# JELEN'S COST AND OPTIMIZATION ENGINEERING

# JELEN'S COST AND OPTIMIZATION ENGINEERING

Third Edition

Edited by

**Kenneth K. Humphreys, PE CCE**

*Executive Director*
*American Association of Cost Engineers*

Sponsored by the American Association of Cost Engineers

**McGraw-Hill, Inc.**

New York   St. Louis   San Francisco   Auckland   Bogotá   Caracas   Hamburg
Lisbon   London   Madrid   Mexico   Milan   Montreal   New Delhi
Paris   San Juan   São Paulo   Singapore   Sydney   Tokyo   Toronto

DEDICATED TO DR. Frederic C. Jelen, and the late Drs. James H. Black and Aaron Rose who were in great measure responsible for the three editions of this book. Their contributions to the cost engineering profession will be a lasting example for many future generations of cost estimators, cost and schedule control professionals and project managers to emulate.

This book was set in Times Roman by Beacon Graphics Corporation.
The editors were B. J. Clark and Eleanor Castellano;
the production supervisor was Friederich W. Schulte.
The cover was designed by Rafael Hernandez.
New drawings were done by Vantage Art.
R. R. Donnelley & Sons Company was printer and binder.

**JELEN'S COST AND OPTIMIZATION ENGINEERING**

2 3 4 5 6 7 8 9 0  DOC  DOC  9 5 4 3 2 1 0

ISBN  0-07-053646-5

**Library of Congress Cataloging-in-Publication Data**

Jelen's cost and optimization engineering. — 3rd ed./edited by
   Kenneth K. Humphreys.
         p.     cm.
   Second ed. published under title: Cost and optimization
engineering. © 1983.
      "Sponsored by the American Association of Cost Engineers."
      Includes bibliographical references.
      ISBN 0-07-053646-5
      1. Costs, Industrial.    2. Cost control.    I. Humphreys, Kenneth
King.    II. American Association of Cost Engineers.    III. Cost and
optimization engineering.
TS167.J45        1991
658.15 — dc20                                    89-77499

# CONTENTS

## 8 Probability, Simulation, Uncertainty 168

## 9 Productivity 196

**12   Linear, Dynamic, and Geometric Programming   317**

**13   Special Mathematical Techniques   344**

## Part III    Cost Estimation, Cost/Schedule Control, and Project Management

### 14    Capital Investment Cost Estimation    361

## Part IV Associated Topics

## Part V   Appendixes

# LIST OF CONTRIBUTORS

D. L. ANDERSON
Brown and Root, Inc.
Houston, Texas

H. JACK BEHRENS
Retired
Culver City, California

KWEKU K. BENTIL
University of Florida
Gainesville, Florida

DR. JAMES H. BLACK
Deceased

DR. ARNOLD H. BOAS
Foster Wheeler Energy Corp.
Livingston, New Jersey

J. R. BRAUWEILER
Northwest Industries
Chicago, Illinois

DR. BRISBANE H. BROWN, JR.
University of Florida
Gainesville, Florida

W. B. HIRSCHMANN
Valuation Consultant
Chicago, Illinois

KENNETH K. HUMPHREYS, PE CCE
Executive Director
American Association of Cost Engineers
Morgantown, West Virginia

T. LYNN HYVONEN
MK-Ferguson Co.
Cleveland, Ohio

WILLIAM A. JANDA, CCE
Fluor Daniel, Inc.
Dallas, Texas

DR. FREDERIC C. JELEN
Retired
Austin, Texas

DR. O. P. KHARBANDA
O. P. Kharbanda and Associates
Bombay, India

DONALD F. McDONALD, JR., CCC
Construction Performance Specialists
Tyler, Texas

DR. JAMES M. NEIL, CCE
MK-Ferguson Co.
Boise, Idaho

HARLEY C. NELSON
Retired
Wiscasset, Maine

RAYMOND A. ROGERS
Brown and Root, Inc.
Houston, Texas

O. JAMES VOGL
Hughes Aircraft Company
Los Angeles, California

DR. CARL L. YAWS
Consultant
Beaumont, Texas

# PREFACE

Cost engineering encompasses all aspects of cost estimating, cost and schedule control, and project management as they apply to engineering work. The cost engineering profession continues to expand in its complexity and importance and has changed significantly since the first edition of this book in 1970.

The third edition has been substantially expanded over its predecessors and incorporates much new material on planning and scheduling, progress control, productivity management, value engineering, and numerous other topics which in 1970 were only beginning to gain attention in the profession. Most of the material from the second edition has been retained, and expanded upon in the third edition.

Dr. Frederic C. Jelen, who edited the first and second editions of the book, has retired, and Dr. James H. Black, who co-edited the second edition and who in fact originally conceived the idea to develop this book in the late 1960s, is deceased. Dr. Aaron Rose began the task of preparing the present edition for the American Association of Cost Engineers, but due to failing health was unable to complete the task and I was asked to assume the job of editor.

For those who use this book and find it to be of value, the credit belongs to Drs. Jelen, Black, and Rose, and to the many contributing authors. However, if the book is found to be lacking in any way, the responsibility is that of the present editor.

Acknowledgment is given to the American Association of Cost Engineers for continuing its sponsorship of the book and to the many sources noted in the text who have given permission to use copyrighted material.

Appreciation is also extended to the Kennedy-Western University, Agoura Hills, California, for their agreement to accept my efforts in preparing this book as meeting the dissertation requirements of their doctoral program in engineering.

Special acknowledgment is given to the Education Board of the American Association of Cost Engineers and its members who were responsible for the preparation of much of the new material appearing in this edition.

Additional thanks go to the following reviewers of this edition: Neil Book, Florida State University; Robert Creese, West Virginia University; David Edelson,

Florida State University; Daniel Loucks, Cornell University; and James Rowings, Iowa State University.

Finally, acknowledgment is given to Marcel Dekker, Inc. for its permission to my undertaking the task of editing this book despite the fact that I am contractually committed to them as an editor and contributing author of a competing series of textbooks.

*Kenneth K. Humphreys*

# JELEN'S COST AND OPTIMIZATION ENGINEERING

# PART
# I

# MATHEMATICS OF COST COMPARISONS

# CHAPTER
# 1

# INTRODUCTION

*F. C. Jelen*
*and K. K. Humphreys*

## 1.1  STATUS OF COST ENGINEERING

Cost is a major consideration in all human activities and gains in importance with increasing technology and the expanding complications of society. Cost is the glue that binds together a wide variety of components into a single structure. Tourists in a strange land may not know a single word in the native language, but they can bargain for their purchases by using simple gestures. Cost is elementary, basic, and understood universally.

As technology and society advance, it becomes necessary to estimate more closely to remain competitive. An estimate based on overdesign may be too high to win the award. If based on underdesign, it may well win the award and end in disaster.

The cost engineering profession is ever-expanding, and its growth is mirrored by the growth of the American Association of Cost Engineers (AACE). The AACE was founded in 1956 and by 1986, 30 years later, it had 6000 members in over 60 countries. Local sections of AACE had been formed in 10 countries, several of which grew into large, independent national organizations in their own right. Worldwide cost engineering and project management conferences and symposia are now regular occurrences and national cost engineering/project management societies exist in numerous countries.

Professional recognition for cost engineers has been established by the certification program of AACE and by international certificates of certification issued by the International Cost Engineering Council (ICEC), a worldwide confederation of cost engineering societies. Under the AACE and ICEC programs individuals who meet certain minimum standards of education and experience and who pass a written examination are recognized as certified cost engineers (CCEs), or if there is a conflict

**3**

with local laws or regulations, as certified cost consultants (CCCs). Over 1000 individuals worldwide have qualified to use the CCE or CCC designation.

## 1.2   SCOPE

In its broadest sense cost engineering is that area of engineering principles where engineering judgment and experience are utilized in the application of scientific principles and techniques to problems of cost estimating, cost control, profitability analysis, project management, and planning and scheduling. In addition, cost engineering embraces the principles of several disciplines and branches into business planning, management science, optimization, operations research, accounting, and economics, among others. Cost engineering covers a wide range of topics as listed in

**TABLE 1.1**
**Component topics in cost engineering**

| Part I.   Supporting skills and knowledge | |
|---|---|
| Computer operations | Time value of money |
| Operations research | Forecasting theory |
| Statistics and probability | Behavioral science/ |
| Basic business and finance |     motivational management |
| Oral and written | Metric-English units |
|     communication |     conversion |

| Part II.   Cost estimating and cost control | |
|---|---|
| Manufacturing | Capital cost |
| Engineering | Maintenance |
| Construction | Revamp |
| Operations | Decommissioning |

| Part III.   Planning and scheduling and project management | |
|---|---|
| Organizational structures | Quality management |
| Integrated project control | Resource management |
| Time management |     Personnel |
|     Scheduling |     Materials |
|     Schedule control |     Equipment and tools |
|     Forecasting | Contract administration |
| Cost management | Societal/legal influences |
|     Budgeting | |
|     Forecasting | |

| Part IV.   Economic analysis and business planning | |
|---|---|
| Value engineering | Profitability |
| Depreciation | Appraisal |
| Comparative studies | Life cycle costing |
| Constructability | Risk analysis |

*Source:* Adapted from the AACE study guide for the certified cost engineer examination.

Table 1.1, which has been taken from a study guide prepared by the AACE for applicants taking the examination for certified cost engineer. The scope of these topics is indicative of the wide breadth of knowledge which a cost engineer must possess.

## 1.3  COST ENGINEERING IN EDUCATION

Individual courses involving costs are offered at the college level in a few disciplines. Industrial engineers have featured costs in their engineering economy course since the inception of industrial engineering. Many civil engineering departments introduce costs in some courses. Chemical engineers include costs in design courses, and the profession has been very active in publications on costs and profitability. Individual courses that include costs are becoming more popular in several disciplines, and although no undergraduate degree in cost engineering is currently available and only a few institutions offer graduate degrees, serious efforts are under way to establish such degree programs. Proposed curricula are shown in Tables 1.2 and 1.3. Table 1.3

**TABLE 1.2**
**Undergraduate curriculum for cost engineering**

| First year | |
|---|---|
| Chemistry and Materials I | Chemistry and Materials II |
| Engineering Graphics | Computer Programming |
| Literature and Composition | History of the Modern World |
| Calculus I | Calculus II |
| Physics I | Physics II |
| Physical Education | Physical Education |
| **Second year** | |
| Calculus III | Differential Equations |
| Physics III | Statics and Dynamics |
| Economics | Literature |
| English Literature | Basic and Cost Accounting |
| Introduction to Cost Engineering | Industrial and Manufacturing Processes |
| **Third year** | |
| Cost Estimation Methods I | Engineering Economics |
| Probability and Statistics | Construction Methods |
| Work Organization and Analysis | Cost Estimation Methods II |
| Strength of Materials | Cost Control and Analysis |
| Humanities | Humanities Elective |
| **Fourth year** | |
| Finance | Project Management |
| Network Planning and Scheduling | Computer Simulation |
| Management Science | Systems Analysis |
| Labor Relations | Humanities Elective |
| Organizational Planning | Law |

*Source:* C. Wolf, *Transactions of the American Association of Cost Engineers,* AACE, Inc.,
Morgantown, W. Va. 1978, pp. 338–341.

**TABLE 1.3**
# Master's degree program in cost engineering

<hr>

Section I:   Summary

<hr>

**A. Undergraduate prerequisite subjects**

Engineering Economics
Basic Cost Estimating
Basic Scheduling
Introduction to Computers

Business Law
Financial Accounting
Technical Writing & Effective Oral
  Communication

**B. Cost engineering required subjects (approximately 12 credits in these subjects)**

Motivational Management
Statutory Aspects of Personnel
  Management
Fundamentals for Business/Project
  Management

Advanced Economic Analysis
Advanced Cost Estimating

**C. Cost engineering elective subjects (minimum of 15 credits from this group of subjects)**

Advanced Computer Subjects
Computer Applications in Industry
Operations Research
  (Operations Management)
Statistics
Value Engineering

Methods and Productivity Improvement
Applied Project Management
Contract Documents
Contract Administration
Special Topics in Cost Engineering

**D. Cost engineering survey or research project (0–4 credits)**

Thesis or graduate report on cost engineering subject

**E. General electives**

Subjects as locally approved to round out degree program

<hr>

Section II:  Course content

<hr>

Capsulized descriptions of the contents for each of the courses identified in Section I follow. It is recognized that each university may use titles and package courses differently from those listed. The important thing is for the total content of the program to incorporate as many as possible of the components noted in the descriptions.

**A. Undergraduate prerequisite subjects**

These subjects are normally considered to be undergraduate level. An incoming candidate for a master's program who does not have college credit for any of these subjects should either add them to his or her total degree program or demonstrate proficiency in them to the satisfaction of the university. In lieu of college credit, the student may be qualified through attendance at company training programs, commercial short courses, self-study programs, or qualification exam.

**1. Engineering Economics.** The equivalent of a 2- to 3-credit course in the fundamentals of economic analysis. It should encompass at least these subjects:

Time value of money
Equivalence
Present value, future value
Rate of return
Depreciation/depletion

Methods for comparison of economic
  alternatives
Break-even analysis
Budgeting and cash flow

**TABLE 1.3** *continued*

---

### Section II: Course content

---

**2. Basic Cost Estimating.** The equivalent of a 3-credit course built around construction or manufacturing cost estimating which provides an introduction to definitive estimating. Includes these subjects:

Labor costs

Determination of work-hour
  requirements

Determination of materials requirements

Materials costs

Determining equipment requirements

Equipment costs

Equipment productivity

Overhead, indirect, and distributable
  costs

Risk costs and contingency

Inflation/escalation

Profit

Practical exercise in pricing a
  construction project or a
  manufactured product

**3. Basic Scheduling.** The equivalent of a 2- to 3-credit course designed to provide basic proficiency in common scheduling techniques. These subjects should be included:

Barchart (Gantt) scheduling

Arrow diagram method (ADM)
  scheduling

Precedence diagram method (PDM)
  scheduling

Time-scaled networks

Line-of-balance charts

**4. Introduction to Computers.** The equivalent of a 3-credit course designed to provide basic computer literacy and programming capability. It should include these subjects:

Computer hardware and software nomenclature and functioning

Programming using a common engineering language

Introduction to a representative group of personal computer commercial software
  programs applicable to cost engineering

**5. Business Law.** The equivalent of a 3-credit overview of the body of law and case studies relating to business organizational forms, contracts, financing sources, documents of title, secured transactions, investments, business taxation, insurance and bonding, bankruptcies and related subjects.

**6. Financial Accounting.** The equivalent of a 3-credit course in basic financial accounting.

**7. Technical Writing & Effective Oral Communication.** During the graduate study program the student must demonstrate competency in both written and oral communication gained through prior formal instruction or experience. Any deficiencies should be remedied through addition of formal classes to the degree program on these subjects.

**B. Cost engineering required subjects**

Subjects in this category are intended to provide a basic background in business or project management and are considered essential for the professional cost engineer. It is estimated that the equivalent of approximately 12 credits of the degree program will be devoted to pursuing these subjects.

**1. Motivational Management.** A course in which the student is introduced to behavioral science theories and their application to effective management of personnel. Typical included subjects are:

Behavioral science theories

Motivators and demotivators of
  personnel

Effective communication

Leadership/management styles

Progressive, corrective discipline

Conflict management

Team building

Participative management

Practical applications through case
  studies

**TABLE 1.3** *continued*

---

### Section II: Course content

---

**2. Statutory Aspects of Personnel Management.** This course is intended to supplement the Motivational Management course by presenting statutory/regulatory factors which affect personnel management, the knowledge of these factors being essential as a "cost avoidance" tool. Included topics are:

| | |
|---|---|
| Unions and collective bargaining | Worker's compensation |
| Equal Employment Opportunity, Affirmative Action, and similar statutory programs | Special hiring/training programs involving disadvantaged population groups |
| Policies involving alien workers | Current legal atmosphere relating to |
| Termination policies and unemployment compensation | special personnel management matters (smoking, sexual |
| Occupational Safety and Health Act | harassment, substance abuse, etc.) |

**3. Fundamentals for Business/Project Management.** This course covers internal organizational forms (functional, task force, and matrix); management under union, merit, and open shop conditions; employee benefit programs; planning and marketing of services; budgeting and cost control; business insurance and bonding; procurement practices; quality control; health, safety, and morale programs; substance abuse control; environmental protection; programs for small, minority-owned, female-owned, and disadvantaged businesses; financial statements and other financial reporting.

**4. Advanced Economic Analysis.** This course is designed to demonstrate and apply principles of economic analysis to a variety of business situations and case studies. Typical of included subjects are:

| | |
|---|---|
| Life cycle cost analysis | Benefit: cost studies |
| Evaluation of purchase, lease, and rental options | Value analysis |
| Profitability studies | Budgeting and cash-flow analysis |
| | Inflation/escalation |

**5. Advanced Cost Estimating.** This course is designed to demonstrate cost-estimating techniques and applications not covered in basic cost estimating. Included subjects are:

| | |
|---|---|
| Approximate estimating techniques | Risk analysis |
| Capital cost estimating | Integration of estimating with cost and schedule control |
| Manufacturing and product cost estimating | |
| Operating cost estimating | Computerized estimating systems |
| Indices | Probabilistic estimating |

### C. Cost engineering elective subjects

Students will select a minimum of 15 credits from this group of subjects.

**1. Advanced Computer Subjects.** This subject area includes the following possibilities:

| | |
|---|---|
| High-level programming languages | Computer modeling |
| Database management | Hardware design |
| Artificial intelligence | Software program development |

**2. Computer Applications in Industry.** This subject area deals with computers and computer programs within a specific industry and can include applications using commercial software and/or development of special software programs.

**TABLE 1.3** *continued*

---

**Section II: Course content**

---

**3. Operations Research (Operations Management).** This subject area includes many topics which may be combined into one or more courses or included as portions of other courses. Potential topics are:

| | |
|---|---|
| Decision theory | Optimization |
| Systems theory | Forecasting |
| Linear and dynamic programming | Statistical quality control |
| Learning curves | Inventory control |
| Queuing | |

**4. Statistics.** A traditional course in statistics which includes these topics:

| | |
|---|---|
| Probability | Sampling |
| Data collection | Variance analysis |
| Frequency distributions | Correlation and regression |

**5. Value Engineering.** Using life cycle costs as a basis for evaluations, this course explains value engineering techniques. A practical exercise in which techniques are applied is normally a major element in this course.

**6. Methods and Productivity Improvement.** This subject area provides special tools to evaluate and improve project performance. Potential included topics are:

Time and motion studies
Constructability, operability, and maintainability analysis
Work sampling and time-lapse photography
Foreman and worker delay reporting
Quality circles and problem-solving teams
Suggestion programs

**7. Applied Project Management.** Topics in this group apply the principles of earlier subjects to a type of project. Potential topics include:

| | |
|---|---|
| Project stakeholders and interests | Security |
| Contract approaches for engineering, construction, and other services | Document control |
| | Startup (if applicable) |
| Organizational structures for project management | The special challenges of the fixed-price contract |
| Insurance and bonding | The special challenges of |
| The impact of governmental regulation | "fast tracking" |
| Risk identification, evaluation, and control | In-process management: |
| | Short-range planning |
| Work breakdown structures and charts of accounts | Constructability analysis |
| | Status reporting |
| Control schedules and budgets | Schedule management |
| Project plans: | Earned-value control |
| Project controls | Analysis, trending, and forecasting |
| Materials management | Change administration |
| Quality management | Contingency management |
| Subcontracting | Management reporting |
| Safety/loss control | Historical data files |
| Environmental protection | Claims administration |
| Construction equipment, tools, and heavy rigging | Project closeout |

**TABLE 1.3** *continued*

---

Section II: Course content

---

**8. Contract Documents.** This is a detailed study of standard (AIA, AGC, U.S. Government, etc.) and selected user-prepared contract forms to include bid forms, contract forms (short and long), general conditions, special or supplementary conditions, drawings, specifications, and addenda. Special attention is given to key contract clauses. Variations of these contract forms for common methods for contracting work, materials, or services are covered.

**9. Contract Administration.** This subject area provides an understanding of the administration of contracts for engineering, construction, other services, equipment, materials, and supplies. Coverage includes roles and responsibilities of contracting parties, organizations for contract administration, contractor/vendor qualification, risk allocation, procurement strategies, incentive contracting, liabilities of contracting parties, change processing, document control, claims, and disputes.

**10. Special Topics in Cost Engineering.** Universities normally have a "Special Topics" course category which enables a student to pursue one or more credits of study in a subject area agreed to between the student and the advisor. Those involving a cost engineering topic may be credited to the electives group.

**D. Cost engineering survey or research project**

A master of science program will generally require a thesis. A master of engineering, construction management, building construction, or comparable program will normally require one or more survey reports. Whatever the requirements, these should be on a cost engineering subject. Credits are normally awarded for thesis work; they may or may not be awarded for survey reports.

**E. General electives**

Remaining hours in the degree program may be devoted to any subject area of interest to the student and approved by the supervising graduate faculty committee.

---

*Source:* American Association of Cost Engineers Education Board, P.O. Box 1557, Morgantown, W. Va. 26507-1557.

*Note:* "Credits" refers to credit-hours. One credit-hour is equivalent to 15 lecture-hours.

includes descriptions of course content. This table was developed by the AACE Education Board.

## 1.4 REQUIRED SKILLS AND KNOWLEDGE OF A COST ENGINEER

The American Association of Cost Engineers Education Board was created in 1982 to coordinate educational activities of AACE. One of the first dilemmas faced by the Education Board was the lack of a definition of a professional cost engineer in terms of skills and knowledge. Such a definition is obviously a prerequisite to development of appropriate degree programs, university courses, continuing education programs, and textbooks such as this.

A survey of cost engineering professionals was therefore conducted to identify and catalog the numerous cost engineering subject areas. "Core" subjects were identified as those whose usage in the profession was occasional to frequent and which were felt by the survey participants to be desirable for professional cost engineers to know.

The AACE Education Board then took these core subjects and expanded them into a series of performance statements to represent the level of proficiency expected in each subject area. The statements, published by AACE for public comment, were revised based on comments received and republished for further comment in 1986. No additional comments were received, indicating general agreement within the profession, and the AACE Board of Directors formally adopted the set of statements as an AACE Recommended Practice in 1987. The final set of statements of the expected level of skill and knowledge of a practicing cost engineer is presented in Table 1.4.

**TABLE 1.4**
**AACE recommended practice: Required skills and knowledge of a cost engineer**

**Section I:   Supporting skills and knowledge**

**Engineering economics**
1. Calculate simple and compound-interest rates and solve interest problems using the basic single payments, uniform series, and gradient formulas.
2. Calculate present value, future value, and equivalent uniform annual value of a cash-flow series.
3. Determine discounted rate of return of a cash-flow series.
4. Evaluate and select the best alternative from two or more alternatives utilizing present value, future value, equivalent uniform annual value, and discounted rate of return.
5. Compare two alternatives using benefit-cost ratio.

**Terminology**
1. Be familiar with cost engineering terms as contained in AACE's *Cost Engineers' Notebook*.

**Computers**
1. For a major computer installation:
   a. Describe at least three methods for inputting data.
   b. Identify at least three peripherals.
   c. Explain the function of the central processing unit.
   d. Identify two methods of mass data storage.
2. Explain current terms as used in data processing, such as:

| | |
|---|---|
| a. Software and hardware | k. Modem |
| b. Alphanumeric | l. Mainframe |
| c. Baud rate | m. Remoting |
| d. Bits, bytes, kilobytes, and megabytes | n. User-friendly |
| e. Cursor | o. Dumb, smart, and interactive terminals |
| f. Daisy wheel, dot matrix, and laser printers | p. Floppy and hard disks |
| g. Emulator | q. Networking |
| h. Fields, records, and files | r. Disk operating system |
| i. Database | s. Turnaround time |
| j. RAM | t. Computer chip |
| | u. Batch and interactive input modes |

3. Name at least three common computer languages.

**TABLE 1.4** *continued*

---

### Section I: Supporting skills and knowledge

---

**Statistics and probability**

1. Given a set of data, determine:
   a. The arithmetic mean
   b. The median
   c. The mode
2. Given a curve of normal distribution and an accompanying table of areas under the curve, determine the probability of the variable:
   a. Being between two given values.
   b. Not being higher than a given number, or lower than that number.
   c. Being exactly equal to a given number.
3. Given a curve of distribution for a variable that is other than normal, draw the resultant cumulative frequency curve. Determine the percent probability of a given value not being overrun or underrun.
4. Given a set of numbers, determine their standard deviation and variance.
5. Handle basic probability exercises involving:
   a. Permutations
   b. Ordered samples
   c. Combinations
   d. Ordered and unordered partitions

**Optimization**

1. Given an optimization goal involving a result $Y$, which is a function of a single variable $X$, use graphical or incremental methods to establish the optimum value of $Y$.
2. Be familiar with nonmathematical analysis/decision techniques used in an operating environment for evaluating options and selection of the one which best meets the objectives.

**Productivity management**

1. Define "productivity." Distinguish between "production" and "productivity."
2. Discuss how the following can influence productivity:

   a. Individual worker skills
   b. Worker attitude
   c. Work-force sociological characteristics
   d. Project demographics
   e. Absenteeism and turnover
   f. Technology employed
   g. Management/supervisor competence
   h. Immediate work area environment
   i. Relationship to other work or contractors
   j. Project geographic location
   k. Job layout
   l. Weather
   m. Overtime
   n. Learning curve
   o. Work rules
   p. Safety program
   q. Materials/tools equipment availability
   r. Crew balance
   s. First-level quality control

3. Describe programs that can be undertaken to improve productivity on a project.
4. Describe time-lapse photography (unattended) and at least one direct observation (attended) type of work sampling system.
5. Describe the application of work sampling in a productivity management program to include its major strengths and weaknesses.

**Human relations/behavioral science**

1. Given a listing of a number of commonly referenced management theories or authors, such as those below, be able to discuss at least three of them.
   a. Abraham Maslow (hierarchy of needs)

**TABLE 1.4** *continued*

---

### Section I: Supporting skills and knowledge

---

    b. Douglas McGregor (theory $X$ and theory $Y$)
    c. Rensis Likert (four model systems)
    d. Chris Argyris (effects of organization life on individuals)
    e. Robert Blake and Jane S. Mouton (managerial grid)
    f. Frederick Herzberg (motivation-hygiene theory)
    g. Koontz and O'Donnell
       The Japanese experience (theory $Z$)
2. Discuss the meaning and provide examples of "participative management."
3. Discuss "motivators" and "demotivators" as related to labor attitudes and productivity.
4. Given a case study situation involving a personnel or productivity problem in a company, analyze the situation and provide appropriate recommendations for solving the problem.
5. Describe the purpose, functioning, and limitations of "quality circles."

**Organizational structures**

1. A large industrial firm includes the functions of administration, finance, research and development, engineering, marketing, and manufacturing. For such a firm:
    a. Draw an organization chart which reflects a hierarchical, vertical, functional organization.
    b. Draw an organization chart for the same company if the intent is to include a manager of projects while still maintaining a vertical, hierarchical organization. In effect the organization is to be "projectized."
    c. Draw an organization chart for the same company if the intent is to maintain centralized functional responsibilities but to manage projects on a matrix basis.
    d. For each of the above, discuss how costs are best reported and controlled.
2. In an organization where projects are managed on a matrix basis, discuss the roles, responsibilities, and methods for its successful operation.
3. Compare the advantages and disadvantages of a vertical, hierarchical organization to a matrix organization.
4. Describe the organization and operation of a project team or task force.

**Measurements**

1. Be able to convert English and metric measurements.

---

### Section II:   Cost estimating

---

**Estimating basics**

1. Describe the differences in the mechanics of compensation for wage and salaried employees. Explain the meanings of "exempt" and "nonexempt" employees.
2. Describe the possible included components of "fringes" or "burden."
3. Describe the basic differences in the handling of base compensation and fringes in a union atmosphere as compared to an open shop atmosphere.
4. Describe the basic mechanics of mandatory government programs that are part of employee benefit packages such as:
    a. Retirement (e.g., social security)
    b. Unemployment insurance
    c. Accident compensation (e.g., worker's compensation)

    In this description identify agency (federal, state or provincial, local) involved in its administration; how rates are established; who pays the bill; what affects rates, if they are variable.

**TABLE 1.4** *continued*

---

### Section II: Cost estimating

---

5. A worker is paid a basic wage rate per regular hour. Be able to convert this to an effective cost per hour due to:
   a. Overtime premium
   b. Other premium pay
   c. Shortened shift time
   d. Travel time
   e. Show-up pay
6. Describe the included elements of materials cost from a supplier's viewpoint and from a buyer's viewpoint.
7. Describe and distinguish among the various estimate categories such as conceptual, approximate, definitive, budget, appropriation, order-of-magnitude, engineer's, contractor's or bidder's, etc. In this discussion refer to applications and relative accuracy.
8. Distinguish between factored and parametric estimating.
9. Understand these terms and their influence on the cost of materials:
   a. Carrying charges
   b. Demurrage
   c. Escalation clauses
   d. Catalog price and market price
   e. FOB
   f. Shrinkage, waste, theft, and damage
10. Identify elements of equipment cost included in:
    a. Ownership costs
    b. Operating costs
11. Know application of these elements of current tax laws with respect to capital equipment:
    a. Investment and other tax credit
    b. Depreciation
12. Given factors of usage and included elements of cost, develop an equivalent cost per hour to be charged for a piece of equipment.
13. Distinguish between the terms "operating life" and "economic life" as applied to equipment.
14. Explain the mechanics of common rental, lease, and purchase contracts for equipment.
15. Given the options of rental, lease, or purchase of equipment and all the economic and use data pertaining to the problem (purchase price, rental price, lease price, down payments, anticipated usage, fixed costs, operating costs, investment tax credit, depreciation, tax rates, time value of money, etc.), determine the most economical option.
16. Explain the difference between "costing" and "pricing."
17. Define and give examples of distributable costs. Describe various methods for handling these in a cost estimate.

#### Industry specific — contractors

1. Describe included components of these costs associated with design engineering work:
   a. Direct salaries
   b. Reimbursable directs
   c. Nonreimbursable directs
   d. Overhead/indirects
   e. Profit
2. Given a unit-price bid form for a construction contract, plus all direct cost data pertaining to bid line items, overhead/indirect costs, and profit markup, calculate the bid unit prices assuming no unbalancing.
3. Explain the mechanics of unit-price bid unbalancing. Explain "front-end loading."
4. Given the construction drawings for a relatively simple structure or system, complete the quantity takeoff of selected items.

**TABLE 1.4** *continued*

---

### Section II: Cost estimating

---

5. Given the quantity takeoff summary for a given civil, electrical, or mechanical work package plus crew composition, labor production rates, labor wage rates, and labor burden, determine the craft labor direct cost for that work package.

**Industry specific — owners**

1. Define contingency and explain how it differs from allowances or management reserve. Discuss a reasonable method for determining the contingency to be included in a budget for a capital project.
2. Discuss the use of factors, ratios, indices, escalation, and inflation in capital cost estimation.
3. Select a process or manufacturing industry with which you are familiar and describe a system by which conceptual estimates of capital cost of a new facility can be made.
4. Given the following scale of operations conversion formula, plus relative outputs/sizes of A and B, location indices, inflation indices, and the value of the "*x*" factor, estimate the cost for a new facility at location B, given actual data for a comparable facility at location A.

$$C_B = C_A \left[ \frac{Q_B}{Q_A} \right]^x$$

5. With respect to a parameter estimating system:
   a. Describe the basic mechanics of such a system.
   b. Given the baseline data for an existing facility plus the standard parameter measures, location indices, and inflation indices, estimate the cost of a comparable proposed facility at another location.
6. Given the estimated fixed and variable components of cost for a manufactured product, determine the break-even price to be charged per unit for a given level of production. Or, given the estimated fixed and variable price components, the desired profit, and the expected price to be charged per unit, determine the minimum level of production and sales required.
7. Define the following with respect to a manufacturing/process facility:

   | | |
   |---|---|
   | Operating/manufacturing cost | Direct cost |
   | General expense | Daily cost |
   | Total product cost | Unit-of-product costs |

8. Distinguish among products, coproducts, and by-products. What is joint costing?
9. Explain the cost factors included in packing and shipping of a product.

---

### Section III:   Cost control

---

**Work breakdown structures and charts of accounts**

1. Given a structure or manufacturing process whose components are generally known, develop a work breakdown structure of four levels. For this purpose the total structure is level 0, major components are level I, subcomponents are level II, and sub-subcomponents are level III. Show at least three components at each level, choosing one of those for subsequent breakdown into three components at the next lower level.
2. Develop a chart of accounts to parallel the work breakdown structure developed above. The format of this code of accounts must be designed to facilitate computer summary of data at each level assuming raw data is input at level III.
3. Define "work package."

**TABLE 1.4** *continued*

---

### Section III: Cost control

---

**Earned value (also called achieved and accomplished value)**

1. Explain each of the following terms and, given data relating to an example project, calculate them:
   a. Budgeted cost of work scheduled (BCWS)
   b. Budgeted cost of work performed (BCWP)
   c. Actual cost of work performed (ACWP)
   d. Budget at completion (BAC)
   e. Estimate at completion (EAC)
   f. Schedule variance (SV)
   g. Cost variance (CV)
2. Be able to explain and handle calculations similar to those of item 1, using work-hour instead of cost data.
3. Given data relating to individual control accounts, determine earned value of each account and determine percent complete of each account plus the percent complete at summary levels.

**Industry specific — contractors**

1. Given contractual provisions concerning reimbursable and nonreimbursable costs on a proposed engineering contract, and also given historical and projected cost data for the company, develop the break-even marker to be applied to direct salaries for current contracts.
2. Explain how engineering tasks (which are system oriented), procurement tasks (which are commodity oriented) and construction tasks (which tend to be area oriented) can be combined to provide integrated engineering-procurement-construction control.

**Industry specific — owners**

1. Apply the line-of-balance method to control of a manufacturing process.

---

### Section IV:  Planning and scheduling

---

**Scheduling basics**

1. Given a series of logic statements relating to the activities in a project, draw the logic diagram for that project.
2. Given an arrow diagram method (ADM) logic diagram and durations for each activity, calculate early start, early finish, late start, late finish, total float, and free float times for all activities. Identify the critical points and minimum project completion time.
3. Given a precedence diagram method (PDM) logic diagram with durations and which includes at least one each finish-start, finish-finish, start-finish, and start-start relationships with lags, calculate early start, early finish, late start, late finish, total float, and free float times for all activities. Identify the critical path(s) and minimum project completion time.
4. Given a summary of all activities in a project network to include durations, early and late start and finish times, and total float, draw a bar chart schedule based on early start of all activities. Show total float of activities where applicable.
5. Discuss the relationship and significance of total float, free float, and independent float in the scheduling of an activity.
6. Compare characteristics of ADM and PDM networks.
7. Convert an ADM type network into a time-scaled network based on early start.
8. For a simple ADM type network with activities whose crew sizes are given, resource level this network within early and late start limits. Draw histograms of worker-loadings for early, late, and leveled configurations.

**TABLE 1.4** *continued*

---

### Section IV: Planning and scheduling

---

**Schedule control**

1. Work packages in a construction or manufacturing project will be some combination of the following types:
   a. Those composed entirely of single, measurable units of work, such as placing concrete (cubic yards) or installation of a component (each).
   b. Those composed of a number of separate, overlapping tasks, each with a different unit of measure. An example would be major pipe installation where the included tasks are hanger installation, pipe erection, valve installation, and welding.
   c. Those composed of easily identified sequential tasks, such as those associated with installation of a piece of mechanical equipment—receipt, setting, alignment, grouting, test.
   d. Those involving level of effort or apportioned effort such as management, administration, quality control, etc.
   e. Those for which start and completion are well defined, but there is no basis for determining intermediate progress. Examples are rotary equipment alignment, testing, and planning.

For each of the above, describe a progress control system than can be established that becomes the baseline for schedule status determination.

**Industry specific—contractors**

1. Describe reasonable methods for planning and statusing engineering work such as drawings, specification writing, etc.
2. Given a "soft logic" work package, such as installation of a pressure piping system, where included activities (e.g., hanger installation, pipe installation, valve installation, and welding) are overlapping, parallel, and have no strict interrelationship, discuss methods for reasonably scheduling this work.

**Industry specific—owners**

1. Distinguish between a batch and a continuous process. How does scheduling for these two types of operation differ?

---

### Section V:  Project management

---

**Contracting**

1. Describe the relationship between risk allocation and contract type.
2. Explain the advantages, disadvantages, and uses of these types of contracts:
   a. Fixed-price contracts
   b. Unit-price contracts
   c. Cost-plus contracts
      (1) Fixed fee
      (2) Incentive fee
      (3) Award fee
3. Describe the contracting roles among the various potential parties: project owner, engineer, project manager, prime contractors, subcontractors.
4. Describe the general contents and purposes of the following documents which may comprise bidding and contract packages:
   a. Invitation to bid or request for proposal
   b. Bid form
   c. Agreement
   d. General conditions
   e. Supplementary or special conditions
   f. Technical specifications
   g. Drawings
   h. Addenda
   i. Modifications

**TABLE 1.4** *continued*

---

**Section V: Project management**

---

5. Distinguish among performance, proprietary, and descriptive specifications.
6. Describe major reasons for changes in a contract and how changes are initiated. Distinguish among formal, constructive, and cardinal changes.
7. Describe the general approach to determination of cost and schedule impact of changes.
8. Distinguish between changes and claims.
9. Distinguish among various types of delay—excusable, nonexcusable, compensatory, and concurrent.

**Integrated project control**

1. Describe the concept of integrated project control to include:
   a. Systems included
   b. The objectives
   c. The basic mechanics of such a system—its structure and information flow

**Industry specific—contractors**

1. Describe the types of bonds that may be required as part of a construction contract.
2. Describe the various types of insurance that may be required as part of a construction contract.
3. Describe the various types of taxes that may be included in a construction contract.
4. Describe examples of "construction plant."
5. Identify typical items included in the category "general expense" in a construction contract.
6. Distinguish between "job (project) overhead" and "general overhead" and provide examples of each.
7. A supplier offers a contractor payment terms such as "2/15 net 30." Given this information, the current time value of money for the contractor, the timing of contractor requests for payment from the owner, and the timing of owner reimbursement of contractor, determine the method of payment that is economically most advantageous.
8. For a construction project explain the difference between and give examples of "permanent materials" and "construction materials and supplies."
9. Calculate the effective cost of "retention" given the terms of the contract and time value of money.
10. Explain "liquidated damages."
11. Explain contractor considerations in the determination of profit markup (risk, competition, desired rate of return, current economic conditions, etc.).

---

**Section VI:   Economic analysis and business planning**

---

**Budgeting and cash flow**

1. Given the items listed, develop the cost "S-curve" for a project.
   a. The schedule for completion of each component account
   b. The estimated rate of incurring costs in each component account
   c. The estimated total amount of indirect costs
   d. The estimated rate of incurring indirect costs
2. Given the information below, develop the cash-flow profile for a project:
   a. The projected expense curve for the project
   b. The projected progress curve for the project
   c. The dollar value of the contract
   d. The payment terms of the contract (frequency, basis of payment, retention, delay between completion of work and payment)

**TABLE 1.4** *continued*

---

**Section VI: Economic analysis and business planning**

---

3. For an investment opportunity calculate the cash flow during an annual period given the following information:

   a. Anticipated expenses
   b. Anticipated receipts
   c. Depreciation or depletion (to include method to be used)

   d. Investment tax credit
   e. Taxation conditions
   f. Resale, residual, or salvage value

**Value engineering**

1. Explain the meanings, using examples if desired, of these four kinds of value that may be associated with an item:

   a. Use value
   b. Esteem value
   c. Exchange value
   d. Cost value

2. The Value Engineering Job Plan is a basic format for organizing and executing the value engineering effort. Although there is no standard labeling of the phases of such a plan, commonly used labels are given here:

   a. Information phase
   b. Creative phase
   c. Judgment phase

   d. Development phase
   e. Recommendation phase

   Briefly describe activity within each of these phases.

3. Describe each of the following problem-solving techniques:

   a. Brainstorming
   b. Checklists
   c. Morphological analysis

   d. Attribute listing
   e. Function analysis

4. Distinguish "value engineering" from "scope control" and "constructability."
5. Distinguish among the terms "lowest life cycle cost," "best quality," and "best value."
6. Explain "life cycle" costs.

---

*Source:* B. D. Dunfield et al., *Cost Engineers' Notebook,* AACE, Inc., P.O. Box 1557, Morgantown, W. Va. 26507-1557. Copyright 1988. Reprinted by permission.

Although no single book can possibly cover all the topics and answer all the questions which are posed in Table 1.4, this text attempts to cover a major portion of them, at least in an introductory manner. The cost engineering master's degree program outlined in Table 1.3 is designed to cover comprehensively the entire scope of the required skills and knowledge document.

# NOMENCLATURE

Nomenclature always presents a problem, particularly in books covering a wide range of topics. The nomenclature used is characterized by being highly delineated. In engineering problems, use of an orderly system of nomenclature is good discipline, and it becomes a necessity in large problems such as those for computer programming. For example, in comparing two machines $A$ and $B$ on a present-value basis $P$, it is not good discipline to write

$$P_A \quad \text{and} \quad P_B$$

If each machine is compared on a 10-year basis, it is much more informative to write

$$P_{A10} \quad \text{and} \quad P_{B10}$$

Indeed, this kind of notation will avoid making a false comparison with, say,

$$P_{A8} \quad \text{and} \quad P_{B10}$$

for the difference in duration is clearly evident.

The nomenclature for a chapter is given immediately following the text of the chapter. If the nomenclature and typography seem to be needlessly exact at times, it merely reflects the modern approach to engineering requirements for good characterization.

Numerical tables in Appendixes 1 through 4 are somewhat extensive and are for quick reference use only. Computers have virtually eliminated the need for tables; the factors can easily be programmed and recalled as needed. Interpolation is eliminated, and a wide choice of factors is available. Numerous modestly priced software packages are readily available for these applications.

# CHAPTER
# 2

# EQUIVALENCE AND COST COMPARISONS

*F.C. Jelen*

## TIME VALUE OF MONEY AND EQUIVALENCE

## 2.1 COMPOUND INTEREST AS AN OPERATOR

Investments are spread over a period of time, and time has a monetary value. An investor expects an investment, with due allowance for dividends, to increase with time in the form of appreciation. The appreciation of 1 year becomes an investment of the next year and in turn should earn appreciation of its own. The increase in investment should be compounded with time, much like the increase of a deposit in a bank, but the time value of money in a practical problem applies without an actual banking transaction.

Suppose a present amount $P$ earns at the rate $i$ per year, expressed as a decimal. Also, let $S$ be the future value at some later time. Then at the end of 1 year

$$S_1 = P + iP = P(1 + i)$$

$P(1 + i)$ becomes the starting value at the beginning of the second year, and at the end of the second year

$$S_2 = P(1 + i) + iP(1 + i) = [P(1 + i)](1 + i) = P(1 + i)^2$$

Repeating for another year gives

$$S_3 = P(1 + i)^2 + iP(1 + i)^2 = [P(1 + i)^2](1 + i) = P(1 + i)^3$$

The form of the relationship is obvious and for $n$ years is

$$S = P(1 + i)^n \tag{2.1}$$

or
$$S = PF_{PS} \tag{2.2}$$

Future value = (present value)(compound-interest factor)

These relationships are the *compound-interest law*. Factors in this book are designated by the letter $F$ with a suitable subscript. The symbol $F_{PS}$ is read "factor for converting $P$ to $S$." It is more completely designated as $F_{PS, i, n}$, where $i$ is the rate of return and $n$ is years. Thus

$$F_{PS, i, n} = (1 + i)^n \tag{2.3}$$

$(1 + i)^n$ or $F_{PS}$ is the *compound-interest factor*, values for which are given in Table A1.1. It should be noted that $P$ is a single payment and that compounding is carried out periodically in steps.

In engineering economy a year is usually taken as a unit period of time. Equations (2.1) and (2.2) are used with $n$ in years and $i$ the yearly rate of return expressed as a decimal, although $i$ is usually stated as a percentage. If compounding occurs $p$ times per year, the relationship is

$$S = P\left(1 + \frac{i}{p}\right)^{np} \tag{2.4}$$

where $np$ is the number of periods and $i/p$ is the rate of return per period, inasmuch as $i$ always refers to the rate of return per year. The last equation can be written

$$S = P\left[\left(1 + \frac{i}{p}\right)^p\right]^n$$

where the expression in brackets is the value of 1 at the end of 1 year. The *increase* in value for 1 year

$$\left(1 + \frac{i}{p}\right)^p - 1 = \text{effective rate of return} \tag{2.5}$$

The *effective rate of return* can be used as if the compounding were done once a year instead of $p$ times per year. For distinction $i$ is referred to as the *nominal rate of return*.

The compound-interest law is used in engineering economy if the time is over 1 year. If the time is less than 1 year, *simple interest* can be used, with interest charged only on the original amount:

$$S = P(1 + ni) \tag{2.6}$$

where $n$ may be a fractional part of a year.

In this book if interest is for 1 or more years plus a fractional part of a year, simple interest will be used for the fractional part and compound interest for the integer part unless otherwise indicated.

**Example 2.1.** Find the simple interest on $10,000 for 2 months at 10 percent per year.

*Solution*

$$S = 10,000[(\tfrac{2}{12})(0.10)] = \$166.67$$

**Example 2.2.** A person borrows $1000 and repays $1050 at the end of 4 months. What interest rate was paid?

*Solution.* By Eq. (2.6),

$$S = 1000[1 + (\tfrac{4}{12}i)] = 1050$$

$$i = 0.15 \text{ or } 15 \text{ percent per year}$$

**Example 2.3.** How long will it take for money to double if compounded annually at 8 percent per year?

*Solution.* By Eq. (2.2),

$$S = PF_{PS,8\%,n} = 2P$$

$$F_{PS,8\%,n} = 2$$

From Table A1.1,

$$F_{PS,8\%,9} = 1.9990$$

That is, money will double in about 9 years at 8 percent per year.

**Example 2.4.** If $10,000 compounds to $14,120 in 4 years, what is the rate of return?

*Solution.* By Eq. (2.1),

$$14,120 = S = P(1 + i)^n = 10,000(1 + i)^4$$

$$(1 + i)^4 = 1.412$$

$$4 \log(1 + i) = \log 1.412 = 0.14983$$

$$\log(1 + i) = 0.03746$$

$$(1 + i) = 1.090$$

$$i = 0.09 \text{ or } 9 \text{ percent per year}$$

**Example 2.5.** Find the value of $10,000 compounded annually at 10 percent for 3 years and 4 months.

*Solution.* By Eq. (2.6) for 4 months,

$$S = 10,000[1 + (\tfrac{4}{12})(0.10)] = \$10,333$$

and, compounding by Eq. (2.2) for 3 years,

$$S = 10,333F_{PS,10\%,3} = 10,333(1.331) = \$13,753$$

Equation (2.1), the compound-interest law, can be written as an expression for $P$, present value, in terms of $S$, future value, and becomes

$$P = \frac{S}{(1 + i)^n} \qquad (2.7)$$

or

$$P = S(1 + i)^{-n} \qquad (2.8)$$

or

$$P = SF_{SP} \qquad (2.9)$$

Present value = (future value) (present-value factor)

The factor $(1 + i)^{-n}$ or $F_{SP}$, which converts $S$ to $P$, is the *single-payment present-value factor*. It will be referred to as the *present-value factor* and is tabulated in Table A1.1. The complete designation of $F_{SP}$ is

$$F_{SP, i, n} = (1 + i)^{-n} \qquad (2.10)$$

It will be noted that $F_{SP}$ and $F_{PS}$ are reciprocal relationships; i.e.,

$$F_{SP} = \frac{1}{F_{PS}} \qquad (2.11)$$

Equations (2.8) and (2.9) show that a future amount is reduced when converted to a present amount. The procedure of reducing a future amount to a present amount is known as *discounting*. Discounting is merely an application of the principle that because of the time value of money, the future is not worth as much as the present when viewed from the present.

Equations (2.1) and (2.8) can be written

$$S = P(1 + i)^n \qquad \textit{(moving with the calendar)}$$

$$P = S(1 + i)^{-n} \qquad \textit{(moving against the calendar)}$$

and can be combined in a single relationship:

$$A_{t2} = A_{t1}(1 + i)^n \qquad (2.12)$$

Amount time 2 = (amount time 1) (compound-interest operator)

where $n$ positive = with the calendar

$n$ negative = against the calendar

The student should become familiar with Eq. (2.12), the *time-value conversion relationship*. The compound-interest factor $(1 + i)^n$ can be regarded as a mathematical operator which moves a unit value $n$ years, with or against the calendar, depending on the sign of $n$.

In engineering economy receipts and disbursements occur at different instants of time. They can be combined only after conversion to a single common instant. The time-value conversion relationship defines the mathematical operator that makes this possible. In solving problems it is convenient to diagram expenditures and receipts as vertical lines positioned along a horizontal line representing time. Expenditures and receipts can point in opposite directions. The problem can be visualized in this manner and each amount moved by a routine technique to some common instant for consideration. The reference instant is called the *focal point* or *focal date*.

**Example 2.6.** A boy is celebrating his sixteenth birthday and will need $3000 on his seventeenth, eighteenth, nineteenth, and twentieth birthdays for his college education. His father agrees to set aside a certain amount now and each year thereafter until the boy is 20 years old; moreover, the contributions will form an arithmetical progression for all years, increasing by 20 percent after the first year. If money is worth 4 percent per year, what is the father's first contribution?

*Solution.* If $x$ is the first contribution, succeeding contributions are $1.2x$, $1.4x$, etc. The diagram is

Take 2 years from now as the focal date. Use the time-value conversion relationship, Eq. (2.12), noting that contributions must equal withdrawals:

$$x(1.04^2) + 1.2x(1.04^1) + 1.4x(1.04^0) + 1.6x(1.04^{-1}) + 1.8x(1.04^{-2})$$
$$= 3000(1.04^1) + 3000(1.04^0) + 3000(1.04^{-1}) + 3000(1.4^{-2})$$
$$x = \$1699 \text{ first contribution}$$

## 2.2  UNIFORM ANNUAL AMOUNT, UNACOST

Consider a uniform end-of-year annual amount $R$ for a period of $n$ years. The diagram is

where $R$ will be referred to as *unacost*. The terms can be replaced by a single present value $P$ by moving all terms to zero time using the time-value conversion relationship, Eq. (2.12):

$$P = R\left[\frac{1}{1 + i} + \frac{1}{(1 + i)^2} + \cdots + \frac{1}{(1 + i)^n}\right] \qquad (2.13)$$

The terms in the brackets can be condensed by the $\Sigma$ (sigma) notation to

$$\sum_{1}^{n} \frac{1}{(1 + i)^n}$$

and represent a geometrical series of $n$ terms with first term $1/(1 + i)$ and ratio $1/(1 + i)$. For a geometric series

$$\text{Sum of geometric series} = \frac{1 - \text{ratio}^{\text{no. of terms}}}{1 - \text{ratio}} \text{ first term} \qquad (2.14)$$

The previous summation becomes

$$\sum_{1}^{n} \frac{1}{(1 + i)^n} = \frac{1 - [1/(1 + i)]^n}{1 - [1/(1 + i)]} \frac{1}{1 + i} = \frac{(1 + i)^n - 1}{i(1 + i)^n} \qquad (2.15)$$

so that Eq. (2.13) is

$$P = R\frac{(1 + i)^n - 1}{i(1 + i)^n} \qquad (2.16)$$

or

$$P = RF_{RP} \qquad (2.17)$$

$$\text{Present value} = (\text{unacost})(\text{unacost present-value factor})$$

$F_{RP}$ must be called by its name, *unacost present-value factor*, to avoid confusion with $F_{SP}$, which is called simply the *present-value factor*. $F_{RP}$ converts $R$ to $P$ and has for its complete designation

$$F_{RP, i, n} = \frac{(1 + i)^n - 1}{i(1 + i)^n} \qquad (2.18)$$

and is tabulated in Table A1.1. It is also known as the *equal-payment-series present-value factor* and as the *annuity present-value factor*.

**Example 2.7.** Find the equivalent value of the following series of receipts as of the end of the third year if money is worth 6 percent per year:

The series has an irregularity at the end of 7 years but with 50 added and subtracted at that time becomes

**Solution.** The present value of all the ascending lines is given by Eq. (2.17), and Eq. (2.9) gives the present value of the single descending line:

$$P = 100F_{RP, 6\%, 10} - 50F_{SP, 6\%, 7}$$

$$P = 100(7.3601) - 50(0.66506) = 702.8 \qquad \text{(zero time)}$$

Move 3 years with the calendar:

$$702.8F_{PS,6\%,3} = 702.8(1.1910) = \$837.00 \qquad \text{(end of third year)}$$

Equation (2.16) written as an expression for $R$ in terms of $P$ becomes

$$R = P\frac{i(1 + i)^n}{(1 + i)^n - 1} \qquad (2.19)$$

$$R = PF_{PR} \qquad (2.20)$$

Unacost = (present value) (capital-recovery factor)

These relationships are very important in engineering economy. The *capital-recovery factor* converts a single zero-time cost to an equivalent uniform end-of-year annual cost, *unacost*. The unacost repays the initial expenditure with interest, hence the name *capital-recovery factor*. The factor has the complete designation

$$F_{PR,i,n} = \frac{i(1 + i)^n}{(1 + i)^n - 1} \qquad (2.21)$$

and is tabulated in Table A1.1. $F_{PR}$ and $F_{RP}$ are reciprocal relationships:

$$F_{PR} = \frac{1}{F_{RP}} \qquad (2.22)$$

Relationships between $R$ and $P$ have been established by Eqs. (2.16) and (2.17). Relationships between $R$ and $S$, the future amount, are easily established. Noting that $P = S(1 + i)^{-n}$ from Eq. (2.8), and placing this expression for $P$ in Eq. (2.16), gives

$$S = R\frac{(1 + i)^n - 1}{i} \qquad (2.23)$$

or

$$S = RF_{RS} \qquad (2.24)$$

Future amount = (unacost) (equal-payment-series future-value factor)

The factor

$$F_{RS,i,n} = \frac{(1 + i)^n - 1}{i} \qquad (2.25)$$

called the *equal-payment-series future-value factor*, is not tabulated in the Appendix but can be obtained from the corresponding $F_{RP}$ factor

$$F_{RS} = F_{RP}(1 + i)^n = F_{RP}F_{PS} \qquad (2.26)$$

Also, Eq. (2.23) can be written

$$R = S\frac{i}{(1 + i)^n - 1} \qquad (2.27)$$

$$R = SF_{SR} \qquad (2.28)$$

Unacost = (future amount) (sinking-fund factor)

The factor

$$F_{SR, i, n} = \frac{i}{(1 + i)^n - 1} \qquad (2.29)$$

called the *sinking-fund factor,* is not tabulated in the Appendix but can be obtained from the corresponding $F_{PR}$ factor

$$F_{SR} = (1 + i)^{-n}F_{PR} = F_{SP}F_{PR}$$

$F_{SR}$ and $F_{RS}$ are reciprocal relationships:

$$F_{SR} = \frac{1}{F_{RS}}$$

Instead of unacost, which is a uniform end-of-year annual amount, one might consider $R_b$, a uniform beginning-of-year annual amount. $R$ and $R_b$ are connected by the relationship

$$R = R_b(1 + i) \qquad (2.30)$$

and substitution of this relationship for $R$ in Eqs. (2.16), (2.19), (2.23), and (2.27) gives various relationships between $R_b$ and $P$ and between $R_b$ and $S$.

So far, four amounts have been introduced:

$P$ = present value, a single amount at zero time

$S$ = future value, a single amount at the end of $n$ years

$R$ = unacost, a uniform end-of-year annual amount for $n$ years

$R_b$ = a uniform beginning-of-year annual amount for $n$ years

The various relationships among these amounts are summarized in Table 2.1. The student should become familiar with the first four items of the table and understand fully the meaning of equivalence. The other relationships in the table are less important.

The symbol $R$ implies an *equivalent* uniform end-of-year amount. For emphasis it may be desirable to designate $R$ as $R_e$, where the subscript $e$ designates end of year, in contrast to $R_b$, for an equivalent beginning-of-the-year amount.

## 2.3  MATHEMATICS OF FINANCE

There are numerous applications of the relationships of Table 2.1 in the mathematics of finance, sometimes called *business mathematics*. Terminology and nomenclature can be different from those in this book, but the principles are the same. *Present worth* and *future worth* are synonymous with and replaced respectively by *present value* and *future value* in this book. The factor $(1 + i)^n$ may be called the *accumulation factor* or *amount of 1* and is sometimes represented by $u^n$. The factor $(1 + i)^{-n}$ is sometimes called the *discount factor* or the *present value of 1* and may be represented by $v^n$. A series of equal payments at equal intervals is called an *annuity*. A periodic payment is called *rent*. An ordinary annuity has the payments at the end of the

**TABLE 2.1**
## Summary of $S$, $P$, $R$, and $R_b$ relationships

| Item | Conversion | Algebraic relationship | Relationship by factor | Name of factor |
|---|---|---|---|---|
| 1 | $P$ to $S$ | $S = P[(1+i)^n]$ | $S = PF_{PS,i,n}$ | Compound-interest factor |
| 2 | $S$ to $P$ | $P = S[(1+i)^{-n}]$ | $P = SF_{SP,i,n}$ | Present-value factor |
| 3 | $R$ to $P$ | $P = R\dfrac{(1+i)^n - 1}{i(1+i)^n}$ | $P = RF_{RP,i,n}$ | Unacost present-value factor |
| 4 | $P$ to $R$ | $R = P\dfrac{i(1+i)^n}{(1+i)^n - 1}$ | $R = PF_{PR,i,n}$ | Capital-recovery factor |
| 5 | $R$ to $S$ | $S = R\dfrac{(1+i)^n - 1}{i}$ | $S = RF_{RS,i,n}$ <br> $S = RF_{RP,i,n}F_{PS,i,n}$ | Equal-payment-series future-value factor |
| 6 | $S$ to $R$ | $R = S\dfrac{i}{(1+i)^n - 1}$ | $R = SF_{SR,i,n}$ <br> $R = SF_{SP,i,n}F_{PR,i,n}$ | Sinking-fund factor |
| 7 | $R_b$ to $P$ | $P = R_b(1+i)\dfrac{(1+i)^n - 1}{i(1+i)^n}$ | $P = R_b(1+i)F_{RP,i,n}$ | |
| 8 | $P$ to $R_b$ | $R_b = \dfrac{P}{1+i}\dfrac{i(1+i)^n}{(1+i)^n - 1}$ | $R_b = \dfrac{P}{1+i}F_{PR,i,n}$ | |
| 9 | $R_b$ to $S$ | $S = R_b(1+i)\dfrac{(1+i)^n - 1}{i}$ | $S = R_b(1+i)F_{RS,i,n}$ <br> $S = R_b(1+i)F_{RP,i,n}F_{PS,i,n}$ | |
| 10 | $S$ to $R_b$ | $R_b = \dfrac{S}{1+i}\dfrac{i}{(1+i)^n - 1}$ | $R_b = S\dfrac{1}{1+i}F_{SR,i,n}$ <br> $R_b = S\dfrac{1}{1+i}F_{SP,i,n}F_{PR,i,n}$ | |

period, and an annuity due has payments at the beginning of the period. Amortization implies the substitution of a number of equal-period payments for a present obligation. A *sinking fund* implies a uniform end-of-year payment which will accumulate to some future value at the time of the last payment. A *deferred annuity* has the first payment delayed in date. Thus if an ordinary annuity has the first payment at the end of the sixth year, it would be deferred 5 years. A *forborne annuity* continues to earn interest after its termination when payments end. A *perpetuity* is an annuity that continues forever. A *balloon payment* is an irregular payment practically always at the end of the term of the payments.

Interest may be compounded more than one period per year; thus for home mortgages the period is 1 month. Elaborate tables accurate to 10 decimals are available, as well as pocket editions. The principal tables are the amount of 1, the (future) amount of 1 per period, and the sinking-fund deposit, which is the end-of-period payment that has a value of 1. Computations can be made from combinations of these tabulations.

Nomenclature has not been standardized, but some common usages with their equivalents as used in this book are $s'' = F_{PS}$, $s_{n\rceil} = F_{RS}$, $1/s_{n\rceil} = F_{SR}$, and $a_{n\rceil} = F_{RP}$.

Some writers use terminology which employs principally capital letters to imitate computer programming. Thus PV and NPV are used for present value or net present value, FV for future value, CPIV for compound-value interest factor (our $F_{PS}$), PVIF for present-value interest factor (our $F_{SP}$), CVIF$_a$ for annuity interest factor (our $F_{RS}$), PVIF$_a$ for present-value interest factor of an annuity (our $F_{RP}$), PMT for payment, and BAL for balloon payment.

Coupon bonds have a *face value* or *denomination,* frequently $1000, to which are attached coupons which become redeemable periodically in the future. The *coupon rate* is an annual rate stated on the bond face. Thus a $1000 10-year bond paying 12 percent per year semiannually would have 20 coupons when issued, each worth $60 at 6-month intervals. The interest rate for the bond fixes the value of the coupons and has no other use. After a bond is issued it will sell at a premium if the bond rate is greater than the going interest rate and, conversely, at a discount. The bond sells so that the *yield to maturity* equals the going interest rate. A buyer pays the present value at the current interest rate for the future coupons and future face value.

**Example 2.8.** A person has $2000 for a down payment on a $10,000 lot. An annual end-of-year payment of $1174.11 for 12 years with money worth 10 percent per year will discharge the debt. However, the person elects to pay $1000 yearly and a balloon payment at the end. Find the balloon payment.

*Solution.* With a balloon payment X the $10,000 is now equivalent to the following time diagram for the payments:

and using Table 2.1, we obtain

$$10,000 = 2000 + 1000F_{RP, 10\%, 12} + XF_{SP, 10\%, 12}$$

$$= 2000 + 1000(6.8137) + X(0.31863)$$

$$X = \$3723.10 \quad \text{(in addition to the regular \$1000 payment)}$$

**Example 2.9.** A person plans to create a forborne annuity by depositing $1000 at the end of the year for 8 years and withdrawing the fund at the end of 14 years from now. If money is worth 10 percent per year, find the accumulated value at the end of the fourteenth year.

*Solution.* The time diagram for the payments is

The present value, by Table 2.1, is

$$P = 1000F_{RP, 10\%, 8} = 1000(5.3349) = 5334.90$$

The future value at the end of 14 years is

$$S = 5334.90 \times F_{PS, 10\%, 14} = 5334.90(3.7975) = \$20,259$$

The problem can also be solved by considering the $1000 paid for 14 years less $1000 paid as annuity for the last 6 years. By Table 2.1, item 5,

$$S = 1000\left[\frac{(1.10)^{14} - 1}{0.10} - \frac{(1.10)^6 - 1}{0.10}\right] = 1000(27.975 - 7.7156) = \$20,259$$

**Example 2.10.** A $1000 face value bond bearing 4 percent per year interest payable semiannually, and due in 6 years, is to be bought so as to yield the purchaser a rate of 8 percent per year. What price should be paid for the bond?

*Solution.* There are 12 semiannual periods involved, with a 2 percent per period coupon rate. The 12 coupons are worth $20 each when due. Also, the purchaser receives the $1000 face value 12 periods hence. The diagram is

The purchaser should pay the present value of these amounts at 8 percent per year, or 4 percent per period.
By Eqs. (2.17) and (2.9),

$$P = 20F_{RP, 4\%, 12} + 1000F_{SP, 4\%, 12}$$

$$= 20(9.3851) + 1000(0.62460) = \$812.30 \text{ purchase price}$$

The bond sells at a discount because the coupon rate is less than the yield rate. If the coupon rate is greater than the yield rate, the bond will sell at a premium.

Note that there are several rates associated with the example. The bond rate is 4 percent per year, the yield to maturity for the purchaser is 8 percent per year, which is the going rate for money, and the bond pays a current rate of 2(20/812.30)(100) or 4.92 percent.

## COST COMPARISONS

## 2.4   COMPARISONS WITH EQUAL DURATIONS

The preceding sections contain all the tools required to make a cost comparison between two machines or alternative investments having equal service lives or durations. The word "machine" is used in a general sense. Two machines generally have cost patterns differing on a time basis, but the costs can be reduced to some common instant and the comparison made on that basis. Any instant can be chosen for the comparison, but the present instant is usually preferred. Hence the comparison is made on the basis of present values for equal durations, which constitutes a fair comparison. The difference between the machines can have any degree of complexity, but the following example will suffice to illustrate the method.

In this and subsequent sections a uniform expense is referred to as a *uniform end-of-year cost*. Actually, the expense may be spread over the year and is merely expressed as an equivalent end-of-year cost. With this understanding, the expense will exist for the year even if the machine is abandoned at the end of the year.

**Example 2.11.** Two machines, each with service life of 5 years, have the following cost comparison. If money is worth 10 percent per year, which machine is more economical?

|  | A | B |
|---|---|---|
| First cost, $ | 25,000 | 15,000 |
| Uniform end-of-year maintenance, $ per year | 2,000 | 4,000 |
| Overhaul, end of third year, $ |  | 3,500 |
| Salvage value, $ | 3,000 |  |
| Benefit from quality control as a uniform end-of-year amount, $ per year | 500 |  |

**Solution.** Construct the cost diagrams for each machine using descending lines for costs and ascending lines for receipts or benefits:

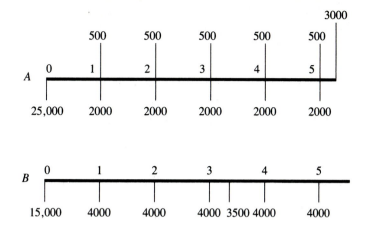

The present value of the costs for the diagrams on a 5-year basis is

$$P_{A5} = 25{,}000 + (2000 - 500)F_{RP,\,10\%,\,5} - 3000F_{SP,\,10\%,\,5}$$

$$= 25{,}000 + 1500(3.7908) - 3000(0.62092)$$

$$= \$28{,}823$$

$$P_{B5} = 15{,}000 + 4000F_{RP,\,10\%,\,5} + 3500F_{SP,\,10\%,\,3}$$

$$= 15{,}000 + 4000(3.7908) + 3500(0.75131)$$

$$= \$32{,}793$$

Machine $A$ is more economical. The cost of machine $A$ at \$28,823 must be qualified by the information that the cost is the present value for 5 years ahead with money worth 10 percent per year.

## 2.5 COMPARISONS WITH UNEQUAL DURATIONS—COMMON DENOMINATOR OF SERVICE LIVES

If two machines have different service lives, a fair comparison can be made only on the basis of equal durations. It would not be equitable to compare the present value of the cost of 2 years of service for machine $A$ with the present value of the cost of 3 years of service for machine $B$. One method of making the comparison is to compare them on the basis of a common denominator of their service lives, here 6 years. The procedure is merely a mathematical convenience without implying that either machine will be repeated.

**Example 2.12.** Two machines have the following cost comparison. If money is worth 10 percent per year, which machine is more economical?

|  | A | B |
|---|---|---|
| First cost, $ | 18,000 | 25,000 |
| Uniform end-of-year maintenance cost, $ per year | 4,000 | 3,000 |
| Salvage value, $ | 500 | 1,500 |
| Service life, years | 2 | 3 |

**Solution.** The cost diagrams of each machine are first reduced to a single present value of the costs:

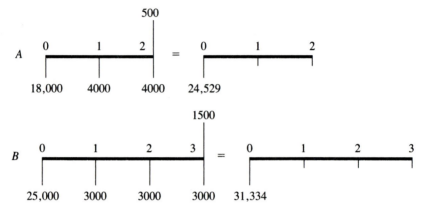

The simplified diagrams are now repeated to obtain 6 years' duration. Note that the present value of machine A is 24,529 as of its *time of installation*.

The present value of each of the preceding diagrams at 10 percent per year is

$$P_{A6} = \$61,555$$

$$P_{B6} = \$54,875$$

Machine B is more economical. The ratio of cost is

$$\frac{A}{B} = \frac{61,555}{54,875} = 1.1217$$

## 2.6   COST COMPARISON BY UNACOST

The method of the preceding section is cumbersome because each problem requires a new duration for the common denominator of the service lives, and the calculation is tedious besides. It is better to use a common denominator for all service lives. One such service life is an equivalent end-of-year uniform annual cost which puts all systems on a 1-year basis.

Unacost fits the requirements exactly. Thus converting any system lasting $n$ years with a present value $P_n$ to an equivalent 1-year cost as of the end of the year results in

$$R = P_n F_{PR,i,n} \tag{2.31}$$

which is a repeat of Eq. (2.20).

Most authors refer to the method simply as the *annual-cost method*. The term *annual cost* is, however, a general term and does not necessarily imply uniformity or the end of the year to all readers. The term *unacost* as used in this book definitely implies uniformity from year to year and the end of the year as part of the definition. Use of the term *unacost* avoids the circumlocutory expression *uniform end-of-year annual amount*.

**Example 2.13.** Solve Example 2.12 on the basis of unacost.

*Solution.* From the first solution,

$$P_{A2} = \$24,529$$

$$P_{B3} = \$31,334$$

By Eq. (2.31),

$$R = P_n F_{PR,i,n}$$

$$R_A = 24,529 F_{PR,10\%,2} = 24,529(0.57619) = \$14,133$$

$$R_B = 31,334 F_{PR,10\%,3} = 31,334(0.40211) = \$12,600$$

Machine $B$ is more economical. The ratio of costs is

$$\frac{A}{B} = \frac{14,133}{12,600} = 1.1217$$

## 2.7   COST COMPARISON BY CAPITALIZED COST

Another common denominator for all service lives is infinity leading to the concept of capitalized cost. Capitalized cost is the present value on an infinite time basis. For a machine costing $P_n$ and lasting $n$ years the present value replacing out to infinity is

$$K = P_n \sum_{x=0}^{\infty} \frac{1}{(1 + i)^{xn}}$$

The $\Sigma$ summation is an infinite series with first term 1 and ratio $1/(1 + i)^n$. The sum is, by Eq. (2.14),

$$K = P_n \frac{1 - [1/(1 + i)^n]^\infty}{1 - [1/(1 + i)^n]} 1 = P_n \frac{(1 + i)^n}{(1 + i)^n - 1} = P_n F_{PK,i,n} \qquad (2.32)$$

where $K$ is the capitalized cost and $F_{PK}$ is the factor that converts a present value to capitalized cost:

Capitalized cost = (present-value basis $n$ years' duration) (capitalized-cost factor)

$F_{PK}$, which converts $P$ to $K$, has the complete designation

$$F_{PK,i,n} = \frac{(1 + i)^n}{(1 + i)^n - 1} \qquad (2.33)$$

and is tabulated in Table A1.1.

The capital-recovery factor $F_{PR}$ and the capitalized-cost factor are related very simply. By Eq. (2.21),

$$F_{PR,i,n} = \frac{i(1 + i)^n}{(1 + i)^n - 1}$$

Comparing this with Eq. (2.33) shows that

$$F_{PR,i,n} = i F_{PK,i,n} \qquad (2.34)$$

Capital-recovery factor = (rate of return) (capitalized-cost factor)

Also, $R$ and $K$ are similarly related:

$$R = iK \qquad (2.35)$$

Unacost = (rate of return) (capitalized cost)

**Example 2.14.** Solve Example 2.12 on the basis of capitalized cost.

*Solution.* From the first solution,

$$P_{A2} = \$24,529$$

$$P_{B3} = \$31,334$$

By Eq. (2.32),

$$K = P_n F_{PK}$$

$$K_A = 24,529 F_{PK,10\%,2} = 24,529(5.7619) = \$141,330$$

$$K_B = 31,334 F_{PK,10\%,3} = 31,334(4.0211) = \$126,000$$

Machine $B$ is more economical. The ratio of costs is

$$\frac{A}{B} = \frac{141,330}{126,000} = 1.1217$$

As a matter of information, it is possible to convert a present value $P_{n1}$ of $n_1$ years' duration to an equivalent present value $P_{n2}$ of $n_2$ years' duration. The unacosts must be equal to each other. Hence applying Eq. (2.20) gives

$$P_{n1} F_{PR, i, n1} = P_{n2} F_{PR, i, n2}$$

$$P_{n2} = P_{n1} \frac{F_{PR, i, n1}}{F_{PR, i, n2}} \tag{2.36}$$

**Example 2.15.** Solve Example 2.12 by the use of Eq. (2.36).

*Solution.* From the first solution,

$$P_A = \text{2-year duration} = \$24{,}529$$

$$P_B = \text{3-year duration} = \$34{,}334$$

Convert the present value of machine $B$ to an equivalent present value for 2 years' duration using Eq. (2.36):

$$P_{B2} = P_{B3} \frac{F_{PR, 10\%, 3}}{F_{PR, 10\%, 2}} = 31{,}334 \frac{0.40211}{0.57619} = \$21{,}867$$

Machine $B$ is more economical. The ratio of costs is

$$\frac{A}{B} = \frac{24{,}529}{21{,}867} = 1.1217$$

Example 2.14 has now been calculated in four ways, all equivalent. The results are summarized in Table 2.2.

## 2.8 EXAMPLES OF COST COMPARISONS

It is desirable to develop expressions for unacost and capitalized cost for various types of costs or receipts. If the present value of the cost for a duration of $n$ years is

**TABLE 2.2**
**Comparison of solutions to Example 2.14**

| Method | Numerical value, $ | Ratio of costs $A/B$ |
|---|---|---|
| Present value on 6-year basis, common denominator of service lives; Example 2.12, first solution | $P_{A6} = 61{,}555$ $P_{B6} = 54{,}875$ | 1.1217 |
| Unacost, equivalent uniform end-of-year annual cost; Example 2.13 | $R_A = 14{,}133$ $R_B = 12{,}600$ | 1.1217 |
| Capitalized cost, present worth for an infinite duration; Example 2.14 | $K_A = 141{,}330$ $K_B = 126{,}000$ | 1.1217 |
| Present value of machine B converted to 2-year duration; Example 2.15 | $P_{A2} = 24{,}529$ $P_{B2} = 21{,}867$ | 1.1217 |

calculated, as from a time diagram, multiplication by the capital-recovery factor $F_{PR}$ and the capitalized-cost factor $F_{PK}$ will give the unacost and the capitalized cost. For example, consider a salvage value $C_{sal}$, a negative cost, due at the end of $n$ years. The present value is

$$-C_{sal}\frac{1}{(1 + i)^n}$$

The equivalent unacost, by Eq. (2.19), is

$$R = -C_{sal}\frac{1}{(1 + i)^n}\frac{i(1 + i)^n}{(1 + i)^n - 1} = -C_{sal}\frac{i}{(1 + i)^n - 1}$$

The capitalized cost, by Eq. (2.33), is

$$K = -C_{sal}\frac{1}{(1 + i)^n}\frac{(1 + i)^n}{(1 + i)^n - 1} = -C_{sal}\frac{1}{(1 + i)^n - 1}$$

Table 2.3 is a summary of a number of relationships. Indeed, every type of cost is covered by the tabulation, inasmuch as all costs can be included as an irregular expense at some time and thus can be accommodated by the $C_{ex}$ and $C_{bx}$ items, which are for irregular costs.

Unacosts can be added together. Capitalized costs can also be added together. For a system

$$R_{SYS} = R_1 + R_2 + R_3 + \cdots$$

$$K_{SYS} = K_1 + K_2 + K_3 + \cdots$$

**Example 2.16.** A new machine costs $8000 and lasts 10 years with negligible salvage value. If money is worth 10 percent per year, how much can be spent now to repair an old machine to extend its life 3 years?

*Solution.* Unacost for a new machine by item 1, Table 2.3, is

$$R_{new} = C_i F_{PR, 10\%, 10} = 8000(0.16275) = \$1302.0$$

For a repair job having the same unacost

$$R_{rep} = 1302.0 = C_i F_{PR, 10\%, 3} = C_i(0.40211)$$

$$C_i = \$3238$$

That is, up to $3238 can be spent for the repair job with money worth 10 percent per year.

**Example 2.17.** If in the preceding example the old machine can be repaired for $2500, what is the savings?

*Solution.* Unacost for a repair job by item 1, Table 2.3, is

$$R_{rep} = C_i F_{PR, 10\%, 3} = 2500(0.40211) = \$1005$$

$$R_{sav} = 1302 - 1005 = \$297$$

That is, $297/year is saved as of the end of each year for 3 years.

**Example 2.18.** A floor requires resurfacing at a cost of $5000 and will last 2 years. If money is worth 10 percent per year, how long must a new floor last if it costs $19,000?

*Solution.* The capitalized cost of resurfacing by item 1, Table 2.3, is

$$K = C_i F_{PK, 10\%, 2} = 5000(5.7619) = \$28,810$$

Assume a service life for the new floor and by repeated trials find the life for which the capitalized cost is $28,810:

$n = 11$:           $K = C_i F_{PK, 10\%, 11} = 19,000(1.5396) = 29,252$

$n = 12$:           $K = C_i F_{PK, 10\%, 12} = 19,000(1.4676) = 27,884$

By interpolation $n = 11.3$ for $K = 28,810$. That is, the new floor must last 11.3 years or longer.

**Example 2.19.** It will cost $5000 to line a stack and obtain a 2-year life. For $9400 a 4-year life can be obtained. At what rate of return are the two investments equal?

*Solution.* Try a 6 percent per year rate of return. Let $R_2$ and $R_4$ be unacost for 2- and 4-year linings. By item 1, Table 2.3,

$$R_2 = C_i F_{PR, 6\%, 2} = 5000(0.54544) = 2727.2$$

$$R_4 = C_i F_{PR, 6\%, 4} = 9400(0.28859) = 2712.7$$

For an 8 percent per year rate of return

$$R_2 = C_i F_{PR, 8\%, 2} = 5000(0.56077) = 2803.9$$

$$R_4 = C_i F_{PR, 8\%, 4} = 9400(0.30192) = 2838.0$$

That is,

At 6 percent:           $R_2 - R_4 = +14.5$

At 8 percent:           $R_2 - R_4 = -34.1$

and by interpolation, $R_2 - R_4 = 0$ at 6.6 percent per year rate of return. Below 6.6 percent the 4-year lining is more economical, and above 6.6 percent the 2-year lining is more economical. The example illustrates that a decision may be reversed depending on the rate of return.

**Example 2.20.** A machine was purchased for $10,000 from a manufacturer who guaranteed the machine for 10 years and also agreed to include 6 percent per year for the time value of money. At the end of 8 years the machine becomes worthless. How much of a refund should the manufacturer pay as of that time?

*First solution.* The unacost acceptable to the purchaser, by item 1, Table 2.3, was

$$R = C_i F_{PR, 6\%, 10} = 10,000(0.13587) = \$1359$$

**TABLE 2.3**
**Summary of unacost and capitalized-cost relationships—no tax**

| Item no. | Item | Present value as a cost for $n$ years' duration | Unacost — Algebraic expression | Unacost — Factor expression | Capitalized cost — Algebraic expression | Capitalized cost — Factor expression |
|---|---|---|---|---|---|---|
| 1 | $C_i$ Initial cost for $n$-year life | $C_i$ | $C_i \dfrac{i(1+i)^n}{(1+i)^n - 1}$ | $C_i F_{PR,i,n}$ | Multiply corresponding algebraic expression for unacost by $\dfrac{1}{i}$ | $C_i F_{PK,i,n}$ |
| 2 | $R$ Uniform end-of-year annual cost | $R \dfrac{(1+i)^n - 1}{i(1+i)^n}$ | $R$ | $R1$ (factor is unity) | | $R\dfrac{1}{i}$ |
| 3 | $R_b$ Uniform beginning-of-year annual cost | $R_b(1+i) \dfrac{(1+i)^n - 1}{i(1+i)^n}$ | $R_b(1+i)$ | $R_b(1+i)$ | | $R_b F_{PK,i,1}$ |
| 4 | $C_{ex}$ Irregular cost at end of $x$th year for article lasting $n$ years | $C_{ex}\dfrac{1}{(1+i)^x}$ | $C_{ex}\dfrac{1}{(1+i)^x} \times \dfrac{i(1+i)^n}{(1+i)^n - 1}$ | $C_{ex}\dfrac{1}{(1+i)^x}F_{PR,i,n}$ | | $C_{ex}\dfrac{1}{(1+i)^x}F_{PK,i,n}$ |

| | | | | | |
|---|---|---|---|---|---|
| 5 | $C_{bx}$ Irregular cost at beginning of xth year for article lasting $n$ years | $C_{bx}\dfrac{1}{(1+i)^{x-1}}$ | $C_{bx}\dfrac{1}{(1+i)^{x-1}} \times \dfrac{i(1+i)^n}{(1+i)^n - 1}$ | $C_{bx}\dfrac{1}{(1+i)^{x-1}} F_{PR,i,n}$ | $C_{bx}\dfrac{1}{(1+i)^{x-1}} F_{PK,i,n}$ |
| 6 | $C_{nd}$ Nondepreciable first cost, e.g., land or an article that lasts forever | $C_{nd}\dfrac{(1+i)^n - 1}{(1+i)^n}$ | $C_{nd}i$ | $C_{nd}i$ | $C_{nd}1$ (factor is unity) |
| 7 | $C_{sal}$ Salvage value at end of $n$th year, a receipt | $-C_{sal}\dfrac{1}{(1+i)^n}$ | $-C_{sal}\dfrac{i}{(1+i)^n - 1}$ | $-C_{sal}\dfrac{1}{(1+i)^n} \times F_{PR,i,n}$ | $-C_{sal}(F_{PK,i,n} - 1)$ or $-C_{sal}\dfrac{1}{(1+i)^n}F_{PK,i,n}$ |

The equivalent time pattern is

The purchaser gave $10,000 for $1359 worth of service as of the end of each year. At the end of 8 years he is owed $1359 1 year away and $1359 2 years away. The immediate value of these is

$$1359(1.06^{-1}) + 1359(1.06^{-2}) = \$2491$$

That is, $2491 would be a fair settlement.

**Second solution.** The purchaser is owed the value of a 2-year machine based on $10,000 for a 10-year machine. For equal unacosts by item 1, Table 2.3,

$$C_{i, 10} F_{PR, 6\%, 10} = C_{i, 2} F_{PR, 6\%, 2}$$

$$10,000(0.13587) = C_{i,2}(0.54544)$$

$$C_{i,2} = \$2491 \quad \text{(as before)}$$

**Third solution.** The purchaser actually received an 8-year machine. The price for an 8-year machine by capitalized cost, item 1, Table 2.3, would be

$$C_{i, 10} F_{PK, 6\%, 10} = C_{i, 8} F_{PK, 6\%, 8}$$

$$10,000(2.2645) = C_{i, 8}(2.6839)$$

$$C_{i, 8} = \$8437$$

The purchaser's overpayment 8 years ago brought up to the present is $(10,000 - 8437)$ $(1.06^8) = \$2491$, as before.

The three solutions illustrate the equivalence of various methods.

**Example 2.21.** Two machines have expenses as follows. If money is worth 10 percent per year, which is more economical and by how much?

|  | A | B |
|---|---|---|
| First cost, $ | 10,000 | 95,000 |
| Uniform end-of-year annual maintenance cost, $ per year | 3,000 | 1,000 |
| Irregular cost, beginning of third year, $ | 4,000 | |
| Irregular cost, beginning of fourth year, $ | 1,500 | |
| Salvage value, $ | 0 | 20,000 |
| Uniform end-of-year savings from quality control, $ per year | 0 | 6,000 |
| Life, years | 4 | 10 |

*Solution.* In a more involved problem the present value can be obtained first, after which conversion to a common-denominator basis can be made.

The present value of 4 years of machine $A$ is

$$P_{A4} = 10,000 + 3000F_{RP, 10\%, 4} + 4000F_{SP, 10\%, 2} + 1500F_{SP, 10\%, 3}$$

$$= 10,000 + 3000(3.1699) + 4000(0.82645) + 1500(0.75131)$$

$$= 23,943$$

for which the unacost, by item 1, Table 2.3, is

$$R_A = 23,943F_{PR, 10\%, 4} = 23,943(0.31547) = \$7553$$

The present value of 10 years of machine $B$ is

$$P_{B10} = 95,000 + (1000 - 6000)F_{RP, 10\%, 10} - 20,000F_{SP, 10\%, 10}$$

$$= 95,000 - 5000(6.1446) - 20,000(0.38554)$$

$$= 56,566$$

for which the unacost is

$$R_B = 56,566F_{PR, 10\%, 10} = 56,566(0.16275) = \$9206$$

Machine $A$ is more economical. The unacost savings is

$$9206 - 7553 = \$1653$$

That is, machine $A$ is more economical by $1653 per year as of the end of each of the 4 years that it is in service.

## 2.9 FINDING THE BEST ALTERNATIVE

Frequently a problem arises in such a way that several alternatives are possible, and the analyst must take precautions so as to arrive at the best answer, not just a better answer. The following problem illustrates the principle.

**Example 2.22.** A new tank costs $10,000 and lasts 5 years. An old tank can be repaired for $1500 to give 1 more year of service, then for $2000 to give 1 more year of service, then for $2500 to give another year of service, etc. If money is worth 10 percent per year, how long should the old tank be repaired?

*Solution.* At zero time, compare the cost of a new tank with the cost of repairing. Using capitalized cost and item 1, Table 2.3, we obtain

$$K_{new} = C_i F_{PK, 10\%, 5} = 10,000(2.6380) = \$26,380$$

$$K_{rep} = C_i F_{PK, 10\%, 1} = 1500(11.00) = \$16,500$$

It is more economical to repair. Repeat the calculation 1 year later:

$$K_{new} = \$26,380 \quad \text{(as above)}$$

$$K_{rep} = 2000F_{PK, 10\%, 1} = 2000(11.00) = \$22,000$$

Again, it is more economical to repair. Repeat the calculation 1 year later, at the end of the second year:

$$K_{\text{new}} = \$26,380 \qquad \text{(as before)}$$

$$K_{\text{rep}} = 2500F_{PK, 10\%, 1} = 2500(11.00) = \$27,500$$

It will not pay to repair. Since the repair cost is rising each year, it follows that the tank should be repaired for 2 years and then replaced with a new tank.

The following solution is incorrect but sometimes offered. Let $K_{r1}, K_{r2}, \ldots$ be the capitalized cost of repairing for $1, 2, \ldots$ years. Taking the present values and converting to a capitalized cost by item 1, Table 2.3, gives

$$K_{r1} = 1500F_{PK, 10\%, 1} = 1500(11.00) = 16,500$$

$$K_{r2} = \left(1500 + \frac{2000}{1.10}\right)F_{PK, 10\%, 2} = 3318.2(5.7619) = 19,119$$

$$K_{r3} = \left(1500 + \frac{2000}{1.10} + \frac{2500}{1.10^2}\right)F_{PK, 10\%, 3} = 5384.3(4.0211) = 21,650$$

Continuing, we have

| Years repair | $K$, $ |
|---|---|
| 1 | 16,500 |
| 2 | 19,119 |
| 3 | 21,650 |
| 4 | 24,096 |
| 5 | 26,457 |

The capitalized cost as found previously for a new tank is $26,380, and the tabulation might be interpreted as a decision to repair for 4 years because only for 5 years' repairing does the capitalized cost of repairing exceed the capitalized cost of a new tank. The reasoning is incorrect, however. The capitalized cost of repairing has been taken by years as a group. The tabulation merely indicates that it is more economical to repair for 4 years than not to repair at all. The group of 4 years includes the savings of the first 2 years, which overcome the loss of the last 2 years.

The following alternative solution is correct. Compare the programs of buying a new tank every 5 years, buying a new tank every 6 years after repairing for 1 year, etc. Take present values and convert to capitalized costs:

$$K_5 = 10,000F_{PK, 10\%, 5} = 10,000(2.6380) = 26,380$$

$$K_6 = \left(10,000 + \frac{1500}{1.10^5}\right)F_{PK, 10\%, 6} = 10,931(2.2961) = 25,099$$

$$K_7 = \left(10,000 + \frac{1500}{1.10^5} + \frac{2000}{1.10^6}\right)F_{PK, 10\%, 7} = 12,060(2.0541) = 24,772$$

Continuing, we obtain

| Total years | Years repair | $K$, $ |
|---|---|---|
| 5 | 0 | 26,380 |
| 6 | 1 | 25,099 |
| 7 | 2 | 24,772 |
| 8 | 3 | 25,010 |
| 9 | 4 | 26,000 |
| 10 | 5 | 26,409 |

The minimum occurs for repairing for 2 years. Repairing for 3 or even 4 years is more economical than not repairing at all, but not the most economical.

The following variation also gives the correct analysis. Consider programs of repairing for various numbers of years and then purchasing a new tank to complete the cycle:

$$K_5 = 10,000 F_{PK, 10\%, 5} = 10,000(2.6380) = 26,380$$

$$K_6 = \left(1500 + \frac{10,000}{1.10}\right) F_{PK, 10\%, 6} = 10,591(2.2961) = 24,318$$

$$K_7 = \left(1500 + \frac{2000}{1.10} + \frac{10,000}{1.10^2}\right) F_{PK, 10\%, 7} = 11,583(2.0541) = 23,793$$

Continuing, we get

| Years repair | $K$, $ |
|---|---|
| 0 | 26,380 |
| 1 | 24,318 |
| 2 | 23,793 |
| 3 | 24,174 |
| 4 | 25,122 |
| 5 | 26,427 |

Again, repairing for 2 years is most economical. Repairing for 3 or 4 years is more economical than not repairing at all, but not the most economical.

## 2.10 CONSIDERATIONS IN COST COMPARISON

In a practical problem all the pertinent costs must be included and the analyst must be on guard to avoid omissions. If alternatives have costs equal in all respects, such costs may be omitted because only differences affect the choice. However, equality in all respects must be strictly true. Thus suppose the comparison is between heat-exchanger tubes lasting 2 and 5 years. The installation cost for each may be the same, but it must be included. In one case the installation cost will be incurred every

2 years, in the other case every 5 years. Conversely, if the cost for cleaning the tubes is the same for each on a yearly basis, this cost can be omitted in the comparison. Of course, inclusion of equal costs is never detrimental and may prevent their being overlooked in subsequent considerations.

It is not necessary that the machines or systems being compared give equal service. Benefits or penalties can include such considerations as extra production, better quality control, or even reduction of pollution. However, savings or losses associated with a machine should be included only if they are real. Downtime for repairs should not be considered a penalty for lost production if the machine can attain the overall production rate even with the downtime. Frequently the downtime of a machine can be a significant consideration.

Every practical comparison contains a number of intangibles. Two machines may have different capacities, although both can meet present demands. No immediate benefit can be given for the larger capacity, but it is an intangible benefit. In a comparison between air cooling and water cooling, air cooling has an intangible value because water is becoming more restricted and relatively more expensive with time. The intangibles can be an annoyance to the analyst. If they can be given a monetary value, they should be included to that extent. At least they can be enumerated so that they are there to be seen. The intangibles can easily swing the decision in a close comparison.

Sometimes such items as service life or future repair costs are not definitely known, but the comparison can be made on the basis of a range. For example, machine $A$ can be calculated on a high, best estimate, and low basis. Also, results can be presented in a form that circumvents the uncertainty. In Example 2.18 the service life of the new floor is not known, but the calculation that the floor would have to last 11.3 years to be economical could facilitate making a decision.

## 2.11  THE TWO TIME FACTORS

It has been emphasized that two time factors are involved in making a cost comparison. They are (1) the instant at which costs are incurred and (2) the duration over which costs extend. Unacost and capitalized cost both have built-in common denominators to accommodate any duration and convert it to a standard duration. If the durations for two machines are the same, both methods are still applicable. The methods are fundamentally present-value methods. Unacost uses a 1-year duration as an end-of-year cost, and capitalized cost uses an infinite duration with a zero-time cost. The methods do not imply respectively that an article will be used 1 year or replaced forever. The 1-year and infinite durations are chosen merely for mathematical convenience. Each method gives the correct comparison for any duration. The methods are readily converted from one to the other by the simple relationship [Eq. (2.35)]

$$R = iK$$

In most problems the method used drops out. Management, however, may understand a unacost figure of, say, $10,000 per year better than $100,000 capitalized cost, although both are equivalent at a 10 percent per year rate of return.

The result of a comparison depends on the time value of money as demonstrated by Example 2.19, where the decision is reversed above and below 6.6 percent. A discussion of what rate of return should be used is not presented here.

A subsequent chapter will discuss the analysis of whole projects or ventures for which all receipts and expenses are known, or at least assigned a definite value. In the present chapter the receipts associated with a particular machine or group of machines are usually not known. Thus in a long assembly line it is not possible to allocate the receipts of the entire operation to individual services or machines. This does not prevent the analyst, however, from using the methods of this chapter for comparing two competing machines. The rate of return used in the calculation is then merely the time value of money that the analyst places on investment for machines performing the indicated service. Generally it will not be the same as the rate of return for the venture.

## 2.12 PROCEDURE FOR COST COMPARISONS

When the time value of money is involved in the comparison of two or more machines or systems, the best procedure is to draw a time diagram of all money values incurred. Then convert each diagram to a present value and the present value for each system to a unacost or capitalized cost, either of which will correct for different durations of the system.

Money put into the system can be displayed by lines below the time base of the time diagram and money received by the system by lines above the time base. The length of the lines is not significant. In this chapter the emphasis is on cost and a cost is treated as a positive amount, with income being a negative amount. Later, in discussing profitability, income will be treated as positive, but this causes no confusion.

The fundamental aspects of the mathematics of cost comparisons have been presented in this chapter. Subsequent chapters will discuss additional considerations such as taxes.

The procedure stressed in this chapter emphasizes the conceptual nature and use of various factors, such as $F_{PR}$, rather than their algebraic form. The best engineering practice in solving a problem involves three distinct steps. Each step should be done independently so that there is no mental block caused by intermingling the steps. The first step is to draw a picture of the problem, here a time diagram. The next step is to set up the problem in a conceptual mathematical form, here the various amounts with the required $F$ factors, to give an equation of value. The third step is to perform the calculations, here solving to obtain the actual numerical values.

## NOMENCLATURE

$A_{t1}$      Amount at time 1, $
$C_{bx}$     Irregular cost at the beginning of the $x$th year, $
$C_{ex}$     Irregular cost at the end of the $x$th year, $
$C_i$      Initial or first cost, $

| | |
|---|---|
| $C_{sal}$ | Salvage value, $ |
| $F$ | Future value by some writers, $ |
| $F$ | Reserved for factors in this book |
| $F_{PK}$ | Factor to convert $P$ to $K$; capitalized cost factor, dimensionless |
| $F_{PR}$ | Factor to convert $P$ to $R$; capital recovery factor, year$^{-1}$ |
| $F_{PS}$ | Factor to convert $P$ to $S$; compound interest factor, dimensionless |
| $F_{RP}$ | Factor to convert $R$ to $P$; unacost present-value factor, years |
| $F_{RS}$ | Factor to convert $R$ to $S$; equal-payment-series future-value factor, years |
| $F_{SP}$ | Factor to convert $S$ to $P$; single-payment present-value factor or present-value factor, dimensionless |
| $F_{SR}$ | Factor to convert $S$ to $R$; sinking-fund factor, year$^{-1}$ |
| FV | Future value |
| $i$ | Rate of return, decimal per year |
| $K$ | Capitalized cost, $ |
| $n$ | Number such as the life of an article, years |
| NPV | Net present value, $ |
| $p$ | Number of periods per year |
| $P$ | Present value, $ |
| PV | Present value, $ |
| PW | Present worth, $; not used |
| $R$ | Uniform end-of-year annual amount; unacost, $ per year |
| $R_b$ | Uniform beginning-of-year annual amount, $ per year |
| $R_e$ | Same as $R$ |
| $S$ | Single future value, $ |
| $u^n$ | $(1 + i)^n$ |
| $v^n$ | $(1 + i)^{-n}$ |

## PROBLEMS

**2.1.** The mortgage on a house is $50,000 payable at 12 percent per year. If the end-of-month payments are $717.35, find how much of the first and second payments go to interest and how much to payment of the principal.

**2.2.** What is the present value of $1000 due in 6 months at 10 percent per year?

**2.3.** A person can discharge a debt by paying $1000 now or $1075 6 months from now. If money is worth 10 percent per year, what is the cash advantage in paying the debt now?

**2.4.** A person borrows $1000 from a credit union. At the end of each month $100 is paid as a principal payment plus interest at 12 percent per year on the principal amount for the month. How much total interest is paid on the loan?

**2.5.** A person signs a note for $1000 due with interest in 8 months at 10 percent per year. Two months later the holder of the note sells it when the current interest rate is 12 percent. What does the note sell for?

**2.6.** A person owes $1000 due in 4 months and $500 due in 1 year. What single payment 6 months from now will discharge the debt if money is worth 8 percent per year?

**2.7.** Find the effective rate of return if money at 8 percent per year is compounded
(a) quarterly  (b) monthly  (c) daily  (d) hourly  (e) continuously

**2.8.** Due now is $5000, and $10,000 is due 6 months from now. What amount will discharge the debt 3 years and 9 months from now if money is worth 12 percent per year?

**2.9.** What will $15,000 amount to in 4 years at 8 percent per year compounded
(*a*) annually?  (*b*) quarterly?

**2.10.** What sum will yield $1850 in 5 years at
(*a*) 6 percent per year?  (*b*) 5 percent per year?

**2.11.** In 1626 Peter Minuit bought Manhattan Island from the Indians for $24 worth of trinkets. What would be the value in 1985 with interest compounded at 6 percent per year?

**2.12.** What is the rate of return if $10,000 in 7 years amounts to
(*a*) $17,138?  (*b*) $16,400?

**2.13.** If money is worth 10 percent per year, find the present value of a $10,000 receipt due in 4 years, plus a $15,000 receipt due in 6 years, plus an $8000 debt due in 9 years, plus a $12,000 receipt due in 10 years.

**2.14.** In Prob. 2.13 find a single debt due in 7 years which would make the receipts and debts equivalent.

**2.15.** What is the present value of a series of equal end-of-year payments of $1000 for 8 years if money is worth 10 percent per year?

**2.16.** A series of end-of-year payments forms a geometric series with first payment $1000, second payment $1210, etc. Find the present value of the payments for 6 years if money is worth 10 percent per year.

**2.17.** What is the present value of a series of equal beginning-of-year payments of $1000 for 8 years if money is worth 10 percent per year?

**2.18.** If $10,000 is owed, what equal yearly payments for 8 years will discharge the debt if money is worth 6 percent per year and payments are made
(*a*) at the end of the year?  (*b*) at the beginning of the year?

**2.19.** A note has a maturity value of $10,000 due in 4 years and 6 months. What is the present value if money is worth 12 percent per year?

**2.20.** A debtor signs a note for $10,000 due 5 years from now, with interest to accumulate at 10 percent per year. The holder of the note sells it 2 years and 6 months from now when money is worth 8 percent per year. What is the note sold for?

**2.21.** If $30,000 is owed and equal payments of $5500 at the end of each year for 8 years will discharge the debt, what is the rate of return?

**2.22.** If $20,000 is owed with money worth 6 percent per year, and equal end-of-year payments of $3000 are made, how long will it take to discharge the debt?

**2.23.** A lot can be bought for $10,000 cash now, or for $4000 cash now plus $7000 cash at the end of 4 years. If money is worth 10 percent per year, which plan is better and by how much?

**2.24.** The population of a city increased from 24,852 to 32,173 in 10 years. Find the average annual rate of increase during the period.

**2.25.** A person bought a house for $50,000 with various closing fees of $800. Three years later the house was sold for $72,000 with $1200 in various fees and a real estate broker's fee of 6 percent of the selling price. What average annual rate of return was realized on the investment?

**2.26.** The sales of a business are $100,000 per year and are increasing at a rate of 20 percent per year. How long will it take for sales to be $1,000,000 per year?

**2.27.** A certain organism can double its population every 6 hours. Assuming no deaths, how long will it take 100 of the species to increase to 1 million?

**2.28.** A house sells for $50,000 and 3 years later for $75,000. Assume that it increases in value at a constant rate. Find
(*a*) what it will sell for 4 years after it sells for $75,000.
(*b*) when it will sell for $250,000.

**2.29.** A person dies and leaves an estate of $100,000, which is invested and earns 8 percent per year. Three children aged 8, 10, and 13 survive the person and at age 18 each child will receive an equal amount from the estate. What is the amount?

**2.30.** A building sells for $100,000, for which $15,000 is paid in cash. A note for $15,000 due in 1 year, a note for $20,000 due in 2 years, and two equal notes payable in 3 and 4 years are given. Money is worth 10 percent per year. What is the maturity value of the notes due in 3 and 4 years?

**2.31.** A person owes $10,000 due 1 year from now, owes $30,000 due 5 years from now, will receive $20,000 3 years from now, and will receive $26,500 8 years from now. At what rate of return will the person be even?

**2.32.** A person pays an ordinary annuity of $1000 per year for 10 years and makes no further payments. If money is worth 10 percent per year, what is the value in the fund at the end of 15 years?

**2.33.** Repeat Prob. 2.32 for an annuity due.

**2.34.** A person deposits $1000 at the end of the year for 20 years and then withdraws a uniform amount at the end of each year for the next 5 years. If money is worth 8 percent per year, what is the amount of each withdrawal?

**2.35.** A person deposits $1000 at the end of each year for 10 years. If money is worth 8 percent per year, find the interest earned by the fund for the sixth year.

**2.36.** The money payment on a house is $300 payable on the first of the month, with money worth 12 percent per year compounded monthly. A person misses the February, March, April, May, and June payments. What single payment can be made on July 1 to become up to date?

**2.37.** Show that the present value of an ordinary annuity can be written

$$P = \frac{1 - v^n}{i} \qquad \text{where } v = (1 + i)^{-1}$$

**2.38.** Find the equivalent cash cost for an automobile that can be purchased for $1000 down and $300 a month for 36 months if money is worth 18 percent per year compounded monthly.

**2.39.** A person sells a lot for $100,000 but for tax purposes takes 29 percent down and the balance in 14 equal payments starting 1 year from now and continuing every 3 months, with money worth 10 percent per year compounded quarterly. Find the amount of the quarterly payments.

**2.40.** A person wishes to accumulate exactly $10,000 by making end-of-year payments of $1000 with money worth 8 percent. How long a period is required? If the last payment is to be at the end of a year and is to make the fund exactly $10,000, what is the amount of the last payment?

**2.41.** A dealer sells an automobile for $9500 less a $1000 allowance for a used automobile. Payments of $276.20 per month are to be made for 36 months starting 1 month from today. What is the monthly rate of interest charged, the nominal rate, and the effective rate?

**2.42.** A person inherits $100,000 but elects to take 10 yearly payments starting 5 years from now. If money is worth 8 percent per year, find the yearly payments.

**2.43.** A debt of $12,000 is to be amortized, with equal end-of-year payments of $2000 and money worth 10 percent per year. Payments will be made only at the end of the year. Find the amount of the last payment.

**2.44.** Lenders sometimes charge points on a loan to avoid exceeding a legal limit on interest rates. Thus for five points the lender would loan only $95 for each $100 mortgage. The borrower borrowing $1000 would receive only $950 but would have to make payments just as if he or she had received $1000. A borrower makes a $100,000 loan payable at the end of each year for 5 years with an interest rate of 12 percent per year but has been charged five points. What is the true interest rate?

**2.45.** A bond has a face value of $1000, pays 6 percent per year semiannually, and becomes due in 4.5 years. Find the purchase price if it is bought to yield
(*a*) 8 percent   (*b*) 4 percent

**2.46.** A bond has a face value of $1000, pays 8 percent per year semiannually, becomes due in 3 years, and is purchased for $1100. What is the yield rate to the purchaser?

**2.47.** A bond with a face value of $1000 pays interest semiannually and has 3 years to go to maturity. A purchaser will net 12 percent per year by buying the bond for $901.65. Find the annual rate appearing on the face of the bond.

**2.48.** A bond with a face value of $1000 pays 8 percent per year semiannually and has 4 years to maturity. Although it should have eight coupons attached, it has only six, the next two due having been removed. A purchaser wishes to earn 10 percent on the money. What should be paid for the bond?

**2.49.** A bond with a face value of $1000 pays 8 percent per year semiannually. It is purchased for $922.78, so that the yield to maturity is 10 percent. Find the time to maturity for the bond.

**2.50.** A machine costs $15,000 and will have a $3000 salvage value at the end of its life in 10 years. A sinking fund is established at 8 percent per year to purchase a replacement.
(*a*) What is the sinking-fund payment?
(*b*) What is the value of the sinking fund at the end of 7 years?
(*c*) How much does the sinking fund increase during the eighth year?

**2.51.** If $10,000 is owed and $1740 is to be paid annually at the end of the year for 8 years to discharge the debt, but the fourth payment is missed, what equal end-of-year payments thereafter will discharge the debt on time?

**2.52.** If $5000 is spent now to avoid $15,000 in repairs 2 years from now, what rate of return is earned by repairing now?

**2.53.** An operation is performed manually at a cost of $10,000 per year. A machine costing $60,000 and lasting 12 years will perform the operation and eliminate the labor cost.
(*a*) What is the rate of return earned by the investment?
(*b*) If in addition the machine is worth $5000 per year because of added production, what is the rate of return on the investment?

**2.54.** A company can produce 1 million units per year on a product which infringes a patent. The company can pay a royalty of $0.08 per unit or invest $200,000 in equipment and incur a cost of $0.05 per unit to avoid the patent. If money is worth 12 percent per year, which course should the company take if only 10 years ahead are to be considered? What are the intangibles?

**2.55.** Three machines are available for an operation. Machine *A* costs $100,000 and lasts 5 years. Machine *B* costs $150,000 and machine *C* costs $200,000. If money is worth 20 percent per year, how long must machines *B* and *C* last to be economically equal to machine *A*?

**2.56.** Two machines have the following costs:

|  | A | B |
|---|---|---|
| First cost, $ | 10,000 | 25,000 |
| Salvage value, $ | 1,100 | 1,500 |
| Uniform end-of-year expense, $ per year | 3,000 | 2,000 |
| Irregular expense, end first year, $ | 1,000 | |
| Irregular expense, end third year, $ | | 2,500 |
| Benefit from quality control, as of end of each year, $ per year | | 600 |
| Life of machine, years | 2 | 5 |

Compare the machines on the following basis:
(a) Present value for some common denominator of the service lives with money worth 8 percent per year.
(b) Present value, basis 4 years, with money worth 8 percent per year.
(c) Unacost with money worth 8 percent per year.
(d) Capitalized cost with money worth 8 percent per year.
(e) The service life of machine B that would be economically equal to machine A with money worth 8 percent per year.
(f) The rate of return at which the two are equal.

**2.57.** Expressions for irregular costs $C_{bx}$ and $C_{ex}$ are given in Table 2.3. Derive the expressions for
(a) $C_{bx}$ and (b) $C_{ex}$.

**2.58.** Derive the expressions for a nondepreciable first cost such as land given in Table 2.3.

**2.59.** A new tank costs $50,000 and lasts 5 years. An old tank can be patched now for $7000 to give one more year of life, then for $9500 to give another year of life, then for $12,000 to give one more year of life, etc. How long should the old tank be patched if money is worth 10 percent per year?

**2.60.** A liquid stream in an oil refinery requires servicing at a cost of $0.04 per barrel in a reactor. Operation is at 10,000 barrels per day for 350 days per year. A new catalyst is available at a cost of $180,000, which would reduce the servicing cost by one-half. If money is worth 8 percent per year, how long will it take for an investment in the new catalyst to pay off?

**2.61.** A pipeline was built 3 years ago to last 6 years. It develops leaks according to the relationship

$$\log N = 0.07T - 2.42 \qquad T > 30$$

where $N$ is the total number of leaks from installation and $T$ is the time in months from installation. It costs $500 to repair a leak. If money is worth 8 percent per year, how much can be spent now for a cathodic system that will reduce leaks by 75 percent?

**2.62.** Plan A is to build an addition having a capacity of 600 Mg/yr at a cost of 3 M$ and 3 years from now to add to the addition at a cost of 2.4 M$ to double production; M is mega-.

Plan B is to build a full-scale plant now for 4.5 M$ with a capacity of 1200 Mg/yr.

The product sells for $3/kg. For plan *A* the variable operating cost is $0.50/kg produced for all the years, and for plan *B* it is $0.40/kg produced for all the years. Sales demand for the years is expected to be

| Year | Mg |
|------|------|
| 1 | 600 |
| 2 | 800 |
| 3 | 1000 |
| 4 | 1200 |
| 5 | 1200 |

If money is worth 10 percent per year, which plan is the more economical?

**2.63.** Which of the following machines is more economical, and by how much per year if money is worth 10 percent per year?

| | *A* | *B* |
|------|------|------|
| First cost, $ | 12,000 | 20,000 |
| Annual expense, end year 1 | 3,000 | — |
| 2 | 3,000 | — |
| 3 | 3,000 | 4,000 |
| 4 | | — |
| 5 | | — |
| Life, years | 3 | 5 |

**2.64.** A machine costs $100,000 and lasts 5 years. Past records show

| Year | Repairs as of beginning of year, k$ | Salvage value, end of year, k$ |
|------|------|------|
| 1 | 0 | 50 |
| 2 | 22 | 20 |
| 3 | 28 | 8 |
| 4 | 38 | 3 |
| 5 | 65 | 2 |

If money is worth 10 percent per year, how long should the machine be kept?

**2.65.** An operation can be performed manually at an end-of-year cost of $72,000/year. How much can be spent for a machine which lasts 10 years if the operating cost is $18,000 at the end of the year and an overhaul costing $24,000 is required at the beginning of the fifth year? Money is worth 10 percent per year.

**2.66.** A machine is bought which will yield a 10 percent per year return based on the following end-of-year savings:

| Year | Savings, $ |
|------|-----------|
| 1 | 10,000 |
| 2 | 15,000 |
| 3 | 17,000 |
| 4 | 12,000 |
| 5 | 10,000 |

The savings for the first 3 years are obtained as predicted, but the machine has to be abandoned with a $5000 value at the end of the third year. What rate of return was actually realized?

**2.67.** It is required to produce 1 million gears per year for a small automobile part. The study will look ahead 12 years, with money worth 14 percent per year.

Plan *A* is to build a production unit in the United States at a cost of 7 M$, with expenses of 2 M$ annually. There will be a 0.5 M$ investment at the end of the sixth year for better pollution control. M is mega-.

Plan *B* is to build a production unit in Asia for 4 M$ having a yearly expense of 1.2 M$, but with a duty of 30 percent of the selling price of that in the United States. At what selling price are the two plans equal?

**2.68.** A company makes 2 million parts per year of a product. The company pays $0.10 per part as a royalty or can circumvent the royalty by investing 300 k$ and incurring a cost of $0.05 per part. Money is worth 10 percent per year, and 8 years of operation is expected. How much is saved or lost per year by paying the royalty?

**2.69.** A machine costs $10,000 and lasts 12 years. If money is worth 10 percent per year, what is the loss on abandonment at the end of 8 years?

**2.70.** Two types of pipe are available for carrying water with costs as follows:

| | A | B |
|------|------|------|
| First cost, $ | 100,000 | 200,000 |
| Annual end-of-year cost, $ per year | 25,000 | 20,000 |
| Salvage value, $ | 0 | 0 |
| Life, years | 15 | 20 |

Type *A* pipe must be repaired at times as part of the annual cost, but the water can become contaminated at these times. Contamination with type *B* pipe can be neglected. If money is worth 6 percent per year, how much benefit must be given to type *B* pipe to make it economically equal to type *A* pipe?

**2.71.** A company manufactures 1 million units per year which sell at $2 less than a competitor's superior product. How much can be spent on research as of the end of this year, assuming a 40 percent chance for success in attaining the extra $2 per unit? Money is worth 10 percent per year, and a period of 8 years beyond this year is to be considered.

**2.72.** A machine which lasts 10 years can be bought for $10,000. Operating expenses for the machine form an arithmetic progression and are $2000 for the first year as an end-of-year expense, $4000 for the next year, etc. At what age should the machine be exchanged for a new one if money is worth 10 percent per year? Ignore the salvage value.

**2.73.** A company predicts and then actually realizes receipts from an operation as follows:

| End of year | Predicted, k$ | Actual, k$ |
|-------------|---------------|------------|
| 1 | 1000 | 1200 |
| 2 | 1300 | 1400 |
| 3 | 1500 | 1000 |
| 4 | 1700 | 300 |
| 5 | 1200 | 100 |
| 6 | 800 | 0 |
| 7 | 400 | 0 |
| 8 | 100 | 0 |

A unit is set up at a cost that will yield 20 percent per year on the basis of the 8-year prediction.

(*a*) What was the actual rate of return?

(*b*) What was the loss on abandonment at the end of the fifth year based on the original 20 percent per year rate of return?

**2.74.** A company owns its office building, which can be sold for $10,000,000 now. Total operating expenses for the building are $600,000 per year, and every 5 years from now a modernization outlay for the building will cost $1,000,000. The company expects to use the building for 20 years, at which time it should sell for $8,000,000. The company can earn 10 percent on money invested in its business. A real estate firm offers to buy the building and lease it back to the company. What is the maximum end-of-year rent that the company could afford to pay?

# REFERENCES

1. Barish, N. N., and S. Kaplan: *Economic Analysis for Engineering and Managerial Decision Making,* 2d ed., McGraw-Hill Book Company, New York, 1978.
2. Canada, J. R.: *Intermediate Economic Analysis for Management and Engineering,* Prentice-Hall, Inc., Englewood Cliffs, N.J., 1971.
3. Cissell, R., and H. Cissell: *Mathematics of Finance,* 4th ed., Houghton Mifflin Company, Boston, 1973.
4. DeGarmo, E. P., J. R. Canada, and W. G. Sullivan: *Engineering Economy,* 7th ed., The Macmillan Publishing Company, New York, 1984.
5. Estes, J. C.: *Handbook of Interest and Annuity Tables,* McGraw-Hill Book Company, New York, 1976.
6. Fleischer, G. A.: *Engineering Economy: Capital Allocation Theory,* Wadsworth, Inc., Monterey, Calif., 1984.
7. Grant, E. L., W. G. Ireson, and R. S. Leavenworth: *Principles of Engineering Economy,* 7th ed., John Wiley and Sons, New York, 1982.
8. Gushee, C. H., ed.: *Financial Compound Interest and Annuity Tables,* 5th ed., Publication No. 376, Financial Publishing Company, Boston, 1970.

9. Humphreys, K. K., ed.: *Project and Cost Engineers' Handbook,* 2d ed., Marcel Dekker, Inc., New York, 1984.

10. Kasner, E.: *Essentials of Engineering Economics,* McGraw-Hill Book Company, New York, 1979.

11. Minrath, W. R.: *Handbook of Business Mathematics,* 2d ed., D. Van Nostrand Company, Princeton, N.J., 1967.

12. Newnan, D. G.: *Engineering Economic Analysis,* 3d ed., Engineering Press, San Jose, Calif., 1983.

13. Ostwald, P. F.: *Cost Estimating for Engineering and Management,* Prentice-Hall, Inc., Englewood Cliffs, N.J., 1974.

14. Riggs, J. L.: *Engineering Economics,* 2d ed., McGraw-Hill Book Company, New York, 1982.

15. Smith, G. W.: *Engineering Economy: Analysis of Capital Expenditures,* 3d ed., Iowa State University Press, Ames, Iowa, 1979.

16. Taylor, G. A.: *Managerial and Engineering Economy,* 3d ed., D. Van Nostrand Company, Inc., New York, 1980.

17. Thorndike, D., and D. S. Benton: *Thorndike Encyclopedia of Banking and Financial Tables,* rev. ed., Warren, Gorham & Lamont, Inc., Boston, 1981.

18. Woods, D. R.: *Financial Decision Making in the Process Industry,* Prentice-Hall, Inc., Englewood Cliffs, N.J., 1975.

# CHAPTER

# 3

# DEPRECIATION AND TAXES: EQUIVALENCE AFTER TAXES

*F. C. Jelen, C. L. Yaws, and K. K. Humphreys*

## DEPRECIATION AND TAXES

## 3.1 NATURE OF DEPRECIATION

In any economic endeavor expenditures are incurred that have various durations. For accounting and tax purposes a sharp demarcation exists between expenditures that are paid for and used up in a single accounting period and expenditures for which the duration exceeds one accounting period. A year is accepted as the period for both accounting and tax purposes; hence there is a sharp distinction between expenditures consumed within 1 year and expenditures which outlast 1 year. Investment in a building will last many years. It would be incorrect to charge the cost of the building entirely to operations for the first year and make no charge for the cost of the building for subsequent years. A means must be found for spreading over the years expenditures that outlast 1 year. An expenditure that decreases in value with time must be apportioned over its life. The term used to describe this loss in value is *depreciation*.

Depreciation arises from two causes: wear and tear and obsolescence. A machine may wear out and eventually be unable to perform its function. A machine may show little wear and tear but become economically inefficient due to technological advancement. The latter is called *obsolescence* and represents a loss in value, just as real as wear and tear. Wear and tear and obsolescence are both recognized by tax rules as depreciation.

Depreciation is an expense. Indeed, the accountant's charge for depreciation is depreciation expense. Depreciation is a fundamental aspect of accounting and cost engineering, which must be considered for manufacturing costs, pricing, taxes, and cost comparisons. A company may use one depreciation method for tax purposes and another method for accounting purposes.

Popular presentations of depreciation can lead the student to narrow interpretations. The popular conception of the existence of a depreciation fund to which payments are made so that the original investment is restored when an article is worn out is not in accord with practice. In this book depreciation is used in its mathematical sense. Depreciation arises when an expenditure outlasts 1 year, and a depreciation schedule is a means of allocating the expenditure as an expense over the years.

A depreciation schedule has two mathematical requirements: the life of the depreciable equipment and the mathematical procedure used in allocating the depreciation expense for each year of life of the equipment.

For U.S. tax purpose the allowable life of equipment is given by the class life system. Table 3.1 is a brief selection of classes taken from *Tax Information on Depreciation,* published by the U.S. Internal Revenue Service. The analyst may choose any service life for internal company use but must use IRS guidelines for tax purposes.

A schedule for allocating depreciation over the years for tax purposes must meet government rules. Prior to 1981 the U.S. government would accept any consistent method, provided the total depreciation at the end of each year did not exceed during the first two-thirds of the useful life the total depreciation that would result if twice straight-line declining-balance depreciation were used.

The use of accelerated depreciation predates the tax situation. An asset, such as a machine, represents two types of cost, the initial depreciable cost and the operating cost. Operating costs tend to be low in the early years and high in the later years; hence the practice arose of using high depreciation in the early years and low depreciation in the later years to keep the sum of depreciation and operating costs more nearly constant. Accelerated depreciation, too, is more in line with the resale value of machinery at intermediate years.

The U.S. tax laws governing depreciation were totally rewritten in 1980 and again in 1986 to mandate certain specific methods of calculating depreciation.

The 1980 law introduced the concept of "recovery property" and a new depreciation system known as ACRS (accelerated cost recovery system). Recovery property is tangible depreciable property placed in service after 1980 and before 1987. ACRS allows the taxpayer to recover the cost of the property over a specified "recovery period," which is determined by the class life of the property.

The 1986 tax reform act provided for a revised form of ACRS known as MACRS (modified accelerated cost recovery system), which applies to tangible depreciable property placed in service after 1986. MACRS effectively mandates the depreciation scheme known as the declining-balance system or, alternately, straight-line depreciation, for most industrial property. Straight-line depreciation is required for real property. MACRS, however, specifies recovery periods as shown in Table 3.1.

For property acquired before 1981 in the United States numerous depreciation methods are acceptable. Many of these same methods continue to be acceptable in

**TABLE 3.1**
## Depreciation class lives and MACRS recovery periods

| | Class life (in years) | MACRS recovery period (in years) |
|---|---|---|
| **Assets used in all businesses** | | |
| Office furniture and equipment | 10 | 7 |
| Information systems, including computers | 6 | 5 |
| Heavy general-purpose trucks | 6 | 5 |
| Railroad cars and locomotives | 15 | 7 |
| Vessels, barges, tugs | 18 | 10 |
| **Depreciable assets used in the following activities** | | |
| Agriculture | 10 | 7 |
| Mining | 10 | 7 |
| Exploration for and production of petroleum | 14 | 7 |
| Petroleum refining | 16 | 10 |
| Manufacture of wood products and furniture | 10 | 7 |
| Manufacture of pulp and paper | 13 | 7 |
| Manufacture of chemical and allied products | 11 | 7 |
| Manufacture of primary nonferrous metals | 14 | 7 |
| Manufacture of machinery | 10 | 7 |
| Manufacture of electronic components, products, and systems | 6 | 5 |
| Manufacture of motor vehicles | 12 | 7 |
| Manufacture of aerospace products | 10 | 7 |
| Ship- and boat-building machinery and equipment | 12 | 7 |
| Railroad machinery and equipment | 14 | 7 |
| Motor transport-freight | 8 | 5 |
| Water transport | 20 | 15 |
| Air transport | 12 | 7 |
| Electric utility hydraulic production plant | 50 | 20 |
| Electric utility nuclear production plant | 20 | 15 |
| Electric utility steam production plant | 28 | 20 |

*Source: Tax Information on Depreciation,* U.S. Internal Revenue Service Publication No. 534, published yearly.

other countries. For this reason and also because some companies use a different depreciation method for accounting purposes from the one used for tax purposes, it is necessary that the student be familiar with all commonly used depreciation techniques. Therefore, problems and examples are included in this text which illustrate all of these various methods.

An additional complication of U.S. tax law is that property placed in service before 1981 which is not yet fully depreciated must continue to be depreciated using whatever method was selected when the property was obtained. Also, in rare cases certain property does not qualify for ACRS or MACRS depreciation and an alternate depreciation method must be selected. Any depreciation method which was generally acceptable prior to 1981 may be used in these cases.

## 3.2 DEPRECIATION TERMS

Suppose an asset has an initial cost $C_i$, a life of $n$ years, and a terminal, or salvage, value $C_{sal}$. Let $D_{f1}, D_{f2}, \ldots$ be the fractional depreciation for each year. The total depreciation, or the *depreciable first cost* $C_d$, is

$$C_d = C_i - C_{sal} \tag{3.1}$$

The depreciation for any arbitrary year, the $m$th year, is

$$D_m = D_{f,m} C_d \tag{3.2}$$

The *book value* of the asset at the end of $m$ years $B_m$ is the original value less the accumulated depreciation expense to that time:

$$B_m = C_i - C_d \sum_{m=1}^{m=m} D_{f,m} \tag{3.3}$$

Book value = initial cost − accumulated depreciation

Subtracting and adding $C_d$ to Eq. (3.3) gives

$$B_m = C_i - C_d + C_d - C_d \sum_1^m D_{f,m} = C_{sal} + C_d \left( 1 - \sum_1^m D_{f,m} \right) \tag{3.4}$$

Book value = salvage value + future depreciation

The book value may bear no relation to the resale value.

When the time value of money is a factor, it becomes necessary to consider the present value of the yearly depreciation.

## 3.3 STRAIGHT-LINE DEPRECIATION (SL)

Allocation of the depreciable cost is uniform for all years in this method:

$$D_{f1} = D_{f2} = \cdots = D_{f,n}$$

or

$$D_f = \frac{1}{n} \tag{3.5}$$

Depreciation for the $m$th year is

$$D_m = \frac{C_d}{n} \tag{3.6}$$

Accumulated depreciation for $m$ years is

$$\sum D_m = C_d \frac{m}{n} \tag{3.7}$$

Book value at the end of $m$ years is

$$B_m = C_i - C_d \frac{m}{n} \tag{3.8}$$

also

$$B_m = C_d\left(1 - \frac{m}{n}\right) + C_{\text{sal}} \tag{3.9}$$

Depreciation remaining for future years is

$$C_d\sum_{m}^{n} D_{f,m} = C_d\sum_{m=m}^{m=n} \frac{1}{n} = C_d(n - m)\frac{1}{n} = C_d\left(1 - \frac{m}{n}\right) \tag{3.10}$$

The present value of $1 of depreciation is

$$F_{\text{SLP},i,n} = \frac{1}{n}\left[\frac{1}{1 + i} + \frac{1}{(1 + i)^2} + \cdots + \frac{1}{(1 + i)^n}\right]$$

where the symbol $F_{\text{SLP},i,n}$ converts straight-line depreciation to a present value. The terms within the brackets are a geometrical series and are summed by Eq. (2.15):

$$F_{\text{SLP},i,n} = \frac{1}{n}\frac{(1 + i)^n - 1}{i(1 + i)^n} = \frac{1}{n}F_{\text{RP},i,n} \tag{3.11}$$

$F_{\text{SLP},i,n}$ is the straight-line present-value factor and is tabulated in Table A1.1 (see Appendix 1).

Table 3.2 is a summary of relationships for various depreciation methods.

## 3.4 SINKING-FUND DEPRECIATION (SF)

This method gives a yearly depreciation schedule which increases systematically with time and hence is opposite to accelerated depreciation. Suppose an annual deposit $R_S$ is made to a sinking fund to restore the depreciable value at the end of $n$ years. Then by item 6, Table 2.1,

$$R_S = C_d\frac{1}{(1 + i)^n - 1} = C_d F_{\text{SP},i,n}F_{\text{PR},i,n} \tag{3.12}$$

The depreciation for any year is the sinking-fund increase for that year, which is the deposit for the year plus interest earned by the fund for the year.

The relationships for sinking-fund depreciation are given in Table 3.2. The method is employed by some governmental agencies and public utilities but is not widely used in industrial operations. It has at times been referred to as *theoretical depreciation* because if the sinking fund bears the same interest rate as that used in an economy study, the book value at any time will be the theoretical value of an asset having the same remaining life and salvage value. This aspect should not be over-emphasized, but it is of interest.

> **Example 3.1.** An asset costs $20,000, lasts 10 years, and has $1000 salvage value. If money is worth 8 percent per year, find the book value at the end of 7 years using sinking-fund depreciation. Compare this to the equivalent cost of an asset having a 3-year life and $1000 salvage value with money worth 8 percent per year.

**TABLE 3.2**
**Relationships for depreciation, $n$-year life***

| Item no. | Method | Annual depreciation for $m$th year | Accumulated depreciation for $m$ years | Present value of $m$ years of depreciation |
|---|---|---|---|---|
| 1 | Straight-line (SL) | $\dfrac{C_i - C_{sal}}{n}$ | $(C_i - C_{sal})\dfrac{m}{n}$ | $\dfrac{C_i - C_{sal}}{n}F_{RP,i,m}$ |
| 2 | Sinking-fund (SF) | $(C_i - C_{sal})F_{SP,i,n}F_{PR,i,n}F_{PS,i,m-1}$ | $(C_i - C_{sal})\dfrac{F_{PS,i,m} - 1}{F_{PS,i,n} - 1}$ | $(C_i - C_{sal})\dfrac{m}{1+i}F_{SP,i,n}F_{PR,i,n}$ |
| 3 | Declining-balance (DB) | $C_i(1 - F_{DB})^{m-1}F_{DB}$ | $C_i[1 - (1 - F_{DB})^m]$ | $C_iF_{DB}\dfrac{1 - \left(\dfrac{1 - F_{DB}}{1 + i}\right)^m}{i + F_{DB}}$ |
| 4 | Sum-of-the-years-digits (SD) | $(C_i - C_{sal})\dfrac{n - m + 1}{0.5n(n + 1)}$ | $(C_i - C_{sal})\dfrac{m}{n(n + 1)}(2n + 1 - m)$ | $\dfrac{2(C_i - C_{sal})}{n(n + 1)i}$ $\times [n - F_{RP,i,m} - (n - m)F_{SP,i,m}]$ |
| 5 | Units-of-production (UP) | $(C_i - C_{sal})\dfrac{M_m}{M}$ | $(C_i - C_{sal})\dfrac{\sum_1^m M_m}{M}$ | $\dfrac{C_i - C_{sal}}{M}\sum_1^m \dfrac{M_m}{(1 + i)^m}$ |

*Book value at end of $m$th year $= C_i$ less accumulated depreciation for $m$ years. Depreciation remaining for $n - m$ future years $=$ book value at end of $m$th year less $C_{sal}$. For declining-balance depreciation, terminal value is not related to salvage value, but total write-off should be to salvage value. At the end of useful life, $m = n$.

*Solution.* The sinking-fund payment, by Eq. (3.12), is

$$R_S = (20,000 - 1000)F_{SP,\,8\%,\,10}F_{PR,\,8\%,\,10} = 19,000(0.46319)(0.14903) = \$1311.6$$

The amount in the sinking fund at the end of 7 years by item 5, Table 2.1, is

$$S = RF_{RP,\,8\%,\,7}F_{PS,\,8\%,\,7} = 1311.6(5.2064)(1.7138) = \$11,703$$

The book value at the end of 7 years, by Eq. (3.3), is

$$B_7 = 20,000 - 11,703 = \$8297$$

This is what a 3-year machine with \$1000 salvage value is worth. Use unacost items 1 and 7, Table 2.3:

$$R_{10} = 20,000F_{PR,\,8\%,\,10} - 1000(1/1.08)^{10}F_{PR,\,8\%,\,10}$$

$$= 20,000(0.14903) - 1000(0.46319)(0.14903) = \$2911.5$$

$$R_3 = 8297F_{PR,\,8\%,\,3} - 1000(1/1.08)^3F_{PR,\,8\%,\,3}$$

$$= 8297(0.38803) - 1000(0.79383)(0.38803) = \$2911.5$$

## 3.5 DECLINING-BALANCE DEPRECIATION (DB)

In this accelerated depreciation method the yearly depreciation is a fixed percentage or fixed fraction of the book value at the beginning of the year. Let $F_{DB}$ be the fixed fraction; then

$$D_1 = C_i F_{DB}$$

$$B_1 = C_i - C_i F_{DB} = C_i(1 - F_{DB})$$

$$D_2 = C_i(1 - F_{DB})F_{DB}$$

$$B_2 = B_1 - D_2 = C_i(1 - F_{DB}) - C_i(1 - F_{DB})F_{DB} = C_i(1 - F_{DB})^2$$

For the $m$th year the relationships are

$$D_m = C_i(1 - F_{DB})^{m-1}F_{DB} \tag{3.13}$$

$$B_m = C_i(1 - F_{DB})^m \tag{3.14}$$

The fixed fraction $F_{DB}$ is limited to $2/n$, where $n$ is the useful life in years. Any value less than $2/n$ may be used. For emphasis in terminology, the declining-balance method with $F_{DB} = 2/n$ can be referred to as *twice the straight-line rate* or *double-declining-balance depreciation*.

## 3.6 DECLINING-BALANCE DEPRECIATION REDUCING TO SALVAGE VALUE

The declining-balance method will not in general reduce the final book value to the salvage value, nor is this requirement necessary, as explained here. If one wishes to choose a declining-balance fraction so that the final book value and salvage are equal, then by Eq. (3.14), for $m = n$

$$B_m = C_{sal} = C_i(1 - F_{DB})^n$$

whence
$$F_{DB} = 1 - \sqrt[n]{\frac{C_{sal}}{C_i}} \qquad (3.15)$$

Equation (3.15) is of historical interest but has little current significance.

## 3.7   SUM-OF-THE-YEARS-DIGITS DEPRECIATION (SD)

This is an accelerated depreciation method having fractional depreciation for which the numerator is the years of life remaining as of the beginning of the year and the denominator is the sum of the years digits for the total life. For example, for a 10-year life $1 + 2 + 3 + \cdots + 10 = 55$, the fractional depreciation for the first year is $10/55$, for the second year $9/55$, etc. For the $m$th year in a total life of $n$ years the factor is

$$F_{SD} = \frac{n - m + 1}{0.5n(n + 1)} \qquad (3.16)$$

The present value of $1 of depreciation is

$$F_{SDP, i, n} = \sum_{m=1}^{m=n} \frac{n - m + 1}{0.5n(n + 1)} \frac{1}{(1 + i)^m} = \frac{n - F_{RP, i, n}}{0.5n(n + 1)i} \qquad (3.17)$$

$F_{SDP, i, n}$ converts sum-of-the-years-digits depreciation to a present value and is the sum-of-the-years-digits present-value factor. It is tabulated in Table A1.1 (see Appendix 1).

The relationships for sum-of-the-years-digits depreciation are given in Table 3.2.

## 3.8   UNITS-OF-PRODUCTION DEPRECIATION (UP)

The methods discussed previously allocate depreciation on a time basis without regard to use. The units-of-production method allocates depreciation on a use basis without regard to time. If $M$ is the total units produced and $M_m$ is the units produced for the $m$th year, the fractional depreciation for the $m$th year is

$$F_{UP, m} = \frac{M_m}{M}$$

The relationships for units-of-production depreciation are given in Table 3.2.

## 3.9   ACCELERATED COST RECOVERY SYSTEM (ACRS)

As stated earlier, in the United States ACRS was mandatory for tax purposes for most tangible depreciable assets placed in service after 1980 and before 1987. Any asset placed in service prior to that time must continue to be depreciated using whatever method was previously applied to that asset (usually one of the methods described in Secs. 3.3 through 3.8 of this chapter).

Under ACRS the allowable deduction is determined by multiplying the cost of the asset by certain specified percentages, depending on the year of purchase and recovery period. Salvage value is not considered in ACRS. To complicate the calculation further, for the years 1981 through 1984, ACRS is approximately equivalent to 150 percent declining-balance depreciation with an eventual switch to straight-line depreciation. For 1985 the calculation is approximately 175 percent declining balance. Then for 1985 and 1986 the calculation becomes 200 percent declining balance with a switch to the sum-of-years-digits method at the optimum point in time. An alternate form of ACRS, based on straight-line depreciation, was also acceptable for the 1981–1986 period.

Because the ACRS method is now of historical interest only, the specific details of applicable deduction percentages and recovery periods are not cited here. These data are published in various readily available sources [1, 2].

## 3.10 MODIFIED ACCELERATED COST RECOVERY SYSTEM (MACRS)

The U.S. Tax Reform Act of 1986 established the MACRS system, which applies to all assets placed in service after December 31, 1986 with very few exceptions [2].

All property depreciated under MACRS is assigned a recovery period and deductions are calculated as specific percentages of the cost of the asset. As with ACRS, salvage value is not considered. The property classes are:

1. *Three-year property,* which is defined as "property with a class life of 4 years or less . . . ."
2. *Five-year property,* which includes property with a class life of more than 4 years but less than 10 years. This category includes computers and office machinery, vehicles, and research equipment.
3. *Seven-year property,* which includes property with a class life of 10 years or more but less than 16 years. This class includes most types of industrial machinery and equipment.
4. *Ten-year property,* which includes items with a class life of 16 to 19 years. This class primarily includes marine vessels.
5. *Fifteen-year property,* which includes items with a class life of 20 to 24 years. Civil infrastructure falls in this category, i.e., roads, municipal wastewater treatment plants, etc.
6. *Twenty-year property,* which includes items with a class life of 25 years or more, such as farm buildings and municipal sewers.
7. *Nonresidential real property.* Straight-line depreciation applies over 31.5 years.
8. *Residential rental property.* Straight-line depreciation applies over 27.5 years.

Under MACRS a one-half year convention generally applies, i.e., all acquisitions are assumed to be made in midyear and for the first tax year the allowable deduction is one-half the allowance for subsequent years.

To calculate MACRS allowances, the declining-balance method of depreciation calculation is used until the optimum year is reached for a switch to the straight-line method. The optimum year is the first year for which the straight-line method will give an equal or greater deduction than does the declining-balance method. Table 3.3 gives the MACRS deduction rates.

As an option, most property may be depreciated over its class life using straight-line depreciation. The one-half year convention, however, continues to apply for the first year. Note that alternate MACRS uses class life, not recovery period, in the calculation.

**TABLE 3.3**
**MACRS deduction rates**

General depreciation system
Applicable depreciation method: 200 or 150 percent
Declining balance switching to straight line
Applicable recovery periods: 3, 5, 7, 10, 15, 20 years
Applicable convention: Half-year

| If the recovery year is: | And the recovery period is: | | | | | |
|---|---|---|---|---|---|---|
| | 3-year | 5-year | 7-year | 10-year | 15-year | 20-year |
| | The depreciation rate is: | | | | | |
| 1 | 33.33 | 20.00 | 14.29 | 10.00 | 5.00 | 3.750 |
| 2 | 44.45 | 32.00 | 24.49 | 18.00 | 9.50 | 7.219 |
| 3 | 14.81 | 19.20 | 17.49 | 14.40 | 8.55 | 6.677 |
| 4 | 7.41 | 11.52 | 12.49 | 11.52 | 7.70 | 6.177 |
| 5 | | 11.52 | 8.93 | 9.22 | 6.93 | 5.713 |
| 6 | | 5.76 | 8.92 | 7.37 | 6.23 | 5.285 |
| 7 | | | 8.93 | 6.55 | 5.90 | 4.888 |
| 8 | | | 4.46 | 6.55 | 5.90 | 4.522 |
| 9 | | | | 6.56 | 5.91 | 4.462 |
| 10 | | | | 6.55 | 5.90 | 4.461 |
| 11 | | | | 3.28 | 5.91 | 4.462 |
| 12 | | | | | 5.90 | 4.461 |
| 13 | | | | | 5.91 | 4.462 |
| 14 | | | | | 5.90 | 4.461 |
| 15 | | | | | 5.91 | 4.462 |
| 16 | | | | | 2.95 | 4.461 |
| 17 | | | | | | 4.462 |
| 18 | | | | | | 4.461 |
| 19 | | | | | | 4.462 |
| 20 | | | | | | 4.461 |
| 21 | | | | | | 2.231 |

*Source:* U.S. Internal Revenue Service: *Tax Information Depreciation,* U.S. Department of the Treasury, Publication No. 534, published yearly.

*Note:* 200 percent declining balance applies to class lives of 10 years or less. 150 percent applies to 15- and 20-year class lives.

MACRS has certain other provisions which are not discussed here and which apply only in unusual circumstances. They are explained in detail in Internal Revenue Service Publication 534 [2].

## 3.11  TAXES AND DEPRECIATION

The federal corporate tax is based on the income after all costs have been deducted. A depreciation cost, accordingly, is of considerable interest to the Treasury Department because it affects the profit tax. Thus the class lives of equipment have been established, as shown in Table 3.1. The regulations are subject to revision at any time, and the most recent information available should be obtained from the Treasury Department in making final computations.

A summary of U.S. federal corporation income taxes is shown in Table 3.4. In determining the influence of depreciation upon the tax, one should recognize that all depreciation costs reduce the tax base and are a tax savings or credit in any year.

In this book it will generally be assumed that the company as a whole is operating at a profit, so that a loss sustained by a single project will reduce the tax base and generate a tax offset for which it should be credited. For a large corporation the tax bracket will be assumed to be above $335,000, so that a 34 percent tax rate prevails. State income taxes should be added if applicable.

Local property taxes and other taxes which are not levied on the basis of profit are just an expense and do not enter the tax considerations of this chapter.

## 3.12  COMPARISON OF METHODS

Table 3.5 shows the yearly depreciation by six depreciation schedules for an asset costing $1100 with a salvage value of $100. In column (4) the declining-balance factor is 0.38096, as calculated from Eq. (3.15), to make the final book value equal to the salvage value. In column (5), with the declining-balance factor at the maximum, here $\frac{2}{5} = 0.4$, the depreciation for the fifth year is limited to $42.6 to prevent the book value from falling below the salvage value.

Note that the declining-balance and the sum-of-the-digits methods have accelerated depreciation at the beginning and high present values. The MACRS method yields the highest present value of all methods because it combines the declining-balance method with a switch to the straight-line method and also does not deduct salvage value. Conversely, sinking-fund depreciation has the highest depreciation per year at the end and the lowest present value.

**TABLE 3.4**
**U.S. federal corporate income tax (1988)**

| | |
|---|---|
| Tax on first $50,000 of taxable income | 15% |
| Tax on next $25,000 of taxable income | 25% |
| Tax on next $25,000 of taxable income | 34% |
| Tax on next $235,000 of taxable income | 39% |
| Tax on taxable income above $335,000 | 34% |

**TABLE 3.5**
**Comparison of depreciation methods***

| | | | **Yearly depreciation, $** | | | |
|---|---|---|---|---|---|---|
| **(1)**<br>**End-of-year** | **(2)**<br>**Straight-**<br>**line** | **(3)**<br>**Sinking-fund**<br>**10% per year** | **(4)**<br>**Declining-**<br>**balance**<br>$F_{DB} = 0.38096$ | **(5)**<br>**Declining-**<br>**balance**<br>$F_{DB} = 0.4$ | **(6)**<br>**Sum-of-the-**<br>**years-digits** | **(7)**<br>**MACRS** |
| 1 | 200.0 | 163.8 | 419.1 | 440.0 | 333.3 | 220.0 |
| 2 | 200.0 | 180.2 | 259.4 | 264.0 | 266.7 | 352.0 |
| 3 | 200.0 | 198.2 | 160.6 | 158.4 | 200.0 | 211.2 |
| 4 | 200.0 | 218.0 | 99.4 | 95.0 | 133.3 | 126.7 |
| 5 | 200.0 | 239.8 | 61.5 | 42.6[†] | 66.7 | 126.7 |
| Total | 1000 | 1000 | 1000 | 1000 | 1000 | 1036.6[‡] |
| | | | **Present value of depreciation at 10% per year** | | | |
| 0 | 758.2 | 744.5 | 822.2 | 828.6 | 806.1 | 850.6 |

*Original cost $1100, salvage value $100, life 5 years.
[†]Brought down to $100 salvage value.
[‡]No salvage value. Additional depreciation of 63.4 claimed in year 6.

As shown in Table 3.5, when using the declining-balance method, care must be taken not to drop the book value below the salvage value. On the other hand, a switch may be required to reach the salvage value. A switch to straight-line depreciation can be made at any time without prior approval. The switch point for the best policy depends on the ratio of salvage value to initial cost, and the useful life.

## 3.13   SPECIAL CONSIDERATIONS

The importance of accelerated depreciation in contributing to profitability lies in the greater present value of larger depreciation allotments in the early years of an economic venture. In practice, many companies use one depreciation method for tax purposes and another for internal use to maintain a better perspective for true costs and pricing. Sometimes a liability account is set up to be drawn against later when delayed taxes may become a burden.

The student should understand depreciation in its true perspective and recognize its fundamental nature, although it is intermingled with taxes and even profitability. Because depreciation is an expense, the manner of allocation over the years has an effect on the profit for individual years. Cash flow, which is essentially the sum of depreciation plus profit, is discussed in a subsequent chapter.

A depreciable asset may not be in service for the time set up for the depreciation program. If a machine becomes inoperable before it has been completely depreciated, there are provisions in the accounting procedure to allow for the unrecovered capital, such as loss on disposal. Conversely, if equipment is useful for longer periods than anticipated, it is not possible to charge additional depreciation for tax purposes. For pricing, however, some kind of cost for wear and tear might very well be included.

If on retirement or abandonment a machine has depreciation available for future years, immediate disposal may give a present-value advantage which can be considered.

**Example 3.2.** A machine has $330,000 book value, $30,000 salvage value, and 3 years of life remaining, and is being depreciated by the straight-line method. If abandoned, it can be taken as an immediate loss on disposal at a 34 percent tax rate. Assume that the machine can be sold for $30,000 at any time. If money is worth 10 percent per year, find the present value of the benefit by abandonment now.

**Solution.** Depreciation is $100,000 per year. This reduces the tax base. Recovery of the salvage value does not affect the tax base. If the machine is retained, the present value after taxes is

$$P_{ret} = 100,000 \times 0.34 \left( \frac{1}{1.10} + \frac{1}{1.10^2} + \frac{1}{1.10^3} \right) + \frac{30,000}{1.10^3} = 84,555 + 22,539$$

$$= \$107,094$$

The present book value is $330,000. Of this, $30,000 recovered now does not affect the tax base; but loss of $330,000 minus $30,000 = $300,000 does affect the tax base. The present value after taxes is

$$P_{aban} = 300,000 \times 0.34 + 30,000 = \$132,000$$

The present-value advantage by abandoning now is

$$132,000 - 107,094 = \$24,906$$

If the machine were part of a group, then it would not be possible to take the loss on disposal now, and the present-value advantage would be simply

$$30,000 - \frac{30,000}{1.10^3} = \$7460$$

## 3.14  DEPLETION

When exhaustible resources are sold, part of the sales realization is a return of capital, and the income tax should adjust for that. Also, it is desirable to have an incentive to encourage exploration for new resources as existing resources are used up.

*Depletion* is a term used to describe the write-off of certain exhaustible natural resources such as minerals, oil and gas, and timber. Depletion applies to production units withdrawn from the property, whereas depreciation is limited to the original cost less the estimated salvage value.

There are two methods for computing depletion: cost depletion and percentage depletion.

In the cost method the value of a depletion unit, say, a ton of ore, is arrived at by calculating the total value to be depleted divided by the tons of ore to be depleted. Deduction for a tax year is the depletion unit multiplied by the number of units sold within the year. Cost depletion can be applied to all types of depleted property and is often called *valuation depletion*.

In the percentage method the depletion allowance for the year is a specified percentage of the "gross income from the property" but must not exceed 50 percent of the taxable income figured without the depletion allowance.

The cost method can always be used, but the percentage method has certain limitations; thus timber is exempt. The deduction should be computed both ways, if applicable, and the larger deduction taken.

Percentage depletion varies from 5 percent to a maximum of 22 percent. Percentage depletion does not apply to oil and gas wells, except for certain small producers subject to limitations.

It should be noted that under cost depletion, when the total cost and accumulated depletion are equal, no further cost depletion is allowed. However, percentage depletion is not limited to original cost, as is ordinarily true with depreciation of assets.

Accounting for depletion can be quite complicated because shifting expenses and production from one year to another can result in a tax advantage. Also, there are complications arising from multiple properties, additions to capital, reevaluation of resources, election to expense or capitalize, production limits for oil and gas wells, and even the disqualification of a small producer who controls a retail outlet properly defined.

**Example 3.3.** A mining property with an estimated 1 megaton (Mt) of ore originally cost $5,000,000. In one year 100 kilotons (kt) of ore are sold for $16/t with expenses of $1,000,000. The percentage depletion allowance is 22 percent, and the tax rate is 34 percent. Which is more advantageous, cost depletion or percentage depletion?

*Solution.* The cost per ton of ore for cost depletion is

$$\frac{5,000,000}{1,000,000} = \$5/t \quad \text{(adjusted basis)}$$

| | Cost depletion, $ | Percentage depletion, $ |
|---|---|---|
| 1. Gross income | 1,600,000 | 1,600,000 |
| 2. Expenses, excluding depletion | 1,000,000 | 1,000,000 |
| 3. Taxable income before depletion | 600,000 | 600,000 |
| 4a. Cost depletion at $5/t | 500,000 | |
| 4b. Percentage depletion at 22% | | 300,000* |
| 5. Actual taxable income (3) − (4) | 100,000 | 300,000 |
| 6. Tax at 34% | 34,000 | 102,000 |
| 7. Cash flow (3) − (6) | 566,000 | 498,000 |

*22% of $1,600,000 is $352,000, but (4b) cannot exceed 50% of (3).

Cost depletion is better than percentage depletion.

**Example 3.4.** A mining property using percentage depletion at 22 percent and a tax rate of 34 percent has gross income and expenses before depletion for two consecutive years as follows, with all amounts in kilodollars:

|                              | Year 1 | Year 2 |
|------------------------------|--------|--------|
| Gross income                 | 1000   | 1000   |
| Expenses, excluding depletion| 400    | 700    |

Is there any advantage in shifting 50 k$ of expenses to year 1, all other things remaining the same?

*Solution*

|                                       | Plan A | | Plan B | |
|---------------------------------------|--------|--------|--------|--------|
|                                       | Year 1 | Year 2 | Year 1 | Year 2 |
| 1. Gross income                       | 1000   | 1000   | 1000   | 1000   |
| 2. Expenses, excluding depletion      | 400    | 700    | 450    | 650    |
| 3. Taxable income before depletion    | 600    | 300    | 550    | 350    |
| 4. Percentage depletion               | 220    | 150*   | 220    | 175*   |
| 5. Actual taxable income              | 380    | 150    | 330    | 175    |
| 6. Tax at 34%                         | 129.2  | 51.3   | 112.2  | 59.5   |
| 7. Cash flow (3) − (6)                | 470.8  | 248.7  | 437.8  | 290.5  |
| 8. Cash flow for 2 years              | 719.5  |        | 728.3  |        |

*Limited to 50% of (3).

Plan *B* is better. One forgoes 470.8 − 437.8 = 33 k$ to gain 290.5 − 248.7 = 48.1 k$ 1 year later, which is a 45.8 percent rate of return.

**Example 3.5.** A mining property is using cost depletion. It was bought for 10 M$ with 1 Mt of estimated ore. After 400,000 t of ore are sold, the mine is then reevaluated and estimated to have 1.2 Mt of ore remaining. The next year 50,000 t of ore are sold. Find the cost depletion for that year.

*Solution.* The original adjusted basis was

$$\frac{10,000,000}{1,000,000} = \$10/t$$

With 400,000 t of ore sold

$$400,000 \times 10 = \$4,000,000 \text{ cost depletion}$$

On the new basis

$$10,000,000 - 4,000,000 = \$6,000,000 \quad \text{(value for 1.2 Mt)}$$

$$\frac{6,000,000}{1,200,000} = \$5/t \quad \text{(new adjusted basis)}$$

$$50,000 \times 5 = \$250,000 \quad \text{(cost depletion for the year)}$$

# COST COMPARISON AFTER TAXES

## 3.15 PRESENT VALUE AFTER TAXES

In Chap. 2 cost comparisons were made on a no-tax basis. In practice, taxes enter and the remainder of this chapter will develop the comparisons of alternatives on an after-tax basis.

One effect introduced by the corporation tax is a reduction of the rate of return. If $i$ is the rate of return before taxes, $t$ the tax rate, and $r$ the rate of return after taxes, all expressed as decimals, then

$$r = i(1 - t) \tag{3.18}$$

If an investment compounds at a rate $i$ before taxes, it will compound at a rate $r$ after taxes.

Income taxes are actually paid quarterly, but in this chapter they will be treated as being paid once a year at the end of the year. Also, in this chapter assets will be regarded as purchased at the first of the year. In actual practice, the half-year MACRS convention would apply for the first year.

Consider now an article that lasts $n$ years with initial cost $C_i$, salvage value $C_{\text{sal}}$, and depreciable cost $C_d$, where

$$C_d = C_i - C_{\text{sal}} \tag{3.19}$$

There is no tax consideration at the time the article is purchased. At the end of the first year depreciation amounting to $D_{f1}C_d$, where $D_{f1}$ is the fractional depreciation, becomes available as an expense recognized for tax purposes. The taxable base is reduced $D_{f1}C_d$ and a savings or reduction in taxes amounting to $D_{f1}C_d t$ is realized. The time-cost diagram is

The present value is

$$P = C_d - C_d t \left[ \frac{D_{f1}}{1 + r} + \frac{D_{f2}}{(1 + r)^2} + \cdots + \frac{D_{fn}}{(1 + r)^n} \right] \tag{3.20}$$

The sum of the terms within the brackets is the present value of $1 of depreciation and depends on the depreciation method selected. Whatever the method used, the equation can be written

$$P = C_d - C_d t \psi = C_d (1 - t\psi) \tag{3.21}$$

where
$$\psi = \sum_1^n \frac{D_{fm}}{(1 + r)^m} \tag{3.22}$$

The $\psi$ factor depends on $r$, $n$, and the method chosen for depreciation. It represents the present value of \$1 of depreciation and can be calculated for any value of $r$ and $n$ for any depreciation method. In particular, for sum-of-the-years-digits and straight-line depreciation

$$\psi_{SD, r, n} = F_{SDP, r, n}$$

$$\psi_{SL, r, n} = F_{SLP, r, n}$$

$F_{SDP}$ converts sum-of-the-years-digits depreciation to a present value, and $F_{SLP}$ converts straight-line depreciation to a present value. Both factors are tabulated in Table A1.1 (see Appendix 1).

Sum-of-the-years-digits depreciation will be used for the remainder of this section, with the understanding that any other method can be used. Equation (3.21) becomes

$$P = C_d(1 - tF_{SDP, r, n}) \tag{3.23}$$

It should be noted that for all depreciation methods

$$\psi_{1r} = \frac{1}{1 + r} = F_{SDP, r, 1} \tag{3.24}$$

$$\psi_{\alpha r} = 0 = F_{SDP, r, \alpha} \tag{3.25}$$

Equation (3.23) can be converted to a unacost or capitalized cost by Eq. (2.20) or (2.32), respectively, giving

$$R = C_d F_{PR, r, n}(1 - tF_{SDP, r, n'}) \tag{3.26}$$

$$K = C_d F_{PK, r, n}(1 - tF_{SDP, r, n'}) \tag{3.27}$$

where $n'$ is used to designate that the life for tax purposes can be different from $n$, the life for an economic study.

It is now possible to duplicate Table 2.3 on an after-tax basis utilizing the method just described for first determining present value. Table 3.6 has been prepared in this manner.

For item 2 in the table $R$ is a uniform end-of-year cost occurring simultaneously with the tax instant. The cost after taxes is $R(1 - t)$ as of the end of each year for $n$ years. The present value, by Eq. (2.16), with $r$ substituted for $i$, is

$$P = R(1 - t)\frac{(1 + r)^n - 1}{r(1 + r)^n} \tag{3.28}$$

For item 3, $R_b$, a uniform beginning-of-year cost, the tax credit $R_b t$ occurs at the end of the year. Combining the two as of the end of the year gives

$$R_b(1 + r) - R_b t = R_b\left(1 - \frac{t}{1 + r}\right)(1 + r) \tag{3.29}$$

which can be converted to a present value by Eq. (2.16) with $r$ substituted for $i$.

**TABLE 3.6**
## Summary of unacost and capitalized-cost relationships with tax

| Item no. | Item | Present value as a cost for $n$-year duration | Unacost — Algebraic expression |
|---|---|---|---|
| 1 | $C_d$ Fully depreciable part of an initial cost for $n$-year life, $n'$ years for tax purposes; $C_d = C_i - C_{sal}$ | $C_d(1 - t\psi_{rn'})$ | $C_d(1 - t\psi_{rn'}) \times \dfrac{r(1 + r)^n}{(1 + r)^n - 1}$ |
| 2 | $R$ Uniform end-of-year annual cost, depreciated fully at instant incurred | $R(1 - t)\dfrac{(1 + r)^n - 1}{r(1 + r)^n}$ | $R(1 - t)$ |
| 3 | $R_b$ Uniform beginning-of-year annual cost, fully depreciated at end of year | $R_b\left(1 - \dfrac{t}{1 + r}\right)(1 + r) \times \dfrac{(1 + r)^n - 1}{r(1 + r)^n}$ | $R_b\left(1 - \dfrac{t}{1 + r}\right)(1 + r)$ |
| 4 | $C_{ex}$ Irregular cost at end of $x$th year for article lasting $n$ years, fully depreciated at instant incurred | $C_{ex}\dfrac{1 - t}{(1 + r)^x}$ or $\dfrac{C_{ex}}{(1 + r)^x} \times [1 - t(1 + r)\psi_{r1}]$ | $C_{ex}\dfrac{1 - t}{(1 + r)^x} \times \dfrac{r(1 + r)^n}{(1 + r)^n - 1}$ |
| 5 | $C_{bx}$ Irregular cost at beginning of $x$th year for article lasting $n$ years, fully depreciated at end of year $x$ | $C_{bx}\left(1 - \dfrac{t}{1 + r}\right) \times \dfrac{1}{(1 + r)^{x-1}}$ or $\dfrac{C_{bx}}{(1 + r)^{x-1}}(1 - t\psi_{r1})$ | $C_{bx}\left(1 - \dfrac{t}{1 + r}\right) \times \dfrac{1}{(1 + r)^{x-1}} \times \dfrac{r(1 + r)^n}{(1 + r)^n - 1}$ |
| 6 | $C_{nd}$ Nondepreciable first cost such as land or an article that lasts forever | $C_{nd}\dfrac{(1 + r)^n - 1}{(1 + r)^n}$ | $C_{nd}r$ |
| 7 | $C_{sal}$ Salvage value at end of $n$th year; treated as a nondepreciable first cost, an expense | $C_{sal}\dfrac{(1 + r)^n - 1}{(1 + r)^n}$ | $C_{sal}r$ |

| Unacost | | Capitalized cost |
|---|---|---|
| **Factor expression for sum-of-the-years-digits depreciation** | **Algebraic expression** | **Factor expression for sum-of-the-years-digits depreciation** |
| $C_d(1 - tF_{SDP,r,n'})F_{PR,r,n}$ | Multiply corresponding algebraic expression for unacost by $\dfrac{1}{r}$ | $C_d(1 - tF_{SDP,r,n'})F_{PK,r,n}$ |
| $R(1 - t)$ | | $R(1 - t)\dfrac{1}{r}$ |
| $R_b(1 - tF_{SDP,r,1})(1 + r)$ | | $R_b(1 - tF_{SDP,r,1})F_{PK,r,1}$ |
| $C_{ex}\dfrac{1}{(1 + r)^x}$ $\times [1 - t(1 + r)F_{SDP,r,1}]F_{PR,r,n}$ | | $C_{ex}\dfrac{1}{(1 + r)^x}$ $\times [1 - t(1 + r)F_{SDP,r,1}]F_{PK,r,n}$ |
| $C_{bx}\dfrac{1 - tF_{SDP,r,1}}{(1 + r)^{x-1}}F_{PR,r,n}$ | | $C_{bx}\dfrac{1 - tF_{SDP,r,1}}{(1 + r)^{x-1}}F_{PK,r,n}$ |
| $C_{nd}r$ | | $C_{nd}$ |
| $C_{sal}r$ | | $C_{sal}$ |

For item 4, $C_{ex}$, an irregular cost occurring at the end of the $x$th year for an article lasting $n$ years, the cost occurs simultaneously with the tax instant. The present value is

$$P = C_{ex}(1 - t)\frac{1}{(1 + r)^x} \tag{3.30}$$

For item 5, $C_{bx}$, an irregular cost occurring at the beginning of the $x$th year, the tax credit $C_{bx}t$ when brought forward to the beginning of the year becomes $C_{bx}t/(1 + r)$. The cost as of the beginning of the year is then reduced to a present value at zero time, giving

$$P = \left(C_{bx} - \frac{C_{bx}t}{1 + r}\right)\frac{1}{(1 + r)^{x-1}} = C_{bx}\left(1 - \frac{t}{1 + r}\right)\frac{1}{(1 + r)^{x-1}} \tag{3.31}$$

If $C_{ex}$ and $C_{bx}$ cannot be written off in 1 year for tax purposes, substitute $F_{\text{SDP},r,n'}$ for $F_{\text{SDP},r,1}$ in items 3 and 4 of Table 3.5.

For item 6, $C_{nd}$, a nondepreciable expense such as land or working capital, an amount $C_{nd}$ is invested now and fully recovered at the end of $n$ years with no tax consideration. The present value is

$$P = C_{nd} - \frac{C_{nd}}{(1 + r)^n} = C_{nd}\frac{(1 + r)^n - 1}{(1 + r)^n} \tag{3.32}$$

Item 7 in Table 3.6 for salvage value $C_{sal}$ is correct in showing the item as positive, an expense. A sum is invested, and although it is fully recovered, the expense arises from the loss of interest.

## 3.16 PROCEDURE FOR COST COMPARISON AFTER TAXES

Costs before taxes must be corrected to the cost after taxes. Some costs will be the same before and after taxes. For example, $10,000 invested in inventory is a swap of money for goods which are later recovered. When invested and recovered, the $10,000 is not subject to tax and is $10,000 before and after tax when invested and when recovered. If a cost occurs at the same instant as the tax payment, the cost after taxes is merely $(1 - t)$ times the cost before taxes. At times, the cost and the tax consideration have different timing. A first depreciable cost is the same before and after taxes at the instant it occurs, as it is merely a swap of money for assets, but each year the depreciation creates a tax benefit that must be recognized. The best procedure is to use a time diagram, include all costs and their timing, include all tax and tax benefits and their timing, and reduce the entire system to a present value after taxes. Finally, convert to an unacost or a capitalized cost to compare with other systems having different service lives. Although all comparisons should be made on an after-tax basis, any value reported back to management should be on a before-tax basis. In any case the before- and after-tax amounts should be well labeled.

**Example 3.6.** Two machines have the following cost comparison. If money is worth 10 percent per year after a 34 percent tax, and sum-of-the-years depreciation is used, which machine is more economical and by how much per year?

|  | A | B |
| --- | --- | --- |
| First cost, $ | 18,000 | 24,000 |
| Uniform end-of-year expense, $/yr | 1,000 | 0 |
| Salvage value, $ | 500 | 0 |
| Service life, years | 2 | 3 |
| Life for tax purposes, years | 5 | 5 |

**Solution.** Refer to Table 3.6 and find the present value of machine A by items 1, 2, and 7 in order:

$$C_d(1 - tF_{SDP, 10\%, 5}) = (18,000 - 500)[1 - 0.34(0.80614)] = \quad 12,703$$

$$R(1 - t)F_{RP, 10\%, 2} = 1000(1 - 0.34)(1.7355) = \quad 1,145$$

$$C_{sal}[1 - (1.10)^2] = 500(1 - 0.82645) = \quad \underline{87}$$

$$P = \$13,935$$

The present value of machine B by item 1, Table 3.6, is

$$C_d(1 - tF_{SDP, 10\%, 5}) = 24,000[1 - 0.34(0.80614)] = \$17,422$$

Convert to unacosts by factor $F_{PR}$:

$$R_A = 13,935F_{PR, 10\%, 2} = 13,935 \times 0.57619 = \$8029 \text{ after tax}$$

$$R_B = 17,422F_{PR, 10\%, 3} = 17,422 \times 0.40211 = \$7006 \text{ after tax}$$

Machine B is cheaper by (8029 − 7006) = \$1023/year on an after-tax basis. The report to management should be that machine B saves

$$\frac{1023}{1 - 0.34} = \$1550/\text{year before tax}$$

**Example 3.7.** A project requires purchase of a machine that costs \$100,000, lasts 10 years, and will be written off in 10 years using sum-of-the-years-digits depreciation. The tax rate is 34 percent. The machine qualifies for an immediate 10 percent investment credit. The operating cost is \$10,000/year at end of year. Inventory worth \$10,000 is purchased at zero time and will be recovered at the end of the project. At the end of 8 years the project will be abandoned and the machine sold for \$15,000, but the machine is part of a group and will continue to be depreciated. If money is worth 10 percent per year after taxes, find the present value of the project.

**Solution.** Depreciation continues for 10 years. The present value of the first cost by Table 3.6, item 1, is

$$C_d(1 - tF_{SDP, 10\%, 10}) = 100,000[1 - 0.34(0.70099)] = \$76,166$$

The present value of the yearly cost by Table 3.5, item 2, is

$$10,000(1 - 0.34)F_{RP, 10\%, 8} = 10,000(0.66)(5.3349) = \$35,210$$

The inventory goes in and out, with no tax consideration. The present value is

$$10,000[1 - (1.10)^{-8}] = \$5335$$

Book value at the end of 8 years with 2 years remaining is

$$\frac{2 + 1}{55}100,000 = \$5445$$

The machine is sold for $15,000, which is $9545 over book value. However, the $5455 is part of the group depreciation and will be recovered in the next 2 years. The entire $15,000 becomes income subject to tax, and the present value is

$$-15,000(1 - 0.34)(1.10)^{-8} = -\$4618$$

The net present value as a cost after taxes is

$$76,166 + 35,210 - 4618 = \$106,758$$

which to the company as a whole would cost a present value of

$$\frac{106,758}{1 - 0.34} = \$161,755 \qquad \text{(before taxes)}$$

If the machine were not part of a group, the $15,000 received at the end of the eighth year would be $9545 subject to tax and $5455 tax free. However, in that case the present value of the first cost after taxes would have to be revised so as not to include the tax benefit from depreciation originally allowed for years 9 and 10.

# NOMENCLATURE

| | |
|---|---|
| $B$ | Book value, $ |
| $C_{bx}$ | Irregular cost at beginning of $x$th year, $ |
| $C_d$ | Depreciable cost, $ |
| $C_{ex}$ | Irregular cost at end of $x$th year, $ |
| $C_i$ | Initial or first cost, $ |
| $C_{nd}$ | Nondepreciable cost, $ |
| $C_{sal}$ | Salvage value, $ |
| $D$ | Depreciation, $ |
| DB | Declining-balance method of depreciation |
| $D_f$ | Fractional depreciation, decimal |
| $F_{DB}$ | Factor for declining-balance depreciation, decimal |
| $F_{PK}$ | Factor to convert $P$ to $K$; capitalized-cost factor, no units |
| $F_{PR}$ | Factor to convert $P$ to $R$; capital-recovery factor, year$^{-1}$ |
| $F_{PS}$ | Factor to convert $P$ to $S$; compound-interest factor, no units |
| $F_{RP}$ | Factor to convert $R$ to $P$; unacost present-value factor, years |
| $F_{SD}$ | Factor for sum-of-the-years-digits depreciation, decimal |

| | |
|---|---|
| $F_{SDP}$ | Present value of $1 for sum-of-the-years-digits depreciation, decimal |
| $F_{SLP}$ | Present value of $1 for straight-line depreciation, decimal |
| $F_{SP}$ | Factor to convert $S$ to $P$; present-value factor, decimal |
| $F_{UP}$ | Factor for units-of-production depreciation, decimal |
| $i$ | Rate of return, generally before taxes, decimal per year |
| $K$ | Capitalized cost, $ |
| $m$ | A number, generally the $m$th year |
| $M$ | Units of production |
| $n$ | A number, generally the service life of an item, years |
| $n'$ | Service life for tax purposes, years |
| $P$ | Present value, $ |
| $r$ | Rate of return, generally after taxes, decimal per year |
| $R$ | Uniform end-of-year annual amount; unacost, $ per year |
| $R_b$ | Uniform beginning-of-year annual amount, $ per year |
| $R_S$ | Sinking-fund deposit, $ per year |
| $S$ | Future value, $ |
| SD | Sum-of-the-years-digits method of depreciation |
| SF | Sinking-fund method of depreciation |
| SL | Straight-line method of depreciation |
| $t$ | Tax rate, decimal |
| UP | Units-of-production method of depreciation |
| $\psi$ | Psi, present value of $1 of depreciation, decimal |

## PROBLEMS

**3.1.** A furniture manufacturer buys special new manufacturing equipment for 1.8 M$ that has an estimated salvage value of 0.3 M$. (M is mega-.) Find
(a) the useful life for tax purposes from tables.
(b) the book value at the end of 4 years for MACRS depreciation.

**3.2.** A box manufacturer uses cutting dies that cost $45,000 with a salvage value of $5000. The dies are expected to produce 32 million boxes before being replaced. If production for four consecutive years is 2, 8, 16, and 6 million boxes, compare the annual depreciation by twice the straight-line rate declining-balance and units-of-production depreciation.

**3.3.** Compare the present value for the depreciation using straight-line, twice the straight-line rate declining-balance with switch at the end of the fourth year, and sum-of-the-years-digits depreciation for an asset valued at $14,000 with $2000 salvage value after 6 years with money worth 12 percent per year.

**3.4.** An asset costs $14,000 and has a useful life of 10 years and salvage value of $2000. Find the declining-balance depreciation factor which will give the same book value at the end of 5 years as does straight-line depreciation.

**3.5.** An asset costs $20,000 with a salvage value of $2000 and has a useful life of 8 years. Find a declining-balance factor which will reduce the book value to $4000 at the end of 5 years.

**3.6.** A variation for the calculation of the sum-of-the-years-digits method called the *remaining life plan* uses a fraction with the numerator being the useful life in years remaining as of the beginning of the year and the denominator being the sum of the numbers representing the remaining useful life. This fraction is applied to the unrecovered cost. An asset costs $12,000 with $2000 salvage value and lasts 5 years. Show that the depreciation for each year by the remaining life plan is the same as that given by the ordinary calculation for the sum-of-the-years-digits method.

**3.7.** A machine for an assembly line costs $50,000 and has an Internal Revenue Service (IRS) class life of 10 years. What is the allowable MACRS deduction for depreciation in each year of the life of the machine?

**3.8.** A machine costs $100,000 with an estimated $10,000 salvage value. It has an estimated operating life of 15,000 h. During the third year it is used 2500 h. Find the depreciation by the UP method.

**3.9.** A company can purchase a machine for $75,000 with $10,000 salvage value and a useful life of 4 years. Annual operating costs are $7000/year, and the machine will generate $40,500/year additional revenue. The income tax rate is 34 percent. Find the net cash flows resulting from the purchase of the machine for each of the 4 years using
(*a*) straight-line depreciation.
(*b*) sum-of-the-years-digits depreciation.

**3.10.** A company has an asset worth $12,000 with a salvage of $2000 at the end of 5 years. It will use sum-of-the-years-digits depreciation for tax purposes and straight-line depreciation for internal use to stockholders. The accountants set up an account called Reserve for Future Taxes from Depreciation, the purpose of which is to set aside revenue from the underpaid (though legal) tax in the early years which can be used as a bookkeeping fund to offset the increase tax in later years. Such a reserve will help to make the apparent profit for each year more uniform despite the variations from straight-line depreciation. Find the transfers in and out of the reserve fund for each of the 5 years if the tax rate is 34 percent.

**3.11.** An accountant imitating sum-of-the-years-digits depreciation suggests a sum-of-the-squared years-digits method. The first year fractional depreciation would be $n^2/\Sigma n^2$, the second year $(n-1)^2/\Sigma n^2$, etc. Set up the fractions for a useful life of 5 years and comment on the method.

**3.12.** The gross income for a mineral property is 1.2 M$ for 1 year, and the income tax is 34 percent. What is the annual cash flow if the annual expenses for the operation, excluding depletion, are
(*a*) 0.8 M$?   (*b*) 0.6 M$?
The depletion allowance is 22 percent. (M is mega-.)

**3.13.** A mining property cost 12 M$ for an estimated 6 Mt of ore. In one year 250 kt of ore are sold for $20/t with expenses of $10/t. If the depletion allowance is 22 percent, which is more advantageous, cost depletion or percentage depletion? The income tax rate is 34 percent. (M is mega-.)

**3.14.** A mining property has a gross income of 6 M$ and an adjusted basis of $3/t for cost depletion. It sells 300 kt and, using cost depletion at an income tax rate of 34 percent, a cash flow of $2.574 M$ is realized. Find the cash flow if a percentage depletion at 22 percent were used. (M is mega-.)

**3.15.** Refer to Example 3.4. Find the minimum amount of expenses that should be shifted from year 2 to year 1 to maximize the cash flow for the 2 years.

**3.16.** A mining company sells ore at $10/t with expenses of $2/t. The income tax rate is 34 percent. Currently the company uses percentage depletion at 22 percent. The company next year will be limited to 15 percent depletion, which will cost the company 322 k$ in cash flow compared to this year for the same production. Find how much the company must charge per ton of ore next year to suffer no reduction in cash flow for the same production.

**3.17.** An individual is in a 33 percent income tax bracket. A $1000 bond paying semi-annually at a rate of 12 percent per year has 3 years to run and is available for $900. What rate of return after taxes does the purchaser realize from the bond?

**3.18.** In Prob. 3.7 a decision is made to depreciate the machine using the alternate MACRS method. What is the depreciation deduction for each year of the life of the machine?

**3.19.** A computer system is purchased at a cost of $35,000. What is the allowable MACRS deduction for each year of the life of the system?

**3.20.** Repeat Example 2.16. Use sum-of-the-years-digits depreciation, with money worth 10 percent per year after a 34 percent tax and a 10-year life for tax purposes for the new machine. Repair expense for tax purposes can be written off in 1 year, like maintenance.

**3.21.** In Prob. 3.20, what is the savings if the old machine can be repaired for $2500?

**3.22.** Repeat Example 2.18. Use sum-of-the-years-digits depreciation with money worth 10 percent per year after a 34 percent tax, a resurfacing job written off in 1 year for tax purposes, and a new floor written off in the years of its life.

**3.23.** Repeat Example 2.19. Use sum-of-the-years-digits depreciation with a 34 percent tax rate. A relining job can be written off in 1 year for tax purposes, a 4-year stack in 4 years.

**3.24.** Repeat Example 2.21. Money is worth 10 percent per year after a 34 percent tax using straight-line depreciation. Machine *A* will be written off in 8 years for tax purposes, machine *B* in 10 years. The irregular expenses can be written off in 1 year. Savings from quality control are treated as ordinary income.

**3.25.** Repeat Prob. 2.53 using straight-line depreciation and a 12-year life for tax purposes for the machine. The tax rate is 34 percent.

**3.26.** Repeat Prob. 2.54. Royalty payments can be treated as a regular expense. The equipment will be capitalized with a 12-year life for tax purposes using straight-line depreciation, with money worth 12 percent after a 34 percent tax.

**3.27.** Repeat Prob. 2.55. The tax rate is 34 percent, all machines will be written off in 10 years using straight-line depreciation, and there is a 20 percent rate of return after taxes.

**3.28.** Repeat Prob. 2.59. The new tank can be written off in 10 years for tax purposes. Use straight-line depreciation and money worth 10 percent per year after a 34 percent tax.

**3.29.** Repeat Prob. 2.60. Money is worth 8 percent after a 34 percent tax. The catalyst can be written off in 5 years using sum-of-the-years-digits depreciation.

**3.30.** Repeat Prob. 2.61. Money is worth 8 percent per year after a 34 percent tax. Repair costs for leaks can be written off in the year incurred. The cathodic protection system can be written off in 4 years using sum-of-the-years-digits depreciation.

**3.31.** Repeat Prob. 2.62. Money is worth 10 percent per year after a 34 percent tax. Use straight-line depreciation. All plans, including additions, are written off in 10 years for tax purposes.

**3.32.** Repeat Prob. 2.63. Money is worth 10 percent per year after a 34 percent tax. Use straight-line depreciation and write off both machines in 6 years for tax purposes.

**3.33.** Repeat Prob. 2.64. Money is worth 10 percent per year after a 34 percent tax. The machine can be written off in 5 years for tax purposes using sum-of-the-years-digits depreciation.

**3.34.** A machine is bought that is written off in 5 years using alternate MACRS depreciation and with no salvage value. The machine will return 10 percent after a 34 percent tax rate based on the following end-of-year savings:

| Year | Savings, $ |
|------|-----------|
| 1 | 10,000 |
| 2 | 15,000 |
| 3 | 17,000 |
| 4 | 12,000 |
| 5 | 10,000 |

The savings for the first 3 years are realized, but the machine must be abandoned at the end of 3 years and sold for $7000. What rate of return was actually realized
(a) if the new machine can be taken as a loss on disposal at the end of 3 years?
(b) if the machine is part of a group and must continue to be depreciated for two additional years?

**3.35.** A machine costs $10,000 and lasts 12 years, with a useful life of 10 years for tax purposes. Money is worth 10 percent per year after a 34 percent tax using sum-of-the-years-digits depreciation. What is the loss on abandonment at the end of 8 years if the machine is junked for no value and is
(a) depreciated as a single-asset account?
(b) depreciated as a multiple-asset account?

**3.36.** Repeat Prob. 2.70. Money is worth 8 percent per year after a 34 percent tax. Use straight-line depreciation and a useful life of 15 years for tax purposes for both kinds of pipe.

**3.37.** Repeat Prob. 2.71. Money is worth 10 percent per year after a 34 percent tax. Use straight-line depreciation for costs that are capitalized. How much can be spent on the research at a 40 percent chance for success if
(a) the research can be written off in 1 year for tax purposes?
(b) the research must be capitalized for a 6-year life for tax purposes?

**3.38.** Repeat Prob. 2.72. Money is worth 10 percent per year after a 34 percent tax. The machine can be written off in 10 years for tax purposes using sum-of-the-years-digits depreciation.

**3.39.** Repeat Prob. 2.74. Money is worth 10 percent per year after a 34 percent tax. The building has already been written off for tax purposes.

**3.40.** A machine cost $252,000 5 years ago and was estimated to have $25,000 salvage value for tax purposes. The life was estimated at 7 years, and sum-of-the-years-digits depreciation has been used. At the present time it will be scrapped and will bring $15,000 on disposal. Money is worth 10 percent after a 34 percent tax rate. What is the value in being allowed to write off the machine in full now compared to holding it on the books for two additional years?

# REFERENCES*

1. Humphreys, K. K. and P. Wellman: *Basic Cost Engineering*, 2d ed., Marcel Dekker, Inc., New York, 1986.
2. U.S. Internal Revenue Service: *Tax Information on Depreciation*, U.S. Department of the Treasury Publication No. 534, published yearly.

---

* See also references for Chap. 2.

# CHAPTER
# 4

# CONTINUOUS INTEREST AND DISCOUNTING

*W. B. Hirschmann and J. R. Brauweiler*

## 4.1 LOGIC FOR CONTINUOUS INTEREST

Interest can be compounded periodically, e.g., annually, semiannually, or even daily, or it can be compounded continuously. Annual discounting is appropriate for handling mortgages, bonds, and similar financial transactions, which require payments or receipts at discrete times. In most businesses, however, transactions occur throughout the year. These circumstances suggest a continuous flow of money, for which continuous compounding and discounting are more realistic to some than are periodic compounding and discounting. This chapter illustrates how continuous interest can cope with a variety of cash flows.

## 4.2 CONTINUOUS INTEREST AS AN OPERATOR

Table 4.1 shows the effect of increasing the number of compounding periods in 1 year, as developed in Sec. 2.1. Note that there is little difference between the factors for monthly and continuous compounding, so that continuous factors could be used with little error to determine mortgage payments, for example.

The future value $S$ of a present amount $P$ with periodic compounding after $n$ years with a nominal decimal interest rate $i$ compounded $p$ times per year is given by Eq. (2.4):

$$S = P\left(1 + \frac{i}{p}\right)^{np} \tag{4.1}$$

**TABLE 4.1**
**Comparison of compounding factors**

| Period | Relationship | For $i = 0.06$ | Factor for $i = 0.06$ |
|--------|-------------|----------------|----------------------|
| Annually | $(1 + i)^1$ | $1.06^1$ | 1.06000 |
| Semiannually | $\left(1 + \dfrac{i}{2}\right)^2$ | $1.03^2$ | 1.06090 |
| Quarterly | $\left(1 + \dfrac{i}{4}\right)^4$ | $1.015^4$ | 1.0613635 |
| Monthly | $\left(1 + \dfrac{i}{12}\right)^{12}$ | $1.005^{12}$ | 1.0616778 |
| Daily | $\left(1 + \dfrac{i}{365}\right)^{365}$ | $1.00016^{365}$ | 1.0618305 |
| Continuously | $e^i$ | $e^{0.06}$ | 1.0618365 |

As $p$ approaches infinity for continuous compounding,

$$\left(1 + \frac{i}{p}\right)^p = \left[\left(1 + \frac{i}{p}\right)^{p/i}\right]^i = e^i \tag{4.2}$$

since the limit of $(1 + 1/x)^x = e$, where $e$ is the Naperian constant 2.71828. From Eqs. (4.1) and (4.2),

$$P = Se^{-in} \tag{4.3}$$

$$S = Pe^{in} \tag{4.4}$$

Thus the factor $e^{in}$ is an operator that moves \$1 $n$ years *with* the calendar at a nominal decimal rate $i$ per year. Similarly, the factor $e^{-in}$ is an operator that moves \$1 $n$ years *against* the calendar at a nominal decimal rate $i$ per year.

Generally there is no confusion between periodic and continuous interest. However, a bar over a letter will be used, when necessary, to emphasize that continuous interest or continuous flow is intended. Thus in keeping with the terminology of Chap. 2,

$$F_{PS,\bar{\imath},n} = e^{in} \tag{4.5}$$

$$F_{SP,\bar{\imath},n} = e^{-in} \tag{4.6}$$

The factor $F_{PS,\bar{\imath},n}$ converts (compounds) \$1 to a future value $n$ years from now at nominal decimal rate $i$ per year and is tabulated in Table A2.1. The factor $F_{SP,\bar{\imath},n}$ is used for discounting and is tabulated in Table A2.2. Continuous compounding gives compact tables since $in$ appears as a product and not separately, as in periodic interest tables of Appendix 1.

Note that for values of $i$ less than 0.01, $e^i = 1 + i$ and that property plus other properties of exponentials permits ready extension of Tables A2.1 and A2.2. For example,

$$e^{10.613} = e^5 e^5 e^{0.61} e^{0.003} = 148.41(148.41)(1.8404)(1.0030) = 40{,}666$$

In this chapter $i$ is the decimal annual rate. The discounting or compounding interval determines the effective annual rate. The effective interest rate is the interest rate which, if used for periodic once-a-year compounding, gives the same end-of-year amount as a more frequent compounding rate. Thus $1 at decimal yearly effective rate $i$ will amount to

$$1 + i_{\text{eff}} \tag{4.7}$$

at the end of the year. With a decimal continuous rate $i_c$ the value of $1 at the end of one year, by Eq. (4.5), is

$$e^{i_c}$$

For the two to be equal (equivalence)

$$e^{i_c} = (1 + i_{\text{eff}})$$

$$i_{\text{eff}} = e^{i_c} - 1 \tag{4.8}$$

$$i_c = \ln(1 + i_{\text{eff}}) \tag{4.9}$$

where ln refers to natural logarithm. Equations (4.8) and (4.9) permit conversion from continuous to *equivalent* effective rates, and the opposite. Some equivalent values are:

5 percent annual =   4.9 percent continuous
10 percent annual =   9.5 percent continuous
20 percent annual = 18.2 percent continuous
30 percent annual = 26.2 percent continuous
40 percent annual = 33.7 percent continuous

Figure 4.1 shows a comparison of equivalent annual and continuous rates. Note that for rates below 15 percent there is very little difference between annual and continuous compounding, but for rates above 30 percent the difference becomes significant.

## 4.3   UNIFORM FLOW

In the previous section compounding and discounting were performed on a single amount. In this section the operations will be performed on a continuous flow. Suppose that an amount flows at the rate $\overline{R}$ per year for $n$ years. Consider a small interval of time $dX$ starting $X$ years from now, as in Fig. 4.2. The flow during this interval is given by rate multiplied by time and is $\overline{R}\,dX$. The present value for this small element of time, from Eq. (4.3), is

$$P_{\text{elem}} = \overline{R}\,dX\,e^{-iX}$$

and for all the elements

$$P = \overline{R}\int_0^n e^{-iX}\,dX = \overline{R}\left[\frac{e^{-iX}}{-i}\right]_0^n = \overline{R}\,\frac{1 - e^{-in}}{i} \tag{4.10}$$

**FIGURE 4.1**
Equivalent annual and continuous rates.

**FIGURE 4.2**
Discounting a uniform flow.

If the preceding relationship is multiplied and divided by $n$, it becomes

$$P = n\bar{R}\left[\frac{1 - e^{-in}}{in}\right] \qquad (4.11)$$

The value $n\bar{R}$ is the *total flow* for the period. The factor within the brackets now appears as a function of $in$ only and can be tabulated compactly. In the terminology of this book

$$F_{\bar{R}P,\bar{i},n} = \frac{1 - e^{-in}}{in} \qquad (4.12)$$

and

$$F_{P\bar{R},\bar{i},n} = \frac{in}{1 - e^{-in}} \qquad (4.13)$$

The factor $F_{\bar{R}P,\bar{i},n}$, which converts $n\bar{R}$ to $P$, is tabulated in Table A2.3, as the evaluation of $(1 - e^{-x})/x$, where $x = in$.

**Example 4.1.** A mine is expected to yield a cash income after taxes of $26,000/year continuously for the next 15 years. If the minimum acceptable rate of return on investment is 12 percent per year after taxes, find the maximum amount that can be economically justified for buying the mine.

**Solution.** By Eq. (4.11) and Table A2.3,

$$P = (n\overline{R})F_{\overline{R}P, 0.12, 15} = 15(26,000)(0.4637) = \$180,843$$

**Example 4.2.** If $1/day is invested as received at 8 percent per year interest, what will the sum be in 15 years?

**Solution.** First, find the present value of the uniform flow of $365/year by Eq. (4.11) and Table A2.3:

$$P = 15(365)F_{\overline{R}P, 0.08, 15} = 15(365)(0.5823) = \$3188$$

Next, convert to a future value by Eq. (4.4) and Table A2.1:

$$S = 3188e^{(0.08)(15)} = 3188(3.3201) = \$10,584$$

**Example 4.3.** Parents plan to save enough money to send their child to college. How much must they invest monthly in 5 percent per year continuous-interest tax-free bonds to accumulate the $20,000 they believe will be needed 17 years hence?

**Solution.** Present value of the $20,000 needed, by Eq. (4.3) and Table A2.2, is

$$P = 20,000e^{-(0.05)(17)} = 20,000(0.4274) = \$8548$$

which, in turn, can be converted to a uniform flow by Eq. (4.11) and Table A2.3:

$$8548 = 17\overline{R}F_{\overline{R}P, 0.05, 17} = 17\overline{R}(0.6736)$$

$$\overline{R} = \$746.47/\text{year}$$

$$\frac{746.47}{12} = \$62.21/\text{month}$$

## 4.4 FLOW CHANGING AT AN EXPONENTIAL RATE

In practice, the cash flow will not be uniform from year to year. If the flow changes at an exponential rate, increasing or decreasing, the use of continuous compounding of interest leads to compact relationships for the present value $P$.

If an initial flow of $R_0$ dollars per year increases continuously at a rate $g$ per year, expressed as a decimal, the rate of flow at any time $X$ by analogy with Eq. (4.4) is

$$R_x = R_0 e^{gx}$$

Consider a small time interval $dX$ starting $X$ years from now, as in Fig. 4.3. The flow for this interval is $R_0 e^{gX} dX$. The present value for this small element of flow from Eq. (4.3) is

$$P_{elem} = R_0 e^{gX} dX e^{-iX} = R_0 e^{(g-i)X} dX$$

and for all the elements

$$P = R_0 \int_0^n e^{(g-i)X} dX = R_0 \frac{e^{(g-i)n} - 1}{g - i}$$

$$= nR_0 \frac{1 - e^{-(i-g)n}}{(i - g)n} = (nR_0)F_{\overline{RP}, \overline{i-g}, n} \tag{4.14}$$

Thus the present value is easily calculated from a knowledge of the initial flow rate and from the factors tabulated in Table A2.3, with $(i-g)n$ replacing $in$ as the argument.

When the rate of increase of $g$ is greater than the interest rate $i$, $(i - g)$ becomes negative and Table A2.3 cannot be used directly, but Eq. (4.14) still applies.

**Example 4.4.** Repeat Example 4.1, but with a forecast that inflation will raise prices 3 percent per year continuously.

*Solution.* By Eq. (4.14) and Table A2.3,

$$P = nR_0 F_{\overline{RP}, 0.12-0.03, 15} = 15(26,000)(0.5487) = \$213,993$$

**Example 4.5.** Repeat Example 4.1 with the condition that the mine will become gradually depleted so that its net income declines at the rate of 5 percent per year.

*Solution.* Here $i = 0.12$ and $g = -0.05$. Thus $i - g = 0.12 - (-0.05) = 0.17$. By Eq. (4.14),

$$P = 15(26,000)F_{\overline{RP}, 0.17, 15} = 15(26,000)(0.3615) = \$140,985$$

**Example 4.6.** Repeat Example 4.1 subject to both an inflation rate of 3 percent per year and a depletion rate of 5 percent per year.

*Solution.* Here $i - g = 0.12 - 0.03 - (-0.05) = 0.14$, and Eq. (4.14) becomes

$$P = 15(26,000)F_{\overline{RP}, 0.14, 15} = 15(26,000)(0.4179) = \$162,981$$

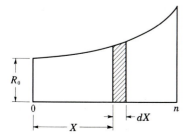

**FIGURE 4.3**
Discounting a flow changing at an exponential rate.

## 4.5 FLOW DECLINING IN A STRAIGHT LINE TO ZERO

Consider Fig. 4.4, in which an initial flow, $R_0$ dollars per year, declines to zero by a straight-line relationship in $n$ years. At time $X$ the flow is $R_x$, and by similar triangles,

$$\frac{R_x}{n - X} = \frac{R_0}{n}$$

or
$$R_x = R_0\left(1 - \frac{X}{n}\right) \qquad (4.15)$$

In a small time interval $dX$ starting $X$ years from now the flow for the interval is

$$R_x\,dX = R_0\left(1 - \frac{X}{n}\right)dX$$

and the present value for this small element of flow, from Eq. (4.3), is

$$P_{\text{elem}} = R_0\left(1 - \frac{X}{n}\right)dX\,e^{-iX}$$

For all the elements

$$P = R_0\int_0^n\left(1 - \frac{X}{n}\right)e^{-iX}\,dX = R_0\int_0^n e^{-iX}\,dX - \frac{R_0}{n}\int_0^n Xe^{-iX}\,dX \qquad (4.16)$$

The first integral on the right has already been evaluated and is

$$R_0\frac{1 - e^{-in}}{i}$$

Tables of integrals show

$$\int Xe^{-aX}\,dX = -\frac{e^{-aX}}{a^2}(aX + 1)$$

so that the second integral on the right of Eq. (4.16) is

$$-\frac{R_0}{n}\left[-\frac{e^{-iX}}{i^2}(iX + 1)\right]_0^n = \frac{R_0}{i}\left(e^{-in} + \frac{e^{-in}}{in} - \frac{1}{in}\right)$$

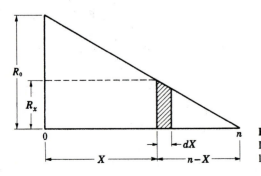

**FIGURE 4.4**
Discounting a flow declining in a straight line to zero.

The combined integrals on the right of Eq. (4.16) become

$$P = \frac{R_0}{i}\left(1 - e^{-in} + e^{-in} + \frac{e^{-in}}{in} - \frac{1}{in}\right) = \frac{R_0}{i}\left(1 - \frac{1 - e^{-in}}{in}\right)$$

The latter can be written

$$P = \frac{nR_0}{2}\frac{2}{in}\left(1 - \frac{1 - e^{-in}}{in}\right) \tag{4.17}$$

The total flow $Q$ is the area of Fig. 4.4 and is $nR_0/2$. Finally, Eq. (4.17) becomes

$$P = Q\left[\frac{2}{in}\left(1 - \frac{1 - e^{-in}}{in}\right)\right] \tag{4.18}$$

A table of discount factors for such a flow is the evaluation of the bracketed terms on the right; i.e.,

$$\frac{2}{x}\left(1 - \frac{1 - e^{-x}}{x}\right) \qquad \text{with } x = in$$

and is tabulated in Table A2.4. This type of flow approximates sum-of-the-years-digits (SD) depreciation and in symbols is

$$F_{\text{SDP},\bar{i},n} = \frac{2}{in}\left(1 - \frac{1 - e^{-in}}{in}\right) \tag{4.19}$$

Table A2.4 is commonly referred to as the *years-digits table*.

**Example 4.7.** A machine costs $150,000 and can be depreciated over 20 years by sum-of-the-years-digits depreciation. Find the present value of the depreciation, before taxes, if the discount rate is 16 percent per year.

*Solution.* By Eqs. (4.18) and (4.19) and Table A2.4,

$$P = QF_{\text{SDP},\bar{i},n} = 150,000F_{\text{SDP},0.16,20} = 150,000(0.4376)$$

$$= \$65,640$$

## 4.6 EQUIVALENT UNIFORM FLOW

In Sec. 4.3 a uniform flow was converted to a present value. The inverse of that procedure, the conversion of a present value to a uniform flow, will be considered in this section. Solving Eq. (4.11) for $\bar{R}$ gives

$$\bar{R} = \frac{P}{n}\frac{1}{(1 - e^{-in})/in} \tag{4.20}$$

which, by Eq. (4.12), becomes

$$\bar{R} = \frac{P}{n}\frac{1}{F_{\bar{R}P,\bar{i},n}} \tag{4.21}$$

Equations (4.20) and (4.21) are important. They permit transformation of a present value $P$ having $n$ years' duration to a uniform flow. $\overline{R}$ will be referred to as a *uniform flow*. $\overline{R}$ can be called the *continuous capital-recovery amount*. It can also be called *unaflow* and is analogous to unacost in periodic compounding.

Equivalent annual flow is important since it permits a comparison when service lives are different. All articles or systems are reduced to a cost for one year expressed as an equivalent annual flow.

**Example 4.8.** A firm has the option of getting a patent license by making either a single payment of $50,000 or royalty payments of $5000/year for the 17-year life of the patent. If the payments can be expensed in either case, and if the firm earns 15 percent per year after taxes, which is the more attractive choice?

**Solution.** Uniform flow for royalty payments is $5000/year, as given. Uniform flow for purchase of a patent, by Eq. (4.21) and Table A2.3, is

$$\overline{R} = \frac{50,000}{17} \frac{1}{F_{\overline{R}P,0.15,17}} = \frac{50,000}{17} \frac{1}{0.3615} = \$8136$$

The annual royalties of $5000 are thus cheaper for this firm. The ratio of costs, purchase to lease, is $8136/5000 = 1.6272$. The calculation assumes that in both cases the tax depreciation will be taken at a uniform rate and will cancel out as affecting both alternatives equally.

**Example 4.9.** A $15,000 mortgage is to be repaid over 20 years at 6 percent per year interest. Find the monthly payments.

**Solution.** By Eq. (4.21) and Table A2.3, unaflow is

$$\overline{R} = \frac{15,000}{20} \frac{1}{F_{\overline{R}P,0.06,20}} = \frac{15,000}{20} \frac{1}{0.5823} = \$1288$$

That is, the flow must be $1288 per year, or

$$\frac{\$1288}{12} = \$107.33/\text{month}$$

## 4.7 CAPITALIZED COST

Capitalized cost, like uniform annual flow, can be used to compare articles or systems having different service lives. It reduces all service lives to a common denominator, i.e., present value on the basis, for mathematical purposes, of service forever.

Consider an article that has an initial cost $C$ and lasts $n$ years. The present value of supplying service forever is

$$P_\infty = Ce^{-i0} + Ce^{-in} + Ce^{-2in} + Ce^{-3in} + \cdots$$

which is an infinite geometrical series with the first term $C$ and the ratio $e^{-in}$. The sum is given by Eq. (2.14), and letting $P_\infty = K$, we have

$$K = \left[\frac{1 - (e^{-in})^\infty}{1 - e^{-in}}\right]C = \frac{1}{1 - e^{-in}}C \qquad (4.22)$$

The bracketed term on the right converts a present value of $n$ years' duration to a capitalized cost; i.e.,

$$K = P_n \frac{1}{1 - e^{-in}} \qquad (4.23)$$

or
$$K = P_n \frac{e^{in}}{e^{in} - 1} \qquad (4.24)$$

where the symbol $P_n$ emphasizes that $P$ is a present value representing $n$ years' duration.

Equations (4.23) and (4.24) are important because they are the basis for using the capitalized-cost concept with continuous interest. The reader is referred to Chap. 2 for a more complete discussion of capitalized cost.

A relationship between capitalized cost $K$ and uniform flow $\overline{R}$ is easily derived. The present value of a uniform flow $\overline{R}$ for $n$ years is, by Eq. (4.11),

$$P_n = n\overline{R} \frac{1 - e^{in}}{in}$$

and the capitalized cost of this present value, by Eq. (4.23), becomes

$$K = n\overline{R} \frac{1 - e^{-in}}{in} \frac{1}{1 - e^{-in}} = \frac{\overline{R}}{i}$$

That is,

$$\overline{R} = iK \qquad (4.25)$$

Equation (4.25) for continuous interest and unaflow is analogous to the corresponding relationship $R = iK$, Eq. (2.35), for periodic interest.

**Example 4.10.** Repeat Example 4.8 on the basis of capitalized cost.

*Solution.* Capitalized cost of the royalty payments, by Eq. (4.25), is

$$K = \frac{\overline{R}}{i} = \frac{5000}{0.15} = 33{,}333$$

Capitalized cost of purchase is given by Eq. (4.24), which, using Table A2.1, becomes

$$K = 50{,}000 \frac{e^{(0.15)(17)}}{e^{(0.15)(17)} - 1} = 50{,}000 \frac{12.807}{12.807 - 1} = 54{,}235$$

It is cheaper to pay the royalties. The ratio of costs, purchase to lease, is $54{,}235/33{,}333 = 1.6271$. This checks the calculation by uniform flow in Example 4.8.

**Example 4.11.** In a given exposure a paint job lasts 4 years and costs $0.20/ft². A supplier offers a new coating which is claimed to last 20 years but costs $0.60/ft². Is it economically attractive to change to the coating which lasts five times as long and costs only three times as much, if money is worth 10 percent? Neglect taxes.

**Solution.** Capitalized costs can be calculated from Eq. (4.24) and Table A2.1 and, for the 4- and 20-year jobs, respectively, are

$$K_4 = \$0.20 \frac{e^{(0.10)(4)}}{e^{(0.10)(4)} - 1} = 0.20 \frac{1.4918}{0.4918} = 0.6067$$

$$K_{20} = \$0.60 \frac{e^{(0.10)(20)}}{e^{(0.10)(20)} - 1} = 0.60 \frac{7.3891}{6.3891} = 0.6939$$

The 4-year coating is more economical. The saving as an annual flow per year per square foot can be obtained from Eq. (4.25) and is

$$\overline{R} = i(K_{20} - K_4) = 0.10(0.6939 - 0.6067) = 0.00872$$

That is, use of the 4-year coating saves $0.00872/year · ft² in comparison with the 20-year coating.

## 4.8 INCOME TAX

The reader is referred to Chap. 3 for a detailed discussion of the effect of income tax using periodic interest. This section is concerned with the inclusion of income tax using continuous interest. A before-tax amount will always transform to an after-tax amount using the expression

$$(\text{After-tax amount}) = (\text{before-tax amount})(1 - t\psi) \qquad (4.26)$$

where $t$ is the decimal tax rate and $\psi$ is the present value of $1 of depreciation. The value for $\psi$ depends on the discounting rate and how it is taken, the depreciation method, the useful life of the article for tax purposes, and the timing of tax payments. If the tax timing and expense or receipt are simultaneous, $\psi = 1$; otherwise $\psi$ will be less than unity.

In practice, income tax is usually paid on a quarterly basis, which agrees neither with yearly periodic discounting and a single end-of-year tax payment nor with continuous discounting and continuous tax payment. However, in practice, there is little error in using approximations. In particular, since comparisons are always being made, simplifying assumptions tend to balance out. For this section it is necessary only to summarize the $\psi$ factor for some circumstances using continuous interest.

For a uniform yearly flow $\overline{R}$, assuming a continuous tax applied simultaneously with the continuous flow,

$$\psi = 1$$

For straight-line depreciation and continuous tax the tax benefit is a uniform flow and the $\psi$ factor is related to Eq. (4.11):

$$\psi = F_{\text{SLP},\bar{i},n} = F_{\overline{RP},\bar{i},n} = \frac{1 - e^{-in}}{in} \tag{4.27}$$

which is tabulated in Table A2.3.

For an approximation to sum-of-the-years-digits depreciation and continuous tax the $\psi$ factor was given by Eq. (4.19) and is tabulated in Table A2.4.

The $\psi$ factor can always be calculated from its definition—the present value of $1 of future depreciation. For example, calculate the $\psi$ factor for sum-of-the-years-digits depreciation using continuous interest but a periodic tax at the end of the year for a useful life of $n$ years. Sum-of-the-years-digits is $n(n + 1)/2$, and the present value is

$$P = \frac{2}{n(n + 1)}[ne^{-i} + (n - 1)e^{-2i} + (n - 2)e^{-3i} + \cdots + e^{-in}] \tag{4.28}$$

The summation can be done by special techniques, giving

$$\psi = \frac{2}{n(n + 1)(e^i - 1)}\left(n - \frac{1 - e^{-in}}{e^i - 1}\right) \tag{4.29}$$

Equation (4.28) is not exactly the same as Eq. (4.19). They both use continuous interest, but Eq. (4.19) uses an approximation to sum-of-the-years-digits depreciation, and they differ in the timing of the tax.

**Example 4.12.** A $1000 investment has an expected life of 20 years and is to be depreciated over a 15-year life using sum-of-the-years-digits depreciation with money worth 10 percent per year after a 34 percent tax. Find (a) the present value of the capital charge after taxes and (b) the equivalent annual flow.

*Solution*
(a) By Eqs. (4.26) and (4.19) and Table A2.4,

$$P = \$1000(1 - 0.34F_{\text{SDP},0.10,15}) = \$1000[1 - 0.34(0.6438)] = \$781.11 \tag{4.30}$$

(b) The after-tax present value given above can be converted to an after-tax uniform flow by Eq. (4.11) and Table A2.3:

$$P = \$781.11 = n\overline{R}\left(\frac{1 - e^{-in}}{in}\right) = 20\overline{R}F_{\overline{PR},0.10,20}$$

$$\$781.11 = 20\overline{R}(0.4323)$$

$$\overline{R} = \$90.34 \quad \text{(after tax)} \tag{4.31}$$

$\overline{R}$ in Eq. (4.31) is figured on an after-tax basis. The $\psi$ factor is 1 for a uniform flow; hence $\overline{R}$ before tax can be calculated from

$$\overline{R}(1 - 0.34) = \$90.34$$

$$\overline{R} = \$136.88/\text{year} \quad \text{uniform annual flow before tax}$$

**TABLE 4.2**
**Summary of relationships for continuous interest**

| Item no. | Item | Description | Algebraic relationship | Factor relationship |
|---|---|---|---|---|
| 1 | $P$ to $S$ | Moves a fixed sum $P$ to another instant of time $n$ years with the calendar | $S = Pe^{in}$ | $S = PF_{PS,\bar{i},n}$ (Table A2.1) |
| 2 | $S$ to $P$ | Moves a fixed sum $S$ to another instant of time $n$ years against the calendar | $P = Se^{-in}$ | $P = SF_{SP,\bar{i},n}$ (Table A2.2) |
| 3 | $\bar{R}$ to $P$ | Converts a uniform flow $\bar{R}$ for $n$ years to present value at the start of the flow | $P = n\bar{R}\dfrac{1 - e^{-in}}{in}$ | $P = n\bar{R}F_{\bar{R}P,\bar{i},n}$ (Table A2.3) |
| 4 | $\bar{R}$ for 1 year to $P$ | Present value of 1 year of uniform flow starting $X$ years hence | $P = \bar{R}e^{-ix}\dfrac{1 - e^i}{i}$ | $P = \bar{R}F_{SP,i,X}F_{RP,\bar{i},1}$ |
| 5 | $P$ of flow changing at an exponential rate for $n$ years | Present value of $R_x = R_0e^{\pm gx}$ for $n$ years | $P = n\bar{R}_0\dfrac{1 - e^{-(i\mp g)n}}{i \mp g}$ | $P = nR_0F_{\bar{R}P,(\bar{i}\mp g),n}$ |

| 6 | $P$ of flow declining in a straight line to zero | Flow goes from $\bar{R}_0$ at zero time to zero in $n$ years; total flow $Q$ is $nR_0/2$ | $P = Q\left[\dfrac{2}{in}\left(1 - \dfrac{1 - e^{-in}}{in}\right)\right]$ | $P = QF_{SDP,\bar{i},n}$ (Table A2.4) |
| 7 | $P$ to $\bar{R}$ | Converts a present value to a uniform flow of $n$ years | $\bar{R} = \dfrac{P}{n}\dfrac{1}{(1 - e^{-in})/in}$ | $R = \dfrac{P}{n}\dfrac{1}{F_{\overline{RP},\bar{i},n}}$ |
| 8 | $\bar{R}$ to $S$ | Converts a uniform flow for $n$ years to a future amount $n$ years hence | $S = n\bar{R}e^{in}\dfrac{1 - e^{-in}}{in}$ | $S = n\bar{R}F_{PS,\bar{i},n}F_{RP,\bar{i},n}$ |
| 9 | $S$ to $\bar{R}$ | Converts a future sum $S$, $n$ years from now, to a uniform flow; sinking-fund payment | $\bar{R} = \dfrac{Se^{-in}}{n}\dfrac{1}{(1 - e^{-in})/in}$ | $R = \dfrac{S}{n}F_{FS,\bar{i},n}\dfrac{1}{F_{RP,\bar{i},n}}$ |
| 10 | $P$ to $K$ | Converts a present value representing $n$ years to a capitalized cost | $K = P\dfrac{e^{in}}{e^{in} - 1}$ | |
| 11 | $K$ to $\bar{R}$ | Converts a capitalized cost to a uniform flow | $\bar{R} = iK$ | |

## 4.9 EQUIVALENCE

The purpose of this chapter, using continuous interest, and that of previous chapters, using periodic interest, has been to demonstrate how to compare two or more systems differing in receipts and expenses, as well as their timing, the evaluation of present value after taxes, and the conversion of the present values to some common denominator of economic life, such as equivalent uniform flow or capitalized cost, to correct for different durations. Comparisons can also be made on the basis of rate of return, discounted cash flow, as developed in the following chapter.

The method to be used for comparing alternatives or ventures and the choice between periodic and continuous interest are left to the analyst. Although an economic specialist may prefer one method, the client is often a manager who is a generalist by necessity. Results must be presented in terms familiar to such a client and must be simple enough to be grasped on the run. Experience shows that if a solution is presented in unfamiliar or seemingly unrealistic terms, it will not be understood; if not understood, it will not be believed; and if not believed, it will not be accepted.

In practice, real problems are complex, not so much in computation as in defining what the cash flows will be. Often 95 percent of the total time in solving a problem is required for determining costs and incomes, the applicable tax and other government regulations, and projecting sales, costs, etc.

A summary of the various relationships using continuous interest is given in Table 4.2. The $\psi$ factor, the present value of $1 of future depreciation, is given in Table 4.3.

One article [5] attempts to coordinate periodic and continuous interest. It shows that both methods of compounding will give equivalent results for a comparison between alternatives if the three following aspects are recognized:

1. Continuous and periodic interest, when equivalent, are not alike numerically. The equivalency is given by Eq. (4.8).
2. A flow of funds may be continuous or periodic, and the nature of the flow must be considered independently of the method used to compound the rate of return.
3. The timing of the tax situation must be consistent in both methods.

A simple example will show what is meant by equivalency and how relationships between continuous and periodic interest can be made compatible.

**Example 4.13.** Derive the relationship for $\psi$ with continuous interest as given by Table 4.3, item 3, from the known relationship with yearly compounding given by Eq. (3.11).

*Solution.* Equation (3.11) can be written

$$\psi = \frac{1}{n} \frac{(1 + i_{\text{eff}})^n - 1}{i_{\text{eff}}(1 + i_{\text{eff}})^n}$$

**TABLE 4.3**
**The $\psi$ factor, the present value of $1 of depreciation**

| Item no. | Type of depreciation | Tax timing | $\psi_{i,n}$ with continuous interest |
|---|---|---|---|
| 1 | Instantaneous | Instantaneous | $1$ |
| 2 | Uniform | Continuous | $\dfrac{1 - e^{-in}}{in}$ (Table A2.3) |
| 3 | Uniform | Periodic, end of year | $\dfrac{1 - e^{-in}}{n(e^i - 1)}$ |
| 4 | Flow declining in a straight line to zero; simulated sum-of-the-years-digits | Continuous | $\dfrac{2}{in}\left(1 - \dfrac{1 - e^{-in}}{in}\right)$ (Table A2.4) |
| 5 | Sum-of-the-years-digits | Periodic, end of year | $\dfrac{2}{n(n + 1)(e^i - 1)}\left(n - \dfrac{1 - e^{-in}}{e^i - 1}\right)$ |
| 6 | Declining balance | Periodic, end of year without switch | $\dfrac{F_{DB}}{e^i - (1 - F_{DB})}\left[1 - \left(\dfrac{1 - F_{DB}}{e^i}\right)^n\right]$ |

Substituting from Eq. (4.8) with $i$ used for $i$ continuous gives

$$\psi = \frac{1}{n}\frac{e^{in} - 1}{(e^i - 1)e^{in}} = \frac{1}{n}\frac{1 - e^{-in}}{e^i - 1}$$

as given in the table. Here the $\psi$ factor is also given by a geometric series which can be summed by Eq. (2.14):

$$\psi = \frac{1}{n}\sum_{n=1}^{n=n} e^{-in} = \frac{1}{n}\frac{1 - e^{-in}}{e^i - 1}$$

## NOMENCLATURE

$C_d$     Depreciable first cost, $

$e$     Naperian constant 2.71828 . . .

$F_{P\overline{R},\bar{\imath},n}$     Factor to convert $P$ to $\overline{R}$ with continuous compounding; reciprocal of $F_{\overline{R}P,\bar{\imath},n}$, year$^{-1}$

$F_{PS,\bar{\imath},n}$     Factor to convert $P$ to $S$ with continuous compounding, $e^{in}$, Table A2.1, dimensionless

$F_{\overline{R}P,\bar{\imath},n}$     Factor to convert $\overline{R}$ to $P$ with continuous discounting; Table A2.3, years

$F_{SP,\bar{\imath},n}$     Factor to convert $S$ to $P$ with continuous discounting, $e^{-in}$, Table A2.2, decimal, dimensionless

| | |
|---|---|
| $F_{\text{SDP},\bar{i},n}$ | Factor to convert a unit total flow declining to zero at a constant rate over $n$ years starting with the reference point and with continuous discounting, decimal, dimensionless, approximates $\psi_{\text{SD}}$ |
| $g$ | Constant in exponential-rate flow change, decimal |
| $i$ | Interest rate, decimal per year; generally nominal rate |
| $i_c$ | Interest rate, specifically emphasizing continuous compounding, decimal per year |
| $i_{\text{eff}}$ | Effective interest rate, decimal per year |
| $i_p$ | Interest rate, specifically emphasizing periodic compounding, decimal per year |
| $K$ | Capitalized cost |
| $n$ | Time, years |
| $n'$ | Time for tax depreciation, years |
| $p$ | Periods per year |
| $P$ | Present value, $ |
| $P_n$ | Present value for $n$ years' duration, $ |
| $Q$ | Total flow, $ |
| $r$ | Nominal rate of return after taxes, decimal per year |
| $\overline{R}$ | Uniform flow, $/year |
| $R_0$ | Initial flow rate, $/year |
| $R_x$ | Flow rate at time $x$, $/year |
| $S$ | Future value, $ |
| SD | Sum-of-the-years-digits depreciation |
| SL | Straight-line depreciation |
| $t$ | Income tax rate, decimal |
| $T$ | Time, years |
| $\psi$ | Factor representing present value of $1 of future depreciation |
| $\psi_{\text{SD}}$ | The $\psi$ factor for sum-of-the-years-digits depreciation |
| $\psi_{\text{SL}}$ | The $\psi$ factor for straight-line depreciation |

## PROBLEMS

**4.1.** Develop a relationship for discounting a flow increasing in a straight line from zero at zero time to $\overline{R}_n$ at time $n$.

**4.2.** Develop a relationship for discounting a flow increasing in a straight line from $\overline{R}_1$ at zero time to $\overline{R}_2$ at time $n$.

**4.3.** Develop a relationship for discounting a series of periodic cash flows of $k$ payments, $Y$ each, at intervals of $n$ years, the first one beginning $n$ years hence.

**4.4.** A firm has a contributory savings plan whereby each employee can set aside 5 percent of his or her gross salary. The firm will match this amount, invest the sums in its capital stock, and reinvest all dividends in capital stock. If an employee's salary is consistently $12,000 per year, how much will the employee accumulate after 20 years if the company's net earnings average 8 percent per year and the stock consistently sells at book value?

**4.5.** What is the average rate of growth of the employee's $600 per year portion of the contribution?

**4.6.** Suppose the employee finds an alternative proposition which promises to double the money every 5 years. Is it better to participate in the savings plan or forgo the company's contribution and invest the employee's contribution in the alternative?

**4.7.** If the parents in Example 4.3 continue their monthly savings during the 4 years their child attends college, i.e., for 21 instead of 17 years, how much must their monthly savings be to permit $3000 per year to be withdrawn uniformly over the 4 years from the seventeenth to the twenty-first birthday?

**4.8.** A new machine costs $8000 and lasts 10 years, using sum-of-the-years-digits depreciation and a 10-year life for tax purposes. If money is worth 10 percent per year after a 34 percent tax, how much can be spent to repair an old machine to extend its life 3 years? The repair job can be written off at once for tax purposes. Compare with Prob. 3.20.

**4.9.** Repeat Example 2.21 using continuous discounting. Money is worth 10 percent per year after a 34 percent tax rate. Use straight-line depreciation. Machine A will be written off in 8 years for tax purposes, machine B in 10 years. Maintenance costs and savings from quality control are uniform flows in the years in which they occur. The salvage value is anticipated and cannot be depreciated for tax purposes:

|                                          | A      | B      |
|------------------------------------------|--------|--------|
| First cost, $                            | 10,000 | 95,000 |
| Maintenance, $ per year                  | 3,000  | 1,000  |
| Extra maintenance, year 3, $             | 4,000  |        |
| Extra maintenance, year 4, $             | 1,500  |        |
| Savings from quality control, $ per year |        | 6,000  |
| Salvage value                            |        | 20,000 |
| Life, years                              | 4      | 10     |

**4.10.** A company completed a plant 10 years ago. It was expected to be serviceable for 25 years, but technical advances and accumulated know-how suggest that obsolesence may have progressed faster than expected, so that it may be profitable to replace it now. Assume, for simplicity, that (1) a new plant would have the same capacity as the old and would produce the same array of products with the same initial revenue for both, so that the advantage of the new plant is reflected only in its lower operating costs; (2) these savings in operating costs are $180,000 per year; (3) depreciation on the old plant is $35,000 per year on a straight-line basis, and present salvage value is zero; (4) the tax rate is 34 percent per year; and (5) any loss on disposal can be taken immediately for tax purposes. If the investment required for the new plant is $1,000,000 and both its economic life and its life for tax purposes are 15 years with sum-of-the-years-digits depreciation, what is the rate of return to be earned by investment in a new plant?

**4.11.** Establish the relationship for the $\psi$ factor using continuous interest and a periodic end-of-year tax given by Table 4.3, item 5.

**4.12.** An item will be depreciated over a 10-year life with money worth 15 percent per year after a 34 percent tax. Use continuous discounting. Compare the $\psi$ factors obtained from Table 4.3, item 4, for simulated sum-of-the-years-digits depreciation and item 5 for periodic end-of-year depreciation. Also, compare $(1 - t\psi)$ with $t = 0.34$ since that is the expression in which $\psi$ will be used.

**4.13.** Let $R_e$ be an equivalent end-of-year amount and $R_c$ a continuous uniform flow during the year, compounded with continuous interest at nominal rate $i$. For equivalency, show that

$$R_e = \frac{R_c}{i}(e^i - 1)$$

**4.14.** Some analysts use a continuous uniform flow for 1 year as if it were a single total flow at midyear, and use simple interest for one-half year to obtain a single end-of-year amount. Let the flow be $1000/year and the periodic interest rate 20 percent per year.
(a) Find the single amount at the end of the year.
(b) The analyst now recalculates the amount using the continuous interest rate equivalent to the periodic rate of 20 percent but continuous flow of the $1000. What single amount is found as of the end of that year?

**4.15.** A manufacturer currently makes a profit $M$ of $100,000/year on an item. Money is worth 8 percent per year, and interest is to be compounded continuously. Twenty years of operation is anticipated, and there is no income tax.
(a) Find the present value of the profit.
(b) Because of increased productivity, the profit per year increases continuously and can be expressed as

$$M = 100,000(2 - e^{-0.1T})$$

where $T$ is time in years. Find the present value of the profit.
(c) Productivity increases as in (b), but competition reduces the profit continuously by 3 percent per year. Find the present value of the profit.

## REFERENCES

1. Hirschmann, W. B.: "Profit from the Learning Curve," *Harvard Business Review,* January–February 1964, pp. 125–139.
2. Hirschmann, W. B., and J. R. Brauweiler: "Investment Analysis: Coping with Change," *Harvard Business Review,* May–June 1965, pp. 62–72.
3. Hirschmann, W. B., and J. R. Brauweiler: "Continuous Discounting for Realistic Investment Analysis," *Chemical Engineering,* July 19, 1965, pp. 210–214.
4. Hirschmann, W. B., and J. R. Brauweiler: "Realistic Investment Analysis, II," *Chemical Engineering,* August 16, 1965, pp. 132–136.
5. Jelen, F. C., and C. L. Yaws: "Unify Interest Compounding," *Hydrocarbon Processing,* vol. 60, April 1981, pp. 223–226.

# CHAPTER
# 5

# PROFITABILITY

*F. C. Jelen and K. K. Humphreys*

## 5.1 NATURE OF PROFITABILITY

In a broad sense profitability is a measure of the total income for a project compared to the total outlay. Profitability can be applied to a short period of time, say, year by year, or to the entire life of a project. It can be applied to a single cost center, an entire project, or an organization as a whole. Profitability always applies to a monetary measure, although intangible benefits and costs must not be ignored. Inherently, a measure of profitability should be adjusted for size, such as the profit per dollar invested, rather than just the profit.

In this chapter criteria for profitability will be developed, followed by a discussion of profitability and some associated topics.

## 5.2 CRITERIA FOR PROFITABILITY

The major classifications for profitability are given in Table 5.1, and there are several variations for some of the entries. The criterion in group I is expressed as time, such as time to recover the investment. Group II uses rate of return as a criterion but does not recognize the timing of cash flows. The discounted-cash-flow rate of return in group IV also uses rate of return as a criterion but does recognize the timing of cash flows. Group III, based on present value, recognizes the timing of cash flows but expresses the profitability as a premium (or deficiency) net present value compared to that at some referenced rate of return.

**TABLE 5.1**
## Criteria for profitability

I. Payout time
    A. Without interest
    B. With interest
    C. Equivalent maximum investment period
    D. Positive equivalent recovery period

II. Return on investment
    A. Return on original investment
    B. Return on average investment

III. Present value
    A. Net present value
    B. Excess present-value index

IV. Discounted-cash-flow rate of return

## 5.3   A SIMPLE PROJECT

The classifications in Table 5.1 will now be applied to a simple project, $A$, having cash flows, as in Table 5.2. The cash flow here is the sum of the profit after taxes plus depreciation and represents an in-pocket amount after taxes. Amounts are in kilodollars. Money going into the project is taken as negative and money coming back from the project as positive, following the customary designations in profitability studies.

Project $A$ involves swapping 1000 k\$ invested at zero time for the after-tax amounts given by years in the last column, starting with 475 k\$ at the end of year 1. The numbers in Table 5.2 alone determine the profitability.

## 5.4   PAYOUT TIME WITHOUT INTEREST

Payout time is also known as *payback time*. It is the time required to reduce the investment to zero.

> **Example 5.1.** Find the payout time without interest for the data in Table 5.2.
> The tabulation on top of page 106 shows that the investment will be reduced to zero between 2 and 3 years, or approximately 2.4 years by interpolation.

**TABLE 5.2**
## Cash flow for project $A$

| Time, end year | After-tax profit, k\$ | Depreciation, k\$ | Cash flow, k\$ |
|---|---|---|---|
| 0 | −1000 | 0 | −1000 |
| 1 | 275 | 200 | 475 |
| 2 | 200 | 200 | 400 |
| 3 | 130 | 200 | 330 |
| 4 | 70 | 200 | 270 |
| 5 | 0 | 200 | 200 |

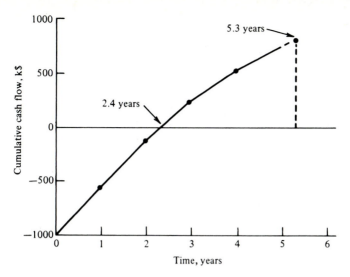

**FIGURE 5.1**
Cumulative cash flow for project *A* of Table 5.2.

| Time, end year | Cumulative cash flow, k$ |
|---|---|
| 0 | −1000 |
| 1 | −525 |
| 2 | −125 |
| 3 | +205 |
| 4 | +475 |
| 5 | +675 |

The values are plotted in Fig. 5.1.

Payout time without interest does not consider working capital, e.g., inventory, since it is assumed that working capital can be recovered at any time and does not affect the time required to get back the fixed investment.

## 5.5  PAYOUT TIME WITH INTEREST

The method allows for a return on investment and is subject to variations. The following variation applies an interest charge only on the remaining fixed investment. A variation also includes an interest charge on working capital but recognizes that working capital itself is fully retrievable at any time.

In practice, payout time with interest is seldom used, and the term *payout time* should mean "without interest" unless interest is specifically mentioned.

**Example 5.2.** Using the data in Table 5.2, find the payout time with interest on the remaining investment at 10 percent per year.

*Solution.* The following tabulation shows that the payout time is between 2 and 3 years or, more exactly, 2.95 years by interpolation:

| (1)<br>End year | (2)<br>Investment<br>for year | (3)<br>Charge on<br>investment<br>for year,<br>10% of (2) | (4)<br>Cash flow | (5)<br>Cash flow<br>after investment<br>charge,<br>(4) − (3) | (6)<br>Cumulative<br>net<br>cash flow |
|---|---|---|---|---|---|
| 0 | | | −1000 | | −1000 |
| 1 | 1000 | 100 | 475 | 375 | −625 |
| 2 | 625 | 62.5 | 400 | 337.5 | −287.5 |
| 3 | 287.5 | 28.75 | 330 | 301.25 | +13.75 |

The investment for each year is the investment for the previous year less the cash flow after the investment charge for the previous year.

## 5.6 ADVANTAGES AND DISADVANTAGES OF PAYOUT TIME

The advantages of payout time are that it is a simple, easily understood method and is commonly used in the business world. The method is particularly adaptable to simple problems, such as replacing an old, inefficient pump that wastes energy with a new, energy-efficient pump. A payout time of 9 months in this case would be an adequate and convincing criterion for purchasing a new pump.

Perhaps the chief disadvantage of payout time is that, as ordinarily used, it gives no credit for the period beyond the payout time.

When payout time is used as a criterion on large projects, it can encourage shoddy investment, favor skimpy design, lead to high start-up costs, and promote a policy-inhibiting perspective. In the end it could lead investors to abandon their investment and become merchandisers for the products of others since no investment assures the minimum payout time — zero.

## 5.7 RETURN ON ORIGINAL INVESTMENT (ROI)

Return on original investment (ROI), also called the *DuPont* or *engineer's method,* is the percentage relationship of the average annual profit (sometimes the average annual cash flow) to the original investment, including nondepreciable items such as working capital. Thus it is given by

$$\frac{\text{Average yearly profit during earning life}}{\text{Original fixed investment} + \text{working capital}}(100) \qquad (5.1)$$

**Example 5.3.** Using the data in Table 5.2, find the return on original investment.

*Solution.* The average profit is

$$\frac{275 + 200 + 130 + 70 + 0}{5} = 135 \text{ k\$/year}$$

By Eq. (5.1), the return on original investment is

$$\frac{135}{1000 + 0}(100) = 13.5 \text{ percent per year}$$

The time value of money is not considered since only the average profit is used, not its timing. The profits from years 1 through 5 could be reversed, and the return on original investment would be the same.

## 5.8 RETURN ON AVERAGE INVESTMENT (RAI)

The return on average investment (RAI) is similar to the return on original investment except that the divisor in Eq. (5.1) is the average outstanding investment.

**Example 5.4.** Using the data in Table 5.2, find the return on average investment.

*Solution.* Average investment is found from the following tabulation:

| Year | Investment, $ |
|------|---------------|
| 1 | 1000 |
| 2 | 1000 − 200 = 800 |
| 3 | 800 − 200 = 600 |
| 4 | 600 − 200 = 400 |
| 5 | 400 − 200 = 200 |
| | Av = 600 |

$$\frac{135}{600}(100) = 22.5 \text{ percent per year}$$

The return on average investment at 22.5 percent is higher, as it must be, than the return on the original investment at 13.5 percent. The two are independent criteria and must be used separately. Comparing the ROI of one system with the RAI of another system would be unfair. The two must not be intermixed in a comparison.

## 5.9 NET PRESENT VALUE (NPV)

The return on original investment and the return on average investment do not reflect the time value of money. Comparing projects on the basis of present value does include the time value of money since all cash flows are referred to zero time before an evaluation is made. The original concept was known as *venture worth* but is now generally referred to as *net present value (NPV)*. Reduction to present value, discounting, is done at a fixed rate, usually the minimum acceptable rate of return on capital. The NPV is then the present value of all inflows less the present value of all outflows. Accordingly, the NPV is a single amount referred to zero time and represents a premium if positive, or a deficiency if negative, at some chosen fixed rate of return.

**Example 5.5.** Using the data in Table 5.2, find the NPV at 10 percent per year and at 25 percent per year.

*Solution.* At 10 percent per year the NPV is

$$\frac{-1000}{1.10^0} + \frac{475}{1.10^1} + \frac{400}{1.10^2} + \frac{330}{1.10^3} + \frac{270}{1.10^4} + \frac{200}{1.10^5} = +318.92$$

At 25 percent per year the NPV is

$$\frac{-1000}{1.25^0} + \frac{475}{1.25^1} + \frac{400}{1.25^2} + \frac{330}{1.25^3} + \frac{270}{1.25^4} + \frac{200}{1.25^5} = -18.91$$

The project has a premium present value of 318.92 k$ at 10 percent per year and a deficiency in present value of 18.91 k$ at 25 percent per year. The negative present value at 25 percent per year does not mean that the project is a loser, but only that it does not pay off at 25 percent per year.

Net present value is not corrected for size. Thus two of the projects would have an NPV of $2 \times 318.92 = 638.8$ k$ at 10 percent per year, but fundamentally the profitability has not changed. This objection can be overcome by using an excess present value index defined as

$$\frac{\text{Present value receipts}}{\text{Present value outlays}}$$

which in this case, at 10 percent per year, would be

$$\frac{1318.92}{1000} = 1.319$$

Net present value does not correct for a difference in the service lives of projects. In the present example at 10 percent per year the NPV for 5 years ahead is 318.92 k$. If the life of the project is to be 10 years, obtained by repeating the project at the end of 5 years, the NPV is changed numerically to

$$318.92 + \frac{318.92}{(1.10)^5} = 516.94 \text{ k\$}$$

If projects have different durations, the NPV should be converted to an annual cost, such as unacost, or a capitalized cost, both of which have built-in factors for different service lives.

**Example 5.6.** Compare project $A$, having an initial investment of 1000 k\$, a 5-year life, and an NPV of 318.92 k\$, with project $W$, having an initial investment of 1500 k\$, a 7-year life, and an NPV of 675 k\$. Both present values are at 10 percent per year. Correct for the difference in size by referring to an initial investment of 1 k\$ and correct for the difference in duration by converting to a premium unacost.

*Solution.* For project $A$

$$\frac{318.92}{1000} F_{PR,\,10\%,\,5} = 0.3192(0.26380) = 0.0842$$

For project $W$

$$\frac{675}{1500} F_{PR,\,10\%,\,7} = 0.4500(0.20541) = 0.0924$$

Thus project $A$ has a premium of 0.0842 k\$/year per 1 k\$ of original investment against 0.0924 for project $W$. Project $W$ is more profitable than project $A$ at 10 percent per year. However, the comparison does depend on the rate of return used, and may even reverse at another rate of return.

One method for correcting for the duration of a project is to evaluate all projects at some fixed duration. Thus one company compares all projects on a 12-year life. The procedure is acceptable for projects lasting 12 years or longer since the time beyond 12 years contributes little to the present value at practical rates of return.

Net present value, or any variation thereof, requires only a direct calculation. No trial-and-error procedure is required since the rate of return is chosen before the calculation is begun. Originally, the avoidance of a trial-and-error calculation was an advantage. Now, however, with the introduction of computers and the ubiquitous programmable calculator, trial-and-error calculations are easily made, and some of the advantage of NPV has been lost.

## 5.10 DISCOUNTED CASH-FLOW RATE OF RETURN (DCFRR)

The *discounted cash-flow rate of return (DCFRR)* is a sophisticated criterion for profitability that corrects for differences in the duration and size of projects, includes the time value of money and the timing of all cash flows, and recognizes that investment changes with time. It is also known as the *interest-rate-of-return method, profitability index (PI)*, *internal rate of return*, and *investor's method*. It includes all cash flows over the entire life of the project and adjusts them to one point fixed in time, usually the original investment or start-up time, using the compound-interest procedure; but the calculated rate of return is independent of the point in time chosen for discounting. A trial-and-error calculation is required to determine the compound-interest rate at which the sum of the time-adjusted cash outflows equals the sum of the time-adjusted inflows. The present value of all cash flows is zero at the DCFRR:

$$0 = \sum_{n=1}^{n=n} \frac{\text{Cash flow for year } n}{(1+r)^n} \tag{5.2}$$

**Example 5.7.** Using the data in Table 5.2, find the DCFRR.

*Solution.* The present value of the cash flows must be zero. Hence

$$P = 0 = \frac{-1000}{(1 + r)^0} + \frac{475}{(1 + r)^1} + \frac{400}{(1 + r)^2} + \frac{330}{(1 + r)^3} + \frac{270}{(1 + r)^4} + \frac{200}{(1 + r)^5}$$

$$P = \begin{cases} +75.16 & \text{for } r = 0.20 \\ -18.91 & \text{for } r = 0.25 \end{cases}$$

By interpolation,

$$P = 0 \qquad \text{for } r = 0.239$$

The DCFRR is 23.9 percent.

The DCFRR on a project is a constant annual rate of return on that part of the investment in the project outstanding each year over its life. The investment base declines over time as returns accumulate to pay interest and retire principal. Therefore, the dollar amount returned each year varies with time, although the rate of return remains constant. In other words, the rate of return is a constant fraction of the nonconstant investment.

The DCFRR method is perhaps the most widely used of all profitability methods. Its wide acceptance is illustrated by the results of a survey of major U.S. corporations which was conducted by the American Association of Cost Engineers. Without exception, the firms responding to the survey used DCFRR as a primary criterion in profitability assessment. While virtually all firms also utilized other profitability measures, DCFRR was the only method used by every firm.

Although DCFRR is widely used and accepted, there are special cases for which it does not yield satisfactory results. Though quite uncommon, situations can occur in which there may be no return at which Eq. (5.2) reduces to zero, or there may be multiple roots and thus more than one rate of return. These cases are the subject of Probs. 5.20 and 5.21.

## 5.11 CASH FLOW

Cash flow is the clue to profitability studies and will be discussed in some detail. Cash flow is simply the movement and timing of cash with respect to a project. Movement can be further defined as in- or out-of-pocket movement. To set up the cash flow for a project, we need to establish all the in- and out-of-pocket amounts and the time at which they occur. Local taxes are easily included. A more complicated task is the evaluation of the corporate income tax which must be included since cash flow is always calculated on an after-tax basis. The income tax is best calculated separately and is included in the cash flow as a single-line item.

Cash flow itself is not a measure of profitability and may merely reflect a high turnover for wear and tear, i.e., depreciation. However, cash flow is used to calculate profitability, as in the DCFRR method, which is a measure of profitability.

Project *D*, now to be discussed, should be studied in detail since it embodies virtually all the principles involved in cash-flow calculations.

A plant will cost 10.4 M$, with payment in equal amounts at 9, 6, 3, and 0 months before start-up taken as zero time. The project will last 5 years, with 0.4 M$ salvage value. The MACRS method of depreciation will be used for tax purposes for a 5-year life. Land at a cost of 1 M$ will be paid for 9 months prior to start-up and is expected to be sold at the end of 5 years for 1.7 M$. Although there will be an internal profit of 0.7 M$ when the land is sold which normally would become subject to the prevailing 34 percent income tax, some allowance for improvements can be anticipated, and the effective tax rate can be taken as 26 percent of the 0.7 M$ gain.

Production, sales, and regular expenses will vary, and the following values can be used as end-of-year amounts:

| Year | Activity as % of full production | Sales, M$ | Regular expenses, M$ |
|------|------|------|------|
| 1 | 33.3 | 5 | 3 |
| 2 | 66.6 | 10 | 5 |
| 3 | 100 | 15 | 7 |
| 4 | 100 | 15 | 7 |
| 5 | 100 | 15 | 7 |

A special expense is classified separately and refers to some rental equipment. It amounts to 0.5 M$/year and must be treated in that manner for tax purposes. However, 3 full years' expense, totaling 1.5 M$, must be paid at start-up time and 0.5 M$ at the end of the fourth and fifth years.

Working capital, consisting chiefly of inventory and cash, will be proportional to sales and will be 1 M$ for full production. Thus 0.33 M$ must be invested at the beginning of the first and second years and 0.34 M$ at the beginning of the third year. Working capital will be recovered in full at the end of the project. Movement of working capital into and out of the project is not subject to income tax. It is merely a bookkeeping exchange of cash and inventory, with no effect on profit.

A start-up cost of 0.7 M$ taken at zero time will be capitalized along with the plant cost.

Part of the project is subject to depletion and can be taken as 25 percent of the sales, with corresponding expenses of 25 percent of the regular expense. The depletion allowance is 22 percent of the annual gross applicable income, limited to 50 percent of the taxable income computed without the deduction for depletion.

Periodic interest compounded on a yearly basis with simple interest for less than 1 year will be used. Income will be taken as positive and outgo as negative. The cash flows for the project and the DCFRR are required.

The calculation for the cash flows is given in Table 5.3. The first seven items are easily calculated from the statements of the project, and they represent the items in the cash flow before the corporate income tax. The income tax is calculated separately in Table 5.4 and becomes item 8 in Table 5.3. It is advantageous to calculate the income tax separately because the income tax can be complicated, and there is no need to destroy the simplicity of Table 5.3 with the details of the tax situation.

TABLE 5.3
**Calculation of cash flow**

| Item | M$ at years from start-up | | | | | | | | |
|---|---|---|---|---|---|---|---|---|---|
| | $-\frac{3}{4}$ | $-\frac{1}{2}$ | $-\frac{1}{4}$ | 0 | 1 | 2 | 3 | 4 | 5 |
| 1. Plant | -2.60 | -2.60 | -2.60 | -2.60 | — | — | — | — | 0.40 |
| 2. Land and salvage | -1.0 | — | — | — | — | — | — | — | 2.10 |
| 3. Sales | — | — | — | — | 5.00 | 10.00 | 15.00 | 15.00 | 15.00 |
| 4. Regular expenses | — | — | — | — | -3.00 | -5.00 | -7.00 | -7.00 | -7.00 |
| 5. Special rental expense | — | — | — | -1.50 | — | — | — | -0.50 | -0.50 |
| 6. Working capital | — | — | — | -0.33 | -0.33 | -0.34 | — | — | 1.00 |
| 7. Start-up expense-capitalized | — | — | — | -0.7 | — | — | — | — | — |
| 8. Income tax from Table 5.4 | — | — | — | — | 0.52 | -0.21 | -1.59 | -1.84 | -1.60 |
| 9. Cash flow — sum of above items | -3.60 | -2.60 | -2.60 | -5.13 | 2.19 | 4.45 | 6.41 | 5.66 | 9.40 |

**TABLE 5.4**
## Calculation of income tax by years

| Item | M$ at end of year | | | | |
|---|---|---|---|---|---|
| | 1 | 2 | 3 | 4 | 5 |
| 1. Sales, not subject to depletion | 3.75 | 7.50 | 11.25 | 11.25 | 11.25 |
| 2. Regular expenses, not subject to depletion | -2.25 | -3.75 | -5.25 | -5.25 | -5.25 |
| 3. Special rental expense, not subject to depletion | -0.50 | -0.50 | -0.50 | -0.50 | -0.50 |
| 4. Depreciable value for project 10.4 + 0.7 = 11.1 | — | — | — | — | — |
| 5. Depreciation factor for year, % | 25.00 | 30.00 | 18.00 | 11.37 | 15.63* |
| 6. Depreciation for year, (4) × (5) | -2.78 | -3.33 | -2.00 | -1.26 | -1.73 |
| 7. Partial base for income tax, (1) + (2) + (3) + (6) | -1.78 | -0.08 | 3.50 | 4.24 | 3.77 |
| 8. Partial income tax at 34% of (7) | 0.61 | 0.03 | -1.19 | -1.44 | -1.28 |
| 9. Sales, subject to depletion | 1.25 | 2.50 | 3.75 | 3.75 | 3.75 |
| 10. Expenses, subject to depletion | -0.75 | -1.25 | -1.75 | -1.75 | -1.75 |
| 11. Calculated taxable income before depletion, (9) + (10) | 0.50 | 1.25 | 2.00 | 2.00 | 2.00 |
| 12. Depletion allowance, 22% of (9), but not over 50% of (11) | 0.25 | 0.55 | 0.83 | 0.83 | 0.83 |
| 13. Partial base for income tax, (11) − (12) | 0.25 | 0.70 | 1.17 | 1.17 | 1.17 |
| 14. Partial income tax at 34% of (13) | -0.09 | -0.24 | -0.40 | -0.40 | -0.40 |
| 15. Income from sale of property and salvage | — | — | — | — | 1.1 |
| 16. Tax on (15) at 26% effective rate on sale of property and 34% on salvage | — | — | — | — | -0.32 |
| 17. Income tax, (8) + (14) + (16) | 0.52 | -0.21 | -1.59 | -1.84 | -1.60 |

*Complete write-off in fifth year.

113

   The income tax calculation given in Table 5.4 includes provision for the regular tax rate of 34 percent, special attention to the sale of the property, and a depletion allowance of 22 percent for part of the project. Its entire gist is to reduce the calculation to a tax for each year to the last line in Table 5.4.

   Depreciation is an expense for tax purposes, although it is not an out-of-pocket outlay at the time it is taken. Similarly, the depletion allowance reduces the tax base, with no actual expenditure being made for the period. Working capital is not part of the tax situation, since it consists merely of funds frozen in the project and ultimately regained. Land which will be sold at a profit at the termination of the project is given special attention for the tax calculation as is income from salvage values.

   The timing for tax purposes may be different from the timing of the actual expenditures. The special rental expense requires an outlay of 1.5 M$ at zero time, but it is a prepaid expense. For tax purposes it is regarded as a 0.5 M$ yearly expense.

   The tax situation can result in an income from savings in taxes for the project for some of the years. During the first year the company as a whole is making money, but the project itself is losing money. The loss by the project reduces the tax which otherwise would be paid by the company. Thus the project gains an income tax credit of 0.52 M$.

   The DCFRR is that rate of return which makes the present value of all cash flows equal to zero at zero time. The cash flows are given on the last line of Table 5.3. Using simple interest for less than 1 year and compound interest for more than 1 year gives

$$0 = -3.60(1 + 0.75r) - 2.60(1 + 0.50r) - 2.60(1 + 0.25r)$$

$$-5.13 + \frac{2.19}{(1 + r)^1} + \frac{4.45}{(1 + r)^2} + \frac{6.41}{(1 + r)^3} + \frac{5.66}{(1 + r)^4} + \frac{9.40}{(1 + r)^5}$$

By trial and error, $r = 0.20$, and the DCFRR for the project is 20 percent.

   A simplified expression for cash flow is sometimes given as

$$\text{Cash flow} = \text{profit after taxes} + \text{depreciation} \qquad (5.3)$$

where depreciation includes the depletion allowance. The expression is not a definition of cash flow, however, nor is it always exact. Consider Tables 5.3 and 5.4 and calculate the cash flow at the end of year 2 using Eq. (5.3). The calculation is

| | |
|---|---|
| Sales | 10.00 M$ |
| Regular expenses | −5.00 |
| Special rental expense | −0.50 |
| Depreciation expense | −3.33 |
| Allowance for depletion | −0.55 |
| Profit before income tax | 0.62 |
| Less income tax at 34% | 0.21 |
| Net profit after taxes | 0.41 |
| Add depreciation | 3.33 |
| Add depletion allowance | 0.55 |
| Calculated cash flow | 4.29 M$ |

which does not agree with the correct value, 4.45 M$, given in Table 5.3. The preceding calculation includes 0.50 M$ for the special rental expense, and it certainly must be considered in the tax and profit, but the expense was actually prepaid and is not a cash outflow at the end of year 2. Hence 0.5 M$ should be added to the cash flow as previously calculated. Again, at the end of year 2 there was an outlay of 0.34 M$ for increased working capital. The outlay is merely money frozen in the project and does not enter the tax or profit consideration, but it is an outlay and should be subtracted from the cash flow calculated. If the two corrections are made, the cash flow at the end of year 2 becomes

$$4.29 + 0.50 - 0.34 = 4.45 \text{ M\$}$$

which agrees with the value given in Table 5.3.

A more compact format for a simple project follows.

**Example 5.8.** A venture requires 1000 k$ now with no salvage value. Receipts and expenses, excluding depreciation, are

| End year | Receipts, k$ | Expenses, k$ |
|----------|--------------|--------------|
| 1 | 1600 | 1000 |
| 2 | 1700 | 1000 |
| 3 | 1140 | 800 |

Working capital of 200 k$ is invested at zero time. MACRS depreciation will be used for tax purposes for a recovery period of 3 years. The income tax rate is 34 percent. Find the DCFRR.

The cash flow is calculated in Table 5.5.

**TABLE 5.5**
**Cash flow for Example 5.8 in k$**

|  | Time 0 | End year 1 | End year 2 | End year 3 | End year 4 |
|---|--------|------------|------------|------------|------------|
| 1. Receipts |  | 1600 | 1700 | 1140 | 0 |
| 2. Expenses, excluding depreciation |  | −1000 | −1000 | −800 | 0 |
| 3. Receipts for tax purposes, (1) + (2) |  | 600 | 700 | 340 | 0 |
| 4. Depreciation |  |  |  |  |  |
|     0.4167(1000) |  | 417 |  |  |  |
|     0.3889(1000) |  |  | 389 |  |  |
|     0.1414(1000) |  |  |  | 141 |  |
|     0.0530(1000) |  |  |  |  | 53 |
| 5. Taxable income, (3) − (4) |  | 183 | 311 | 199 | −53 |
| 6. Tax at 34% |  | 62.2 | 105.74 | 67.66 | −18.02 |
| 7. Fixed capital | −1000 |  |  | 0 |  |
| 8. Working capital | −200 |  |  | 200 |  |
| 9. Cash flow, (3) − (6) + (7) + (8) | −1200 | 537.8 | 594.26 | 472.34 | 18.02 |

**Solution.** For the present value of the cash flow to be zero

$$P = 0 = -1200 + \frac{537.8}{(1 + r)} + \frac{594.26}{(1 + r)^2} + \frac{472.34}{(1 + r)^3} + \frac{18.02}{(1 + r)^4}$$

$$P = \begin{cases} +58.51 & \text{for } 0.14 \\ -17.80 & \text{for } 0.16 \end{cases}$$

By interpolation,

$$P = 0 \qquad r = 0.153$$

The DCFRR is 15.3 percent.

## 5.12   THREE PROJECTS COMPARED

Project $A$, as given in Table 5.2, features a declining income. Table 5.6 repeats project $A$ together with project $B$, which has a steady income, and project $C$, which has an increasing income. All amounts are in kilodollars. All projects have an initial investment of $-1000$ k$ and employ straight-line depreciation. Table 5.6 summarizes the calculations for income taken as discrete amounts at the end of the year, and interest is compounded periodically at the end of the year. Profitability calculated by four common methods — return on original investment, payout time without interest, DCFRR, and NPV — is compared for the three projects. The rank varies with the

**TABLE 5.6**
**Effect of economic yardsticks on project ranking***

| Time, end year | Project $A$, declining income | | Project $B$, steady income | | Project $C$, increasing income | |
| --- | --- | --- | --- | --- | --- | --- |
| | Cash flow | Profit after taxes | Cash flow | Profit after taxes | Cash flow | Profit after taxes |
| — | $-1000$ | | $-1000$ | | $-1000$ | |
| 1 | 475 | 275 | 355 | 155 | 200 | 0 |
| 2 | 400 | 200 | 355 | 155 | 300 | 100 |
| 3 | 330 | 130 | 355 | 155 | 400 | 200 |
| 4 | 270 | 70 | 355 | 155 | 450 | 250 |
| 5 | 200 | 0 | 355 | 155 | 490 | 290 |
| | | Av = 135 | | Av = 155 | | Av = 168 |
| | | Rank | | Rank | | Rank |
| Return on original investment | 13.5 | 3 | 15.5 | 2 | 16.8 | 1 |
| Payout time, years | 2.38 | 1 | 2.82 | 2 | 3.22 | 3 |
| Discounted cash flow | 23.9% | 1 | 22.8% | 2 | 20.8% | 3 |
| Present value at 10% | 319 | 3 | 346 | 1 | 342 | 2 |

*Depreciation in all cases is $200/year.

criterion used for profitability. It will be shown in Sec. 5.14 that a single criterion is not adequate as an absolute measure for judging profitability.

## 5.13 CONTINUOUS INTEREST

Periodic interest has been used so far in this chapter to avoid confusion for the reader who understands no other method. The use of continuous interest in profitability studies is becoming more common, and the reader is referred to Chap. 4 for a discussion of its development and use. Recalculation of projects *A*, *B*, and *C* of Table 5.6 on the basis of continuous interest gives the following result when the rate expressed is the nominal rate:

|  | *A* | Rank | *B* | Rank | *C* | Rank |
|---|---|---|---|---|---|---|
| Discounted cash flow | 21.5% | 1 | 20.5% | 2 | 18.9% | 3 |
| Present value at 10% | 304 | 3 | 328 | 1 | 322 | 2 |

The rank for both discounted cash flow and present worth at 10 percent with continuous compounding is the same in this case as the rank with discrete compounding (cf. Table 5.6). Generally discrete and continuous compounding give the same order of rank.

To facilitate the computation with continuous interest, the more commonly used discount or present-value factors are tabulated in Appendix 2. Table A2.1 covers cash effects occurring at a point in time before time zero, the present. Table A2.2 covers cash effects occurring at a point in time after time zero. Table A2.3 covers a cash flow occurring uniformly over time. The table lists factors which discount $1 cash flow to its present value at the start of the flow. Table A2.4 covers a cash flow declining to zero at a uniform rate discounted to a present value at the start of the flow. Tax credit generated by sum-of-the-years-digits depreciation is an approximation of such a cash flow. Factors for most other special growth and decline patterns can be developed from these tables.

## 5.14 INADEQUACY OF THE SINGLE CRITERION

The question arises, which criterion for judging profitability is best? Before answering this question, we need to recognize that even a single criterion is not unique. Consider three ventures, *A*, *B*, and *C*, having end-of-year cash flows as follows, with all amounts in kilodollars:

|  | Zero time | Year 1 | Year 2 | Year 3 |
|---|---|---|---|---|
| *A* | −1000 | 750 | 390 | 180 |
| *B* | −1000 | 350 | 470 | 660 |
| *C* | −1000 | 533 | 467 | 400 |

Each of these ventures has a DCFRR of 20 percent:

$$\text{Venture } A: \; -1000 + \frac{750}{(1.20)^1} + \frac{390}{(1.20)^2} + \frac{180}{(1.20)^3} = 0$$

$$\text{Venture } B: \; -1000 + \frac{350}{(1.20)^1} + \frac{470}{(1.20)^2} + \frac{660}{(1.20)^3} = 0$$

$$\text{Venture } C: \; -1000 + \frac{533}{(1.20)^1} + \frac{467}{(1.20)^2} + \frac{400}{(1.20)^3} = 0$$

Ventures $A$, $B$, and $C$ require a 1000 k$ investment, last 3 years, and give a cal-culated 20 percent per year DCFRR. However, they differ in the pattern of the cash flows. Obviously, the single criterion, 20 percent per year, has washed out the indi-viduality of the ventures.

Table 5.7 gives a breakdown of the three ventures on a year-by-year basis. The difference between the three ventures becomes apparent. All have 1000 k$ invested in the first year, but the investments remaining for years 2 and 3 are different, as shown by items 4 and 7 in the tabulation. Figure 5.2 shows the investment as a func-tion of time. Venture $A$ can be classified under "accelerated reduction of investment" since the investment is reduced most in the early years. This implies, too, the need for accelerated reinvestment if the initial investment is to be maintained. Venture $B$ can be classified under "retarded reduction of investment" since the investment is re-duced most in the later years. This also implies a delay in generating funds for rein-vestment. Venture $C$ is intermediate and reduces the investment uniformly from year to year. Each venture has its advantages and disadvantages, and the only purpose here is to show that the difference among the ventures shows up in an analysis on a year-to-year basis.

A single criterion is not unique, and there are an infinite number of cash-flow patterns that will produce the same numerical value for the criterion. A single crite-rion is like an average, and it wipes out the individuality of a venture.

**TABLE 5.7**
**Details of ventures $A$, $B$, and $C$ at a 20 percent per year rate of return in k$**

|  | A | B | C |
|---|---|---|---|
| 1. Investment at zero time | 1000 | 1000 | 1000 |
| 2. Less income end year 1 | −750 | −350 | −533 |
| 3. Plus 20% of item 1 | 200 | 200 | 200 |
| 4. Investment beginning year 2 | 450 | 850 | 667 |
| 5. Less income end year 2 | −390 | −470 | −467 |
| 6. Plus 20% of item 4 | 90 | 170 | 133 |
| 7. Investment beginning year 3 | 150 | 550 | 333 |
| 8. Less income end year 3 | −180 | −660 | −400 |
| 9. Plus 20% of item 7 | 30 | 110 | 67 |
| 10. Investment end year 3 | 0 | 0 | 0 |

**FIGURE 5.2**
Investment as a function of time.

The most popular single criterion is DCFRR as stated earlier. Other popular criteria include payout time, return on original investment, and NPV. In practice, an analyst will not use any single criterion but will consider, in addition, the individuality of the project, such as by breaking it down to a year-by-year analysis; review what the project means to the company in the future; analyze how uncertainty affects the project's profitability; and consider a host of other issues. There is no single best criterion since no one criterion alone can reveal all the details.

## 5.15 PROJECTS FOR PROFITABILITY STUDIES

A checklist of the major cost and revenue elements of a typical project is shown in Table 5.8, with the items arranged roughly by time of occurrence. Table 5.9 lists the common project types subject to profitability studies. The mathematical treatment is the same for all. The comparisons can vary—e.g., abandonment or neglect versus

**TABLE 5.8**
**Time, cost, and revenue elements of a typical project**

1. Costs incurred before construction
2. Fixed capital
3. Start-up costs
4. Working capital
5. Subsequent investment
6. Sales price and volume trends
7. Annual expenses
8. Tax effects
9. Salvage value

**TABLE 5.9**
**Project types for profitability studies**

1. Abandonment
2. Process modification and cost reduction
3. Replacement and displacement
4. Expansion
5. New process, product, or market
6. Purchased know-how vs. internal development
7. Joint venture
8. Lease vs. buy
9. Make vs. buy

**TABLE 5.10**
**Some important characteristics**
**of projects**

1. Sunk costs
2. Differing lives
3. Cost sharing
4. Cost savings
5. Quality improvement
6. New business
7. Special financing
8. Risk and uncertainties

shutting down — but the principles do not vary. The difficult task is to arrive at numerical values for all the pertinent cost items involved and not to overlook any applicable costs or benefits.

Table 5.10 lists some underlying characteristics on which the project types depend. Sunk costs are costs incurred in the past and generally do not enter a profitability analysis since profitability studies are concerned with the future. However, the tax situation associated with them may influence the future and should not be omitted. For example, if a piece of equipment is abandoned without any value the original cost is a sunk cost and does not enter a replacement analysis. However, the depreciation on its book value and its timing cannot be ignored since present and future tax offsets may be involved. For differing lives, criteria such as unacost, capitalized cost, and DCFRR make fair comparisons, but comparisons based on straight present value can be misleading unless the difference in service lives has been compensated for. Cost-sharing projects can introduce difficult decisions about how such costs are to be allocated.

**Example 5.9.** The laboratory must decide whether to buy or lease a highly specialized instrument which is undergoing considerable technological advancement.

A new instrument can be bought for $50,000. It has a technological life of 4 years and can be disposed of for $2000 to be taken as regular income. It is fully depreciable in 4 years using sum-of-the-years-digits depreciation for tax purposes. The maintenance cost at the end of each year is expected to be $6000 per year. If the instrument is owned, it can be modified and used in connection with other instruments, and this fact is given a value of $8000/year, as of the end of the year.

A rented instrument costs $15,000/year, payable at the beginning of the year with free servicing. Rental is on an annual basis, renewable each year at the lessee's option. Renting gives the advantage of deferring buying a year at a time in a period of high technological advancement, which is judged to be worth $20,000 on a present basis for a 4-year period.

If money is worth 10 percent per year after a 34 percent tax, should the instrument be bought or rented?

*Solution.* Unacost, the equivalent uniform end-of-year cost, on a purchased-instrument basis (Table 3.5, items 1, 4, and 2 in order, and Table A1.1) is

$$C_d(1 - tF_{\text{SDP}, 10\%, 4})F_{\text{PR}, 10\%, 4} = 50,000[1 - 0.34(0.83013)]0.31547 = \$11,322$$

The disposal income, negative cost, is

$$-C_{ex}\frac{1-t}{(1.10)^4}F_{PR,10\%,4} = -2000\frac{1-0.34}{1.4641}0.31547 = -284$$

$$R(1-t) = (6000-8000)(1-0.52) = \underline{-1320}$$

$$R = \$9718$$

Unacost for a rented instrument (by Table 3.5, items 3 and 1 in order, and Table A1.1) is

$$R_b\left(1-\frac{t}{1+r}\right)(1+r) = 15,000\left(1-\frac{0.34}{1.10}\right)1.10 = \$11,400$$

Delay value treated as a first cost lasting 4 years with an instant tax benefit is

$$-20,000(1-t)F_{PR,10\%,4} = -20,000(1-0.34)(0.31547) = \underline{-4,163}$$

$$R = \$7237$$

In this example it is more economical to rent. Note that the credit given to the rented article for its value in deferring a purchase until later is the determining factor in the analysis. It is difficult to assign this value. When renting on a longer-term basis without annual renewal options, the financing effects of the rental contract must be recognized as equivalent to financing with debt. In most of these cases purchase is more economical than renting.

## 5.16 CLASSIFICATION SYSTEM FOR ECONOMIC APPRAISING

Several economic evaluations are normally required during the development of a project. They arise when a decision for a different order of magnitude of capital and effort is required, when significant new data become available, or when new alternatives are conceived.

The AACE has prepared a classification system dividing these studies into four classes, depending on the amount of information available [3]: (A) final, (B) intermediate, (C) preliminary, and (D) screening.

A *final* study (class A) is primarily used in preparing appropriation requests. It is the best available evaluation at the time of its preparation for management action. However, the cost of preparing the appraisal does not warrant its use unless positive action is contemplated.

An *intermediate* study (class B) is used to summarize a complete or nearly complete project. It may be used to prepare requests for bids, as well as to prepare appropriation requests when urgency warrants.

A *preliminary* study (class C) is of a quality adequate to guide existing development work, to decide between processing alternates, and to determine relative economic importance of processing steps for programming development work.

A *screening* study (class D), which requires the least amount of information of the four classes, is used to screen proposed projects and to rank them in order of economic attractiveness. It is also used in estimating whether proposals are economically reasonable.

## 5.17 BENEFIT-COST ANALYSIS

A benefit-cost analysis is a special type of decision making and is closely related to decisions on profitability. The method is most applicable to public works projects, but it can arise in private enterprise for such projects as building a baseball field at a plant for the use of company personnel. The main feature of benefit-cost analysis is that the benefits are highly subjective and usually involve social, political, and special-interest ramifications.

Table 5.11 summarizes a benefit-cost procedure, Table 5.12 is a checklist for project costs, and Table 5.13 is a checklist for project benefits and disbenefits. The principles involved in a benefit-cost analysis are elementary to a cost engineer, but analysts involved in this work must be highly specialized. The intimate contact with the public's interest required by this work demands that the analyst be responsive to public opinion and expert in evaluating intangible benefits. Benefit-cost analysis usually involves the public interest, a matter subject to considerable subjective evaluation. The analyst must be an expert in evaluating intangible benefits and be adept and fair in judging the value of the public interest.

> **Example 5.10.** A canal can be built from city $D$ to city $E$ at a cost of 10 M\$, and the annual cost of maintenance and operation will be 0.2 M\$. Annual favorable consequences to the general public are estimated at 3.62 M\$, offset by adverse consequences of 0.5 M\$ to a small segment of the population. The canal is expected to last 10 years, and the current interest rate is 6 percent per year. The annual cost of capital recovery will be 1 M\$.
> (a) Calculate the $B/C$ ratio if the disbenefits are taken as negative benefits.
> (b) Change the disbenefits to costs and recalculate the $B/C$ ratio.
> (c) Find the excess of benefits over costs.
>
> **Solution.** Part $(a)$, converting to present values, is
>
> $B$: $\qquad (3.62 - 0.5)F_{RP,6\%,10} = 3.12(7.3601) = 22.964$
>
> $C$: $\qquad 10.0 + (1.0 + 0.2)F_{RP,6\%,10} = 10.0 + 1.2(7.3601) = 18.832$
>
> $$B/C = 22.964/18.832 = 1.22$$
>
> Part $(b)$ is
>
> $B$: $\qquad 3.620F_{RP,6\%,10} = 3.62(7.3601) = 26.644$
>
> $C$: $\qquad 10.0 + (1.0 + 0.2 + 0.5)F_{RP,6\%,10} = 10.0 + 1.7(7.3601) = 22.512$
>
> $$B/C = 26.644/22.512 = 1.18$$
>
> Part $(c)$ is
>
> $$B - C = 22.964 - 18.832 = 4.132 \text{ M\$} \qquad \text{present value}$$
>
> $$B - C = 26.644 - 22.512 = 4.132 \text{ M\$} \qquad \text{present value}$$

**TABLE 5.11**
## Summary of the benefit-cost analysis procedure

1. Define the set of feasible, mutually exclusive public-sector alternatives to be compared
2. Define the objectives to be used
3. Develop the cost-savings and benefit-disbenefit profiles for each alternative
4. Specify the interest rate to be used
5. Specify the measure of merit to be used
6. Compare the alternatives using the measure of merit
7. Perform supplementary analyses
8. Select the preferred alternative

**TABLE 5.12**
## Checklist for project costs and savings

I. Project costs (positive costs or disbursements)
  A. Capital outlay
    1. Land, rights-of-way, easements, severance damages, or water-rights acquisition
    2. Equipment and buildings
    3. Design
    4. Filing fees
    5. Construction
    6. Relocation costs
    7. Engineering
    8. Working capital
  B. Annual expenses
    1. Operating costs
    2. Taxes forgone
    3. Maintenance costs
    4. Replacement costs
    5. Depreciation or amortization costs
    6. Administrative costs

II. Project savings (negative costs or receipts)
  A. Capital savings
    1. Residual or salvage values
    2. Paid-up rents, royalties, or license fees
  B. Annual savings
    1. Sale of any products or publications
    2. Sale of any services
    3. Toll revenues
    4. Fines collected
    5. Increased taxes resulting from property appreciation and increased business activity
    6. Charges to users, such as running royalties, rentals, or license fees

**TABLE 5.13**
## Checklist for project benefits and disbenefits

I. Benefits
  A. Primary or direct benefits
    1. Reduced damage to health
    2. Reduced damage to property
    3. Reduced damage to aesthetic qualities

**TABLE 5.13** *continued*

    4. Reduced time requirement for performance

    5. Reduced operating cost of equipment

    6. Reduced insurance costs

    7. Increased convenience

    8. Increased safety

    9. Increased direct employment (increased income)

    10. Increased land or other property values

    11. Increased satisfaction or enjoyment

    12. Value of increased production

    13. Preservation of areas of unique natural beauty and scenic, historical, or scientific interest

  B. Secondary or indirect benefits

    1. Increased availability of health care facilities

    2. Increased availability of property for intended use

    3. Increased recreational use

    4. Increased indirect (service) employment

    5. Increased income from broader new uses (e.g., low-flow augmentation for water quality control might achieve navigational uses)

    6. Increased income from reduced interruptions in secondary activities (e.g., flood control might reduce wage losses caused by flooded factories)

    7. Increased income from forward production linkages (e.g., increased cotton production resulting from flood control projects would lead to increased employment to process the cotton into products)

    8. Increased income from backward production linkages (e.g., increased cotton production resulting from flood control projects would lead to increased employment in farm machinery, fertilizer, and pesticide industries)

II. Disbenefits

  A. Primary or direct disbenefits

    1. Any private investment to produce or utilize the project output (e.g., new machinery to utilize water cleaned up in a water pollution project)

    2. Any operating costs associated with private investment to produce or utilize the project output (e.g., the operating costs of schools required by population increases resulting from the project)

    3. Increased costs from project construction (e.g., increased cost of transportation around a reservoir being built for flood control)

    4. Any damage to health, property, or aesthetic qualities

    5. Any increase in the time requirements or operating costs of equipment

    6. Any decrease in property values, satisfaction, or enjoyment

    7. Any loss in production

    8. Any destruction of areas of natural beauty and scenic, historical, or scientific interest

  B. Secondary or indirect disbenefits

    1. Increased costs for, or reduction in, public services

    2. Increased damage or hazards created by the project

    3. Decreases in land values or natural resources

    4. Decrease in business activity

    5. Any decrease in availability of health care facilities, or availability of the property for its intended use or for recreational use

    6. Decrease in employment

    7. Decreased income as a result of interruptions in secondary activities because of project operations

    8. Loss in income and employment from the loss in forward or backward production linkages

## 5.18   LIFE CYCLE COST

Life cycle cost is not fundamentally an index of profitability. It is a concept used for making decisions between alternatives and is a term used principally by the government for expressing the total cost of an article or system. A life cycle cost indicates that the first cost alone is not enough to evaluate fully an article or system but that all costs incurred over the life cycle must be considered. The concept is nothing new in cost engineering, but an analyst should be aware of its growing importance in contacts with the government. General Services Administration Bulletin No. FPMR E-153 suggests that all federal agencies abandon the traditional practice of procuring supplies and services on the basis of lowest competitive bids and substitute the criterion of lowest life cycle cost.

Life cycle costs can be compared, after allowing for a difference in service lives, as absolute measures of cost. The life cycle cost emphasizes two principles. First, all costs throughout the life cycle must be included. Second, no costs should be excluded because they might be regarded as common costs. Thus life cycle costs can be judged in their entirety, yet with inclusion of all details.

The engineer has an inherent disposition to reduce a problem to its basic costs. The final decision will be based on the pertinent costs, ignoring common costs to some extent. Life cycle costing will lead to the same decision, and the extra detail may have advantages. The analysis is less restricted and is not tempered by considerations of what an article is being compared with; hence it offers the best opportunity for avoiding errors or omissions. A checklist for life cycle costing is given in Table 5.14.

## 5.19   COST PER UNIT OR COST-EFFECTIVENESS

Sometimes it is not possible to measure return, service, or improvement in terms of dollars. Typical examples are gasoline octane number improvement, effectiveness of a weapon, and effectiveness of an antipollution device. It is possible, however, to measure unit costs, or cost-effectiveness. This method is used widely in pollution and defense projects when the outputs or improvements cannot be measured in dollars.

**TABLE 5.14**
**Checklist for life cycle costing**

| | |
|---|---|
| Purchase or manufacturing cost | Benefits and penalties for quality |
| Transportation cost | Salvage value |
| Installation cost | Distribution expense |
| Direct costs | General and administrative |
| Indirect costs | expense |
| Maintenance | Conformity with trends |
| Inventory for materials | Safety and ecological |
| Inventory for parts | considerations |
| Periodic overhauls | Governmental rules and |
| Supervision | regulations |
| Service reliability | Uncertainty and risk |

The results are measured in terms such as dollars per unit of output or per unit of improvement.

**Example 5.11.** If $51/year is spent on control measures, the deposition of sulfuric acid droplets can be reduced from 94 to 78 kg/100 m$^2$. Find the cost-effectiveness in dollars per percentage decrease in the weight of acid deposition per 100 m$^2$/year.

*Solution*

$$\frac{94 - 78}{94} \, 100 = 17 \text{ percent decrease}$$

$$\text{Cost-effectiveness} = \frac{\text{annual cost control for 100 m}^2}{\text{percentage decrease in deposition}}$$

$$= \frac{51}{17} = \$3/\text{year for 100 m}^2 \text{ for a 1 percent reduction}$$

## 5.20  REVENUE REQUIREMENT METHOD

In the electric utility industry a special technique of profitability analysis and project comparison known as *revenue requirement* is widely used. In essence this technique compares investments on the basis of the amount of net revenue required in order to meet a prescribed return rate and has evolved out of the fact that regulatory bodies generally specify the rate of return on investment which electric utilities are permitted to earn. The technique for calculating revenue requirement will not be shown here because of its highly specialized area of application. However, one should note that the revenue requirement calculation is complementary to the DCFRR technique and can be reconciled with it.

The revenue requirement method is difficult to apply to major projects because of its inability to cope with interest during construction, varying project lives, and some forms of tax credits. DCFRR does not have these limitations and thus is recommended in favor of the revenue requirement method.

## 5.21  COMPARISON OF ALTERNATIVES

A very common situation which occurs in business is the selection of an investment, or more than one investment, from a group of mutually exclusive alternatives. Since the goal is usually to achieve some specified minimum acceptable rate of return (MARR) on investment and also to maximize the return, comparison of alternatives on some consistent basis is necessary.

The NPV and DCFRR techniques can readily identify and eliminate those alternatives which do not meet the MARR requirement, but if not applied correctly, they can lead to selection of a less than optimum choice from among those alternatives which remain.

The proper methodology is to arrange the alternatives in increasing order of cost or initial investment, and then to calculate the NPV or DCFRR. If the investment does not meet the required MARR, it is rejected and the next higher-cost alternative is considered, etc. Once an alternative is found which meets the specified MARR criterion, it becomes the basis for comparison with all other higher-cost alternatives on an incremental basis.

In comparing two alternatives $A$ and $B$, consider that the cost of $B$ equals the cost of $A$ plus some increment; i.e.,

$$C_B = C_A + C_{B-A}$$

If only the cost of $A$ and $B$ are considered, both alternatives may meet the specified rate of return criteria, but the incremental cost may not; i.e., the added investment to choose the more expensive alternative may actually be a money loser. Thus the investment analysis must be performed on the incremental costs and revenues of the alternatives.

There is considerable debate in the literature on the best method of performing incremental analysis, with some authors insisting that the NPV method is adequate to discern which alternative is best, whereas others favor applying the DCFRR method on the incremental differences between the alternatives [17, 18, 20].

If the objective is simply to meet the MARR criterion with any investment, the NPV method suffices and whichever alternative has the highest positive NPV is the alternative of choice. If the objective is to maximize return, recognizing that the incremental investment could be invested in something else, conceivably a multiple of the lowest-cost acceptable alternative, the DCFRR method should be used to determine the return rate of the incremental investment.

## 5.22 SENSITIVITY ANALYSIS

In evaluating a project, a large number of estimates may be made. It becomes necessary to recognize how sensitive the final estimate is to the individual component estimates. Knowing this, one can concentrate on the components having the greatest effect.

For example, the total cost of cooling for a unit in a plant includes a component cost for inhibitors for the cooling water. The cost of the inhibitors can vary twofold depending on operations. However, this cost is only 1 percent of the total cost at most. Clearly, the total cost is insensitive to the inhibitor cost. Sensitivity analysis is becoming an important tool in cost analysis; thus it is an integral part of linear programming (see Chap. 12).

Sensitivity analysis is not the same as uncertainty analysis. In the preceding case the cost for the inhibitors can vary twofold, a high degree of uncertainty, but the total cost, as noted, is insensitive to the inhibitor cost. Sensitivity analysis is discussed quantitatively in Chap. 8.

## 5.23 UNCERTAINTY AND RISK ANALYSIS

Modern times are dominated by rapid change and instability, creating an atmosphere of uncertainty and risk. Frequently the terms *uncertainty* and *risk* are used as equivalents, but the purist views them differently. Risk is a measure of reaction to uncertainty. For example, a new office copier is introduced at one-third the present price, but whether it is built well enough to be worth buying is uncertain. A large company using many copiers can buy one at little risk. An individual who uses only one copier would take a much greater risk. The uncertainty about the copier is the same in each case, but the risk to the buyers is different.

A profitability study always involves some uncertainty since not all the components that go into the study are exactly known. Most are estimates. The trend now is to demand attention to and inclusion of uncertainty. Suppose the cost for a project is reported as 10 M$. That value is the expected value only; it must now be supplemented with a measure of the uncertainty. What is the probability that the cost will be 12 M$ or more, or between 11 M$ and 13 M$? Management will even want to know how sensitive the uncertainty of the project is to the uncertainty of the component elements. Methods for attacking these problems on a quantitative basis are discussed in Chap. 8.

## 5.24 PITFALLS IN PROFITABILITY ANALYSIS

Profitability is a concern of high-level management but requires conscious attention at all levels of an organization if it is to be attained. A list of pitfalls in profitability analysis is given in Table 5.15, and comments on some of the items follow.

Profit and profitability are different concepts. Accounting principles determine the measure of profit. *Profit* is fundamentally a short-term evaluation—e.g., an income statement for 1 year—and is therefore amenable to distortion. It is possible for a company to show a profit for 1 year when in truth it is losing money over the long run. The recognition of time as a fundamental constituent of profitability is paramount, and *profitability* is a long-term concept.

**TABLE 5.15**
**List of pitfalls in profitability analysis**

| | |
|---|---|
| Profitability is a long-term concept | Allocation problems |
| A single criterion is inadequate |   Utility costs |
| Circumvent the criterion |   Transfer pricing |
| Estimation of cash flow |   Residual and scrap value |
|   Sales volume |   Overhead |
|   Sales price | Influence of government |
|   Raw materials costs |   Subsidies |
|   Raw materials availability |   Control of wages and materials |
| Operating costs |   Safety, ecology, and pollution |
| Investment |   Regulating power |
| Depreciation |   Social factors |
| Inflation | Inventory pricing |
| Replacement frequency and cash flow | Uncertainty |
| Evolutionary factor | Productivity rate gain |
|   Position at end of project |   Diffusion |
|   Availability and commitment of funds |   Fringe benefits |
| Quality control |   Personnel factors |
| Reliability of production | Reviews and postmortems |

Profitability can be evaluated only *after* time has elapsed. Profitabilities for the future are only estimates. The automobile industry showed a profit for the late 1970s, but it failed to tool up for the oncoming demand for small, economical automobiles. The year 1980 was a disaster for profits and revealed in retrospect that the profits of the late 1970s were false because the investment necessary for change and modernization had been disregarded. Again, nothing should be given an ultimate judgment on a short-term basis, and even a short-term venture should be evaluated on the basis of its long-term effect since it will influence the future beyond its termination.

A single criterion is inadequate for a full evaluation of profitability, as discussed in Sec. 5.14.

A common error in comparing profitability estimates is the failure to use a common denominator of service lives, e.g., using NPV when there is a difference in the duration of projects.

Once a fixed value for a criterion is established, it becomes a record to be challenged and beaten. Demanding a short payout time will encourage cheap investment, which actually may be more costly in the end. Criteria and their usage must be flexible.

The most important estimates in the evaluation of profitability are the predictions of cash flows, with sales volume and sales price the most critical factors. The possibility of disastrous change in raw materials costs and availability is becoming commonplace in a world that seems to be running short of everything.

Operating costs are more critical than investment since the former are repetitive and the latter is made only once. The depreciation schedule used in a profitability study that considers the time value of money will affect the calculated profitability. The effect of inflation is discussed in Chap. 6.

Some restraint should be used when comparing cash flows. Although this is not a definition, cash flow is principally the sum of profit plus depreciation. A high cash flow may merely signify a high depreciation expense, i.e., a high cost for wear and tear of equipment caused by frequent replacement.

All activities become part of an evolutionary process. A decision made today will influence the physical plant and marketing position of a company after the economic life of the equipment has expired. The company's position at the end of the project transcends the merely physical aspect of the project, such as its salvage value.

Future improvement in quality control must be anticipated since a competitor may offer a superior product. Reliability of production may be a factor, involving attention to inventories from raw materials to finished goods and investment in spare parts and spare equipment.

Allocation of overhead costs can have an important influence on apparent costs and profitability of individual products and services. The problem is one of the most acute in cost engineering and is discussed in Chap. 17.

The influence of government is omnipresent. Plants have closed because they were unable to meet antipollution requirements. Although laws cannot be made retroactive, plants will have to meet future requirements. Whether government interferes as a guardian of antipollution, social justice, or military preparedness, a cost exists whatever its origin. Costs tend to increase with time. It is necessary to recognize the increasing influence of government and to make judgments that are in tune with anticipated governmental action. Swimming with the political tide is better than drowning in it.

Inventory pricing can have an effect on short-term profits, as discussed in Chap. 17. Uncertainty and risk analysis are developed in Chap. 8 since an application of probabilities and distributions is involved.

Productivity commands considerable attention and is discussed in Chap. 9, particularly the dissipation of productivity by a factor called *diffusion*.

Whether a current expense is charged as such or capitalized and written off as a future expense is important to apparent current profitability. Capitalizing expenses will increase current profits, but at the expense of the profitability picture in the future.

During periods of inflation there is a temptation to capitalize expenses. A fixed dollars amount, written off in part or wholly as a future expense, will appear to be less at a future time when the value of the dollar is less.

Reviews and postmortems can be instructive but are seldom made, because there is a widespread unwillingness to search for causes of mistakes.

Profitability must emphasize the long term versus the short term. Emphasis on the short term leads to a policy of "milking the franchise" and patching. Maintenance, research, modernization, expansion, and necessary costs are reduced or eliminated. Short-term profitability may appear enhanced, but real profitability may be destroyed.

A company must have a policy for profitability. Single criteria are useful tools, but success is more an attitude than a devotion to mathematical criteria. It is the uncertainties and intangibles that will dominate eventual profitability.

The essential ingredients for profitability for a manufacturing company are as follows:

1. Profitability should be judged on a long-term basis.
2. Operations must be as efficient as possible, recognizing that technology is always in flux. Modernize or pass into history.
3. The diffusion effect of peripheral activities should be held to a minimum.

Real growth is measured by the proportion of necessary personnel employed with respect to the total available population. A conflict exists here. Operating with the highest technical efficiency possible while avoiding diffusion effects will tend to reduce personnel. A progressive organization faces a constant demand for growth in markets and activities just to stay even.

The only certainty in a profitability study is that profitability evolves in unexpected ways. Decisions should try to be in tune with the times and anticipate the future.

## NOMENCLATURE

| | |
|---|---|
| B/C | Benefit-cost ratio, dimensionless |
| $C_d$ | Depreciable first cost, $ |
| $C_{ex}$ | Irregular cost at end of $x$th year, $ |
| DCF | Discounted cash flow, $ |
| DCFRR | Discounted cash-flow rate of return, decimal or percent per year |
| $F_{PR}$ | Factor to convert $P$ to $R$, capital recovery factor, year$^{-1}$ |

| | |
|---|---|
| $F_{RP}$ | Factor to convert $R$ to $P$, years |
| $F_{SDP}$ | Present value of $1 for sum-of-the-years-digits depreciation, decimal |
| $i$ | Rate of return, decimal per year |
| MARR | Minimum acceptable rate of return, decimal or percent per year |
| $n$ | Number, such as life of an article, years |
| NPV | Net present value, $ |
| $P$ | Present value, $ |
| $r$ | Rate of return, generally after taxes, decimal per year |
| $R$ | Uniform end-of-year annual amount, unacost, $ per year |
| $R_b$ | Uniform beginning-of-year annual amount, $ per year |
| RAI | Return on average investment, decimal or percent per year |
| ROI | Return on original investment, decimal or percent per year |
| $t$ | Tax rate, decimal |

## PROBLEMS

**5.1.** A venture has cash flows after taxes, excluding working capital and depreciation, as follows:

| Time, years | Amount, M$ |
|---|---|
| 0 | −5 |
| 1 | 1.2 |
| 2 | 1.3 |
| 3 | 1.4 |
| 4 | 1.5 |
| 5 | 1.1 |
| 6 | 0.9 |
| 7 | 0.5 |

Working capital is 1.5 M$ invested at zero time. Depreciation is constant at 1 M$ for 5 years. Find the payout time without interest.

**5.2.** In Prob. 5.1 find the payout time with interest at 10 percent per year (*a*) excluding working capital, (*b*) including working capital.

**5.3.** In Prob. 5.1 find the net present value (NPV) at 10 percent per year on a 5-year life and the discounted cash-flow rate of return (DCFRR) on a 5-year life.

**5.4.** Calculate the results for projects *A*, *B*, and *C* in Table 5.6, assuming that incomes occur uniformly throughout the year, rather than at the end of the year, using continuous compounding of interest.

**5.5.** It is necessary to produce 1.1 million units per year of an automobile part, looking 12 years ahead and with money worth 14 percent per year compounded periodically and with no tax.

Plan *A* is to build a plant in the United States at a cost of 7 M$, with end-of-year expenses at 2 M$/year. It is expected that 0.5 M$ will have to be invested for pollution control, but this can be delayed until the end of the fifth year.

Plan B is to build a plant outside the United States for 4 M$, with end-of-year expenses at 1.2 M$/year, but there will be a duty charge of 30 percent of the selling price in the United States.

Find the selling price in the United States at which the two plans are equal.

**5.6.** Two plans are available for constructing a plant. Income will be the same for both plans for all the years and cancels out in the analysis. There is no tax consideration, and a period 8 years ahead is the basis for comparison, with money worth 12 percent per year using periodic interest.

Plan *A* is to build a half-size plant now at a cost of 1 M$ with an operating cost of 200 k$ as of the end of the year for 4 years. Expansion costing 0.9 M$ will occur at the end of year 4. End-of-year operating costs will then become 350 k$/year.

Plan *B* is to build a full-scale plant now at a cost of 1.5 M$, with an end-of-year operating cost of 250 k$.

Which plan should be chosen?

**5.7.** An investment of $20,000 will be made 2 years from now in a project lasting 4 years. It will permit before-tax savings of $10,000 per year for 4 years. Sum-of-the-years-digits depreciation will be used for an 8-year life for tax purposes. Salvage value at the end of year 4 will be the undepreciated balance. Money is worth 12 percent per year after a 46 percent combined federal and state tax rate. Use periodic interest and find the present value as of today.

**5.8.** A company can choose between two plans. Plan *A* is to build a plant now, having a capacity of 600 Mg/year at a cost of 1 M$, and 3 years from now add to the plant at a cost of 800 k$ to double the production capacity.

Plan *B* is to build a full-scale plant now for 1.5 M$ having a capacity of 1.2 Mkg/f160year

The product sells for $1/kg. For plan *A* the operating cost is $0.5/kg, and for plan *B* it is $0.4/kg, in both cases for all the years. Sales demand for the years is expected to be

| Year | Mg |
|------|------|
| 1 | 600 |
| 2 | 800 |
| 3 | 1000 |
| 4 | 1200 |
| 5 | 1200 |

If money is worth 10 percent per year and there is no tax consideration, which plan is more economical based on present value using periodic interest?

**5.9.** The investment for a new project is

| | |
|---|---|
| Fixed depreciable capital at zero time, k$ | 250 |
| Start-up costs at zero time, capitalized, k$ | 175 |
| Working capital at zero time, k$ | 130 |
| Investment credit for tax purposes at zero time, k$ | 17.5 |
| Project life, years | 20 |
| Tax rate, %, combined federal and state | 46 |
| End-of-year operating cost, k$ | 813 |
| End-of-year insurance, etc., as percent of total investment, including start-up | 2.5 |
| End-of-year sales, k$ | 1000 |
| Depreciation method | sum-of-the-years-digits |

Use periodic interest and find the DCFRR.

**5.10.** A venture requires an investment of 10 M$, 15 percent of which is nondepreciable land. Net annual end-of-year income is 2.2 M$, excluding depreciation. The plant will last 10 years, with straight-line depreciation for tax purposes. The combined tax rate is 46 percent. Find the DCFRR using periodic interest.

**5.11.** A project requires a depreciable investment of 1.5 M$ having no salvage value and 0.5 M$ for nondepreciable land. The depreciation will be over 4 years, with a combined tax at 46 percent. Yearly depreciation is

| End year | Fractional depreciations |
|----------|--------------------------|
| 1 | 0.4 |
| 2 | 0.3 |
| 3 | 0.2 |
| 4 | 0.1 |

Sales realization and expenses, before depreciation, will be

| End year | Sales, M$ | Expenses, M$ |
|----------|-----------|--------------|
| 1 | 0.3 | 0.2 |
| 2 | 0.75 | 0.25 |
| 3 | 0.9 | 0.3 |
| 4 | 0.9 | 0.3 |
| 5 | 0.9 | 0.3 |

Find the DCFRR of the project using periodic interest.

**5.12.** A new facility costs $100,000, $40,000 of which is for land and $60,000 for construction, and has a 20-year economic life. Sales volume is estimated at $150,000/year in the third and subsequent years. The annual cash return after taxes, including depreciation, is $9000/year at full production. Working capital is $15,000 at full production. In the first and second years volume is estimated at one-third and two-thirds of capacity, respectively. Assume that the working capital and annual cash return are proportional to volume. The salvage value is negligible. Find the present value at 0, 4, and 6 percent and the DCFRR using periodic interest.

**5.13.** A project requires an immediate investment of 1 M$, 22 percent of which is for working capital. The profit-and-loss statement for a year will show a 100 k$ net profit after taxes for each of the 10 years of the project life. The economic life and tax life are the same. Use straight-line depreciation and a 46 percent combined tax.
(*a*)  Find the DCFRR.
(*b*)  Find the payout time without interest.

**5.14.** The investment for a new project is

| | |
|---|---|
| Fixed capital | |
|    Depreciable portion, k$ | 250 |
|    Expensed proportion, k$ | <u>175</u> |
| | 425 |
| Working capital, k$ | <u>130</u> |
| | 555 |

The depreciable and expensed portions are spent uniformly from $-1$ to 0 years, and working capital is spent at zero time. A 7 percent investment credit is allowed on the depreciable portion taken at zero time. The project life is 20 years and the annual before-tax margin is $400,000, excluding depreciation. The combined tax rate is 46 percent. Use periodic interest and 20 year sum-of-the-years-digits depreciation. Find the DCFRR for the project.

**5.15.** A project will have a construction cost of 4 M$ occurring uniformly 1 year prior to start-up. Another 1 M$ for construction must be paid at start-up. Start-up costs occurring uniformly during the first year at 0.5 M$ will be capitalized. Salvage at the end of year 4 will be 0.2 M$. The life for economic and tax purposes will be 4 years using straight-line depreciation and a combined 46 percent tax. Other data in megadollars are as follows:

| | Time 0 | End year 1 | End year 2 | End year 3 | End year 4 |
|---|---|---|---|---|---|
| Sales | | 3.0 | 3.5 | 3.8 | 2.8 |
| Expenses before depreciation | | 0.6 | 0.8 | 0.7 | 0.5 |
| Value of inventory | 0.8 | 1.0 | 0.9 | 0.8 | 0 |

Find the DCFRR using periodic interest.

**5.16.** Project $A$ requires a capital investment at zero time of 1 M$ which will be written off in 5 years for tax purposes using twice the straight-line declining-balance depreciation method for the first 3 years with a switch to straight-line depreciation for the last 2 years. The project lasts 5 years, and the combined tax rate is 46 percent. Income taken at the end of the year, before taxes and depreciation, will be

| Year | k$ |
|---|---|
| 1 | 600 |
| 2 | 700 |
| 3 | 700 |
| 4 | 700 |
| 5 | 600 |

Project $B$, which competes with project $A$, shows a 20 percent per year DCFRR with periodic interest. Is project $A$ more acceptable than project $B$?

**5.17.** A venture will have the following financial considerations in kilodollars:

| | Time 0 | End year 1 | End year 2 | End year 3 | End year 4 |
|---|---|---|---|---|---|
| Sales | | 1500 | 1800 | 1700 | 1400 |
| Expenses before depreciation | | 300 | 400 | 350 | 250 |
| Value of inventory | 400 | 500 | 450 | 425 | 0 |

The investment at the start is 2000 k$, having 100 k$ salvage value to be written off in 4 years using sum-of-the-years-digits depreciation. The combined tax rate is 46 percent. Find the DCFRR using periodic interest.

**5.18.** The Jones Company is considering the replacement of 5-year-old machines with some improved units which recently became available. Management is concerned with the high operating cost of the existing machines but expects to replace all of them in 5 years when a new system will be installed. From the following information, determine whether or not the equipment should be replaced now. Money is worth 9 percent per year after a combined 46 percent tax. Use continuous-interest and straight-line depreciation:

|  | Old machines | New machines |
| --- | --- | --- |
|  | 24,000 | 17,000 |
| Number of machines | 6 | 5 |
| Cost of each machine, $ | 4,200 | 7,800 |
| Depreciated value at present each (tax books), $ | 2,100 | |
| Trade-in value on new machines, $ | 1,500 | |
| Market value of new machine in 5 years, $ | | 3,000 |
| Variable costs of operation, annual, $ | 24,000 | 17,000 |

**5.19.** Show that the rate of return at which the equivalent uniform annual cost (unacost) or capitalized cost is zero is the DCFRR.

**5.20.** When an investment occurs in the future rather than at zero time, the DCFRR may have more than one value, or there may be no value.

A project requires the expenditure of 15 M$ 5 years from now and will have an annual cash flow of 1 M$/year starting 1 year from now and continuing through year 20. Find two DCFRRs for the project.

**5.21.** Repeat Prob. 5.20, changing the expenditure 5 years from now to 5 M$, and show that there is no finite rate of return for the project.

**5.22.** An airport costing $750,000 is to be built. The annual cost of capital recovery for the investment in land, buildings, and runways will be $75,000, and the annual maintenance cost will be $100,000. Annual favorable consequences to the general public will be $625,000, and annual adverse consequences to a small segment of the public will be $50,000. The prevailing interest rate is 8 percent, and the project life is 10 years.
(*a*) Consider the adverse consequences as a cost and find the benefit-cost ratio.
(*b*) Find the benefit-cost ratio if the adverse consequences are considered a reduction in benefits.
(*c*) Find the excess of benefits over costs.

**5.23.** Total environmental damage from iron and steel operations is 528 M$/year. By spending 687 M$/year on control measures, the damage could be reduced by 496 M$/year. Find the cost-effectiveness as measured by the dollars of control cost per dollar of damage reduction.

**5.24.** It costs 35 M$ to install a limestone-scrubbing unit on a 500-MW power plant to meet environmental standards for sodium dioxide removal from stack gas. It would cost 60 M$ to install a sodium carbonate scrubbing unit on a 750-MW power plant to perform the same service. Find the cost-effectiveness for each alternative in dollars per kilowatt and state which system is more cost-effective based on the investment cost.

# REFERENCES

1. Abrams, H. J.: "Economics of Investment Criteria, Part I," *Chemical Engineer* (London), September 1970, pp. CE252–CE259.
2. Agarwal, J. C., and I. V. Klumpar: "Profitability, Sensitivity, and Risk Analysis for Project Engineers," *Chemical Engineering,* vol. 29, September 1975, pp. 66–72.
3. American Association of Cost Engineers: "Classification System for Economic Appraisals," *Cost Engineers' Notebook,* 1963.
4. Comber, N. H., and A. K. Biswas: *Evaluation of Environmental Intangibles,* Genera Press, Bronxville, N.Y., 1973.
5. Federal Register: "Cost-Effectiveness Analysis," *Appendix A,* vol. 38, no. 174, September 10, 1973, pp. 24639–24640.
6. Holland, F. A., F. A. Watson, and J. K. Wilkinson: "Probability Techniques for Estimates of Profitability," *Chemical Engineering,* January 7, 1974, pp. 105–108.
7. Jelen, F. C.: "Analyze Ventures This Way," *Petroleum Refiner* (now *Hydrocarbon Processing*), March 1964, pp. 125–128.
8. Jelen, F. C.: "Pitfalls in Profitability Analysis," *Hydrocarbon Processing,* January 1976, pp. 111–115.
9. Jelen, F. C., and C. L. Yaws: "How Energy Affects Life Cycle Costs," *Hydrocarbon Processing,* July 1977, pp. 89–91.
10. Jelen, F. C., and C. L. Yaws: "Project Cash Flow Description and Interpretation," *Hydrocarbon Processing,* March 1978, pp. 77–81.
11. Levin, R. I., and C. A. Kirkpatrick: *Quantitative Approaches to Management,* 3d ed., McGraw-Hill Book Company, New York, 1975.
12. Mishan, E. J.: *Cost-Benefit Analysis,* Frederick A. Praeger, Inc., New York, 1976.
13. Newnan, D. G.: *Engineering Economic Analysis,* 3d ed., Engineering Press, Inc., San Jose, Calif., 1988.
14. Park, W. R., and D. E. Jackson: *Cost Engineering Analysis,* 2d ed., John Wiley & Sons, Inc., New York, 1984.
15. Phung, D. L.: "Cost Analysis Methodologies," *Cost Engineering,* May–June 1980, pp. 139–143.
16. Reul, R. I.: "Algorithms vs. Concepts," *American Association of Cost Engineers Bulletin,* vol. 7, June 1965, pp. 46–49.
17. Riggs, J. L.: *Engineering Economics,* 2d ed., McGraw-Hill Book Company, New York, 1982.
18. Rose, A.: "Tutorial: Incremental Analysis in Capital Budgeting," *Cost Engineering,* October 1988, pp. 13–15.
19. Strickland, T. H., and E. C. Grady: "Capital Budgeting in Project Evaluation: Part I," *Hydrocarbon Processing,* March 1981, pp. 179–204.
20. Theusen, H., G. Fabrycky, and G. J. Thuesen: *Engineering Economy,* 7th ed., Prentice-Hall, Inc., Englewood Cliffs, NJ, 1989.
21. Thorne, H. C.: "Post Installation Appraisals: What American Oil Looks For," *Petroleum Refiner* (now *Hydrocarbon Processing*), March 1965, pp. 109–111.
22. White, J. A., M. H. Agee, and K. E. Case: *Principles of Engineering Economic Analysis,* John Wiley & Sons, Inc., New York, 1977.
23. Zinn, C. D., and W. G. Lesso: "Analyze Risks with This Method," *Hydrocarbon Processing,* December 1979, pp. 74–79.

CHAPTER

# 6

# INFLATION

*F. C. Jelen*

## 6.1  COST COMPARISONS UNDER INFLATION

Inflation is a factor in cost evaluation and cost comparisons. This chapter is not concerned with the causes of inflation, the social, economic, and political consequences, or the problem of mitigating inflation. Its purpose is to develop the mathematical relationships so that inflation can be included along with rate of return, taxes, and productivity gain.

When inflation is a factor, costs change each year as inflation increases or decreases. Although a varying inflation rate can be accommodated, the rate is, at best, a prediction. It will be assumed that the inflation rate is constant and that all costs and revenues are equally affected. A cost of $A$ now will become $A(1 + d)$ 1 year from now, $A(1 + d)^2$ 2 years from now, etc., where $d$ is the decay or inflation rate per year expressed as a decimal. The relationship is

$$A_{t2} = A_{t1}(1 + d)^n \tag{6.1}$$

where $A_{t1}$ is a cost now and $A_{t2}$ is the cost $n$ years from now.

## 6.2  UNABURDEN

When there is no inflation unacost, a uniform end-of-year cost is a satisfactory criterion for comparing different systems. With inflation a satisfactory end-of-year criterion is not a uniform cost in dollars but becomes a uniform burden or cost, called *unaburden*, that increases each year by the factor $(1 + d)$. Thus the burden will be $R_D$ in dollars at the end of the first year and $R_D(1 + d)^{n-1}$ in dollars at the end of the $n$th year. The tax offsets from depreciation are not affected by inflation since depreciation is based on the original cost.

**137**

Suppose an article now has a depreciable first cost $C_d$ and lasts $n$ years. The present value of the first cost and its tax credits is given in Eq. (3.21) and is

$$P = C_d(1 - t\psi) \tag{6.2}$$

The problem is to find a unaburden $R_D$ which on an after-tax basis is equivalent to the value given in Eq. (6.2). The present value of the end-of-year costs, where $R_D$ is the actual cost at the end of the first year, is

$$P = \sum_1^n R_D \frac{(1 + d)^{n-1}}{(1 + r)^n} \tag{6.3}$$

Equation (6.3) is a geometric series with the ratio $(1 + d)/(1 + r)$ and first term $R_D/(1 + r)$, and having $n$ terms. It can be evaluated by Eq. (2.14) and is

$$P = R_D \frac{(1 + r)^n - (1 + d)^n}{(1 + r)^n} \frac{1}{(r - d)} \tag{6.4}$$

For equal present values Eqs. (6.2) and (6.4) are equal, leading to

$$R_D = C_d(1 - t\psi_{r,n'}) \frac{(1 + r)^n}{(1 + r)^n - (1 + d)^n}(r - d) \tag{6.5}$$

Equation (6.5) is the relationship that converts a depreciable first cost $C_d$ into a unaburden $R_D$. Note that the life for tax purposes $n'$ can be different from the economic life $n$. In Eq. (6.5) the unaburden is calculated on an after-tax basis, but it is possible to refer to a before-tax unaburden as well.

To emphasize, unaburden $R_D$ is the actual value at the end of the first year and increases by the factor $(1 + d)$ each year. The equivalent value at zero time is $R_D/(1 + d)$. No confusion should result between $R_D$ and $R_D/(1 + d)$. Either would be satisfactory as a definition, but unaburden is defined as the actual cost at the end of the first year.

Unaburden is an important concept. It permits rate of return, tax rate, and inflation rate to be included independently. It allows articles and systems with different durations to be compared. Unaburden corrects for different durations, as can be shown by reworking Eqs. (6.2) and (6.3) based on a duration of $2n$ years. Equation (6.2) becomes

$$P = C_d(1 - t\psi) + \frac{C_d(1 + d)^n}{(1 + r)^n}(1 - t\psi) \tag{6.6}$$

and Eq. (6.3) becomes

$$P = \sum_1^{2n} R_D \frac{(1 + d)^{n-1}}{(1 + r)^n} \tag{6.7}$$

Setting the two equal and executing the algebraic operations gives Eq. (6.5) as before.

## 6.3 UNABURDEN FOR VARIOUS COSTS

Using the previous procedure, relationships for other types of costs can be derived. A summary is given in Table 6.1. In using the table, notice that all costs are entered as their present cost. The relationships will correct for the expected change in cost with inflation when the costs are actually incurred. For item 7 $t$ is the tax rate for the increase in salvage value arising from inflation over the expected salvage value without inflation. Note that $t$ is usually the same rate as the tax rate for regular income.

Normally, the rate of return $r$ will be greater than the inflation rate $d$. Table 6.1 also gives the correct relationships if the inflation rate is greater than the rate of return.

A special case arises when the inflation rate is equal to the rate of return, i.e., when $r = d$. Some of the relationships in Table 6.1 assume the form 0/0, but evaluation by the calculus leads to the following for the items in the table taken consecutively where $r = d$:

$$R_D = \frac{C_d}{n}(1 - t\psi_{r,n'})(1 + r) \tag{6.8}$$

$$R_D = R(1 - t)(1 + r) \tag{6.9}$$

$$R_D = R_b(1 - t\psi_{r1})(1 + r) \tag{6.10}$$

**TABLE 6.1**
**Summary of unaburden relationships**

| Item no. | Item | | Unaburden |
|---|---|---|---|
| 1 | $C_d$ | Depreciable first cost* | $C_d(1 - t\psi_{m'})\dfrac{(1 + r)^n}{(1 + r)^n - (1 + d)^n}(r - d)$ |
| 2 | $R$ | Uniform end-of-year burden* | $R(1 - t)(1 + d)$ |
| 3 | $R_b$ | Uniform beginning-of-year burden* | $R_b(1 - t\psi_{r1})(1 + r)$ |
| 4 | $C_{ex}$ | Irregular cost at end of $x$th year* | $C_{ex}(1 - t)(1 + d)^x\dfrac{(1 + r)^{n-x}}{(1 + r)^n - (1 + d)^n}(r - d)$ |
| 5 | $C_{bx}$ | Irregular cost at beginning of $x$th year* | $C_{bx}(1 - t\psi_{r1})(1 + d)^{x-1}\dfrac{(1 + r)^{n-x+1}}{(1 + r)^n - (1 + d)^n}(r - d)$ |
| 6 | $C_{nd}$ | Nondepreciable first cost* | $C_{nd}(r - d)$ |
| 7 | $C_{sal}$ | Salvage value* | $C_{sal}\left[1 + t\dfrac{(1 + d)^n - 1}{(1 + r)^n - (1 + d)^n}\right](r - d)$ |

* All costs are calculated in terms of the present cost.

$$R_D = \frac{C_{ex}}{n}(1 - t)(1 + r) \qquad (6.11)$$

$$R_D = \frac{C_{bx}}{n}(1 - t\psi_{r1})(1 + r) \qquad (6.12)$$

$$R_D = C_{nd}(0) = 0 \qquad (6.13)$$

$$R_D = C_{sal}\frac{t}{n}\frac{(1 + r)^n - 1}{(1 + r)^n}(1 + r) \qquad (6.14)$$

**Example 6.1.** A machine lasts 2 years with negligible salvage value; it can be written off in 1 year for tax purposes and can be purchased now for $20,000. How much can be spent now for a new machine which lasts 10 years with negligible salvage value and can be written off in 10 years for tax purposes using sum-of-the-years-digits depreciation when money is worth 10 percent per year after a 34 percent tax, if the inflation rate is (a) 4 percent per year, (b) 10 percent per year, and (c) 12 percent per year?

**Solution**
(a) For 4 percent per year inflation use item 1, Table 6.1. Unaburden for the 2-year machine with $\psi = F_{SDP, 10\%, 1} = 0.90909$ is

$$R_D = 20,000[1 - 0.34(0.90909)]\frac{(1.10)^2}{(1.10)^2 - (1.04)^2}(0.10 - 0.04)$$

$$= \$7,812$$

For the same unaburden for the 10-year machine, with $\psi = F_{SDP, 10\%, 10} = 0.70099$,

$$7812 = C_d[1 - 0.34(0.70099)]\frac{(1.10)^{10}}{(1.10)^{10} - (1.04)^{10}}(0.10 - 0.04)$$

$$C_d = \$73,390$$

(b) For 10 percent per year inflation use Eq. (6.8). For the 2-year machine

$$R_D = \frac{20,000}{2}[1 - 0.34(0.90909)](1.10) = \$7,600$$

For the same unaburden for the 10-year machine

$$7600 = \frac{C_d}{10}[1 - 0.34(0.70099)](1.10)$$

$$C_d = \$90,700$$

(c) For a 12 percent per year inflation rate use item 1, Table 6.1:

$$R_D = 20,000[1 - 0.34(0.90909)]\frac{(1.10)^2}{(1.10)^2 - (1.12)^2}(0.10 - 0.12)$$

$$= \$7,531$$

For the same unaburden for the 10-year machine

$$7531 = C_d[1 - 0.34(0.70099)]\frac{(1.10)^{10}}{(1.10)^{10} - (1.12)^{10}}(0.10 - 0.12)$$

$$C_d = \$97,600$$

**Example 6.2.** A new tank costs \$10,000, lasts 3 years, and belongs to a group depreciated in 5 years with sum-of-the-years-digits depreciation. If money is worth 10 percent per year after a combined 46 percent tax and the inflation rate is 4 percent per year, how much can be spent for repairing the tank at the end of each year as an equal burden so as to obtain a 5-year life?

*Solution.* Unaburden for a new tank lasting 3 years, by item 1, Table 6.1, with $\psi = F_{\text{SDP, 10\%, 5}} = 0.80614$, is

$$R_D = 10,000[1 - 0.46(0.80614)]\frac{(1.10)^3}{(1.10)^3 - (1.04)^3}(0.10 - 0.04)$$

$$= \$2437.5$$

Unaburden for a new tank, plus repairs, is given by items 1 and 2, Table 6.1, and for equal economy is

$$2437.5 = 10,000[1 - 0.46(0.80614)]\frac{(1.10)^5}{(1.10)^5 - (1.04)^5}(0.10 - 0.04)$$

$$+ R(1 - 0.46)(1.04)$$

$$R = \$1592 \qquad \text{present dollars per year before taxes}$$

That is, \$1592(1.04) is the actual cost at the end of the first year, \$1592(1.04)^2 is the actual cost at the end of the second year, etc.

## 6.4  CAPITALIZED COST

Capitalized cost can be used when inflation is a factor in comparing systems, and it will make a correct adjustment for different service lives. To convert a unaburden $R_D$ to a capitalized cost $K$, evaluate Eq. (6.3) out to infinity:

$$K = \sum_1^\infty R_D\frac{(1 + d)^{n-1}}{(1 + r)^n} \tag{6.15}$$

which can be summed similarly to Eq. (6.3), and if $r > d$, becomes

$$K = \frac{R_D}{r - d} \qquad \text{or} \qquad R_D = (r - d)K \tag{6.16}$$

Thus for $r > d$ Table 6.1 can be used to obtain the capitalized cost by dividing each unaburden by $(r - d)$.

If the inflation rate is greater than the rate of return, then a concept called the *extended capitalized cost* is obtained from Table 6.1 by dividing the unaburden by

$(r - d)$, as before, and then taking the negative of that expression. The extended capitalized cost remains positive and is just as suitable for the circumstance $d > r$ as is the regular capitalized cost for the circumstance $r > d$. For details, see the original articles [1].

For the special case in which the rate of return and the inflation rate are equal, Eqs. (6.8) through (6.14) based on unaburden must be used.

## 6.5 INCLUSION OF PRODUCTIVITY

If the productivity gain is equal to $p$ each year, expressed as a decimal, costs will reduce by a factor $(1 - p)$ each year due to productivity alone. With both productivity and inflation, costs will change by the factor $(1 - p)(1 + d)$ each year. The intriguing problem of including all four rates—rate of return, tax rate, inflation rate, and productivity gain rate—has been evaluated using capitalized cost as the criterion for equivalence.

The reader is referred to the original articles [5] for details of the method, which includes all kinds of costs. The capitalized cost for a depreciable first cost $C_d$ is given by

$$K = C_d(1 - t\psi_{r,n'})\frac{(1 + r)^n}{(1 + r)^n - (1 - p)^n(1 + d)^n} \tag{6.17}$$

**Example 6.3.** An article that lasts 1 year can be purchased for $10,000 and written off in 1 year. How much can be paid for an article that lasts 5 years if it can be written off in 3 years for tax purposes using straight-line depreciation when the rate of return is 16 percent per year after a combined 46 percent tax and the productivity gain rate is 3 percent per year with an inflation rate of 10 percent per year?

**Solution.** Capitalized cost for the 1-year article, by Eq. (6.17), with $\psi = F_{SLP, 16\%, 1} = 0.86207$, is

$$K = 10,000[1 - 0.46(0.86207)]\frac{1.16}{1.16 - [(0.97)(1.10)]} = \$75,269$$

For the same capitalized cost for the 5-year article, with $\psi = F_{SLP, 16\%, 3} = 0.74863$,

$$75,269 = C_d[1 - 0.46(0.74863)]\frac{(1.16)^5}{(1.16)^5 - [(0.97)^5(1.10)^5]}$$

$$C_d = \$39,210$$

That is, $39,210 could be paid for the 5-year article.

This solution is correct because the same productivity gain rate applies to the 1-year and 5-year articles. Now and in the future it will be possible to pay 3.921 times as much for the 5-year article as for the 1-year article. However, if the productivity gain rates for the articles are different, the ratio of justifiable costs will change with time. The reader should refer to the original reference.

It is interesting to rework the example for different combinations of the productivity gain rate $p$ and the inflation rate $d$, with the following results:

| Productivity rate, $p$ | Inflation rate, $d$ | Justifiable first cost for 5-year article |
|---|---|---|
| 0 | 0 | $34,959 |
| 0.03 | 0 | 33,218 |
| 0 | 0.10 | 41,500 |
| 0.03 | 0.10 | 39,210 |

The tabulation shows that productivity and inflation have opposing effects. Inflation adds to what can be paid for the longer-lasting article by buying into the future at today's cost. Productivity gain penalizes the longer-lasting article because the purchaser will be delayed in entering the market to obtain the reduced cost resulting from technological advancement.

The inflation rate as deduced from cost indices is the *net* inflation rate after the productivity gain rate. Under the present circumstances it is adequate to use the net inflation rate as a combination rate and omit the productivity gain rate, except for a few cases of new technology in which the productivity gain rate might be significant.

## 6.6 ALLOWANCE FOR INFLATION

Inflation has another serious consequence: It causes companies to reinvest funds on an after-tax basis just to stay even. Thus depreciation is based on the original investment and not on the current cost. The depreciation expense for accounting purposes is less than the true current cost. Consequently, the tax base is higher than it should be and taxes are overpaid. The true profits are less than those indicated. The inadequate allowance for depreciation is not the sole culprit. Inventory increases in value even if it remains the same physically and reinvestment of money is required just to stay even.

The concept of an *allowance for inflation* has been proposed as an accounting term to recognize the hidden costs created by inflation [3]. The hidden cost of inflation is particularly insidious since it must be met on an after-tax basis, and at the present 34 percent U.S. tax rate a profit of over $1.50 before taxes is required to meet a $1 inflation cost. The hidden cost of inflation becomes a double burden and strikes hardest at those industries having high long-term investments—the same industries that need the most help in fighting inflation. Reference 3 shows how to calculate the allowance for inflation on a yearly basis; furthermore, it emphasizes that the current use of net profit after taxes is really a profit after taxes but *before* inflation. The true profit is the profit after taxes and after inflation.

Deflation can be calculated using the mathematical equations of this chapter by giving $d$, the inflation rate, a negative value. However, deflation may be of historical interest only.

In engineering economy and cost comparisons future inflation is assumed and therefore must be predicted. There is no excuse for omitting inflation merely because it cannot be predicted precisely. Indeed, many contracts now contain various escalation clauses in order to remain realistic in terms of the changing dollar.

Inflation in the last analysis is not a matter of changing prices, but rather of changing currency, and is one of the most important current topics in cost engineering.

## NOMENCLATURE

| | |
|---|---|
| $A_{t1}$ | Amount at time 1, $ |
| $C_{bx}$ | Irregular cost at beginning of year $x$, $ |
| $C_d$ | Depreciable first cost, $ |
| $C_{ex}$ | Irregular cost at end of year $x$, $ |
| $C_{nd}$ | Nondepreciable cost, $ |
| $C_{sal}$ | Salvage value, $ |
| $d$ | Inflation rate, decay of $, decimal per year |
| $F_{SDP}$ | Present value of $1 of future sum-of-the-years-digits depreciation, decimal |
| $F_{SLP}$ | Present value of $1 of future straight-line depreciation, decimal |
| $K$ | Capitalized cost, $ |
| $n$ | A number, such as number of years for economic analysis |
| $n'$ | Number of years of life for tax purposes |
| $p$ | Technological advancement rate, decimal per year |
| $P$ | Present value, $ |
| $r$ | Rate of return after taxes, decimal per year |
| $R$ | Uniform end-of-year annual amount without inflation, $/year |
| $R_D$ | Unaburden, a uniform end-of-year burden with inflation, $/year |
| $t$ | Tax rate, decimal |
| $\psi$ | Present value of $1 of future depreciation, decimal |

## PROBLEMS

**6.1.** An article costs $20,000 and lasts 1 year. How much can be spent for an article that lasts 3 years if money is worth 20 percent per year after a 34 percent tax, with a 10 percent per year inflation rate and a 3-year life with straight-line depreciation?

**6.2.** A machine costs $100,000, lasts 2 years, and is written off in 3 years with straight-line depreciation for tax purposes. Using unaburden, find how much can be spent for a machine that lasts 5 years and is written off in 3 years using straight-line depreciation for tax purposes if the rate of return is 16 percent per year after a 34 percent tax and the inflation rate is 10 percent per year.

**6.3.** A factory floor can be resurfaced for $10,000 and lasts 1 year. Money is worth 16 percent per year after a 34 percent tax, straight-line depreciation will be used, and the inflation rate is 10 percent per year. How long must a special resurfacing job last if it costs $40,000 and can be written off in 3 years for tax purposes?

**6.4.** Use the data of Prob. 2.63. Money is worth 16 percent per year after a 34 percent tax, and the inflation rate is 10 percent per year. All costs are given in terms of the present dollar. Use a 3-year life for each machine for tax purposes and straight-line depreciation. Which machine is cheaper, and by how much in before-tax dollars?

**6.5.** Show on a time diagram how Eqs. (6.6) and (6.7) are related. Evaluate the present values and show the algebraic calculations leading to Eq. (6.5), thereby demonstrating that unaburden $R_D$ is independent of duration.

**6.6.** Derive the relationship for item 2, Table 6.1.

**6.7.** Derive the relationship for item 3, Table 6.1.

**6.8.** Derive the relationship for item 4, Table 6.1.

**6.9.** Derive the relationship for item 5, Table 6.1.

**6.10.** Derive the relationship for item 6, Table 6.1.

**6.11.** Derive the relationship for item 7, Table 6.1.

**6.12.** Derive Eq. (6.8) from item 1, Table 6.1, for the special case $r = d$.

**6.13.** Maintenance repairs for a machine cost $10,000 at the beginning of the year and are written off in full at the end of the year, and the repairs have a 1-year life. If a new machine costs $37,620, lasts 5 years, and is written off in 5 years using straight-line depreciation, and if money is worth 16 percent per year after a combined 46 percent tax, at what inflation rate would the 5-year machine be justified?

**6.14.** Sometimes a shortcut period is used in which the desired rate of return is added to the inflation rate and the problem worked out using the combined rate. The method thus reduces to a single rate which can be evaluated by referring to Chap. 3. Rework Prob. 6.1 on this basis using a single rate of return of 30 percent (20 + 10 = 30 percent) and find how much can be spent for the 3-year article. Do you agree that the rate of return and the inflation rate can be combined?

**6.15.** Rework Prob. 6.14 on the basis of a 30 percent rate of return, 20 percent over the inflation rate, but including the 10 percent inflation rate using the method given in this chapter. Find how much can be spent for the 3-year article.

**6.16.** Repeat Prob. 6.2 on the basis of capitalized cost. Note that the method used, whether based on unaburden or capitalized cost, disappears.

**6.17.** In using Eq. (6.17) for a system, the rate of return $r$, the tax rate $t$, and the inflation rate $d$ should be given the same value throughout the system. Does this require that the productivity gain rate $p$ be constant throughout the system?

# REFERENCES

1. Jelen, F. C.: "Consider Inflation in Comparative Cost Analyses," *Chemical Engineering,* May 1956, pp. 164–169; June 1956, pp. 247–252.
2. Jelen, F. C.: "Remember All Three in Cost Analyses," *Chemical Engineering,* January 27, 1958, pp. 123–128.
3. Jelen, F. C.: "Include Inflation in Cost Accounting," *Petroleum Refiner* (now *Hydrocarbon Processing*), July 1959, pp. 113–116.
4. Jelen, F. C., and M. S. Cole.: "Can New Equipment Cost Less?" *Hydrocarbon Processing,* July 1971, pp. 97–100.
5. Jelen, F. C., and M. S. Cole.: "Methods for Economic Analysis," *Hydrocarbon Processing,* Part 1, July 1974, pp. 133–139; Part 2, September 1974, pp. 227–233; Part 3, October 1974, pp. 161–163.

# CHAPTER 7

# BREAK-EVEN
# AND
# MINIMUM-COST
# ANALYSIS

*F. C. Jelen*

## 7.1  BREAK-EVEN ANALYSIS

*Break-even analysis* is a method for presenting costs and profits in a form designed to aid interpretation and analysis. The presentation can be graphical or tabular. Typically, this information presents profits as a function of a percentage of plant utilization, reveals plants where utilization operations are showing no profit and just breaking even, and determines the point at which it will pay to suspend operations.

As a simple example, consider a plant having *fixed costs* $C_F$ in dollars per year which do not change with the production rate, such as depreciation and local taxes. Other costs will be variable and will change with production, such as materials and labor. Suppose each unit of production has a cost of $V$\$ per unit, where $V$ is the *variable cost per unit of production*. Then, if a linear relationship exists, the variable cost itself will be $NV$, where $N$ is the number of units per production. Note that $V$ is the cost per unit of production and may be constant and that $NV$ is the actual variable cost. If $C_T$ is the total annual cost, then

$$C_T = NV + C_F \tag{7.1}$$

In Fig. 7.1 $C_T$ is a straight line because $V$ is assumed to be independent of $N$, and $C_F$ is constant. Other situations are considered subsequently in which the variable cost per unit of production $V$ depends on $N$, and the variable cost and total cost become curved.

**149**

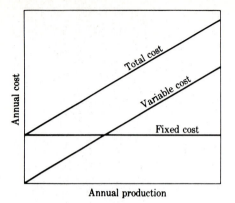

FIGURE 7.1
Diagram of relations between costs and production.

Fixed costs include expenditures which are independent of production rate, such as (1) indirect plant costs including depreciation, (2) management expense, and (3) selling expense.

The variable cost is made up of items such as materials and labor. In general the variable cost per unit of production is essentially constant. However, some variable costs per unit of production depend on factors such as power, which is sold on a declining-rate scale, or labor costs, which may increase per unit of production at high production rates because of inefficient labor or overtime.

## 7.2 ECONOMIC-PRODUCTION CHARTS

The economic-production chart is a geometric chart which utilizes cost and price data and reduces them to a simple visual model. The economic-production, or break-even, chart was proposed by Rautenstrausch [6] and can be utilized in either graphical or analytical form to evaluate an economic operation. The charts make visible the interrelationship of fixed costs, variable costs, sales revenue, and profits. The cost engineer can evaluate the overall economy of the operation from the charts and determine the relative effects of changes in the components.

The mathematical relationships take a very simple form if the following assumptions are made for the purpose of this section:

**1.** $V$ is constant; hence the variable costs $NV$ are linearly dependent on production.
**2.** Fixed costs are independent of production.
**3.** There are no financial costs.
**4.** There is no income other than that from operations.
**5.** All units are sold at the same price per unit.
**6.** All units produced are sold.

If $S$ is the net sales realization *per unit* and $Z_g$ the gross profit, then using the expression for $C_T$ obtained from Eq. (7.1), we have

$$Z_g = NS - C_T = NS - (NV + C_F) \tag{7.2}$$

If $Z_t$ is net profit, i.e., profit after taxes, and the decimal tax rate is $t$, then

$$Z_t = Z_g(1 - t) \tag{7.3}$$

The various relationships all plot as straight lines against $N$, units of production, and are shown in Fig. 7.2. The point at which the sales line crosses the total-cost line is called the *break-even point* and is the point below which operation results in a loss and above which it yields a profit.

The break-even point can be found mathematically by setting $Z_g = 0$ in Eq. (7.2):

$$N \text{ break-even} = \frac{C_F}{S - V} \tag{7.4}$$

This point is not a function of the tax rate since no profit is obtained at this production level.

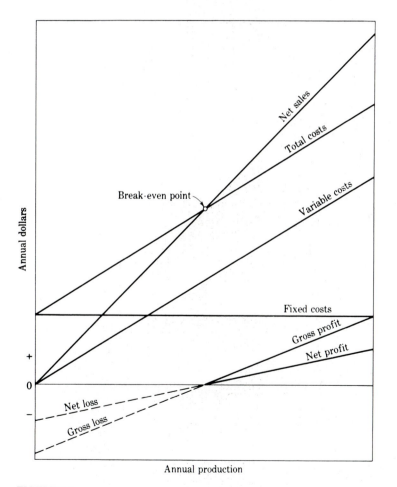

**FIGURE 7.2**
Economic-production chart.

The quantitative effects of various selling prices can be visualized from the chart by sketching a separate net-sales line for each different unit sales price, and the procedure can be used for other variables. The charts can be utilized for any period of time and for any combination of products provided the pertinent costs are prorated.

## 7.3 ECONOMIC-PRODUCTION CHART ABOVE 100 PERCENT CAPACITY

Usually a project operates at maximum efficiency when producing at 100 percent capacity. However, even if the variable cost per unit $V$ increases at operations above 100 percent of capacity, the total profit probably will continue to increase. These points are illustrated in Fig. 7.3. If the two parts of the plot are considered as two steps, all the previous relations apply for each step, the proper terms being included

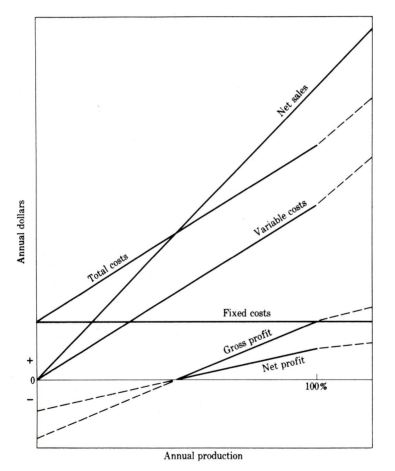

**FIGURE 7.3**
Economic-production chart for operations above 100 percent capacity.

for operations above 100 percent of capacity, here denoted by primes. For example, gross profit can be written

$$Z_g = (N + N')S - (NV + N'V' + C_F) \tag{7.5}$$

where $N'$ and $V'$ are the number of production units above 100 percent capacity and the associated variable cost per unit and $N$ and $V$ are the corresponding values at 100 percent capacity.

**Example 7.1.** A project can produce 12,000 units per year at 100 percent of capacity. Fixed costs are \$10,000/year. The variable cost per unit is \$3 up to 100 percent of capacity and \$3.30 above 100 percent of capacity. All units are sold at a net selling price of \$5. The tax rate is 34 percent. Find the break-even point and the profit per year after taxes at 80 and 120 percent of production capacity.

*Solution.* The break-even point is given by Eq. (7.3):

$$N = \frac{C_F}{S - V} = \frac{10,000}{5 - 3} = 5000 \text{ units per year}$$

The gross profit at 80 percent of capacity is given by Eq. (7.2):

$$Z_g = 0.80(12,000)\,(5) - [0.80(12,000)\,(3) + 10,000] = \$9200/\text{year}$$

which after a 34 percent tax is

$$Z_t = 9200(1 - 0.34) = \$6072/\text{year}$$

The gross profit at 120 percent of capacity is given by Eq. (7.5):

$$Z_g = 1.20(12,000)\,(5) - [12,000(3) + 0.2(12,000)\,(3.30) + 10,000] = \$18,080/\text{year}$$

which after a 34 percent tax is

$$Z_t = 18,080(1 - 0.34) = \$11,933/\text{year}$$

**Example 7.2.** The *average* net sales realization for a unit of product in dollars for a production of $M$ kilounits per year is given by $2000 - 40M$. Find the net sales realization at a production rate of 2500/year and the incremental sales income at that production rate.

*Solution.* Let $N = 1000M$, where $N$ represents one unit. Then the average net sales is

$$2000 - 40M = 2000 - 40\frac{N}{1000} = 2000 - 0.04N$$

For $N = 2500$

$$\frac{\sum NV}{N} = 2000 - 0.04N$$

$$\sum NV = 2000N - 0.04N^2 = 2000(2500) - 0.04(2500)^2$$

$$= \$4,750,000 \text{ net sales realization}$$

The incremental sales realization is

$$\frac{d}{dN}(2000N - 0.04N^2) = 2000 - 0.08N = 2000 - 0.08(2500) = \$1800$$

That is, each additional unit $N$ brings in \$1800.

## 7.4   ECONOMIC-PRODUCTION CHART FOR DUMPING

The principles discussed in connection with Fig. 7.3 apply to the practice of *dumping,* which occurs when a manufacturer sells $N$ units of production at one sales price, $S$ per unit, but because demand is not sufficient to take all the production, he or she sells $N''$ units at a lower price, $S''$ per unit. The manufacturer obtains greater sales and keeps the plant operating at higher output. The diagram is similar to Fig. 7.3 and shows a break in the sales line, gross profit, and net profit at $N$. The gross profit is given by

$$Z_g = NS + N''S'' - [(N + N'')V + C_F] \tag{7.6}$$

**Example 7.3.** For the data given in Example 7.1 the manufacturer can sell only 80 percent of rated capacity at \$5 per unit. How much should the manufacturer charge for additional units if production is brought up to 100 percent of capacity and profits after taxes are increased by an additional \$1000?

*Solution.* The net profit after taxes at 80 percent of production was \$6072. The gross profit at 100 percent of production will be

$$\frac{6072 + 1000}{1 - 0.34} = 10,715$$

The manufacturer will sell $0.80(12,000) = 9600$ units at \$5 and $0.20(12,000) = 2400$ units at $S''$. By Eq. (7.6),

$$10,715 = 9600(5) + 2400S'' - [(9600 + 2400)(3) + 10,000]$$

$$S'' = \$3.63 \text{ per unit}$$

## 7.5   NONLINEAR ECONOMIC-PRODUCTION CHARTS

In practice, it is necessary to consider the analysis of economic-production charts in which some entries vary with production in a nonlinear manner. Costs which change disproportionately with increased production are classified as *regulated costs,* here denoted by $NR$. Included in this category are labor, maintenance, and utilities. The following example illustrates a general case.

**Example 7.4.** A plant operation for 1 year has fixed costs $C_F$, variable costs $NV$, regulated costs $NR$, and sales realization $NS$, as follows, where $N$ is the operating rate in percent:

$$C_F = 20,000$$

$$NV = 500N$$

$$\begin{cases} NR = 15,000 + 30N^{1.5} & N > 0 \\ NR = 0 & N = 0 \end{cases}$$

$$NS = 2000N - 7.5N^2$$

A tabulation and a plot of various items are shown in Table 7.1 and Fig. 7.4.

At the break-even point there is no profit:

$$0 = Z_g = NS - (NV + NR + C_F)$$

$$= 2000N - 7.5N^2 - (500N + 15,000 + 30N^{1.5} + 20,000)$$

$$= 1500N - 30N^{1.5} - 7.5N^2 - 35,000$$

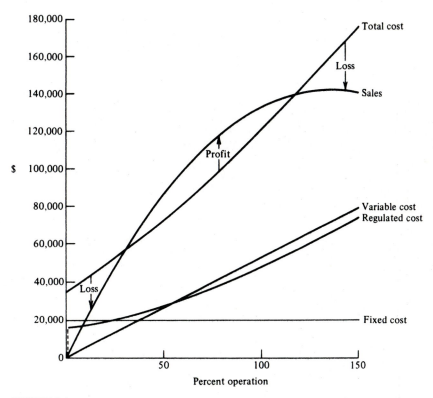

**FIGURE 7.4**
Chart for Example 7.4.

TABLE 7.1
Tabulation for Example 7.4, dollar basis, 1 year

| (1)<br>Line no. | (2)<br>N<br>percent<br>operation | (3)<br>$C_F$<br>fixed<br>cost | (4)<br>NV<br>variable<br>cost | (5)<br>NR<br>regulated<br>cost | (6)<br>Total cost,<br>(3) + (4) + (5) | (7)<br>NS<br>sales<br>realization | (8)<br>Profit<br>before<br>taxes,<br>(7) − (6) |
|---|---|---|---|---|---|---|---|
| 1 | 0 | 20,000 | 0 | 0 | 20,000 | 0 | −20,000 |
| 2 | 0.1 | 20,000 | 50 | 15,001 | 35,051 | 200 | −34,851 |
| 3 | 5 | 20,000 | 2,500 | 15,335 | 37,835 | 9,813 | −28,022 |
| 4 | 10 | 20,000 | 5,000 | 15,949 | 40,949 | 19,250 | −21,699 |
| 5 | 11.4 | 20,000 | 5,700 | 16,155 | 41,855 | 21,825 | −20,030 |
| 6 | 20 | 20,000 | 10,000 | 17,683 | 47,683 | 37,000 | −10,683 |
| 7 | 30 | 20,000 | 15,000 | 19,930 | 54,930 | 53,250 | −1,680 |
| 8 | 32.1 | 20,000 | 16,050 | 20,456 | 56,506 | 56,472 | 0 |
| 9 | 40 | 20,000 | 20,000 | 22,589 | 62,589 | 68,000 | 5,411 |
| 10 | 50 | 20,000 | 25,000 | 25,606 | 70,606 | 81,250 | 10,644 |
| 11 | 60 | 20,000 | 30,000 | 28,943 | 78,943 | 93,000 | 14,057 |
| 12 | 70 | 20,000 | 35,000 | 32,570 | 87,570 | 103,250 | 15,680 |
| 13 | 74.2 | 20,000 | 37,100 | 34,175 | 91,275 | 107,108 | 15,833 |
| 14 | 80 | 20,000 | 40,000 | 36,466 | 96,466 | 112,000 | 15,534 |
| 15 | 90 | 20,000 | 45,000 | 40,614 | 105,614 | 119,250 | 13,636 |
| 16 | 100 | 20,000 | 50,000 | 45,000 | 115,000 | 125,000 | 10,000 |
| 17 | 110 | 20,000 | 55,000 | 49,610 | 124,610 | 129,250 | 4,640 |
| 18 | 116.8 | 20,000 | 58,400 | 52,869 | 131,269 | 131,283 | 0 |
| 19 | 120 | 20,000 | 60,000 | 54,436 | 134,436 | 132,000 | −2,436 |
| 20 | 130 | 20,000 | 65,000 | 59,467 | 144,467 | 133,250 | −11,217 |
| 21 | 138.5 | 20,000 | 69,250 | 63,899 | 153,149 | 133,133 | −20,016 |
| 22 | 140 | 20,000 | 70,000 | 64,695 | 154,695 | 133,000 | −21,695 |
| 23 | 150 | 20,000 | 75,000 | 70,114 | 165,114 | 131,250 | −33,864 |

There is a lower break-even point at $N = 32.1$, as read from the tabulation, the graph, or by a trial-and-error solution of the last equation. It is the lowest production rate that will not result in a loss.

There is also an upper break-even point at $N = 116.8$ known as a *profit limit point*. A production level *above* this point will result in a loss.

The maximum profit is \$15,833 and occurs at $N = 74.2$. It can be obtained from the tabulation, the figure, by maximizing the right side of the last equation, or by differentiating the last equation, setting it equal to zero, and solving for $N$ by trial and error.

Another concept of interest is the *shutdown point*. At times it may be more economical to shut down completely and absorb the fixed costs rather than operate. In this plant operation there is a shutdown point at $N = 11.4$. Operation below this point will result in a loss larger than the \$20,000 fixed cost at shutdown.

There is a second shutdown point at $N = 138.5$. It is an upper shutdown point; i.e., it would be more economical to shut down than to operate *above* that point.

There are other calculations of interest. The maximum average unit profit is

$$\frac{Z_g}{N} = \frac{1500N - 30N^{1.5} - 7.5N^2 - 35,000}{N}$$

$$= 1500 - 30N^{0.5} - 7.5N - \frac{35,000}{N}$$

Differentiating and setting equal to zero gives

$$0 = -15N^{-0.5} - 7.5 + 35,000N^{-2}$$

for which $N = 61.0$ and $Z_g/N = \$234.42$.
The minimum average unit cost is

$$\frac{C_F + NV + NR}{N} = \frac{20,000 + 500N + 15,000 + 30N^{1.5}}{N}$$

$$= \frac{35,000}{N} + 500 + 30N^{0.5}$$

This has a minimum value of \$1096.90 at $N = 176$, which obviously is outside the range of interest in this problem.

The minimum average variable cost is

$$\frac{NV}{N} = \frac{500N}{N} = \$500$$

and is a constant for all values of $N$.

The minimum average regulated cost is

$$\frac{NR}{N} = \frac{15,000 + 30N^{1.5}}{N} = \frac{15,000}{N} + 30N^{0.5}$$

which has a minimum value of 450 at $N = 100$.

The maxima, minima, and averages can mean different things and should not be confused. Thus maximum profit does not correspond to maximum average profit.

However, there is one similarity that should be noted. Unit incremental net sales revenue and unit incremental total costs are equal at maximum profit, and the equality is not accidental. Thus

$$\frac{\Delta NS}{\Delta N} = \frac{d}{dN}(2000N - 7.5N^2) = 2000 - 15N$$

and is equal to 887 for $N = 74.2$ at maximum profit. Also,

$$\frac{\Delta(C_F + NV + RN)}{\Delta N} = \frac{d}{dN}(20,000 + 500N + 15,000 + 30N^{1.5})$$

$$= 500 + 45N^{0.5}$$

and is also equal to 887 for $N = 74.2$.

The incremental expressions are also known as *marginal expressions*. Thus the marginal cost is the additional cost that results from producing one more unit at a certain production level.

The unit incremental, or marginal, cost is an important criterion in analyzing production charts and can be used to allocate production between competing production facilities.


## 7.6 ONE-PRODUCT MULTIPLE-MACHINE EXAMPLE

Up to this point the minimum-cost analyses were concerned only with the level at which a facility making one product should be operated. Letting the term *machine* signify a process or plant or any single producing unit, consider now two or more machines being used to make the same product with different cost relationships. The problem is to select the level at which each machine should operate so that overall production costs will be minimal. The following example* has been modified from one originally presented by Schweyer [7].

**Example 7.5.** Total annual costs for three machines, $A$, $B$, and $C$, producing $N_A$, $N_B$, and $N_C$ annual units, are

$$C_{TA} = N_A V_A + C_{FA} = 0.06N_A^{1.1} + 1200 \tag{7.7}$$

$$C_{TB} = N_B V_B + C_{FB} = 0.02N_B^{1.3} + 1000 \tag{7.8}$$

$$C_{TC} = N_C V_C + C_{FC} = 0.002N_C^{1.6} + 600 \tag{7.9}$$

If 400 annual production units are made, how should production be allotted to the three machines?

---

* With permission from Van Nostrand Reinhold Company, New York.

*Solution.* The total cost for all three machines is obtained by adding Eqs. (7.7), (7.8), and (7.9):

$$C_T = 0.06N_A^{1.1} + 0.02N_B^{1.3} + 0.002N_C^{1.6} + 2800 \qquad (7.10)$$

If $N = 400$ is the total production from all three plants, then

$$400 = N_A + N_B + N_C \qquad (7.11)$$

The problem now is to minimize Eq. (7.10) subject to the restraint, Eq. (7.11). Using the method of Lagrange multipliers as developed in Chap. 10, the Lagrange expression for the problem at hand is

$$LE = 0.06N_A^{1.1} + 0.02N_B^{1.3} + 0.002N_C^{1.6} + 2800 + \lambda(N_A + N_B + N_C - 400)$$

The partial derivatives are set equal to zero for a minimum:

$$\frac{\partial LE}{\partial N_A} = 0.066N_A^{0.1} + \lambda = 0 \qquad (7.12)$$

$$\frac{\partial LE}{\partial N_B} = 0.026N_B^{0.3} + \lambda = 0 \qquad (7.13)$$

$$\frac{\partial LE}{\partial N_C} = 0.0032N_C^{0.6} + \lambda = 0 \qquad (7.14)$$

$$\frac{\partial LE}{\partial \lambda} = N_A + N_B + N_C - 400 = 0 \qquad (7.15)$$

These equations show that

$$0.066N_A^{0.1} = 0.026N_B^{0.3} = 0.0032N_C^{0.6} \qquad (7.16)$$

Comparing with the partial derivatives of Eqs. (7.7) to (7.9) gives

$$\frac{\partial C_{TA}}{\partial N_A} = \frac{\partial C_{TB}}{\partial N_B} = \frac{\partial C_{TC}}{\partial N_C} \qquad (7.17)$$

These results indicate that minimum total costs in a one-product multiple-machine program are achieved when the incremental, or marginal, costs are the same for all machines. Figure 7.5 illustrates the variation of the incremental costs with production. Optimum operation is obtained by drawing a horizontal line (equal incremental costs) and reading the value of the abscissa at the intersection of the horizontal line and each of the incremental-cost curves. The horizontal line is adjusted in position until the sum of the abscissas corresponds to the total production desired.

An analytic solution can be made by trial and error by assuming, say, $N_C$ and calculating $N_A$ and $N_B$ from Eq. (7.16) and repeating until here $N_A + N_B + N_C = 400$. This leads to $N_A = 39$, $N_B = 75$, and $N_C = 285$, which checks with the indicated graphical solution.

The treatment given in this section will give the correct answer if the incremental curves shown in Fig. 7.5 are regular; i.e., among other things they are continuous and single-valued. In practice, the curves do resemble those of Fig. 7.5 and the method is generally applicable.

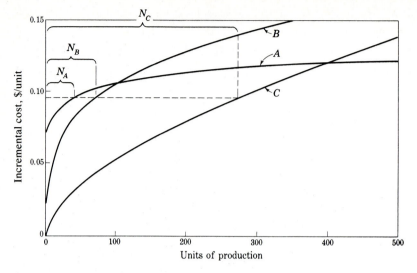

**FIGURE 7.5**
Incremental costs for Example 7.5.

## 7.7 MULTIPLE-PRODUCTS ONE-MACHINE EXAMPLE

The analysis up to this point implies that production involves one product. A more elaborate problem occurs when there are two or more activities, i.e., a mixture of products, which compete for production time on a machine. When there are multiple activities and all the relationships vary in a linear, or first-order, manner, linear programming as developed in Chap. 12 may be used. For cases in which one or more relationships are nonlinear, an analysis can be made using the principles of economic-production charts. The following example* is modified from one given by Schweyer [7].

**Example 7.6.** A machine can make product $A$ at 1000 per hour or product $B$ at 2000 per hour. The available total production time is 2400 h/year. Annual fixed costs are $80,000. Other data are

| Product | Average variable cost per unit, $ per unit | Average net sales per unit, $ per unit |
|---------|------------------------------|----------------------------|
| $A$ | $6N_A \times 10^{-8} + 0.2$ | 0.50 |
| $B$ | 0.15 | 0.20 |

---

*With permission from Van Nostrand Reinhold Company, New York.

where $N_A$ is the yearly production units of product $A$. Find the allocation of products which gives the maximum gross profit per year.

*Solution.* The total annual cost is

$$C_T = NV + C_F = 6N_A^2 \times 10^{-8} + 0.2N_A + 0.15N_B + 80,000 \qquad (7.18)$$

where $N_B$ is the annual production of product $B$. The total revenue $S_T$ is

$$S_T = 0.5N_A + 0.2N_B \qquad (7.19)$$

and the gross profit $Z_g$ is the difference between Eqs. (7.19) and (7.18):

$$Z_g = 0.3N_A + 0.05N_B - 6N_A^2 \times 10^{-8} - 80,000 \qquad (7.20)$$

The profit $Z_g$ is to be maximized subject to the following restraint equation, which limits total hours for products $A$ and $B$ to 2400 per year:

$$\frac{N_A}{1000} + \frac{N_B}{2000} = 2400 \qquad (7.21)$$

or

$$2N_A + N_B - 4,800,000 = 0 \qquad (7.22)$$

The Lagrange expression is

$$\text{LE} = 0.3N_A + 0.05N_B - 6N_A^2 \times 10^{-8} - 80,000 + \lambda(2N_A + N_B - 4,800,000) \quad (7.23)$$

The partial derivatives are set equal to zero for a maximum:

$$\frac{\partial \text{LE}}{\partial N_A} = 0.3 - 12 \times 10^{-8}N_A + 2\lambda = 0 \qquad (7.24)$$

$$\frac{\partial \text{LE}}{\partial N_B} = 0.05 + \lambda = 0 \qquad (7.25)$$

$$\frac{\partial \text{LE}}{\partial \lambda} = 2N_A + N_B - 4,800,000 = 0 \qquad (7.26)$$

whence $N_A = 1,667,000$ units per year and $N_B = 1,467,000$ units per year.

The example can also be solved using the incremental approach. Equations (7.18) for total costs and (7.19) for total revenue as written involve $N_A$ and $N_B$, but $N_B$ can be eliminated by Eq. (7.22), giving equations for $C_T$ and $S_T$ as a function of $N_A$ only. The derivatives of these equations must be equal at the maximum profit, as shown in the previous section. Setting

$$\frac{dC_T}{dN_A} = \frac{dS_T}{dN_A}$$

will lead to the value of $N_A$.

The example can be solved analytically by trial and error by assuming, say, $N_A$, obtaining $N_B$ from Eq. (7.22), and calculating and tabulating the total profit $Z_g$ from Eq. (7.20). Repeating for various values of $N_A$ will establish the maximum profit.

## 7.8 CRITIQUE OF THE USE OF BREAK-EVEN AND MINIMUM-COST ANALYSIS

Break-even analysis is a valuable tool. In general it is well illustrated by Example 7.4, which was analyzed by a tabulation, a chart, and supplemental calculations and analyses. By showing variations in the curves as the parameters are modified, a single chart can display a range of circumstances, and multiple charts permit wide variations for study. The technique is easily understood and is a good means for presentation.

Utilization of the functional equation for maximum profit is a good introduction to optimization. Other criteria for profitability have been developed in Chap. 5. One of the most serious limitations of the simple economic evaluations presented in this chapter is lack of consideration for the time value of money.

Another simplifying assumption involves the projection of sales prices at different production levels. In practice, probabilities must be utilized to predict the future values of the variables. A general comment on mathematical models seems appropriate at this point.

Mathematical models are an abstraction of a real system. In reality, the systems are much more complex than any mathematical model which could be built. Therefore, it is the task of the analyst to select the significant aspects of the real system and ignore those aspects which make a negligible contribution at the level of accuracy desired.

The analyst must always compromise between constructing the most accurate model and keeping it simple enough to have a manageable solution. The computer has made possible the development of very complex models the solution of which would not have been feasible by analytical or other manual means. Thus it has increased the degree of accuracy with which real systems can now be simulated.

## NOMENCLATURE

$C_F$     Fixed cost, \$

$C_T$     Total cost, \$

LE     Lagrange expression

$N$     Number of units of production

$N'$     Number of units of production above 100 percent of capacity

$N''$     Number of units of production dumped

$R$     Regulated cost per unit of production, \$ per unit

$S$     Net realizable sales value per unit sold, \$ per unit

$S''$     Net realizable sales value per unit dumped, \$ per unit

$S_T$     Total realizable sales value, \$

$t$     Tax rate, decimal

$V$     Variable cost per unit of production, \$ per unit

$V'$     Variable cost per unit of production above 100 percent of capacity, \$ per unit

$Z_g$     Gross profit before tax, \$

$Z_t$     Net profit after tax, \$
$\lambda$     Lagrange multiplier

## PROBLEMS

**7.1.** Waste water can be processed for pollution control by two methods, A and B. Pertinent costs are

|                              | A       | B       |
| ---------------------------- | ------- | ------- |
| Labor, \$/year               | 70,000  | 95,000  |
| Depreciation, \$/year        | 50,000  | 28,000  |
| Chemicals, \$/m$^3$ treated  | 0.15    | 0.10    |

How much water must be treated per year for the two methods to be equal?

**7.2.** The average variable cost for a unit of product in dollars for a production of $N$ million units per year is given by

$$1.75 + 0.40N - 0.1N^2$$

Find the variable cost at a production rate of 2.5 million units of production per year. What is the incremental variable cost per unit of production at that production rate?

**7.3.** A boiler plant now recovers 5000 kg of fly ash per day at a collection cost of \$3/t and a disposal cost by dumping in the ocean at \$2/t. A plant costing \$175,000 with a life of 20 years must be built to recover an additional 6000 kg of fly ash at the same collection cost per metric ton. The plant would process the old and new ash collections. If the plant were not built, the new ash would have to be dumped in the ocean. Find the required selling price per metric ton of fly ash if all the fly ash processed in the new plant could be sold and the processing cost is \$1.80/t.

**7.4.** A plant operating at 80 percent of capacity has fixed costs of \$55,000/year, direct costs of \$120,000/year, and sales of \$200,000/year. What is the break-even point?

**7.5.** An office intercom system will cost $1000 + 150N$ dollars for $N$ connections. The system will last 8 years, with a maintenance cost of \$20/year for each connection. A telephone company will charge \$4/month for each intercom outlet by using existing telephone connections. How many connections are necessary to justify the intercom system?

**7.6.** A company has fixed costs of \$300,000/year. At 75 percent of capacity, the variable cost is \$800,000 and the sales realization is \$1,500,000. On a single break-even chart, show the effect of increasing and decreasing the selling price per unit by 10 percent. What is the break-even point for (a) the selling price as is, (b) the selling price increased by 10 percent, (c) the selling price decreased by 10 percent, and (d) extra profit when the selling price is decreased by 10 percent, permitting capacity production as opposed to the present rate at the present selling price?

**7.7.** A company now sells 1000 units per week at a profit of \$20,000/week, with fixed costs at \$10,000/week and variable costs at \$5000/week. If the fixed costs increase by 15 percent, by what fraction must the selling price be raised to maintain the existing profit if output is the same?

**7.8.** A company with fixed costs of $200,000 and a break-even point of 60 percent for net sales of $400,000 will show what profit before taxes at 100 percent of capacity? If the profits tax is 34 percent, what is the total capitalized value, or market value, if prospective purchasers desire an 18 percent return after taxes based on full-capacity operations? Does the depreciation charge affect the capitalized value?

**7.9.** The profit at 65 percent of capacity is 50 percent greater than the profit at 50 percent of capacity. Fixed costs are $100,000, and net sales are 300 percent greater than the variable cost at both operating rates. Find the profit at 100 percent of capacity.

**7.10.** A meat-waste processing plant is designed for operations of 20,000 kg/day at 100 percent of capacity when the average variable costs are $0.48/kg. At 75 percent of capacity, the average variable costs are $0.56/kg. The fixed costs are $1040 per operating day.
(a) Compute the unit costs for operations at both 75 and 100 percent of capacity.
(b) What is the incremental cost in dollars per day per kilogram in going from 75 percent to 100 percent of capacity?
(c) What is the average unit incremental cost in going from the lower to the higher capacity?

**7.11.** If the ratio of direct costs to sales dollars is 0.4 and the fixed costs are $250,000 annually for a product selling at $0.50/kg, what is the unit cost of the product when annual net sales are (a) $1,250,000 and (b) at a maximum capacity of 4,000,000 kg?

**7.12.** For net sales of a company amounting to $600,000 annually, when the fixed costs are $350,000 and the direct costs are 35 percent of the net sales dollars:
(a) What is the gross profit?
(b) What is the break-even point in terms of sales dollars?
(c) What sales are required for a profit of $80,000?

**7.13.** Under normal conditions an operation with a capacity of 10,000 units annually has a selling price of $100, which represents $15 in profit, $30 in fixed costs, and $55 in variable costs. Demand drops to 2000 units annually. Fixed costs can be reduced 10 percent below normal under these conditions and 25 percent below normal if the plant shuts down.
(a) Should the plant remain open or shut down?
(b) How low can the selling price go before the plant should shut down when demand is 2000 units annually?

**7.14.** A manufacturer can produce a part for $5 per part, with fixed annual costs of $15,000. He can buy the part for $8 for the first 10,000 and for $4 for all additional parts used in a year. At what production rate would it be cheaper to buy the part?

**7.15.** A hardware store pays $10 for a tool and plans to run an advertising campaign costing $500 for the allocated space. The number of sales expected at various prices is

| Expected sales | Selling price |
|---|---|
| 100 | $20 |
| 200 | $16 |
| 500 | $15 |
| 1000 | $12 |

Find the selling price that will give the largest profit.

**7.16.** In Prob. 7.15 assume that the number to be sold at each selling price is proportional to the square root of the amount spent on advertising. Find the selling price and advertising cost that will maximize the profit, and find the maximum profit.

**7.17.** An operation rated at 10,000 units per year has fixed costs of $8000/year, variable costs per unit at $5 for production up to 10,000 units per year, variable costs per unit at $6 for production above 10,000 units per year, and a selling price of $9 for each unit of production. Find the break-even point and the profit after a 34 percent tax at 100 percent and at 150 percent production capacity.

**7.18.** In the previous problem assume that all units above 10,000 per year must be sold at $4 to fulfill a prior legal contract. At what production rate will the operation cease to make a profit and show a profit limit point?

**7.19.** A study of a plant operation having annual fixed costs of $100,000/year has variable costs and net sales expressed as

$$NV = N^3 \times 10^{-7} - 0.0012N^2 + 21.6N$$

$$NS = 76N - 0.0036N^2$$

where $N$ represents tons and costs are for 1 year.
(a) Find the break-even point and the profit limit point.
(b) Find the maximum annual profit and the production rate at which it occurs.
(c) Find the average profit and the incremental profit at a production rate of 9000 t/year.

**7.20.** Economic analysis of a plant operation has shown the following empirical relationships, where $V$ is a variable cost per unit and $S$ is net sales realization per unit, with $N$ being units of production:

$$\text{Average sales price} = A - BN^2 \quad \$/\text{kg}$$

$$\text{Average variable cost} = CN^3 + DN^2 \quad \$/\text{kg}$$

If $C_F$ is the total constant annual fixed costs and $A$, $B$, $C$, and $D$ above are constants, develop equations to evaluate $N$ for the following in terms of $A$, $B$, $C$, $D$, and $C_F$:
(a) The break-even point
(b) The production rate for maximum profit
(c) The maximum profit per kilogram
(d) The minimum unit cost if $N$ differs from the value in (c)
(e) The marginal cost, or incremental cost

**7.21.** The marginal or incremental revenue for a part can be expressed as

$$100 - 0.02N$$

where $N$ is the number of parts produced per period. Total costs, including fixed costs, can be expressed as

$$2 \times 10^{-4}N^2 + 10,000$$

Find (a) the break-even point, (b) the maximum profit and the production rate at which it occurs, and (c) the minimum average cost and the production rate at which it occurs.

**7.22.** The average revenue for a quantity $Q$ per kilogram is given by

$$100 - 0.01Q$$

and the total cost for $Q$ per period is given by

$$50Q + 30,000$$

Determine (a) the marginal revenue, (b) the marginal cost, and (c) the point of maximum profit and the maximum profit per period.

**7.23.** Machine $A$ has a capacity of 20,000 units per period, is operating at 30 percent of capacity, and has total costs per period expressed as

$$C_T = 300N + 4N^2 + 12,000$$

where $N$ is the percent of operating capacity.

Machine $B$ has a capacity of 10,000 units per period, is operating at 55 percent of capacity, and has total costs per period expressed as

$$C_T = 75N + 2N^2 + 8000$$

where $N$ is the percent of operating capacity.

(a) If one more unit per period is to be produced, to which machine should it be assigned, i.e., which will have the smaller added cost?

(b) How should the current production be allotted between the machines for minimum cost?

(c) In (b), what will be the incremental cost for each machine?

(d) Find a total production such that the machines share the production load equally and have equal incremental costs. This is equivalent to finding a point at which the machines share the load equally and also have a minimum cost.

**7.24.** A machine is available for 8 h/day to make products $A$ and $B$. The machine can make 800 units of product $A$ exclusively or 2400 units of product $B$ exclusively or a mixture of both. Product $A$ sells for \$11 per unit and product $B$ for \$7 per unit. The machine has a fixed cost of \$100/8-h day. The variable costs per 8-h day for $N_A$ and $N_B$ units are

$$3N_A + 0.002N_A^2$$

$$5N_B + 50$$

Find the maximum profit possible per day if all units can be sold, and show that the incremental costs for $A$ and $B$ are equal at the optimum mix when an equivalent comparison is used.

**7.25.** A one-product, two-mix operation is being considered with a limit of \$100,000 per time period. Labor $A$ costs \$2000 per unit of labor per period, and feedstock $B$ costs \$1000/t. The amount of labor and material used determines the production $Q$, which in tons per period is given by

$$Q = 5AB$$

(a) What amounts should be spent for labor and materials, and what is the optimum production?

(b) Show how the Lagrange multiplier solves the problem.

# REFERENCES

1. Brighton, E. F., and J. L. Pappas: *Managerial Economics*, 2d ed., The Dryden Press, Hinsdale, Ill., 1976.
2. Kasner, E.: *Essentials of Engineering Economics*, McGraw-Hill Book Company, New York, 1979.
3. Levin, R. I., and C. A. Kirkpatrick: *Quantitative Approaches to Management*, 4th ed., McGraw-Hill Book Company, New York, 1978.
4. Newnan, D. G.: *Engineering Economic Analysis*, 3d ed., Engineering Press, Inc., San Jose, Calif., 1988.

5. Park, W. R., and D. E. Jackson: *Cost Engineering Analysis,* 2d ed., John Wiley & Sons, Inc., New York, 1984.
6. Rautenstrausch, W.: *Economics of Business Enterprise,* John Wiley & Sons, Inc., New York, 1939.
7. Schweyer, H. E.: *Analytic Models for Managerial and Engineering Economics,* Van Nostrand Reinhold Company, New York, 1964.
8. Taylor, G. A.: *Managerial and Engineering Economy,* 3d ed., D. Van Nostrand Company, Inc., New York, 1980.

# CHAPTER
# 8

# PROBABILITY, SIMULATION, UNCERTAINTY

*F. C. Jelen*

## DISCRETE FREQUENCY DISTRIBUTIONS

## 8.1 DISTRIBUTION FUNCTIONS

Many practical problems involve uncertainty and are probabilistic as opposed to deterministic. It is no longer sufficient to report that a new venture is expected to have a capital investment of 10 M\$. It may be necessary to give some measure indicating that it may cost 12 M\$ or more. The components going into the estimate have uncertainty and a range of values. The final estimate has an expected value, but also has uncertainty and a range of possible values. The analysis of such problems depends on the distribution that makes up the average or expected value and is part of the mathematics of probability and statistics. The purpose of this chapter is to review those aspects of the mathematics which are essential for an understanding of the application to cost engineering.

Consider the case of throwing dice and noting the total shown. Table 8.1 and Fig. 8.1 give data computed for one, two, and three dice, and the probabilities $P(X)$ for the occurrence of various sums.

The three distributions shown are examples of a *discrete distribution*, for only integer values can appear; i.e., the sums are not a continuous variable. The probabilities in Table 8.1 are easily calculated. Thus in the case of two dice there are 36

**TABLE 8.1**
**Frequency distribution with dice**

| Sum of dice X | One die | | Two dice | | | Three dice | |
|---|---|---|---|---|---|---|---|
| | Number of ways for appearance | Probability P(X) of X | Number of ways for appearance | Probability P(X) of X | Cumulative probability ΣP(X) of X for Sec. 8.3 | Number of ways for appearance | Probability P(X) of X |
| 1 | 1 | 0.1667 | | | | | |
| 2 | 1 | 0.1667 | 1 | 0.0278 | 0.0278 | | |
| 3 | 1 | 0.1667 | 2 | 0.0556 | 0.0833 | 1 | 0.0046 |
| 4 | 1 | 0.1667 | 3 | 0.0833 | 0.1667 | 3 | 0.0139 |
| 5 | 1 | 0.1667 | 4 | 0.1111 | 0.2778 | 6 | 0.0278 |
| 6 | 1 | 0.1667 | 5 | 0.1389 | 0.4167 | 10 | 0.0463 |
| 7 | 6 | | 6 | 0.1667 | 0.5833 | 15 | 0.0694 |
| 8 | | | 5 | 0.1389 | 0.7222 | 21 | 0.0972 |
| 9 | | | 4 | 0.1111 | 0.8333 | 25 | 0.1157 |
| 10 | | | 3 | 0.0833 | 0.9167 | 27 | 0.1250 |
| 11 | | | 2 | 0.0556 | 0.9722 | 27 | 0.1250 |
| 12 | | | 1 | 0.0278 | 1.0000 | 25 | 0.1157 |
| 13 | | | 36 | | | 21 | 0.0972 |
| 14 | | | | | | 15 | 0.0694 |
| 15 | | | | | | 10 | 0.0463 |
| 16 | | | | | | 6 | 0.0278 |
| 17 | | | | | | 3 | 0.0139 |
| 18 | | | | | | 1 | 0.0046 |
| | | | | | | 216 | |

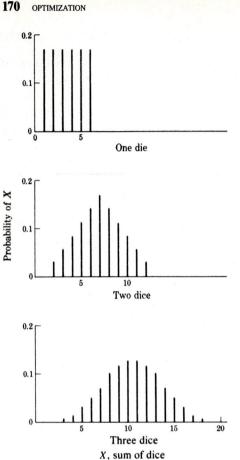

**FIGURE 8.1**
Distributions of sums with dice.

equally likely ways the dice can turn up, and the sum 6 can be obtained in five ways: 1 — 5, 2 — 4, 3 — 3, 4 — 2, and 5 — 1. The probability for the sum 6 is 5/36 = 0.1389.

The distributions shown are symmetrical, but in practice, distributions may be nonsymmetrical, or skewed.

## 8.2 PROPERTIES OF DISCRETE-DISTRIBUTION FUNCTIONS

An inspection of Fig. 8.1 shows differences, and it is desirable to develop parameters which show these differences. Only two are considered here, the mean and the variance.

The *mean* is the arithmetic mean of the distribution or a number of samples taken from it. If $X_i$ is the value of the $i$th in a total of $N$ equally likely events, the mean $\mu$ is

$$\mu = \frac{\sum\limits_{i=1}^{i=N} X_i}{N} \tag{8.1}$$

or in terms of the probabilities

$$\mu = \sum XP(X) \tag{8.2}$$

where $X$ is the value of the event and $P(X)$ is the probability. The summation is over all values of $X$.

Suppose one die is thrown; then by Eq. (8.1), the average is

$$\mu = \frac{1 + 2 + 3 + 4 + 5 + 6}{6} = 3.5$$

For two dice the mean is

$$\mu = \frac{1(2) + 2(3) + 3(4) + \cdots + 3(10) + 2(11) + 1(12)}{36} = 7.0$$

a result which can also be obtained from Eq. (8.2):

$$\mu = 2(0.0278) + 3(0.0556) + \cdots + 11(0.0556) + 12(0.0278) = 7.0$$

For three dice the mean is 10.5.

A very important concept in connection with probability is the *expectation*, which is identical to the average of the occurrences but is used in a very broad sense. The expectation of $X$ is

$$E(X) = \sum X(PX) \tag{8.3}$$

Expected value of $X$ = sum of (value of $X$)(probability of $X$ occurring)

**Example 8.1.** One die is thrown, and the player receives twice the face value if odd and one-half the face value if even. Find the expectation.

*Solution.* By Eq. (8.3),

$$E(X) = 2(1)\left(\tfrac{1}{6}\right) + \tfrac{1}{2}(2)\left(\tfrac{1}{6}\right) + 2(3)\left(\tfrac{1}{6}\right) + \tfrac{1}{2}(4)\left(\tfrac{1}{6}\right) + 2(5)\left(\tfrac{1}{6}\right) + \tfrac{1}{2}(6)\left(\tfrac{1}{6}\right) = 4$$

The second difference between distributions is shown by the *variance*, which is a measure of how much the distribution varies from the average. A small variance means that the distribution is peaked; a large variance, that it is spread out. Two distributions may have the same average but differ drastically in the variance or the way they are spread out. The variance $\sigma^2$ is defined as

$$\sigma^2 = \frac{\sum\limits_{1}^{N} (X_i - \mu)^2}{N - 1} \tag{8.4}$$

and $\sigma$ itself is known as the *standard deviation*.

**Example 8.2.** The life of a tool in hours before breakage is given by the following data obtained for 50 tools. Tabulate the frequency distribution of the tool life and find the average tool life and its variance:

| | | | | |
|------|------|------|------|------|
| 28.1 | 31.6 | 29.9 | 31.4 | 31.6 |
| 29.9 | 33.7 | 30.4 | 33.9 | 31.5 |
| 30.9 | 26.5 | 33.6 | 31.7 | 36.4 |
| 32.6 | 30.9 | 29.3 | 33.7 | 32.8 |
| 27.7 | 30.5 | 30.8 | 34.4 | 33.5 |
| 29.1 | 37.8 | 30.4 | 33.3 | 30.5 |
| 30.5 | 32.0 | 26.5 | 29.6 | 31.4 |
| 29.6 | 29.3 | 29.8 | 34.4 | 34.4 |
| 30.4 | 31.0 | 29.8 | 38.8 | 38.9 |
| 27.7 | 26.7 | 31.6 | 28.3 | 26.5 |

The data are grouped and condensed in Table 8.2. The average value is given by Eq. (8.1).

*Solution*

$$\mu = \frac{4(26.45) + 2(27.45) + \cdots + 1(37.45) + 2(38.45)}{50} = 31.21 \text{ h}$$

and the variance, by Eq. (8.4), is

$$\sigma^2 = \frac{4(26.45 - 31.21)^2 + 2(27.45 - 31.21)^2 + \cdots + 2(38.45 - 31.21)^2}{50 - 1}$$

$$= 8.26$$

**TABLE 8.2**
**Tabulation for Example 8.2**

| Tool life, h | Midpoint X | Number of tools with this life | Probability P(X) of X |
|--------------|------------|-------------------------------|------------------------|
| 26.0–26.9 | 26.45 | 4 | 0.08 |
| 27.0–27.9 | 27.45 | 2 | 0.04 |
| 28.0–28.9 | 28.45 | 2 | 0.04 |
| 29.0–29.9 | 29.45 | 9 | 0.18 |
| 30.0–30.9 | 30.45 | 9 | 0.18 |
| 31.0–31.9 | 31.45 | 8 | 0.16 |
| 32.0–32.9 | 32.45 | 3 | 0.06 |
| 33.0–33.9 | 33.45 | 6 | 0.12 |
| 34.0–34.9 | 34.45 | 3 | 0.06 |
| 35.0–35.9 | 35.45 | 0 | 0.00 |
| 36.0–36.9 | 36.45 | 1 | 0.02 |
| 37.0–37.9 | 37.45 | 1 | 0.02 |
| 38.0–38.9 | 38.45 | 2 | 0.04 |
| | | 50 | |

## 8.3 CUMULATIVE-DISTRIBUTION FUNCTION

A frequency distribution function can be shown as a plot of cumulative values known as a *cumulative-distribution function*. The plots are useful in certain applications. Figure 8.2 is a plot of cumulative probability of the value $X$ or less for two dice with data obtained from Table 8.1.

It is emphasized that any frequency function may be changed to a probability function by changing the area under the function equal to unity. The cumulative-probability distribution function is, accordingly, unity when taken in the limit.

Cumulative functions may be used to find the probability for a random variable $X$ lying between two limits $A$ and $B$. Thus if $A < B$, the probability is

$$P(A < X \le B) = \sum_{i=A+1}^{B} P(X_i) = \sum_{i=0}^{B} P(X_i) - \sum_{i=0}^{A} P(X_i) \tag{8.5}$$

**Example 8.3.** Find the probability of throwing a 4, 5, 6, or 7 with two dice. Here the range from above is 3 to 7 inclusive.

***Solution.*** By Eq. (8.5) and Fig. 8.2,

$$P(3 < X \le 7) = \sum_{0}^{7} P(X) - \sum_{0}^{3} P(X) = 0.583 - 0.083 = 0.500$$

The same result can be obtained, of course, by adding $P(4)$, $P(5)$, $P(6)$, and $P(7)$.

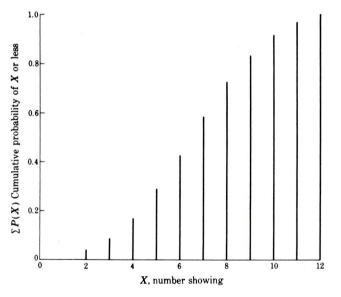

**FIGURE 8.2**
Cumulative probability with two dice.

## 8.4 BINOMIAL DISTRIBUTION

The *binomial distribution* applies to events that can take on only two values, such as the head or tail for a tossed coin, or accept or reject for an object. The relationship is

$$P(X) = \frac{N!}{X!(N-X)!} p^X (1-p)^{N-X} \tag{8.6}$$

where $P(X)$ is the probability of *exactly* $X$ occurrences in $N$ trials and $p$ is the probability for success in one trial. In the case of an unbiased coin $p$ and $1-p$ are both 0.5, but in most problems $p$ will not be 0.5. The distribution is symmetrical if and only if $p = 0.5$.

The mean and variance of a binomial distribution, respectively, are

$$\mu = Np \tag{8.7}$$

$$\sigma^2 = Np(1-p) \tag{8.8}$$

The binomial distribution assumes that the trials are independent. For sampling from a small population without replacement, the binomial distribution should be replaced by the hypergeometric distribution [2], which is

$$P(X) = \frac{\binom{H}{X}\binom{M-H}{W-X}}{\binom{M}{W}} \tag{8.9}$$

where
$$\binom{H}{X} = \frac{H!}{X!(H-X)!}$$

$P(X)$ is the probability of obtaining exactly $X$ objects having a certain property by withdrawing $W$ objects without replacement from a population containing $M$ objects, $H$ of which possess the property.

A generalization of the binomial distribution leads to the multinomial distribution. Instead of having two results $A$ and not $A$ with probabilities $p$ and $1-p$, there are now $k$ results possible, each with probability $P_1, P_2, \ldots, P_k$. The probability of getting exactly $X_1, X_2, \ldots, X_k$ occurrence of $A_1, A_2, \ldots, A_k$ is

$$P(X_1, X_2, \ldots, X_k) = \frac{N!}{X_1!, \ldots, X_k!} P_1^{X_1}, \ldots, P_k^{X_k} \quad \text{with } N = \sum_1^k X_i \tag{8.10}$$

## 8.5 POISSON DISTRIBUTION

The Poisson distribution is of great importance and is

$$P(X) = \frac{e^{-\mu}\mu^X}{X!} \tag{8.11}$$

where $P(X)$ = probability of exactly $X$ occurrences

$e$ = Naperian constant 2.71828...

$\mu$ = expected or average number of occurrences

The distribution is applicable only when the events occur completely at random and the number that occurs is small compared to the potential number that could occur.

The mean and variance of the Poisson function are both $\mu$. Tables are available for the Poisson function and the accumulated Poisson function.

**Example 8.4.** At 3 P.M. telephone calls arrive at the company switchboard at the rate of 120/h. Find the probability that exactly 0, 1, and 2 calls arrive between 3:00 and 3:01.

*Solution.* Take 1 min as the period of time. Then

$$\mu = \frac{120}{60} = 2 \text{ calls per minute expected}$$

By Eq. (8.11),

$$P(0) = \frac{e^{-2}2^0}{0!} = 0.1353$$

$$P(1) = \frac{e^{-2}2^1}{1!} = 0.2707$$

$$P(2) = \frac{e^{-2}2^2}{2!} = 0.2707$$

## CONTINUOUS FREQUENCY DISTRIBUTIONS

## 8.6 CONTINUOUS DISTRIBUTIONS

Continuous distributions arise in practice particularly with measurement as opposed to counting. Although continuous distributions can be reduced to discrete distributions by grouping, much as in Example 8.2, frequently there are advantages in using the continuous property. Essentially this amounts to substituting the integral sign for the summation sign. With this understanding all the principles that apply for discrete distributions are easily transferred to continuous distributions.

One distribution that must be clearly understood is the use of probability density in place of probability. If a function is continuous, the chance for an event to have exactly the value 3.00... is zero. There is an infinity of possible results, and the chance of exactly 3 is $1/\infty = 0$. However, it is possible to consider a probability density in the neighborhood of 3, with units of probability per unit interval, which when multiplied by an interval length gives the probability for occurrence within the interval.

The symbol $P_d(X)$ will be used for the probability density at the point $X$, with the understanding that it must be multiplied by some interval of $X$ to convert it to a probability. Generally $P_d(X)$ depends on $X$ and is a function of $X$:

$$P_d(X) = F(X) \tag{8.12}$$

By analogy with discrete functions, the mean or expected value is

$$\mu = E(X) = \frac{\displaystyle\int XP_d(X)\, dX}{\displaystyle\int P_d(X)\, dX} \tag{8.13}$$

and the variance is

$$\sigma^2 = \frac{\displaystyle\int (X - \mu)^2 P_d(X)\, dX}{\displaystyle\int P_d(X)\, dX} \tag{8.14}$$

All continuous distribution functions used in this chapter are so expressed that the area under the probability-density function is unity and the cumulative probability, as for discrete functions, is unity in the limit. That is,

$$\int_{-\infty}^{+\infty} P_d(X)\, dX = 1 \tag{8.15}$$

## 8.7 CUMULATIVE-DISTRIBUTION FUNCTION

The cumulative-distribution function for a continuous distribution is obtained by integration and is

$$P_c(A) = \int_{-\infty}^{A} P_d(X)\, dX \tag{8.16}$$

where $P_c(A)$ is the cumulative probability of $X$ having the value $A$ or less.

The probability that $X$ will lie between $A$ and $B$, where $A < B$, is

$$P_c(A \leq X \leq B) = \int_{A}^{B} P_d(X)\, dX = \int_{-\infty}^{B} P_d(X)\, dX - \int_{-\infty}^{A} P_d(X)\, dX$$

$$P_c(A \leq X \leq B) = P_c(B) - P_c(A) \tag{8.17}$$

The cumulative-distribution function is the integral of the probability-density function. Hence the derivative of the cumulative-distribution function is the probability-density function at the point.

## 8.8  RECTANGULAR OR UNIFORM DISTRIBUTION

Suppose a distribution is uniform over the range 0 to $a$. Then the probability density is

$$P_d(X) = \begin{cases} \dfrac{1}{a} & 0 \le X \le a \\ 0 & \text{elsewhere} \end{cases} \qquad (8.18)$$

The condition for Eq. (8.15) is met:

$$\int_{-\infty}^{+\infty} P_d(X)\, dX = \int_0^a \frac{1}{a}\, dX = 1$$

The cumulative probability for $X$ or less, by Eq. (8.16), is

$$P_c(X) = \int_0^X \frac{1}{a}\, dX = \frac{X}{a} \qquad 0 \le X \le a \qquad (8.19)$$

A plot for the probability-density and the cumulative-probability functions is given in Fig. 8.3.

**Example 8.5.**  A continuous-distribution function is uniform in the range of 0 to 5. Find the average value and the variance.

*Solution.*  Here $a = 5$, and by Eq. (8.18),

$$P_d(X) = 0.20 \qquad 0 \le X \le 5 \qquad (8.20)$$

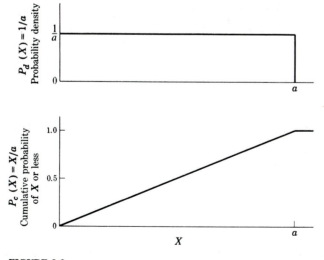

**FIGURE 8.3**
Rectangular or uniform distribution.

Substitution in Eq. (8.13) gives the mean value:

$$\mu = \frac{\int_0^5 X(0.20)\,dX}{\int_0^5 0.20\,dX} = \frac{0.20(\frac{25}{2} - 0)}{0.20(5 - 0)} = 2.5$$

which is obvious. The variance, by Eq. (8.14), is

$$\sigma^2 = \frac{\int_0^5 (X - 2.5)^2 0.20\,dX}{\int_0^5 0.20\,dX} = 0.20\left(\int_0^5 X^2\,dX - 5\int_0^5 X\,dX + 6.25\int_0^5 dX\right)$$

$$\sigma^2 = 0.20[\tfrac{125}{3} - 5(\tfrac{25}{2}) + 31.25] = 2.083$$

## 8.9 EXPONENTIAL DISTRIBUTION

The probability-density function for the distribution is

$$P_d(X) = \begin{cases} ae^{-aX} & X \geq 0 \\ 0 & X < 0 \end{cases} \tag{8.21}$$

The condition for Eq. (8.15) is met:

$$\int_0^\infty ae^{-aX}\,dX = -\frac{a}{a}\left|e^{-aX}\right|_0^\infty = 1$$

The cumulative probability for $X$ or less, by Eq. (8.16), is

$$P_c(X) = \int_0^X ae^{-aX}\,dX = 1 - e^{-aX} \tag{8.22}$$

The mean of the exponential density distribution is $1/a$ and the variance is $1/a^2$. The probability-density function and the cumulative probability of $X$ or less are shown in Fig. 8.4.

There is a connection between the Poisson distribution and the exponential distribution. For example, in queuing problems if the arrival rate, in arrivals per unit time period, follows a Poisson distribution with $\lambda$ average arrivals per period, then by Eq. (8.11),

$$P(X) = \frac{e^{-\lambda}(\lambda)^X}{X!}$$

It can be shown that the time between arrivals has an exponential distribution with the following probability density:

$$P_d(T_a) = \lambda e^{-\lambda T}$$

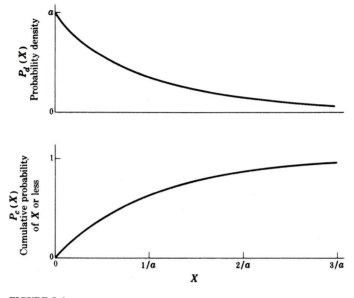

**FIGURE 8.4**
Exponential distribution.

where $T_a$ is the time between arrivals measured in periods $T$. The cumulative probability becomes the time between arrivals of $T_a$ or less and is

$$P_c(T_a) = 1 - e^{-\lambda T}$$

## 8.10 NORMAL DISTRIBUTION IN STANDARD FORM

The normal distribution written in standard form using $Z$ for the independent variable has the following probability-density function:

$$P_d(Z) = \frac{1}{\sqrt{2\pi}} e^{-Z^2/2} \tag{8.23}$$

The area under the curve is unity; the curve is centered on $Z = 0$, is symmetrical, and has a mean equal to zero and a variance equal to unity in the standard form.

The cumulative probability of $Z$ or less is given in Table A3.1. The probability-density and cumulative functions are shown in Fig. 8.5.

The normal distribution was originally derived from theoretical considerations and has been amply verified by experimental data. It is now recognized that its usefulness is derived from the fact that even when the values in a population are not normally distributed if large groups of samples are drawn from the population, the averages of the groups will be normally distributed. For the same reason, if an experimental measurement is subject to a large number of independent causes for variation, the observations will be normally distributed.

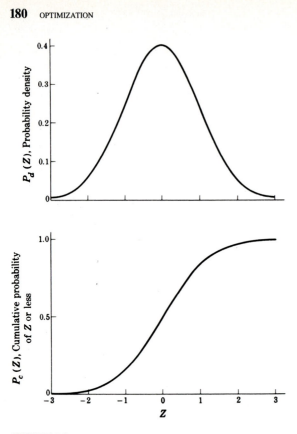

**FIGURE 8.5**
Standard normal distribution.

## 8.11 NORMAL DISTRIBUTION WITH ARBITRARY PARAMETERS

Usually a normal distribution has a mean value that is not equal to zero; nor will the variance be unity. The probability density generally will be

$$P_d(X) = \frac{1}{\sigma\sqrt{2\pi}} \exp - \frac{(X - \mu)^2}{2\sigma^2} \tag{8.24}$$

where $\mu$ is the mean and $\sigma^2$ is the variance.

If a transformation is made with

$$Z = \frac{X - \mu}{\sigma} \tag{8.25}$$

the normal curve is transformed to a standard normal curve with mean value zero and variance equal to unity. The technique, accordingly, is to transform actual conditions to the standard normal distribution using Eq. (8.25), work out the problem in the standard domain, and revert the answer to the actual domain by Eq. (8.25).

**Example 8.6.** Measurements on a production-control test give a mean value of 85 with a variance of 17.4. Find the probability for obtaining a value from 80 to 88. Also, find the value which will be exceeded only 10 percent of the time.

*Solution.* Transformation to the standard domain, by Eq. (8.25), gives

$$
Z = \begin{cases} \dfrac{80 - 85}{\sqrt{17.4}} = -1.20 & \text{for } X = 80 \\[2ex] \dfrac{88 - 85}{\sqrt{17.4}} = +0.72 & \text{for } X = 88 \end{cases}
$$

From Table A3.1,

$$Z = -1.20 \qquad P_c(Z) = 0.1151$$

$$Z = +0.72 \qquad P_c(Z) = 0.7642$$

$$P_c(80 \le X \le 88) = P_c(-1.20 \le Z \le 0.72)$$

By Eq. (8.17),

$$P_c(-1.20 \le Z \le 0.72) = 0.7642 - 0.1151 = 0.6491$$

The probability is 0.649 that $Z$ will lie between $-1.20$ and $+0.72$, which is the same probability that $X$ will lie between 80 and 88.

For the second part of the example, note that if $X$ is exceeded 10 percent of the time, 90 percent of the time the value will be $X$ or less. By Table A3.1,

$$\text{For } P_c(Z) = 0.900: \quad Z = +1.28$$

By Eq. (8.25),

$$1.28 = \frac{X - 85}{\sqrt{17.4}}$$

$$X = 90.34 = \text{value of } X, \text{ which will be exceeded 10 percent of the time}$$

## SIMULATION

## 8.12   MONTE CARLO METHOD

Problems with uncertainty can be quite complicated, and a solution by direct analytical means may become exceedingly difficult if not impossible. The most general and basic method is to resort to simulation. The number of calculations can become large, but the modern computer is a valuable aid. The application known as the *Monte Carlo method* generates the uncertainty within the problem in conformance with the probabilities involved. A large number of calculations are made to establish averages, distribution within the averages, and any other data sought.

First, it is necessary to understand the use of random numbers in the Monte Carlo method. Numbers are generated at random in order to be as free as possible from any pattern or bias. They can be generated by mechanical or electrical means, but now they are generated mathematically as needed in computers and programmable pocket calculators. Table A3.2 gives 1000 random numbers. The numbers can be read horizontally, vertically, diagonally, at random, and by any method which is not biased or prejudiced.

Any distribution, whether empirical or theoretical, discrete or continuous, can be simulated with random numbers. It is necessary to convert the distribution to a cumulative probability of $X$ or less and then make an assignment of random numbers.

**Example 8.7.** Show how the Poisson distribution for a mean value of 2 can be simulated from a table of random numbers.

**Solution.** From Eq. (8.11),

$$P(X) = \frac{e^{-2}2^X}{X!}$$

A tabulation is given in Table 8.3. If three random numbers are chosen with 000 representing 1000, then 135 of these numbers can be assigned to $P(0)$, 271 to $P(1)$, etc., to simulate the frequency of the occurrence of these events. The assignment of numbers is easily and systematically made from the cumulative probabilities, as shown in the last column of the tabulation.

**TABLE 8.3**
**Assignment of random numbers for the Poisson distribution with mean = 2**

| $X$ | $P(X)$ | $\Sigma P(X)$ | Numbers assigned, inclusive |
|---|---|---|---|
| 0 | 0.1353 | 0.135 | 001–135 |
| 1 | 0.2707 | 0.406 | 136–406 |
| 2 | 0.2707 | 0.677 | 407–677 |
| 3 | 0.1804 | 0.857 | 678–857 |
| 4 | 0.0902 | 0.947 | 858–947 |
| 5 | 0.0361 | 0.983 | 948–983 |
| 6 | 0.0120 | 0.995 | 984–995 |
| 7 | 0.0034 | 0.999 | 996–999 |
| 8 | 0.0009 | 1.000 | 000–000 |
| 9 | 0.0002 | | |

The distribution can now be simulated by generating random numbers and converting them to $X$. Thus, using the first three numbers of each group of five taken from Table A3.2, $X$ is generated as in the following tabulation:

| Random no. | X | Random no. | X | Random no. | X |
|------------|---|------------|---|------------|---|
| 141 | 1 | 265 | 1 | 897 | 4 |
| 582 | 2 | 749 | 3 | 592 | 2 |
| 821 | 3 | 086 | 0 | 328 | 1 |
| 481 | 2 | 745 | 3 | 841 | 3 |
| 442 | 2 | 109 | 0 | 665 | 2 |
| 456 | 2 | 566 | 2 | 346 | 1 |
| 724 | 3 | 700 | 3 | 063 | 0 |

The average $X$ is 1.90, which is reasonably close to the true value 2.0 considering the small number of trials.

Example 8.8 can be evaluated without Monte Carlo simulation but illustrates the method.

**Example 8.8.** Assemblies $A$ and $B$ become one product $C$. $A$ and $B$ each contain defects on a statistical basis as follows:

| Defects | No. of times occurring in $A$ | No. of times occurring in $B$ |
|---------|-------------------------------|-------------------------------|
| 0 | 5 | 2 |
| 1 | 5 | 3 |
| 2 | 15 | 5 |
| 3 | 30 | 10 |
| 4 | 20 | 20 |
| 5 | 10 | 40 |
| 6 | 5 | 10 |
| 7 | 5 | 5 |
| 8 | 3 | 3 |
| 9 | 2 | 2 |
|   | 100 | 100 |

The number of defects in $C$ is the product of the defects in $A$ and $B$. Using the Monte Carlo method, find the expected number of defects in $C$.

*Solution.* Table 8.4 shows the assignment of random numbers to simulate the defects in $A$ and $B$. Random numbers are now taken in groups of four, say, the first four numbers in column 1 of Table A3.2. The first two numbers are for the simulation of defects in $A$, the second two numbers for the defects in $B$. The calculation is given in Table 8.5. The last column is the cumulative average defects in $C$ obtained by dividing the cumulative defects by the cumulative number of trials. The average after 20 trials is 13.3, whereas the correct value is 16.7. Agreement will improve with the number of trials.

**TABLE 8.4**
**Assignment of random numbers in Example 8.8**

| | A | | | B | | |
|---|---|---|---|---|---|---|
| Defects | P(X) | ΣP(X) | Random numbers | P(X) | ΣP(X) | Random numbers |
| 0 | 0.05 | 0.05 | 01–05 | 0.02 | 0.02 | 01–02 |
| 1 | 0.05 | 0.10 | 06–10 | 0.03 | 0.05 | 03–05 |
| 2 | 0.15 | 0.25 | 11–25 | 0.05 | 0.10 | 06–10 |
| 3 | 0.30 | 0.55 | 26–55 | 0.10 | 0.20 | 11–20 |
| 4 | 0.20 | 0.75 | 56–75 | 0.20 | 0.40 | 21–40 |
| 5 | 0.10 | 0.85 | 76–85 | 0.40 | 0.80 | 41–80 |
| 6 | 0.05 | 0.90 | 86–90 | 0.10 | 0.90 | 81–90 |
| 7 | 0.05 | 0.95 | 91–95 | 0.05 | 0.95 | 91–95 |
| 8 | 0.03 | 0.98 | 96–98 | 0.03 | 0.98 | 96–98 |
| 9 | 0.02 | 1.00 | 99–00 | 0.02 | 1.00 | 99–00 |

**TABLE 8.5**
**Calculation of average defects in $C$ for Example 8.8**

| (1) Trial no. | (2) Random number | (3) Defects in $A$ | (4) Defects in $B$ | (5) Defects in $C$ $(3) \times (4)$ | (6) Cumulative defects in $C$ | (7) Average defects in $C$ |
|---|---|---|---|---|---|---|
| 1 | 14 15 | 2 | 3 | 6 | 6 | 6.0 |
| 2 | 58 20 | 4 | 3 | 12 | 18 | 9.0 |
| 3 | 82 14 | 5 | 3 | 15 | 33 | 11.0 |
| ... | ...... | . | . | ... | ... | ... |
| 19 | 59 82 | 4 | 6 | 24 | 256 | 13.5 |
| 20 | 18 57 | 2 | 5 | 10 | 266 | 13.3 |

# UNCERTAINTY AND RISK ANALYSIS

## 8.13 NORMAL DISTRIBUTION AND UNCERTAINTY

In practice, a single estimate for a project is desired, along with an analysis of how the actual amount might vary from the expected value. The concept is known by various names, such as *uncertainty analysis* and *risk analysis*. The problem is to combine the uncertainties in the individual components to express uncertainty in the overall project. In this section it is assumed that the uncertainty in each component is normally distributed. The following section shows a method that can be used if normal distributions do not prevail.

If the components are normally distributed, the variances are additive if they are weighted for their sensitivity to the overall effect. Suppose the estimate for a project is Z and it is made up of estimates for A, B, and C with variances of $\sigma_A^2$, $\sigma_B^2$, and $\sigma_C^2$. The expected value of Z is obtained from a calculation of Z using the expected values for A, B, and C. The sensitivity of Z to the individual components is given by $(\partial Z/\partial A)$, $(\partial Z/\partial B)$, and $(\partial Z/\partial C)$. Thus $(\partial Z/\partial A)$ is a measure of $\Delta Z/\Delta A$, which is the change in Z for a unit change in A, all other uncertainties not included. The variance for Z is given by

$$\sigma_Z^2 = \left(\frac{\partial Z}{\partial A}\right)^2 \sigma_A^2 + \left(\frac{\partial Z}{\partial B}\right)^2 \sigma_B^2 + \left(\frac{\partial Z}{\partial C}\right)^2 \sigma_C^2 \tag{8.26}$$

The partial derivatives squared are the weighting factors for the individual variances.

The variance for a component can be obtained from Eq. (8.4). However, if only a few estimates for a component are available, then a high value and a low value, i.e., the range, can be used. The standard deviation, the square root of the variance, can be taken as

$$\sigma = \frac{\text{high value} - \text{low value}}{8} \tag{8.27}$$

**Example 8.9.** The total cost $C_T$ for a project in megadollars is given by

$$C_T = 3X^{1.2}Y^{-0.5} + 10Y + F \tag{8.28}$$

X is estimated from six values: 2.5, 1.6, 1.5, 3.0, 1.4, and 2.0. Y is estimated from a high of 2.5 and a low of 0.5. F is certain, with a value of 0.5.
(a) Find the expected value for $C_T$, its sensitivity to X and Y, and how the actual value for $C_T$ might vary from the expected value. Normal distributions are assumed.
(b) Find the variance for $C_T$.

*Solution*

(a) The mean for X is given by Eq. (8.1):

$$\frac{2.5 + 1.6 + 1.5 + 3.0 + 1.4 + 2.0}{6} = 2.0$$

The variance for X is given by Eq. (8.4):

$$\sigma_X^2 = \frac{(2.5 - 2)^2 + (1.6 - 2)^2 + \cdots + (2.0 - 2)^2}{6 - 1} = 0.404$$

The mean for Y is $(2.5 + 0.50) \div 2 = 1.5$ and the variance, by Eq. (8.27), is

$$\sigma_Y^2 = \left(\frac{2.5 - 0.5}{8}\right)^2 = 0.063$$

The mean value for F is 0.5, and the variance is zero.

The expected value for $C_T$ is obtained from Eq. (8.28) using the mean, or expected value, of the components:

$$C_T = 3(2)^{1.2}(1.5)^{-0.5} + 10(1.5) + 0.5 = 21.127$$

The sensitivity of $C_T$ to $X$ and $Y$ is obtained from Eq. (8.28):

$$\frac{\partial C_T}{\partial X} = 3(1.2)X^{0.2}Y^{-0.5} = 3.6(2)^{0.2}(1.5)^{-0.5} = 3.376$$

$$\frac{\partial C_T}{\partial Y} = 3(X)^{1.2}(-0.5)Y^{-1.5} + 10 = -1.5(2)^{1.2}(1.5)^{-1.5} + 10 = 8.124$$

This is equivalent to saying that $C_T$ changes 3.376 times as fast as $X$ for a unit change in $X$ and 8.124 times as fast as $Y$ for a unit change in $Y$. In other words, a 1 percent change in $X$ ($\Delta X = 0.02$) would cause a change in $C_T$ amounting to $3.376 \times 0.02 = 0.06752$, which on a percentage basis is $(0.06752)(100) \div 21.127$, or 0.320 percent. Similarly, for $Y$, a 1 percent change in $Y$ (at $Y = 1.5$) gives a 0.577 percent change in $C_T$:

$$\frac{(\partial C_T/\partial Y)Y}{C_T} = \frac{(8.124)(1.5)}{21.127} = 0.577 \text{ percent}$$

The total cost $C_T$ is more sensitive to the estimate for $Y$ than $X$, since a 1 percent error in $X$ causes a 0.320 percent error in $C_T$, but a 1 percent error in $Y$ causes a 0.577 percent error in $C_T$. That completes the sensitivity analysis.

(b) Next, find the variance for $C_T$, which by Eq. (8.26), is

$$\sigma^2 = (3.376)^2(0.404) + (8.124)^2(0.063) = 8.763$$

$$\sigma = 2.96$$

At this point it is known that the expected value for $C_T$ is 21.127 and the standard deviation is 2.96. Transform this to a standard normal distribution by Eq. (8.25), which here becomes

$$Z = \frac{C_T - 21.127}{2.96}$$

Assume values for $Z$, and from Table A3.1 the following tabulation results:

| Z | Cumulative probability for Z or less | $C_T$ |
|---|---|---|
| −1.65 | 0.05 | 16.2 |
| −1.28 | 0.10 | 17.3 |
| −0.84 | 0.20 | 18.6 |
| −0.52 | 0.30 | 19.6 |
| −0.25 | 0.40 | 20.4 |
| 0 | 0.50 | 21.1 |
| +0.25 | 0.60 | 21.9 |
| +0.52 | 0.70 | 22.7 |
| +0.84 | 0.80 | 23.6 |
| +1.28 | 0.90 | 24.9 |
| +1.65 | 0.95 | 26.0 |

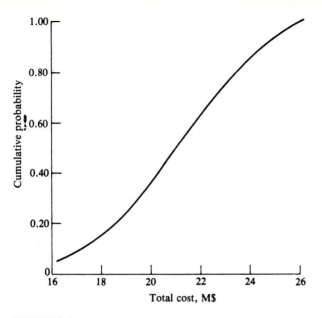

**FIGURE 8.6**
Plot for Example 8.9.

A plot of the last two columns gives the cumulative probability of $C_T$ or less and is shown in Fig. 8.6. The plot shows that the probability that the project will cost 20 M\$ or less is 0.35 or 35 percent. The probability of a cost of 24 M\$ or less is 83 percent. The probability of a cost between 20 M\$ and 24 M\$ is $83 - 35 = 48$ percent.

## 8.14 SKEWED DISTRIBUTIONS AND UNCERTAINTY

In practice, estimates and uncertainties will not follow a normal distribution or even be symmetrical about the mean; a cost for a venture has a greater chance of being 50 percent higher than the estimated cost than it does of being 50 percent lower. In the previous section normal distributions were assumed for the advantage of using relatively simple mathematical analysis. If the distributions are not normal, the general approach is to employ simulation using the Monte Carlo technique and a computer. Skewed distributions can be represented by mathematical formulations such as the *beta-density function* and the *beta distribution,* but the approach outlined here requires no mathematical representation of the distributions. Indeed, all kinds of distributions can be intermixed in the same problem.

As an illustration, let a net present value, NPV, be represented by the following relationship:

$$\text{NPV} = A + \frac{B}{(1 + r)} + \frac{C}{(1 + r)^2} + \frac{D}{(1 + r)^3} \tag{8.29}$$

Suppose that $B$, an expected cash flow at the end of year 1, instead of following a normal distribution, as in Fig. 8.5, has a peak at 750 k$ but is unsymmetrical or skewed, with a greater chance for values below 750 k$ than above. The graph of Fig. 8.7 is accepted as fitting the situation. It is peaked at 750 k$ but there is a greater chance for smaller values than for larger values. The ordinate scale here is arbitrary and need not be indicated. The range of $B$ is 400 k$ to 900 k$.

First, divide the area into several smaller areas which need not be equally spaced. By some convenient means, which will usually be graphical, estimate the area, width, and center height of each element such that the height multiplied by the base will be the area. The dimensions for the area are arbitrary. The procedure leads to Table 8.6, with assignment of random numbers following the Monte Carlo method of Sec. 8.12. Thus elementary area III is represented by 510 k$, with a probability of 0.0868 and random numbers 048–134. Similar tabulations are made for other components. The expected value of $B$ is obtained from Eq. (8.3) and is

$$E(X) = \sum X(PX) = 420(0.0128) + 460(0.0340)$$

$$+ 510(0.0868) + \cdots + 870(0.0791) = 682$$

The expected value for $B$ is 682, not 750 because the distribution is not symmetrical about 750.

The expected value for the NPV and its distribution is now computed by the Monte Carlo method applied to Eq. (8.29). Three random numbers are generated which by Table 8.6 give a value for $B$; thus if 264 came up, then $B$ would have the value 630. Single values for $A$, $C$, and $D$ would be generated and Eq. (8.29) would give a value for the NPV. The procedure would be repeated thousands of times and in the end the printout would give the average NPV, which would be the expected NPV.

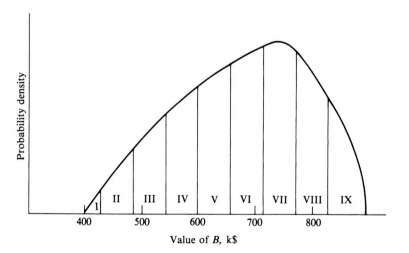

**FIGURE 8.7**
Distribution of the value of $B$.

**TABLE 8.6**
**Calculations from Fig. 8.7**

| Element | Area | Width | Height | Abscissa where height occurs | Fraction of total area | Cumulative fraction of total area | Random numbers |
|---------|------|-------|--------|------------------------------|------------------------|-----------------------------------|----------------|
| I | 3.0 | 2 | 1.5 | 420 | 0.0128 | 0.013 | 001–013 |
| II | 8.0 | 2 | 4 | 460 | 0.0340 | 0.047 | 014–047 |
| III | 20.4 | 3 | 6.8 | 510 | 0.0868 | 0.134 | 048–134 |
| IV | 29.4 | 3 | 9.8 | 570 | 0.1251 | 0.259 | 135–259 |
| V | 36.0 | 3 | 12.0 | 630 | 0.1532 | 0.412 | 260–412 |
| VI | 41.1 | 3 | 13.7 | 690 | 0.1749 | 0.587 | 413–587 |
| VII | 44.1 | 3 | 14.7 | 750 | 0.1877 | 0.775 | 588–775 |
| VIII | 34.4 | 3 | 11.5 | 810 | 0.1464 | 0.921 | 776–921 |
| IX | 18.6 | 3 | 6.2 | 870 | 0.0791 | 1.000 | 922–000 |
| | 235.0 | | | | 1.0000 | | |

It would also print out how the calculated NPV values occurred in various ranges. For example, it might show that the expected NPV is 15 k$ and that 10 percent of the values were in the range $-40$ to $-20$, 25 percent of the values were in the range $-20$ to $-0$, etc.

A cumulative probability curve such as Fig. 8.6 could then be plotted and would be just as useful. The curve would not be centered on the ordinate value 0.5 since the distributions are skewed and nonsymmetrical.

## 8.15 LEAST SQUARES

Frequently the problem arises of fitting data or observations to an equation. The only application considered here is the fitting of data to a straight line. The problem is related to probability and the normal distribution applied to the propagation of errors. Not all the data can be made to fall on a straight line. The amount by which each datum fails to fall on a straight line is referred to as the *error*. According to the theory of probability, the best straight line is that for which the variance of the data from the straight line is a minimum. In other words, the sum of the squares of the errors about the line is a minimum; hence the name *least squares* is used.

Let $X_i$ be a value of the independent variable for which $Y_i$ is a known observation. The equation of a straight line is

$$Y = sX + b \tag{8.30}$$

The error for $Y_i$, its vertical distance from the straight line, is

$$\text{Error} = Y_i - Y = Y_i - (sX_i + b)$$

For the sum of the squares of the error to be a minimum it is necessary to minimize the expression

$$\sum \text{error}^2 = \sum (Y - sX - b)^2$$

where the $i$ designation has been dropped, since it is understood that the summation is for all the observed points. For a minimum (see Chap. 10)

$$\frac{\partial \Sigma}{\partial s} = 2 \sum (Y - sX - b)(-X) = 0$$

$$\frac{\partial \Sigma}{\partial b} = 2 \sum (Y - sX - b)(-1) = 0$$

The two relationships are

$$\sum XY = s \sum X^2 + b \sum X$$

$$\sum Y = s \sum X + \sum b = s \sum X + Nb$$

where $N$ is the number of observations.

The last two equations are known as the *normal equations*. Solved together, they give

$$s = \frac{N \sum XY - \sum X \sum Y}{N \sum X^2 - \left(\sum X\right)^2} \tag{8.31}$$

$$b = \frac{\sum X^2 \sum Y - \sum X \sum XY}{N \sum X^2 - \left(\sum X\right)^2} \tag{8.32}$$

The last two equations give the value of $s$ and $b$ for the best straight line. The calculations are easily made by tabulation.

An alternative calculation for $b$ is based on summing Eq. (8.30) for the $N$ pairs of $X$, $Y$ data, giving

$$b = \frac{\sum Y}{N} - s \frac{\sum X}{N} \tag{8.33}$$

If the relationship is in the form

$$Y' = K(X')^s$$

there results, by taking logarithms,

$$\log Y' = s \log X' + \log K$$

By substituting $Y$ for $\log Y'$, $X$ for $\log X'$, and $b$ for $\log K$, the equation reduces to a straight line and can be treated by the method of this section.

A calculation by the method of least squares is given in the next chapter in Example 9.11.

# NOMENCLATURE

| | |
|---|---|
| $a$ | Constant |
| $A$ | One of $k$ kinds of events in the multinomial distribution |
| $b$ | Constant in the equation for a straight line |
| $e$ | Naperian constant 2.71828 . . . |
| $E(X)$ | Expected value of $X$ |
| $H$ | Number of objects in population with desired property; used in hypergeometric distribution |
| $k$ | Number of kinds of events in the multinomial distribution |
| $K$ | Constant |
| $M$ | Number of objects in the population |
| $N$ | Number of trials or data |
| NPV | Net present value |
| $p$ | Probability for success in a single trial in the binomial distribution |
| $P(X)$ | Probability for event with value $X$ |
| $P_c(X)$ | Cumulative probability of $X$ or less |
| $P_d(X)$ | Probability density at $X$, probability per unit interval |
| $s$ | Slope of a straight line |
| $T$ | Time in periods |
| $T_a$ | Time between arrivals, periods |
| $W$ | Number of objects withdrawn from population in a hypergeometric distribution |
| $X$ | Value of an event |
| $Z$ | Independent variable in the standard normal distribution measured from the mean value |
| $\lambda$ | Average arrivals per period |
| $\mu$ | Mean or average value |
| $\sigma$ | Standard deviation |
| $\sigma^2$ | Variance |

# PROBLEMS

**8.1.** Estimates for a project in megadollars are: $A$: 1.62, $B$: 2.84, $C$: 1.50, $D$: 3.00, $E$: 2.05, $F$: 2.00, $G$: 2.34, $H$: 1.67, $I$: 2.18, $J$: 2.67. Find the mean value, the expected value, and the variance.

**8.2.** Repeat Prob. 8.1 weighting the estimates as follows: $A$: 1.5, $B$: 1, $C$: 1, $D$: 2.0, $E$: 0.5, $F$: 0.7, $G$: 2.0, $H$: 1.5, $I$: 0.6, $J$: 1.5.

**8.3.** The hours required to perform a certain task by different workers are as follows, with time in hours and minutes:

| | | | | | |
|---|---|---|---|---|---|
| 1:56 | 1:22 | 2:16 | 1:54 | 2:36 | 2:04 |
| 2:22 | 2:27 | 0:50 | 1:51 | 2:50 | 1:11 |
| 1:42 | 2:13 | 1:20 | 2:31 | 1:47 | 2:09 |
| 2:07 | 1:37 | 2:22 | 1:44 | 2:06 | 2:06 |
| 1:50 | 1:54 | 2:02 | 1:04 | 2:58 | 1:42 |
| 2:21 | 2:18 | 2:43 | 2:16 | 1:27 | 1:32 |

Take 15-min intervals, such as 0:45 to 1:00, 1:01 to 1:15, etc.

(a) Make a histogram of the data, i.e., a plot of numbers in the group against work-hours.

(b) Using the midtime for each group, find the mean and the variance.

**8.4.** For the grouping from Prob. 8.3 make a plot of the cumulative probability of the distribution. If an arbitrary worker is chosen, what is the probability that

(a) The time will be less than 1:30 h?

(b) The time will be more than 2 h but less than 2:30 h?

**8.5.** The probability of making a successful bid is 10 percent for a construction company. If the company makes 10 bids, find the probability that

(a) Two will be successful.

(b) None will be successful.

(c) No more than two will be successful.

(d) At least two will be successful.

Use a binomial distribution.

**8.6.** A maintenance crew consists of nine people, of whom six are workers and three are key-workers. On a certain day seven people report for work. Find the probability that of the two absentees

(a) Both are key-workers.

(b) One is a key-worker.

(c) Neither is a key-worker.

Use a hypergeometric distribution.

**8.7.** A drafting department has 30 persons, 5 of whom are inferior. A job requires three persons chosen at random. Find the probability that one or more of the persons will be inferior. Use a hypergeometric distribution.

**8.8.** Show that the mean value of the Poisson distribution, Eq. (8.11), is $\mu$.

**8.9.** A maintenance department receives a call for help on the average of 14 times in an 8-h shift. Find the probability that in a 1.5-h period it will receive

(a) no calls   (b) two calls   (c) at least two calls

Use a Poisson distribution.

**8.10.** Given

$$F(X) = \begin{cases} 0.25e^{-x/4} & x \geq 0 \\ 0 & x < 0 \end{cases}$$

(a) Find the mean and variance of the function.

(b) Find the cumulative distribution function.

**8.11.** Use tables available in handbooks and textbooks on probability and statistics which give the probability, density, and ordinate for the standard normal distribution. Given a normal distribution with a mean value of 1.5 and a variance of 4, tabulate and draw the normal density function.

**8.12.** A construction company finds that actual costs against its estimates are normally distributed and that for $1 of estimate the variance is 0.04. If it estimates $300,000 on a project, what is the probability that it will cost more than $350,000?

**8.13.** Actual costs are expected to be normally distributed from an estimate. If the estimate is $100,000 and there is a 20 percent probability that the actual cost is not greater than $115,000, find the probability that the actual cost will not be greater than $125,000.

**8.14.** A company estimates $100,000 with a standard deviation of $22,000. How much should the company bid in order to be

(a) 95 percent sure that the cost will not exceed the bid?

(b) 99 percent sure?

**8.15.** The time needed to complete a project is expected to be 12 months normally distributed with a variance of 2 months$^2$. Find the probability that the project will take no longer than 10.5 months.

**8.16.** A company pays a \$10,000 penalty for each day a project exceeds 40 days to complete, up to a limit of 5 days' penalty. The company expects to complete the project in 30 days, normally distributed, with a variance of 100 days$^2$. Find the expectation of the penalty for not completing the project in 40 days.

**8.17.** A company is bidding on repairing machines for a fixed cost per month. It costs \$125,000 to repair a machine, and the number of repair jobs required will follow a Poisson distribution, with 0.7 machine per month expected. Find the expected cost per month.

**8.18.** Using the data of Prob. 8.3 and 15-min intervals, assign random numbers from 01 to 00 for the probabilities of the time required.

**8.19.** Process A follows a normal distribution, with a mean time of 1 h and a variance of 0.25 h$^2$. Plot the cumulative probability of X hours or less against X. Show how to assign two-digit random numbers to processing time X.

**8.20.** Process B follows an exponential distribution, as in Sec. 8.9, with a mean time of 2 h. Plot the cumulative probability of X hours or less against X. Show how to assign two-digit random numbers to processing time X.

**8.21.** A product can be manufactured with three sequential independent processes, A, B, and C. Process C has a constant time of 4 h. Processes A and B are described in Probs. 8.19 and 8.20. Using the Monte Carlo method, find the average time required to produce five units.

**8.22.** At a certain bus stop at the same time each morning a bus arrives with an average of two empty seats, following a Poisson distribution. The number of passengers at the bus stop averages two, following a Poisson distribution. Using the Monte Carlo method, find the percentage of times that the bus leaves passengers stranded, assuming no passengers are allowed to stand in the bus.

**8.23.** Solve Prob. 8.21 by using expectancies instead of a Monte Carlo solution.

**8.24.** The cost $C_T$ of a project in megadollars depends on two inputs, X and Y, according to the relationship

$$C_T = 7XY^{1.2} + X^2 + 1.4$$

Estimates for X and Y are normally distributed, with X having a mean value of 1 and a variance of 0.5, and Y having a mean value of 2 and a variance of 1. Find the expected cost and the probability that it will lie between the expected cost and 20 percent over the expected cost. Show a cumulative distribution curve for the cost.

**8.25.** The cash flows for a project in kilodollars are

| End year | Cash flow |
|----------|-----------|
| 0 | −1000 A |
| 1 | +750 B |
| 2 | +390 C |
| 3 | +180 D |

The cost at zero time is certain since it is a turnkey cost. The cash flow at the end of year 1 is obtained from six estimates: 720, 800, 780, 820, 700, and 680. Cash flow at the end of year 2 is estimated twice as 500 and 280. Cash flow at the end of year 3 is obtained from two estimates: 250 and 110.

(a) Find the expected NPV for the project at an 18 percent rate of return.

(b) How sensitive is the NPV to estimates for $B$, $C$, and $D$?

(c) Find the probability that the NPV will be between $+30$ k\$ and $+50$ k\$.

**8.26.** The cost in megadollars for a project is composed of two costs, $A$ and $B$, added together. The costs and probabilities are as follows:

| A | | B | |
|---|---|---|---|
| Cost | Probability | Cost | Probability |
| 1.0 | 0.05 | 2.1 | 0.05 |
| 1.1 | 0.10 | 2.2 | 0.15 |
| 1.2 | 0.20 | 2.3 | 0.25 |
| 1.3 | 0.40 | 2.4 | 0.35 |
| 1.4 | 0.25 | 2.5 | 0.20 |

Using a Monte Carlo technique, find the expected cost for the project. How would you obtain a distribution of the project cost?

**8.27.** An electric power company reports to its management for each 30-day period and combines all data by intervals of one 8-h shift. On the average it goes down one 8-h shift in a 24-h day every 90 days because of failures. Find the distribution for failures of 0 through 6 during a 90-day period if the distribution of failures follows (a) a Poisson distribution, (b) a binomial distribution.

**8.28.** The following table shows the observed values of two variables, $X$ and $Y$:

| $X$ | 6 | 5 | 8 | 8 | 7 | 6 | 10 | 4 | 9 | 7 |
|---|---|---|---|---|---|---|---|---|---|---|
| $Y$ | 8 | 7 | 7 | 10 | 5 | 8 | 10 | 6 | 8 | 6 |

(a) Construct a scatter diagram.

(b) Find a straight-line relationship between $Y$ and $X$ using the method of least squares.

**8.29.** Solve Prob. 8.26 by using expectancies instead of a Monte Carlo solution.

# REFERENCES

1. Blum, J. R., and J. I. Rosenblatt: *Probability and Statistics,* W. B. Saunders Company, Philadelphia, 1972.
2. Brownlee, K. A.: *Statistical Theory and Methodology in Science and Engineering,* 2d ed., John Wiley & Sons, Inc., New York, 1965.
3. Campbell, D. W.: "Risk Analysis," *AACE Bulletin* (now *Cost Engineering*), August–October 1971, pp. 8–11.
4. Chalmers, B. J.: *Understanding Statistics,* Marcel Dekker, Inc., New York, 1987.
5. Chatfield, C.: *Statistics for Technology,* 3d ed., Methuen, Inc., New York, 1983.
6. Cheremisinoff, N. P.: *Practical Statistics for Engineers and Scientists,* Technomic Publishing, Lancaster, PA, 1987.

7. Gray, H. L., and P. L. Odell: *Probability for Practicing Engineers,* Barnes & Noble, Inc., New York, 1970.
8. Hogg, R. V., and E. A. Tanis: *Probability and Statistical Inference,* The Macmillan Company, New York, 1977.
9. Holland, F. A., F. A. Watson, and J. K. Wilkinson, "Sensitivity Analysis of Project Profitabilities," "Statistical Techniques Improve Decision-Making," "Probability Techniques for Estimates of Profitability," "Estimating Profitability When Uncertainties Exist," and "Numerical Measures of Risk," in H. Popper (ed.), *Modern Cost Engineering,* McGraw-Hill Book Company, New York, 1979, pp. 29, 41, 47, 53, 61.
10. Koehn, E.: "Estimating with Probabilistic Unit Costs and Quantities," *Cost Engineering,* March–April 1978, pp. 63–66.
11. Ostle, B., and R. W. Mensing: *Statistics in Research,* 3d ed., The Iowa State University Press, Ames, 1975.
12. Quirin, W. L.: *Probability and Statistics,* Harper & Row, Publishers, Incorporated, New York, 1978.
13. Wadsworth, G. P., and J. G. Bryan: *Applications of Probability and Random Variables,* 2d ed., McGraw-Hill Book Company, New York, 1974.
14. Woodroofe, M.: *Probability with Applications,* McGraw-Hill Book Company, New York, 1978.

# CHAPTER 9

# PRODUCTIVITY

*H. J. Behrens, O. J. Vogl,*
*F. C. Jelen, K. K. Humphreys,*
*J. M. Neil, and*
*B. H. Brown, Jr.*

## DISPLACEMENT

## 9.1 DISPLACEMENT VERSUS REPLACEMENT

*Replacement* will be used to refer to a situation in which an asset is worn out or cannot physically perform its intended use and must be exchanged for another article soon, if not now. *Displacement* will be used to refer to a situation in which an asset can perform its function, at least to some degree, but can be exchanged for another asset for an overall economic gain. One distinguishing difference is that a decision regarding displacement can be ignored or delayed, perhaps for years.

A pure replacement problem, i.e., one based solely on the inability of an existing machine to function, can be analyzed by the methods of Chaps. 2 and 3 for periodic interest and Chap. 4 for continuous interest. The analyst should always regard replacement problems with a broad perspective, for the time when one asset must be replaced may introduce an opportunity for the displacement of other assets.

The possibility of displacement arises from either loss of efficiency or technological advancement. Loss of efficiency has several causes, such as increased maintenance expense. Technological advancement creates new machines that produce a superior product at less cost.

**196**

The displacement problem is a key factor in industrial leadership since it is an important part of the productivity problem. Companies which merely confront replacement problems as they occur generally lag behind. Those having an active displacement policy move forward.

## 9.2   ONE YEAR MORE OF THE EXISTENT

Let an existing machine be called the *existent* and the new machine the *displacer*. A decision to purchase the displacer depends on a comparison of the cost of keeping the existent in operation against that of having the displacer in operation. The comparison can be made on the basis of unacost or capitalized cost, and as for all cost comparisons, only differences need be considered. Costs that are equal in all respects for the two machines can be omitted because they have no effect on the decision.

In general operating costs and other disadvantages of the existent become progressively greater with time, and only 1 year ahead for the existent need be considered because it will make its best showing in the coming year. Exceptions to this rule will be discussed later.

The cost of keeping the existent in service 1 year more is readily calculated. Suppose the existing machine can be sold now for $C_{sal,0}$, the present realizable salvage value, and next year for $C_{sal,1}$, the realizable salvage value at the end of 1 year. The present realizable salvage value may not be the same as the book value and indeed may be considerably less. The difference between the book value and the salvage value is known as the *sunk cost*. The sunk cost has no influence on the decision except for the tax situation. A sunk cost is revealed but not created by displacement. It represents an inadequate depreciation expense in previous years and cannot be corrected by any policy of future years. The decision insofar as the existent is concerned for 1 year more is to sell the machine now for its realizable salvage value $C_{sal,0}$ or to keep the machine and incur an operating cost $C_{e1}$ as of the end of the year and recover the salvage value at that time, $C_{sal,1}$. These amounts must, of course, be used with the tax situation included.

For the tax situation let $B_0$ be the book value now and $B_1$ that 1 year from now. Suppose the machine is sold now and the tax credit is available immediately. The loss on disposal is $B_0 - C_{sal,0}$ and the savings in taxes is $t(B_0 - C_{sal,0})$ where $t$ is the decimal tax rate. If the machine is not sold now, the receipts that are avoided must be regarded as an expense, which on an after-tax basis is

$$C_{sal,0} + t(B_0 - C_{sal,0}) \tag{9.1}$$

At the end of the year depreciation amounting to $D_1$ can be taken, $C_{sal,1}$ is received as salvage value, and the loss on disposal is $B_1 - C_{sal,1}$. Altogether the after-tax receipt is

$$tD_1 + C_{sal,1} + t(B_1 - C_{sal,1})$$

Noting that $B_1 = B_0 - D_1$, the last relationship reduces to

$$C_{sal,1} + t(B_0 - C_{sal,1}) \tag{9.2}$$

Let $C_{el}$ be the operating cost for the coming year as of the end of the year, subject to a tax credit as of the end of the year. Then the after-tax cost is

$$C_{el}(1 - t) \tag{9.3}$$

Combine Eq. (9.1), referring to the beginning of the year, with Eqs. (9.2) and (9.3), referring to the end of the year. The cost diagram and the present value, basis 1 year, are

$$P_1 = C_{sal,0} + t(B_0 - C_{sal,0}) - \frac{C_{sal,1} + t(B_0 - C_{sal,1}) - C_{el}(1 - t)}{1 + r} \tag{9.4}$$

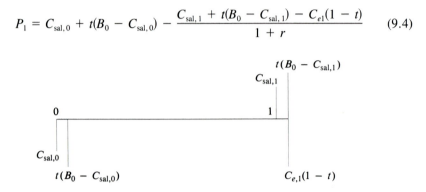

This can be converted to a unacost by Eq. (2.20), with $F_{PR,r,1} = 1 + r$:

$$R = P_1(1 + r)$$

and to a capitalized cost by Eq. (2.32), with $F_{PK,r,1} = (1 + r)/r$:

$$K = P_1\frac{1 + r}{r}$$

Unacost or capitalized cost for the displacer can be obtained in the conventional manner, as by Table 3.5, and a comparison with the existent will show which is more economical.

**Example 9.1.** Four years ago a machine cost $23,000. It has been depreciated on a 10-year life with no salvage value using straight-line depreciation. If sold now, it will bring $2000 with immediate tax credit for any loss on disposal. If sold at the end of the year, it will bring $1500 with immediate tax credit for any loss on disposal. Operating costs for next year are $3800 as of the end of the year with immediate tax credit.

A new machine will cost $50,000 with a 12-year life. Depreciate for a 7 year recovery period for tax purposes using MACRS depreciation. The operating cost will be $3000 as of the end of each year with $6000 per year savings due to better quality control, both subject to an immediate tax credit.

If money is worth 10 percent per year after a 34 percent tax rate, should the displacer be purchased?

**Solution.** The book value now of the existent is

$$B_0 = \tfrac{6}{10}(23,000)$$

$$= \$13,800$$

The loss on disposal at zero time is

$$B_0 - C_{sal,0} = 13,800 - 2000 = \$11,800$$

and the tax credit is

$$t(11,800) = 0.34(11,800) = \$4012$$

The depreciation plus loss on disposal at the end of the year is

$$B_0 - C_{sal,1} = 13,800 - 1500 = \$12,300$$

and the tax credit is

$$t(12,300) = 0.34(12,300) = \$4182$$

The after-tax cost of the end-of-year expense is

$$C_{el}(1 - t) = 3800(1 - 0.34) = \$2508$$

The cost diagram is

The present value at 10 percent per year is

$$P_1 = 2000 + 4012 - \frac{1500 + 4182 - 2508}{1.10} = \$3127$$

and unacost for 1 year more of the existent is

$$R = 3127(1.10) = \$3440$$

Unacost for the displacer, by Tables 2.1, 3.3, and 3.6, is

$$C_d\left(1 - t\sum \frac{D_x}{(1 + 0.10)^x}\right)F_{PR,10\%,12} = 50,000\left[1 - 0.34\left(\frac{0.1429}{1.1} + \frac{0.2429}{(1.1)^2} + \frac{0.1749}{(1.1)^3}\right.\right.$$

$$\left.\left. + \frac{0.1249}{(1.1)^4} + \frac{0.0893}{(1.1)^5} + \frac{0.0892}{(1.1)^6} + \frac{0.0893}{(1.1)^7} + \frac{0.0446}{(1.1)^8}\right)\right](0.14676) = 5542$$

$$R(1 - t) = (3000 - 6000)(1 - 0.34) = \underline{-1980}$$

$$R \text{ displacer} = \$3562$$

It will be more economical to retain the existent. The unacost is \$3440 against \$3562 on an after-tax basis.

If the item being displaced is part of a group for depreciation purposes, the tax situation changes. The depreciation will be taken on the group, whether or not the item is displaced. For example, a loss on disposal cannot be taken as an immediate tax advantage.

**Example 9.2.** Repeat Example 9.1, considering the machine as part of a group for depreciation purposes.

*Solution.* If sold now, the machine would bring $2000 subject to tax. The book value is not a factor since that will remain on the books for future depreciation. The cash flow lost now by retaining the machine is

$$2000(1 - t) = 2000(1 - 0.34) = \$1320$$

At the end of the year it would bring $1500 subject to tax. Any cash flow because of depreciation during the year is the same whether or not the machine is sold. At the end of the year there will also be a $3800 operating cost subject to tax. The end-of-year cost after taxes is

$$(-1500 + 3800)(1 - 0.34) = \$1518$$

The present value after taxes at 10 percent per year is

$$P_1 = 1320 + \frac{1518}{1.10} = \$2700$$

and unacost for one more year of the existent is

$$R = 2700(1.10) = \$2970$$

As before, the unacost after tax for the displacer is $3562.

Unacost for the existent is less than in the previous example since loss of an immediate tax benefit from the depreciation is not a factor.

# 9.3  MORE THAN 1 YEAR OF THE EXISTENT

It was assumed in the preceding section that 1 year would be the most favorable period for the existent. The treatment of the existent for more than 1 year ahead will now be analyzed, followed by some examples.

Refer to Example 9.1. In addition, let the sales value at the end of the second year be $900 and the operating cost for the second year as of the end of the year $C_{e2}$ be $1400. The tax timing is the same as before. Now, find the unacost for the existent based on retention for 2 years.

At zero time the expense incurred by retaining the machine is the same as before, and on an after-tax basis is

$$C_{\text{sal}, 0} + t(B_0 - C_{\text{sal}, 0}) = 2000 + 4012 = \$6012$$

The end-of-first-year expense is $3800, as before. Depreciation for the year is

$$\tfrac{1}{10}(23{,}000) = \$2300$$

The expense for the end of the first year after taxes is

$$3800(1 - t) - 2300t = 2508 - 782 = \$1726$$

The end-of-second-year expense is $1400. Depreciation for the second year is also $2300. The book value at the end of the second year is

$$B_2 = B_0 - D_1 - D_2 = 13{,}800 - 2300 - 2300 = \$9200$$

where $D_1$ and $D_2$ are first- and second-year depreciation. The loss on disposal at the end of the second year is

$$B_2 - C_{sal,2} = 9200 - 900 = \$8300$$

The expense for the end of the second year after a 0.34 decimal tax rate is

$$1400(1 - t) - 2000t - 900 - 8300t = -\$3478$$

The cost diagram on an after-tax basis, by symbols and numbers, is:

The present value is

$$P_2 = 6012 + \frac{1726}{1.10} - \frac{3478}{1.10^2} = \$4707$$

and the unacost is

$$R = P_2 F_{PR, 10\%, 2} = 4707(0.57619) = \$2712$$

This is less than the unacost based on 1 year more, $3440, because there is a sharp drop in operating expenses for the second year. Normally, with operating costs advancing, the unacost for 1 year more will be the minimum.

The example demonstrates how more than 1 year is treated with taxes included. The tax timing must be considered whether 1 or more years ahead is investigated, but the analyst should be able to compute the unacost or capitalized cost for any tax timing and duration of the existent.

Cases in which multiyear service for the existent are more favorable than that for just 1 more year of service are usually evident from inspection. In practice, they arise principally from two causes: (1) net realizable salvage value may decline sharply for the first year, and (2) benefits of an overhaul may extend more than 1 year. Examples of each type follow (on a no-tax basis to simplify the presentation).

**Example 9.3.** An improperly designed pump has an excessive power cost. The salvage value now is $1000, declining to $250 next year and to $100 the following year. The operating cost for 1 year more is $700 and $800 for the following year as of the end of each year. A displacer costing $3800 will last 5 years with negligible salvage value. The operating expense will be uniform at $400/year as of the end of the year. If money is worth 8 percent per year, should replacement be made now?

*Solution.* Capitalized cost of the existent, basis 1 year more of service, by Table 2.3, is

$$\left(1000 - \frac{250}{1.08}\right) F_{PK,8\%,1} = 768.52(13.500) = 10,375$$

$$700\frac{1}{1.08} F_{PK,8\%,1} = \frac{700}{1.08} 13.50 = \underline{8,750}$$

$$K_1 = 19,125$$

Capitalized cost of the existent, basis 2 years of service, by Table 2.3, is

$$\left(1000 - \frac{100}{(1.08)^2}\right) F_{PK,8\%,2} = 914.26(7.0096) = 6409$$

$$700\frac{1}{1.08} F_{PK,8\%,2} = \frac{700}{1.08} 7.0096 = 4543$$

$$800\frac{1}{1.08^2} F_{PK,8\%,2} = \frac{800}{1.1664} 7.0096 = \underline{4808}$$

$$K_2 = 15,760$$

Capitalized cost of the displacer, by Table 2.3,

$$(3800 - 0) F_{PK,8\%,5} = 3800(3.1307) = 11,897$$

$$400\frac{1}{0.08} = \underline{5,000}$$

Displacer $K = 16,897$

Based on only 1 more year of service, the existent is more expensive than the displacer: $K_1 = \$19,125$ against $16,897; but it would be incorrect to replace now. If the existent is kept for 2 years, despite its rise in operating cost during the second year, it will be more economical than the displacer: $K_2 = \$15,760$ against $16,897.

**Example 9.4.** A tank with negligible salvage value now or later can be repaired for $3000 to extend its life 1 year, after which $1000/year at the beginning of each year will extend its life 1 more year. A new tank costs $14,500, lasts 10 years with negligible salvage value, and has no annual expense. If money is worth 8 percent per year, should a new tank be purchased now?

**Solution.** Use unacost and Table 2.3. The cost of the existent, for 1 year of service, is

$$3000 F_{PR,\,8\%,\,1} = 3000(1.0800) = R_1 = \$3240$$

The unacost of the existent, for 2 years of service, is

$$3000 F_{PR,\,8\%,\,2} = 3000(0.56077) = \quad 1682.3$$

$$1000\,\frac{1}{1.08}\,F_{PR,\,8\%,\,2} = \frac{1000}{1.08}(0.56077) = \quad \underline{519.2}$$

$$R_2 = \$2201.5$$

The unacost of the existent, for 3 years of service, is

$$3000 F_{PR,\,8\%,\,3} = 3000(0.38803) = \quad 1164.1$$

$$1000\,\frac{1}{1.08}\,F_{PR,\,8\%,\,3} = \frac{1000}{1.08}(0.38803) = \quad 359.3$$

$$1000\,\frac{1}{(1.08)^2}\,F_{PR,\,8\%,\,3} = \frac{1000}{1.1664}(0.38803) = \quad \underline{332.7}$$

$$R_3 = \$1856.1$$

The unacost of the displacer, for 10 years of service, is

$$14{,}500 F_{PR,\,8\%,\,10} = 14{,}500(0.14903) = \$2160.9$$

Based on 1 or 2 more years of service, the existent compares unfavorably with the displacer, but it is more economical based on a 3-year period.

## 9.4   UNIFORM-GRADIENT SERIES

In many practical problems the operating expenses of a machine vary from year to year. If the variation from year to year is systematic, the present value of the expenses for a number of years can be computed by an algebraic relationship. A particularly important case arises when the annual cost increases by $G$ as of the end of each year, with the first-year end-of-year cost zero. The costs as of the end of the year form the following sequence for $n$ years:

$$0, G, 2G, 3G, \ldots, (n - 1)G$$

The series is known as a *uniform-gradient series,* and $G$ is the uniform gradient. The present value of this series for $n$ years is

$$P = G\left[\frac{0}{1 + i} + \frac{1}{(1 - i)^2} + \frac{2}{(1 + i)^3} + \cdots + \frac{n - 1}{(1 + i)^n}\right] \tag{9.5}$$

The series is classified as a recurring series of the second order, which can be summed algebraically by special techniques developed in Chap. 13. The series for $G = 1$ is referred to as the *uniform-gradient-series present-value factor* with the symbol $F_{GP,\,i,\,n}$. Mathematically, the sum is

$$F_{GP,\,i,\,n} = \frac{1}{i}\left[\frac{(1 + i)^n - 1}{i(1 + i)^n} - \frac{n}{(1 + i)^n}\right] \tag{9.6}$$

or
$$F_{GP,i,n} = \frac{F_{RP,i,n} - nF_{SP,i,n}}{i} \qquad (9.7)$$

The $F_{GP,i,n}$ factor is tabulated in Table A1.1.

Usually the gradient occurs along with a uniform annual cost, but the two are easily segregated, as shown in the following example.

**Example 9.5.** A series of costs at the end of each year increases uniformly by $100/year and is $1000 for the first year. If money is worth 8 percent per year, what is the present value of 12 years of costs?

**Solution.** The present value is

$$P = \frac{1000}{1.08} + \frac{1100}{1.08^2} + \frac{1200}{1.08^3} + \cdots + \frac{2100}{1.08^{12}}$$

which can be broken down into two series:

$$P = 1000 \left( \frac{1}{1.08} + \frac{1}{1.08^2} + \cdots + \frac{1}{1.08^{12}} \right)$$

$$+ 100 \left( \frac{0}{1.08} + \frac{1}{1.08^2} + \frac{2}{1.08^3} + \cdots + \frac{11}{1.08^{12}} \right)$$

or in sigma notation:

$$P = 1000 \sum_1^{12} \left( \frac{1}{1.08} \right)^n + 100 \sum_1^{12} \frac{x-1}{1.08^x}$$

The two sigma series on the right are, respectively, the present value of a uniform annual series, Eq. (2.17), and a uniform-gradient series, Eq. (9.7). The combined present values are

$$P = 1000 F_{RP,8\%,12} + 100 F_{GP,8\%,12}$$
$$= 1000(7.5361) + 100(34.634) = \$11,000$$

## 9.5 BEST POLICY WITH A UNIFORM-GRADIENT COST

A practical and instructive case involves an item or machine that incurs an operating inefficiency at a constant rate. Consider the simplified situation for an automobile which costs $12,000 when new and has a salvage value at the end of each year amounting to 75 percent of the salvage value at the beginning of the year; i.e.,

$$C_{sal} = 12,000(0.75^n)$$

where $n$ is any year, and the yearly operating expenses of the automobile increase by $300 per year. The fixed expenses common to all the years need not be considered. The variable expenses are

$$C_{en} = (n - 1)(300)$$

where $C_{en}$ is an expense at the end of year $n$. Money is worth 6 percent per year with no tax.

If an automobile is kept $n$ years, the present value of all the costs is

$$P_n = 12{,}000 - \frac{C_{sal}}{1.06^n} + 300\left(\frac{0}{1.06} + \frac{1}{1.06^2} + \frac{2}{1.06^3} + \cdots + \frac{n-1}{1.06^n}\right)$$

Substituting the value for $C_{sal}$, and $F_{GP, 6\%, n}$ for the series in parentheses, we have

$$P_n = 12{,}000 - \frac{12{,}000(0.75^n)}{1.06^n} + 300 F_{GP, 6\%, n}$$

Finally, converting to a unacost by Eq. (2.20) gives

$$R = P_n F_{PR} = F_{PR, 6\%, n}\left[12{,}000 - \frac{12{,}000(0.75^n)}{1.06^n} + 300 F_{GP, 6\%, n}\right]$$

After obtaining the values in Table A1.1 for $1/1.06^n = F_{PS, 6\%, n}$ and for $F_{GP, 6\%, n}$, the last relationship yields the following:

| Years automobile retained | Unacost, $/year |
|---|---|
| 1 | 3720 |
| 2 | 3414 |
| 3 | 3186 |
| 4 | 3024 |
| 5 | 2910 |
| 6 | 2832 |
| 7 | 2790 |
| 8 | 2769.6 |
| 9 | 2769.9 |
| 10 | 2787 |

The minimum unacost occurs at 8 years. That is, starting out with a new automobile, the optimum time to trade is at the end of 8 years for the schedule of costs as given.

The same conclusion is reached using the incremental method of Sec. 9.2. The used automobile makes its best showing on the basis of 1 year ahead. Consider the situation at the end of the seventh year. The salvage value then is $1602, dropping 1 year later to $1200. Operating costs for the coming year are $2100 as of the end of the year. The equivalent cost as of the end of the year for just 1 more year of service is

$$1602(1.06) - 1200 + 2100 = \$2598$$

Inasmuch as only 1 year is involved, $2598, being an end-of-year amount, is also a unacost. From the previous tabulation the unacost for keeping a car 7 years is $2790/year. That is, trading at the end of 7 years amounts to a $2790 unacost, whereas keeping the automobile another year amounts to a unacost of $2598. The automobile should be kept another year.

Repeat the calculations at the end of the eighth year. The unacost for 1 year more is

$$1200(1.06) - 900 + 2400 = \$2772$$

whereas from the previous tabulation the unacost for operating for only 8 years is $2769.60. The ninth year will not pay, confirming the previous conclusion.

The preceding analysis was based on the assumption that a new automobile was purchased. Now, consider the situation in which a 2-year-old car is bought, subject to all the previous conditions. The yearly costs will form the series

$$600, 900, 1200, 1500, \ldots$$

which can be broken down to a constant amount of 600 and the series

$$0, 300, 600, 900, \ldots$$

A tabulation similar to the previous tabulation can be made. The first cost is $6750, and the constant operating cost of $600/year must be included. The tabulation for starting with a 2-year-old automobile is

| Years retained | Unacost, $/year |
|---|---|
| 1 | 2694 |
| 2 | 2583 |
| 3 | 2496 |
| 4 | 2487 |
| 5 | 2484 |
| 6 | 2499 |
| 7 | 2532 |

The optimum policy when starting out with a 2-year-old automobile is to trade 5 years later when the automobile is 7 years old. This is not the same conclusion that was reached when starting out with a new automobile, which should be traded when it is 8 years old. Also, the minimum unacost starting with a 2-year-old automobile is $2484, in contrast to $2770 when starting with a new automobile. Proceeding in this manner for all the alternatives gives the data in Table 9.1.

**TABLE 9.1**
**Optimum policy for an automobile**

| Age of automobile when purchased, year | Optimum no. of years retained | Age when traded, years | Unacost, $/year |
|---|---|---|---|
| 0 | 8 | 8 | 2769 |
| 1 | 7 | 8 | 2598 |
| 2 | 5 | 7 | 2484 |
| 3 | 3 | 6 | 2412 |
| 4 | 1 | 5 | 2376 |
| 5 | 1 | 6 | 2382 |
| 6 | 1 | 7 | 2466 |

For the figures given the overall optimum or minimum is \$2376/year, obtained by starting out with a 4-year-old automobile and keeping it 1 year. It has been assumed that a used automobile can be purchased for the same amount as its trade-in value. A different set of figures will lead to a different result, but the example is illustrative. If the yearly expenses and salvage value are not systematic, it will be necessary to calculate the present values by discounting each year individually without the benefit of a systematic relationship; but the principles remain the same.

In practice, not everyone can start out with a 4-year-old automobile and follow the optimum policy. In an industrial problem, however, it is not always necessary to start at the beginning of the line. Suppose a process can use a rough ore, semirefined ore, or refined ore, all of which can be made available in any quantity at different prices. All subsequent processers can choose refined ore if that is the most economical choice.

## 9.6 DELAY VALUE OF THE EXISTENT

An existent can be displaced by a new machine merely because it acquires operating inefficiency with time that has nothing to do with technological advancement. The problem is illustrated by the optimum policy for buying an automobile in the preceding section. On the other hand, an existent can suffer no operating inefficiency with time and yet be displaced because of technological advancement of a machine now available. Usually both factors apply. The analyst compares the existent for its unacost now to the unacost of the new machine.

Future technological advancement cannot be purchased in the present. However, the expectation of future technological advancement cannot be ignored. Indeed, its value should be included as a reduction in the unacost or capitalized cost of the existent. By keeping the existent, it becomes possible eventually to obtain a later model of the new machine which presumably will be even more technologically advanced. Technological advancement works in two ways. The technological advancement which has already been made works against the existent, but the possibility of future advancement works for it. This latter factor has generally been overlooked. For new types of assets undergoing rapid technological advancement the delaying value of the existent may be significant. An exact mathematical evaluation is not possible because the result depends on the model, but the situation has been mentioned in the literature [3]. However, the delay value of the existent should not become a policy for procrastination since if a company falls too far behind, it may not be in a position to catch up.

## 9.7 A MATHEMATICAL MODEL
## FOR TECHNOLOGICAL ADVANCEMENT

Consider a situation in which technological advancement alone makes displacement possible; i.e., the existing machine suffers no loss of efficiency with time. The mathematical model for the machine will be based on two assumptions:

**1.** Each year a new model will become available, costing $1 - p$ of the cost of the model of the preceding year, where $p$ is the technological advancement rate expressed as a decimal. If the cost of a new machine is $C_i$ now, 1 year from now it will be $C_i(1 - p)$, 2 years from now $C_i(1 - p)^2$, etc.

2. Some of the operating costs of the machine will likewise incorporate technological advancement. Let $R_p$ denote the yearly end-of-year operating expense for this year's model that is subject to technological advancement. Next year's model will have a comparable expense $R_p(1 - p)$, the following model $R_p(1 - p)^2$, etc. Uniform yearly expenses that are not affected by technological advancement do not vary from model to model and need not be considered. When a machine is bought, the yearly expense $R_p$ remains constant until a new machine is purchased $n$ years later, which will reduce the yearly expense to $R_p(1 - p)^n$.

These two assumptions permit evaluation of the expression for the economic life of a machine for various technological advancement rates. If a machine is bought now and replaced every $n$ years, the pertinent capitalized cost of the machine is the present value after taxes taken to infinity for (1) the first costs occurring every $n$ years and (2) the yearly operating cost occurring every year and changing every $n$ years. This is

$$K = C_i(1 - t\psi_{rn}) + C_i(1 - t\psi_{rn})\frac{(1 - p)^n}{(1 + r)^n}$$

$$+ C_i(1 - t\psi_{rn})\frac{(1 - p)^{2n}}{(1 + r)^{2n}} + \cdots + \sum_{1}^{n} \frac{R_p(1 - t)}{(1 + r)^x}$$

$$+ \sum_{n+1}^{2n} \frac{R_p(1 - t)(1 - p)^n}{(1 + r)^x} + \sum_{2n+1}^{3n} \frac{R_p(1 - t)(1 - p)^{2n}}{(1 + r)^x} + \cdots \quad (9.8)$$

Although Eq. (9.8) appears formidable, it reduces to

$$\frac{K}{C_i} = \frac{(1 + r)^n(1 - t\psi_{rn}) + (R_p/C_i)\{(1 - t)[(1 + r)^n - 1]/r\}}{(1 + r)^n - (1 - p)^n} \quad (9.9)$$

where $r$ = decimal rate of return after taxes

$t$ = decimal tax rate

$\psi$ = factor associated with depreciation (see Sec. 3.15)

Multiplication of Eq. (9.9) by $r$ will give an equivalent expression for unacost.

The inclusion of all four rates — rate of return, tax rate, inflation rate, and productivity rate — in comparing alternatives was discussed in Sec. 6.5. Under existing conditions the effect of inflation on increasing costs overwhelms the effect of technological advancement on decreasing costs, so that the mathematical relationships involving technological advancement alone are not realistic.

## 9.8 THE MAPI MODEL

The Machinery and Applied Products Institute (MAPI) has made a strong effort to promote the comparison of existing machines with current machines as a guide for modernization. The system has undergone several revisions and is offered principally in worksheet form with accompanying charts.

The MAPI model is actually based on a new machine that acquires an operating inefficiency at a uniform rate. Applications are made using worksheets and charts. Charts are available for different depreciation methods or for expensing fully in 1 year, as well as for making 1-year or multiyear comparisons. The charts have several fixed parameters built in, such as tax rate, ratio of borrowed to total capital, and interest rate on borrowed capital. If other values of the parameters are required, algebraic relationships must be used, but they are complicated. The presentation uses picturesque terminology such as *defender, challenger,* and *adverse minimum,* but the method seems to have little current usage.

## 9.9 GROUP DISPLACEMENT

Some assets exist in multiples and fail on an individual basis, light bulbs being a classic example. Such assets can be replaced at a high cost per unit of replacement or displaced as a group at a low cost per unit. There is an optimum policy for the minimum cost obtained by replacing on an individual basis for a certain time and then displacing the entire group.

Suppose that 1000 light bulbs have the distribution of service lives given in Table 9.2. Let 1000 light bulbs be installed; assume that bulbs are replaced at the beginning of each week and that installation on an individual basis is $1 per bulb and on a group basis is $0.45 per bulb. At zero time 1000 bulbs are installed at a cost of $450.

The number of bulbs that must be installed at the beginning of each week on a replacement basis is given by Table 9.3. An *R* following a number indicates that replacement is required. Thus column 2 shows that in accordance with the probabilities of Table 9.2, 150 replacements will be required at the beginning of period 2, 200 at the beginning of period 3, etc. Column 3 of Table 9.3 shows that in accordance with the probabilities of Table 9.2, the 150 bulbs installed at the beginning of period 2 will require 22 replacements at the beginning of period 3, 30 at the beginning of period 4,

**TABLE 9.2**
**Distribution of service lives for 1000 light bulbs**

| Period week no. | Failures during week | Probability of failure during week |
|---|---|---|
| 1 | 150 | 0.15 |
| 2 | 200 | 0.20 |
| 3 | 300 | 0.30 |
| 4 | 250 | 0.25 |
| 5 | 100 | 0.10 |

**TABLE 9.3**
**Replacement of light bulbs**

| (1) Period | (2) | (3) | (4) | (5) | (6) | (7) | (13) | (14) | (15) |
|---|---|---|---|---|---|---|---|---|---|
| 1 | 1000 | | | | | | | | |
| 2 | 150R | 150 | | | | | | | |
| 3 | 200R | 22R | 222 | | | | | | |
| 4 | 300R | 30R | 33R | 363 | | | | | |
| 5 | 250R | 45R | 44R | 54R | 393 | | | | |
| 6 | 100R | 38R | 67R | 73R | 59R | 337 | | | |
| 7 | | 15R | 56R | 109R | 79R | 51R | | | |
| 8 | | | 22R | 91R | 118R | 67R | | | |
| 9 | | | | 36R | 98R | 101R | | | |
| 10 | | | | | 39R | 84R | | | |
| 11 | | | | | | 34R | | | |
| 12 | | | | | | | 339 | | |
| 13 | | | | | | | 51R | 340 | |
| 14 | | | | | | | 68R | 51R | 339 |
| 15 | | | | | | | 102R | 68R | 51R |
| 16 | | | | | | | 85R | 102R | 68R |
| 17 | | | | | | | 34R | 85R | 102R |
| 18 | | | | | | | | 34R | 85R |
| 19 | | | | | | | | | 34R |

etc. Continuing, column 4 shows that 222 replacements will be required at the beginning of period 3, with replacements as indicated.

Table 9.3 shows that the replacements per period overshoot, oscillate, and stabilize at 339. The number can be obtained independently. The average life of the bulbs is

$$\frac{1(150) + 2(200) + 3(300) + 4(250) + 5(100)}{150 + 200 + 300 + 250 + 100} = 2.95 \text{ weeks per bulb}$$

The reciprocal is the bulbs per week for 1 bulb installed. For 1000 bulbs installed

$$\frac{1000}{2.95} = 339 \text{ bulbs per week}$$

Table 9.4 shows how the minimum-cost program is determined. The final column is the cumulative dollars divided by the cumulative periods and is the dollars per period. The minimum cost is obtained by displacing the bulbs as a group every third period with intermediate replacements of 150 and 222 bulbs. If no group displacement is made, the cost is $339 per period, as determined from the equilibrium replacement value of Table 9.3.

**TABLE 9.4**
**Optimum displacement schedule for light bulbs**

| Period | Bulbs installed | | Dollars this period | Cumulative $ | Dollars per period average |
| | As a group | As replacements | | | |
|---|---|---|---|---|---|
| 1 | 1000 | | 450 | 450 | 450 |
| 2 | | 150 | 150 | 600 | 300 |
| 3 | | 222 | 222 | 822 | 274 |
| 4 | | 363 | 363 | 1185 | 296 |
| 5 | | 393 | 393 | 1578 | 316 |
| 6 | | 337 | 337 | 1915 | 319 |
| Indefinitely | | 339 | 339 | | 339 |

# THE LEARNING CURVE

## 9.10 PRACTICE IMPROVES PERFORMANCE

It is a fundamental human characteristic that a person engaged in a repetitive task will improve his or her performance. This distinguishing quality is one manifestation of optimization, a human trait, to be discussed in Chap. 10. In a broad sense, this quest for improvement is the basis of technological advancement which passes endlessly from generation to generation.

In a large-scale repetitive project improvement occurs smoothly and continuously. The phenomenon is real and has a specific application in cost analysis, cost estimating, or profitability studies related to examination of future costs and confidence level in an analysis. Improved performance in manufacturing was developed as an analytical tool first by the aircraft industry, which still leads in the application, but the principle is now recognized and used by many other types of manufacturers.

The learning curve as developed in these sections is a powerful tool for predicting, guiding, and encouraging increases in productivity.

## 9.11 THE LEARNING CURVE AND LEARNING-CURVE FUNCTION

Figure 9.1 shows the effort required, expressed as direct work-hours per pound, plotted against the cumulative plane number for the Century series aircraft from government World War II production data released as part of the Aeronautical Material Planning Report (AMPR) data. The curve is the composite data for eight types of fighter planes, with several series in some types, produced by four manufacturers.

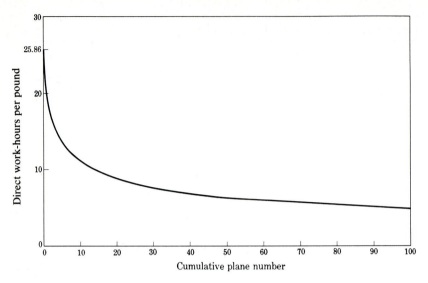

**FIGURE 9.1**

Industry average unit curve for the Century series aircraft. (*AMPR data.*)

The curve is typical and shows a progressive improvement in productivity, but at a diminishing rate. This suggests an exponential relationship between productivity and cumulative production. Figure 9.2 is a plot of the data on log-log graph paper, where the data plot as a straight line. This suggests a relationship of the form

$$E_N = KN^s \tag{9.10}$$

where $E_N$ = effort per unit of production, such as work-hours, required to produce the $N$th unit

$K$ = constant, numerically theoretical effort required to produce the first unit

$s$ = slope constant

The *slope constant s* is negative because the effort per unit decreases with production.

The effort per unit $E_N$ can be expressed in any convenient units. The effort can be converted to dollars, and eventually will be in many cases. Nevertheless, the use of dollars directly in the learning-curve function should be undertaken with care because the dollar is subject to considerable inflation and use of the dollar may mask the true reduction of effort per unit with production.

Relationship (9.10) will plot as a straight line on log-log graph paper. Taking logarithms of both sides, we obtain

$$\log E_N = s \log N + \log K$$

which in terms of $Y = \log E_N$, $X = \log N$, and $b = \log K$ has the form

$$Y = sX + b$$

the equation of a straight line.

**FIGURE 9.2**

Log-log plot of the industry average unit curve for the Century series aircraft. (*AMPR data.*)

A *learning curve,* sometimes called a *manufacturing progress curve* or simply *progress curve,* is merely a representation of the decrease in effort per unit required for a repetitive manufacturing operation. Figures 9.1 and 9.2 are both learning curves. Equation (9.10) is a *learning-curve function.* Not all data will fit the simple relationship of Eq. (9.10), and some effort has been made to use more refined relationships in specific cases. Nevertheless, Eq. (9.10) has surprising validity for manufacturing operations and is the only relationship considered in this chapter. Moreover, a log-log plot for the relationship will be used exclusively for the obvious advantage of producing a straight-line plot. A log-log representation also has the advantage of condensing a wide range of variables.

Although Eq. (9.10) will be referred to as the *learning-curve function* and Fig. 9.2 as the *learning curve,* strict usage perhaps requires that they be referred to as the *exponential learning-curve function* and *exponential learning curve,* respectively.

## 9.12 PROPERTIES OF THE LEARNING-CURVE FUNCTION

Equation (9.10) embodies many important properties. It implies a constant fractional or percentage reduction in effort for doubled (or tripled, etc.) production. For example, for any fixed value for $s$, Eq. (9.10) gives

$$E_1 = K(1^s)$$

$$E_2 = K(2^s)$$

$$\frac{E_2}{E_1} = \frac{K(2^s)}{K(1^s)} = 2^s$$

also
$$E_2 = K(2^s)$$

$$E_4 = K(4^s)$$

$$\frac{E_4}{E_2} = \frac{K(4^s)}{K(2^s)} = 2^s$$

Similarly, $E_8/E_4 = 2^s$, etc. Every time production is doubled, the effort per unit required is a constant $2^s$ of what it was. Inasmuch as $s$ is negative, the effort per unit required decreases with production.

It is common practice to express the learning-curve function in terms of the gain for double production. Thus a 90 percent learning-curve function requires only 90 percent of the effort per unit every time production is doubled.

Let $L_D$ be the decimal ratio of effort, per production unit, required for doubled production. Then by Eq. (9.10),

$$L_D = \frac{E_{2N}}{E_N} = \frac{K(2N)^s}{K(N)^s} = 2^s$$

Taking logarithms, we have

$$\log L_D = s \, \log 2 \tag{9.11}$$

or
$$s = \frac{\log L_D}{\log 2} \tag{9.12}$$

Equation (9.12) establishes a relationship between $s$, the slope constant, and $L_D$, the *decimal learning ratio*. Table 9.5 gives some corresponding values. Because of the occurrence of negative logarithms, it is sometimes more convenient to use $L_P$, the *per-*

**TABLE 9.5**
**Conversion of the decimal learning ratio to a slope constant**

| Decimal ratio of effort per unit for doubled production $L_D$ | Slope constant $s$ |
|---|---|
| 1.00 | 0 |
| 0.95 | −0.0740 |
| 0.90 | −0.1520 |
| 0.85 | −0.2345 |
| 0.80 | −0.3219 |
| 0.75 | −0.4150 |
| 0.70 | −0.5146 |
| 0.65 | −0.6215 |
| 0.60 | −0.7370 |
| 0.55 | −0.8625 |
| 0.50 | −1.0000 |

*centage learning ratio,* which by definition is $100L_D$. If $L_P/100$ is substituted for $L_D$ in Eq. (9.12), it becomes

$$s = \frac{\log(L_P/100)}{\log 2} = \frac{\log L_P - 2}{\log 2}$$

which in terms of minus $s$, a positive number, is

$$-s = \frac{2 - \log L_P}{\log 2} \tag{9.13}$$

**Example 9.6.** If 846.2 work-hours is required for the third production unit and 783.0 for the fifth unit, find the percentage learning ratio and the work-hours required for the second, fourth, tenth, and twentieth units.

***Solution.*** By Eq. (9.10),

$$E_3 = 846.2 = K(3^s)$$

$$E_5 = 783.0 = K(5^s)$$

By division,

$$\frac{846.2}{783.0} = 1.0807 = \left(\frac{3}{5}\right)^s = 0.6^s$$

$$\log 1.0807 = 0.03371 = s \log 0.6 = s(9.77815 - 10) = s(-0.22185)$$

$$s = \frac{0.03371}{-0.22185} = -0.1520$$

By Eq. (9.13),

$$-(-0.1520) = \frac{2 - \log L_P}{0.30103}$$

$$\log L_P = 1.95424$$

$$L_P = 90.0 \text{ percent} = \text{percentage learning ratio}$$

Using the data for $N = 3$ and $s = -0.1520$ in Eq. (9.10), we obtain

$$846.2 = K(3^{-0.1520})$$

$$\log 846.2 = \log K - 0.1520 \log 3$$

$$2.9278 = \log K - 0.07252$$

$$\log K = 3.0000$$

$$K = 1000$$

so that the learning-curve function is

$$E_N = 1000N^{-0.1520}$$

The effort required for any unit can now be calculated directly. Thus for the twentieth unit

$$E_{20} = 1000(20^{-0.1520})$$

$$\log E_{20} = \log 1000 - 0.1520 \log 20$$

$$= 3.0000 - 0.1520(1.30103) = 2.80224$$

$$E_{20} = 634.2$$

A tabulation for other units is

| | |
|---|---|
| $E_1$ | 1000.0 |
| $E_2$ | 900.0 |
| $E_4$ | 810.0 |
| $E_5$ | 783.0 |
| $E_{10}$ | 704.7 |
| $E_{20}$ | 634.2 |

Note that

$$\frac{E_2}{E_1} = \frac{E_4}{E_2} = \frac{E_{10}}{E_5} = \frac{E_{20}}{E_{10}} = 0.900$$

**Example 9.7.** Every time production is tripled, the unit work-hours required are reduced by 20 percent. Find the percentage learning ratio.

*Solution.* It will require 80 percent, or 0.80, for the ratio of effort per unit for tripled production. By Eq. (9.10),

$$\frac{E_{3N}}{E_N} = 0.80 = \frac{K(3N)^s}{K(N^s)} = 3^s$$

$$s = \frac{\log 0.80}{\log 3} = \frac{9.90309 - 10}{0.47712} = \frac{-0.09691}{0.47712} = -0.20311$$

By Eq. (9.13), for doubled production

$$-(-0.20311) = \frac{2 - \log L_P}{\log 2}$$

$$\log L_P = 1.93886$$

$$L_P = 86.86 \text{ percent} = \text{percentage learning ratio}$$

Another important property of the learning-curve function is that when plotted on log-log graph paper, the slope of the line is numerically equal to $s$, the slope constant.

**Example 9.8.** If 423 work-hours is required for the third production unit and 372 for the seventh, find the slope constant by graphical means.

**FIGURE 9.3**
Plot for Example 9.8.

*Solution.* The points are plotted in Fig. 9.3. The slope of the line from measurements, before reproduction, is

$$\text{Slope} = \frac{\Delta Y}{\Delta X} = \frac{-11.4 \text{ mm}}{75.0 \text{ mm}} = -0.152 \text{ slope constant}$$

The modulus of the vertical and horizontal log scales must be the same when using the method, as in the example.

## 9.13   CUMULATIVE VALUES

Equation (9.10) gives instantaneous values of effort per unit and thus the work-hours for the $N$th unit.

For the total effort for $N$ units from 1 through $N$ the *cumulative effort* required, $E_T$, is

$$E_T = E_1 + E_2 + \cdots + E_N = \sum_{N=1}^{N=N} E_N \tag{9.14}$$

An exact summation can be made by summing term by term, which can be tedious; from tables, if available; or by approximation relationships.

An approximation is obtained by treating Eq. (9.10) as a continuous function instead of a discrete function and integrating with a suitable change of limits:

$$E_T \cong K \int_{0.5}^{N+0.5} N^s \, dN = K \left[ \frac{N^{s+1}}{s+1} \right]_{0.5}^{N+0.5}$$

$$\cong \frac{K}{s+1} [(N+0.5)^{s+1} - 0.5^{s+1}] \tag{9.15}$$

The approximation improves as $N$ increases. If $N$ is very large compared to 0.5, Eq. (9.15) reduces to

$$E_T = \frac{K}{s+1} N^{s+1} \tag{9.16}$$

The cumulative average unit effort required is the cumulative effort required for the first $N$ units divided by the cumulative number of units $N$; i.e.,

$$\frac{E_T}{N} = \frac{\sum_{N=1}^{N=N} E_N}{N}$$

which from Eqs. (9.15) and (9.16), respectively, can be approximated by

$$\frac{E_T}{N} \cong \frac{K}{N(s + 1)}[(N + 0.5)^{s+1} - 0.5^{s+1}] \tag{9.17}$$

$$\frac{E_T}{N} \cong \frac{K}{s + 1}N^s \tag{9.18}$$

**Example 9.9.** If 846.2 work-hours is required for the third production unit and 783.0 for the fifth, find the total work-hours required for the fifteenth through twentieth units and the average work-hours for the six units.

**Solution.** From the solution to Example 9.6 the slope constant $s$ is $-0.1520$ and $K = 1000$. The work-hours required for the first 14 units from Eq. (9.15), with $s + 1 = -0.1520 + 1 = 0.8480$, is

$$E_{T, 14} = \frac{1000}{0.8480}[(14 + 0.5)^{0.8480} - 0.5^{0.8480}]$$

$$= \frac{1000}{0.8480}(9.6570 - 0.5556) = 10{,}733 \text{ work-hours}$$

For the first 20 units, by the same equation,

$$E_{T, 20} = \frac{1000}{0.8480}[(20 + 0.5)^{0.8480} - 0.5^{0.8480}]$$

$$= \frac{1000}{0.8480}(12.953 - 0.556) = 14{,}619 \text{ work-hours}$$

Net work-hours for the fifteenth to twentieth units is

$$E_{T, 20} - E_{T, 14} = 14{,}619 - 10{,}733 = 3886 \text{ work-hours}$$

The average for the group of six is

$$\frac{3886}{6} = 647.7 \text{ work-hours per unit}$$

The exact answer is 648.0. The approximation equation gives a close answer in this example because $N$ is reasonably large, and moreover, the effect of the early units is wiped out.

# 9.14   THE LEARNING CURVE
# AND ECONOMIC EVALUATION

The learning curve should be considered in making economic evaluations for the future and can be used as a criterion for judging the achievement actually realized. The time value of money becomes an ingredient of the analysis when looking into the future.

**Example 9.10.** A company has just received an order for four sophisticated space gadgets. The purchaser will take delivery of one unit starting 1 year hence and one unit at the end of each of the succeeding 3 years. He will pay for each unit immediately upon receipt and will not take delivery ahead of time. The seller, however, can make the units ahead of time, if he prefers, and store them at no cost for later delivery.

The only pertinent cost in making these units is labor at $10/h. All units made in the *same* year are subject to an 80 percent learning ratio for the hours required, and the first unit requires 100,000 h of labor. Learning occurs only within a year and is not carried over from one year to another.

Consider the labor costs to be paid as of the end of the year. If money is worth 16 percent per year after a 34 percent tax, decide which of the following two plans is more economical:
(a)   Build four units the first year.
(b)   Build one unit each of the 4 years.

*Solution.* Revenue and its timing are the same for either plan and need not be considered. The object is to choose the plan with the lower present value of cost after taxes. For an 80 percent learning ratio $s = -0.3219$ from Table 9.5. The first unit costs $10(100,000) = \$1,000,000$. The cost of units made in the same year with learning is

| | | |
|---|---|---|
| First: | $1,000,000(1^{-0.3219}) =$ | $1,000,000$ |
| Second: | $1,000,000(2^{-0.3219}) =$ | $800,000$ |
| Third: | $1,000,000(3^{-0.3219}) =$ | $702,104$ |
| Fourth: | $1,000,000(4^{-0.3219}) =$ | $640,000$ |
| | Total $=$ | $3,142,104$ |

The present value (PV) of the cost after taxes for all units made in the first year is

$$PV = (1 - 0.34)\frac{3,142,000}{1.16} = \$1,787,700$$

The present value of the cost after taxes for the four units made in successive years is

$$PV = (1 - 0.34)\left(\frac{1,000,000}{1.16} + \frac{1,000,000}{1.16^2} + \frac{1,000,000}{1.16^3} + \frac{1,000,000}{1.16^4}\right) = \$1,846,800$$

It will pay to make the four units the first year. The gain from the learning curve offsets the advantage in deferring the labor costs by the alternate plan. Making four the first year, however, is not necessarily the best possible plan. There are other combinations, such as three the first year and one the last year.

## 9.15 THE LEARNING-CURVE FUNCTION FROM SINGLE-UNIT DATA

For a given set of data Eq. (9.10) is established by determining the best values for $K$ and $s$. If the data are all in the form giving the effort required for single, individual units of production, the data can be plotted on log-log graph paper and the best line drawn by eye. When the best line has been established, any two points on the line can be used to determine, graphically or analytically, the slope of the line (the slope constant) and $K$, which is numerically the intercept on the ordinate $N = 1$. The graphical method is rapid but depends on judgment, and can become inaccurate if the points are badly scattered.

The most accurate method for determining the best straight line is application of the method of least squares described in the previous chapter.

## 9.16 THE LEARNING-CURVE FUNCTION FROM GROUPED DATA

In practice, the effort required to produce single, individual units may not be known, but instead only the average effort to produce a group, or lot, of units. Data for such a circumstance are given in Table 9.6. A total of 35 units has been made, but the precise number of work-hours required for any single unit is not known. Before the work-hours can be plotted, it is necessary to take each group and convert it to a point. This point within a group is some unit number, not necessarily an integer, with which the average work-hour can be associated. It is a balance or focus point and a single value to be paired with the average effort per unit for the group. It is loosely referred to as the *lot midpoint*, but *lot equivalent point* is perhaps more descriptive.

Several methods are available for determining the lot equivalent point. The *lot-midpoint method* uses the arithmetical mean of the first and last unit numbers in the lot. For example, for a lot of 6 units, the equivalent point is at $(1 + 6)/2 = 3.5$, or 3.5 units from the last unit of the preceding lot.

The *rule-of-thumb lot-midpoint method* places the lot equivalent point by means of the following rules:

First lot:      Less than 10, take half of the last unit number. Ten or more, take one-third of the last unit number.

Subsequent lots:      Take one-half of the number of units in the lot.

**TABLE 9.6**
**Work-hours for consecutive lots**

| Lot no. | Quantity in lot | Work-hours for lot | Average work-hours per unit for lot |
|---|---|---|---|
| 1 | 3 | 621.76 | 207.24 |
| 2 | 4 | 357.72 | 89.43 |
| 3 | 8 | 428.72 | 53.43 |
| 4 | 10 | 449.80 | 44.98 |
| 5 | 10 | 368.10 | 36.81 |

For the data in Table 9.6

Lot 1:                                    $\frac{3}{2} = 1.5$ lot equivalent point

Lot 2:                                    $\frac{4}{2} = 2$

or two units from unit 3, the last unit in the preceding lot. Thus $3 + 2 = 5$, or unit 5.

Lot 3:                                    $\frac{8}{2} = 4$

or four units from unit 7, the last unit in the preceding lot, thus $7 + 4$, or unit 11. The complete tabulation is given in Table 9.7.

The *geometric-mean lot-midpoint method* uses the geometric mean to determine the equivalent point of each lot. For a lot of five the geometrical mean is

$$\sqrt[5]{1(2)(3)(4)(5)} = 2.6052$$

which is added to the last unit number of the preceding group. Some geometric means are given in Table 9.8.

The *true lot-midpoint method* is the most accurate method for determining the lot equivalent point but the most difficult to calculate. It involves two separate steps. In the first step an approximate method is used for the data to establish an approximate value for $s$, the slope constant. The approximate value is then used in the second step to find the true equivalent point for the lots. The data so generated are then used with the method of least squares to establish a revised but accurate value of $s$.

**TABLE 9.7**
**Data of Table 9.6 after adjustment by the rule-of-thumb method**

| Lot. no. | Quantity in lot | Assigned to unit | Average work-hours per unit for lot |
|---|---|---|---|
| 1 | 3 | 1.5 | 207.24 |
| 2 | 4 | 5 | 89.43 |
| 3 | 8 | 11 | 53.43 |
| 4 | 10 | 20 | 44.98 |
| 5 | 10 | 30 | 36.81 |

**TABLE 9.8**
**Geometric means for lot sizes**

| No. in lot | Geometric mean | No. in lot | Geometric mean |
|---|---|---|---|
| 1 | 1.0000 | 9 | 4.1472 |
| 2 | 1.4142 | 10 | 4.5287 |
| 3 | 1.8171 | 15 | 6.4235 |
| 4 | 2.2134 | 20 | 8.3044 |
| 5 | 2.6052 | 25 | 10.177 |
| 6 | 2.9938 | 30 | 12.045 |
| 7 | 3.3800 | 40 | 15.768 |
| 8 | 3.7644 | 50 | 19.483 |

As an example, consider the second lot in Table 9.6. It will be shown below that an approximate value for $s$ for the data in Table 9.6 is $s = -0.57616$. The second lot contains the fourth, fifth, sixth, and seventh units, and for $K = 1$, Eq. (9.1) gives

$$E_4 = 1(4^{-0.57616}) = 0.44990$$

$$E_5 = 1(5^{-0.57616}) = 0.39562$$

$$E_6 = 1(6^{-0.57616}) = 0.35617$$

$$E_7 = 1(7^{-0.57616}) = \underline{0.32590}$$

$$\text{Total} = 1.52759$$

$$\text{Av} = \frac{1.52759}{4} = 0.38190$$

The individual unit which will have an effort of 0.38190 with $K = 1$ and $s = -0.57616$, as before, is

$$0.38190 = 1N^{-0.57616}$$

$$N = 5.3159 = \text{true lot equivalent point}$$

Thus for the second lot an average of 89.43 work-hours per unit is assigned to unit 5.3159. Other lots are treated similarly. The points so obtained can be subject to a least-squares calculation to determine the best values of $s$ and $K$.

If tables of cumulative values are available for values of $s$ reasonably close to the approximate value, the numerical work for the true lot-midpoint method can be simplified.

**Example 9.11.** Using the data in Table 9.6 and the rule-of-thumb lot-midpoint method, determine the learning-curve function that represents the data.

*Solution.* Application of the rule-of-thumb lot-midpoint method to the group data leads to the five points given by the last two columns of Table 9.7. The points are plotted in Fig. 9.4.

A line put through the points by eye is shown. Using the extreme points on the line, the slope as determined by measurement, before reproduction, is $\Delta Y/\Delta X = -73.6/148.5 = -0.496$.

The percentage learning ratio corresponding to the slope can be obtained from Eq. (9.13):

$$-(-0.496) = \frac{2 - \log L_P}{\log 2}$$

$$\log L_P = 1.8507$$

$$L_P = 70.91 \text{ percent} = \text{percentage learning ratio}$$

and the value for $K$ is the intercept on the axis $N = 1$ and is about 204. Thus the line drawn by eye gives a 70.91 percent learning ratio and the following learning-curve function:

$$E_N = 204N^{-0.496}$$

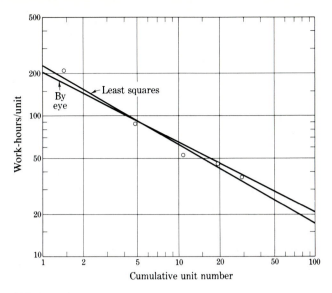

**FIGURE 9.4**
Plot of data and learning curves for Example 9.11.

The data can also be subjected to a least-squares analysis, described in the previous chapter, and shown for the five points in Table 9.9:

$$s = \frac{5\sum XY - \sum X \sum Y}{5\sum X^2 - \left(\sum X\right)^2} = \frac{5(8.03466) - 4.69460(9.21440)}{5(5.47862) - 4.69460^2} = -0.57616$$

$$\log K = \frac{\sum Y}{5} - s\frac{\sum X}{5} = \frac{9.21440}{5} + 0.57616\frac{4.69460}{5} = 2.38385$$

$$K = 242.02$$

The learning-curve function resulting from the least-squares computation is

$$E_N = 242.02N^{-0.57616}$$

**TABLE 9.9**
**Least-squares calculation for Example 9.11**

| N assigned | Effort $E_N$, work-h | X log N | Y log E | XY | $X^2$ |
|---|---|---|---|---|---|
| 1.5 | 207.24 | 0.17609 | 2.31647 | 0.40791 | 0.03101 |
| 5 | 89.43 | 0.69897 | 1.95148 | 1.36403 | 0.48856 |
| 11 | 53.39 | 1.04139 | 1.72746 | 1.79896 | 1.08449 |
| 20 | 44.98 | 1.30103 | 1.65302 | 2.15063 | 1.69268 |
| 30 | 36.81 | 1.47712 | 1.56597 | 2.31313 | 2.18188 |
| | | 4.69460 | 9.21440 | 8.03466 | 5.47862 |

and the slope constant of $-0.57616$ corresponds to a 67.075 percent learning ratio. The straight line represented by the preceding relationship is plotted in Fig. 9.4.

Although the slope constant, $-0.57616$, is not exact because the rule-of-thumb lot-midpoint method was followed, the true lot-midpoint method can now be used, as indicated previously, to establish an even better representation for the learning-curve function.

## 9.17 DEVELOPMENT OF LEARNING-CURVE TABLES

Tables can be readily constructed for use with the learning-curve function. The decimal or percentage learning ratio is a parameter and the unit number is a variable. The three most important items to tabulate are

$E_N$ = effort for the $N$th unit taking unity as the effort for the first unit = $N^s$

$$E_T = \sum_1^N E_N = \text{cumulative effort for units 1 through } N$$

$$\frac{E_T}{N} = \frac{\sum_1^N E_N}{N} = \text{cumulative average unit effort for units 1 through } N$$

Tables can be programmed easily and printed out by a computer for any value of $L_D$, the decimal learning ratio. An abbreviated tabulation for $L_D = 0.90$, 0.80, and 0.70 is given in Table A4.1.

## CONSTRUCTABILITY ANALYSIS

## 9.18 DEFINITION

A definition of *constructability* does not appear in the dictionary. The dictionary has the word, "construct*i*ble," but defines it only as an adjective form of "construct." The Construction Industry Institute has adopted the spelling utilizing the "a" and for their purposes has defined it as:

> Constructability is the optimum use of construction knowledge and experience in planning, design, procurement, and field operations to achieve overall project objectives.

A common view of constructability is that it only involves:

Determining more efficient methods of construction after mobilization of field forces.

Allowing construction to review engineering documents periodically during the design phase.

Assigning construction personnel to the engineering office during design.

A modularization or preassembly program.

Although each of these are part of constructability, they are only a part. Only through the effective and timely integration of construction input into planning and design as well as field operations will the potential benefits of constructability be achieved.

The planning/execution phases for a typical major industrial project involve conceptual engineering, detailed engineering, procurement, construction, and start-up. Constructability considerations must appear during the conceptual stage at the same time as operability, reliability, and maintainability considerations must surface. Constructability analysis continues through the remaining phases. Figure 9.5 relates the potential for influencing final project cost against time. Note how decisions made early in the planning and execution of the project have a high influence as compared to those made later, thus the need for constructability analysis to be considered in the earliest stages of planning.

Unfortunately, constructability is not a subject found in common textbooks on engineering and construction. It is taught in very few schools and has become the subject of significant research only recently by the Construction Industry Institute. This is not to say that constructability has not been practiced by owners and contractors — it has, but each company has approached it in the company's own way and too few of them have any documented programs.

## 9.19 CONSTRUCTABILITY OBJECTIVES

Constructability analyses and decisions are directed toward the construction phase. Basically the intent is to save construction time and cost without compromising any other project objectives such as quality, reliability, operability, maintainability, durability, and appearance. Thus constructability decisions are oriented toward:

Reducing total construction time by creating conditions which maximize potential for more concurrent (rather than sequential) construction and minimize rework and wasted time.

Reducing work-hour requirements by creating conditions which promote better productivity or creating designs which demand less labor.

Reducing costs of construction equipment (and tools) by reducing requirements for such equipment, creating conditions which promote more efficient use of that equipment, and minimizing need for high-cost, special purpose equipment.

Reducing materials costs through more efficient design, use of less costly materials, and creation of conditions which minimize waste.

Who performs constructability analysis? Obviously, it must be someone familiar with the construction process. In a design office a common practice is to colocate construction personnel with the design engineering staff. But this is costly and cannot always be afforded. Actually all lead discipline engineers should have some back-

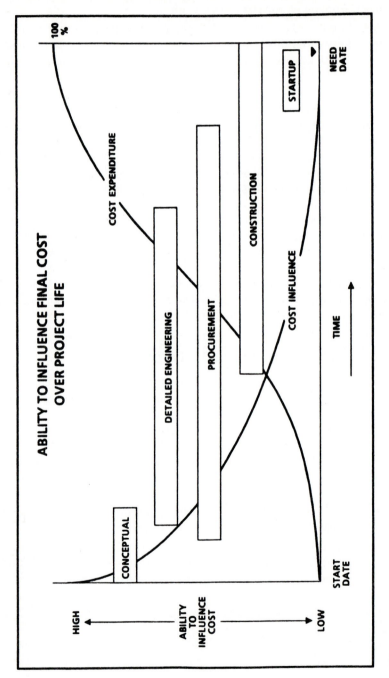

FIGURE 9.5

226

ground in field construction, so that they can provide this talent when field construction personnel are not available. And constructability should be a part of the training program for every design engineer. Of course, constructability analysis related to construction strategies is performed by the field personnel who will be implementing those strategies.

## 9.20 CONSTRUCTABILITY CONCEPTS

As with many subjects, constructability is best understood if a structure can be established to guide users in its application. One can use "concepts" and "applications" as two levels of detail — a concept is a general guideline or philosophy; an application is a specific example. A number of concepts and example applications are given below. The concepts are not necessarily all-inclusive and there is some overlapping among those listed. Obviously there are many examples of each concept that can be added.

**CONCEPT 1. RECOGNIZE THAT START-UP AND CONSTRUCTION DRIVE ENGINEERING AND PROCUREMENT SCHEDULING.** Nonavailability of design drawings, construction materials, or permanent equipment when needed are two of the major contributors to poor construction productivity and higher costs. All participants in the project must plan and schedule their individual efforts to meet the needs of others. The ultimate need date in the project is the owner's need date. Start-up is scheduled to meet that date. Construction is planned to meet start-up need dates. Procurement is planned to meet construction need dates and engineering need dates (e.g., designs of vendor-engineered items). Engineering must be scheduled to meet both procurement and construction need dates.

**CONCEPT 2. USE CONTRACTING AND MANAGEMENT APPROACHES THAT PROMOTE CONSTRUCTION EFFICIENCY.** On engineering-procurement-construction projects executed with overlapping of those phases it is essential that strong single management of the total effort be available from the outset of conceptual engineering. Further, construction contract packages must be of a quality to permit fixed-price bidding as a means to reduce or eliminate the problems associated with changes.

**CONCEPT 3. EMPHASIZE STANDARDIZATION AND REPETITION.** Standardization and repetition maximize application of the learning curve to the work force, permit volume buying of materials, and simplify purchasing and warehousing. While this may cause some overdesign, the added cost of materials is usually offset by the other savings. Examples include standardization of beam sizes in flooring, standardization of foundation sizes, and standardization of foundation bolt sizes.

**CONCEPT 4. TAKE MAXIMUM ADVANTAGE OF READILY AVAILABLE, OFF-THE-SHELF MATERIALS AND EQUIPMENT.** A tremendous variety of commercial materials and equipment is available on the world market. During the design phases it makes little sense to go through the expense of designing some new component when a comparable one that has already been designed and tested is available. During construction maximum use should be made of available work-saving devices. In the tools category there are automatic welding machines, nail guns, etc. In the materials category, for concrete there are commercial forms and a complete family of high chairs, clips, and other gadgets to speed the placement and tying of resteel. Comparable items are available for carpentry, electrical, and other work. Design and construction supervisors must keep up with the market for such items via trade shows and catalogs.

**CONCEPT 5. CHOOSE CONFIGURATIONS THAT SIMPLIFY ERECTION OR INSTALLATION.** Constructability analysts continually envision themselves as the constructors and ask themselves if the design concept being considered is the easiest to build. Will some modification be much easier to build, yet be just as functional? Take, e.g., a simple steel utility bridge over a roadway. One design might call for the two verticals to be erected and the horizontal to be bolted to the sides of the two verticals. An optional design would be to place the horizontal on top of the verticals for bolting. The latter choice requires much simpler positioning. Another example is the specifying of wiring harnesses with termination hardware already installed or with plug-type connectors.

**CONCEPT 6. CREATE DESIGNS WHICH PROMOTE ACCESSIBILITY AND PROVIDE ADEQUATE SPACE FOR CONSTRUCTION PERSONNEL, MATERIAL, AND EQUIPMENT.** The tendency for design personnel to compress facility size as a materials cost-cutting measure must be balanced against the higher labor costs, inefficient use of construction equipment, and material damage that can accompany congestion and interferences. Also, it must be remembered that a design which eases the work of the constructors will usually contribute to ease of maintenance.

**CONCEPT 7. ADAPT DESIGNS AND STRATEGIES TO PROJECT SITUATION.** Each locality has its own weather patterns, locally available resources, and population characteristics. If construction is to be in an area with a very short construction season or limited labor availability, maximum use of factory-assembled modules and components designed for rapid assembly is in order. Scheduling of the construction also must be built around the weather — e.g., some buildings may be erected early to provide protected work space for later construction. Also consider local labor and specialty contracting capabilities — select designs which best utilize these capabilities as they will be less costly than imported capabilities. If the local population is notorious for poor productivity or quality, maximize use of remote, off-site fabrication. In a union environment consider jurisdictional rules and wage scales when selecting a design approach.

**CONCEPT 8. USE REALISTIC SPECIFICATIONS.** Designs must be buildable. The imposition of specifications with unjustifiable tolerances will add many work-hours to the project and contribute to many quality control controversies. For example, the imposition of nuclear quality specifications on ordinary construction can be overkill. Designers are also cautioned against specifying an expensive, hard-to-install material when another is far more economical. For example, PVC conduit is lighter, more flexible, and more easily worked with than rigid conduit. Designers must learn to challenge each specification.

**CONCEPT 9. ESTABLISH CONDITIONS THAT PROMOTE WORKER EFFICIENCY.** Worker productivity is dependent on many factors, one of which is the efficiency of the site. Efficient site layout involves good planning of access roads, offices, construction roads, laydown areas, fabrication yards, and other construction support facilities for the purpose of minimizing travel time and congestion. Efficient site layout should be considered as early as the conceptual design stage since the siting of the permanent structures directly affects the layout potential. Other actions to promote efficiency include dust control, good construction area drainage, weed control in laydown areas, and regular cleanup of scrap and debris.

**CONCEPT 10. PERFORM WORK WHEN AND WHERE IT IS MOST EFFICIENTLY ACCOMPLISHED.** Once a design has been accepted, there still may be options as to when and where the construction or erection work is accomplished. Affecting this decision are such factors as weather, availability of skills, pay scales of workers involved, available space, schedule constraints, and safety considerations. Complicated process equipment is often best assembled in modules at a factory, requiring only positioning and connecting at the project site. This takes advantage of the factory environment which generally has more skilled workers, better productivity, and better working conditions. Pay scales also may be more favorable. Module assembly is accomplished in parallel with other construction work and this permits field construction schedules to be shortened. On-site prefabrication yards for forms, resteel cages, and piping spools allow such work to be accomplished under the best conditions, and if such yards are provided weather cover, permit this work to continue during inclement weather. Work on the ground can be accomplished more efficiently and safely than in the air. For example, insulated components can be at least partially insulated while on the ground with only the finish work left for the air.

**CONCEPT 11. ORGANIZE TO MINIMIZE UNSCHEDULED ACTIVITY.** Redoing work or standing around waiting for instructions, tools, materials, or equipment are common types of unscheduled activity—they drive up costs. A foreman who spends much of the workday trying to find needed materials is also performing unscheduled work for a foreman since the foreman is hired to supervise a crew and without this supervision the crew is inefficient. Use of detailed work package planning and adoption of a philosophy of never starting on a work package until materials and equipment availability is assured are techniques to minimize waiting time. Use of separate crews for materials pickup and spotting at point of use is a technique to keep the foreman with the crew.

**CONCEPT 12. INCORPORATE FLEXIBILITY FOR FIELD MANAGERS.** Construction is a labor-intensive operation subject to all the problems associated with that group plus the vagaries of weather, materials deliveries, equipment availability, and other factors. Field managers must not be overly constrained through imposition of arbitrary schedules, work sequences, or other conditions by higher managers.

## 9.21  ESSENTIAL ELEMENTS

Three elements must be present if a constructability program is to realize full potential. First, constructability must be looked upon as a program requiring proactive attention. The often-held view that constructability is "review" of designs by someone familiar with construction is totally wrong. By the time designs are ready for review, it may be too late to change anything or such changes will be costly. Instead, those with knowledge of construction must jointly participate with the other interested parties—owner, engineer, and operator—to "brainstorm" concepts and approaches before they are committed to a drawing. In other words, constructability is a component of planning that must be included in all phases.

Second, constructability is a "team" effort. Only if the interests of owner, engineer, constructor, and operator are jointly represented in all decisions will the optimum solution be realized. Reduction of construction costs is certainly an important objective, but doing so must not compromise other needs.

Finally, constructability must have management commitment and support. The time and resources needed for such a program must be made available.

## VALUE ENGINEERING

## 9.22  GENERAL DESCRIPTION

Value engineering (VE) is a proactive function targeted at the design itself. It is a multidiscipline, systematic activity which has as its objective the development of a design of a facility or item which will yield least life cycle costs or provide greatest value while also meeting all functional, safety, quality, operability, maintainability, durability, and other criteria established for it. The Society of American Value Engineers is an active organization that is well known through its appropriate acronym, SAVE.

As stated above, the term *value engineering* is associated with a project or product *under development*. On the other hand, *value analysis* is the term to be applied to a value study of something that has already been designed or produced. Although many individuals speak of VE studies during the construction phase of a project, these really should be called *value analysis* studies. A third term is *value management*. This is an all-encompassing term that includes all forms of value study. Thus it includes *value engineering, value analysis,* and *constructability analysis.*

True VE is a very structured and complicated process—considerable training and experience is required for full competency. However, simply understanding these principles can be extremely beneficial to a cost engineer since they are applicable to any number of situations in both private and business lives. Cost-effectiveness is a goal of every cost engineer; VE is a powerful tool which can promote that goal.

Although one may tend to think of cost and value as synonymous, this is not always true. There are at least four types of value that may be associated with an item:

**COST VALUE.** This is the amount of money that must be spent to produce or procure an item.

**EXCHANGE VALUE.** This is the value of an item on the open market should one try to sell or trade it. The *exchange value* of a new car drops in comparison to its *cost value* as soon as the new owner takes the car from the showroom.

**USE VALUE.** This is the value of an item to the user because of the functions or services it provides. A well-maintained, basic, old car may have little *exchange value* but just as much *use value* to the owner as a newer, fancier car. That *use value* would equal the *cost value* of equally reliable replacement transportation.

**ESTEEM VALUE.** This is a value "in the eye of the beholder" or a consequential value derived from some investment. A piece of jewelry has a definite *cost value*, its *exchange value* is usually less than cost, it has little or no *use value*, but it has an *esteem value* to an individual who loves jewelry. In effect, its *esteem value* to its owner must at least equal its *cost value* or that individual would not buy it. In the case of an office building a very austere structure may satisfy basic needs. However, most businesses strive to operate in attractive buildings with pleasant interiors in order to attract clients and motivate employees. The value of the increased business and greater employee productivity is the *esteem value* which the owner hopes will exceed the difference in cost between a basic structure and the more attractive one.

All forms of value enter into VE studies. Value engineering is concerned with the life cycle costs of the facility or item. Following are some categories of cost that may be associated with a constructed facility:

Feasibility studies
Engineering
Real estate
Permits and licenses
Special consultant fees
Construction
Legal fees
Furnishings
Operating costs
Maintenance costs
Insurance
Demolition costs
Rebuild/upgrade costs
Interest payments
Time-value-of-money on committed capital

## 9.23 FORMAL VALUE ENGINEERING

A formal VE program within a design organization calls for the creation of a specially trained team to evolve and critically evaluate alternative design concepts. Formation of such teams should be recognized in the design contract as a reimbursable expense, a situation only rarely found. Lacking this, the principles of VE can still be practiced. Basically VE is built on the principle that two or more heads are better than one and there may be a better approach. In other words, do not leave design decisions totally within the hands of a single person since that person's experience may be limited and not lead to the most cost-effective solution. Hence the idea is to "brainstorm" ideas before putting pencil to paper. For this purpose utilize any available design, construction, or operating personnel whose experience might bring forth a good idea. Although this effort in itself consumes time, it utilizes personnel from locally available staff and can be expected to return savings far in excess of investment.

The VE team should be comprised of people from different disciplines in order to get a broad range of ideas. The leader of the group needs to establish a time and place that the group can meet without interruption. The leader establishes the rules for brainstorming sessions and provides general direction, but must not be dictatorial, impose his or her ideas on the group, or otherwise constrain the open exchange of ideas. The leader will act as the recorder. The keys to good brainstorming are (1) a well-selected group in the proper frame of mind, (2) an atmosphere of openness, and (3) a firm rule of *no criticism* of other's ideas. A group member should feel comfortable saying whatever comes into his or her mind even though the idea may appear ridiculous to some. The reason is that a far-out idea may stimulate someone else to think of an idea that would not have come forth without the stimulation of the first idea. This is why four people working in a group can produce more ideas together than four people working independently. The size of the brainstorming group can vary depending on the magnitude of the problem, but groups of three to five are normal.

Formal VE follows a structured sequence of actions. The various authorities on VE group these into phases, but, depending on the authority, from four to eight phases may be listed. Whatever their categorization, however, the total included actions are essentially the same. Listed here, under six headings, are the phases:

**1. INFORMATION-GATHERING PHASE.** In this phase all information comprising the basis for the project is assembled. Included are such things as performance requirements, special owner requirements, applicable laws and regulations, schedule requirements, etc. *What are the operational and other requirements we must satisfy?*

**2. FUNCTIONAL ANALYSIS PHASE.** Next, the basic and secondary functions to be provided by the facility are identified. *What functions are involved in meeting the requirements?* Functions are described in a generic fashion using an action verb and a noun to invite visualization of various options. For example, "transport coal" leads to options of conveyor belts, trucks, rail, etc. Other typical verb–noun combinations are:

Transmit load
Illuminate room

Convey liquid
Heat building

A basic function is primary function of the item under study. A secondary function is anything else. The basic function of a lamp is to provide illumination; a secondary function is to decorate the room.

Questions to be asked by the team in this phase include:

What basic functions are involved in meeting facility/product requirements?
Which secondary functions are associated? Which are essential?
How does each accomplish the function?
What is the worth of value of a function?

**3. CREATIVE PHASE.** Different ideas for performing the various basic or secondary functions are identified and listed based on the functional analysis. For example, heating requirements for a structure can be satisfied by any number of options (heat pump, gas furnace, oil burner, space heaters, fireplace, electric baseboard heaters, etc.). *In how many ways can a function be performed?* It is not uncommon to have 5 to 25 ideas proposed.

It is worthwhile in this phase to itemize obvious advantages and disadvantages of each idea as a basis for initial screening in the analytical phase.

**4. ANALYTICAL PHASE.** Some authorities recommend that only the group leader from the creative phase continue in the analytical phase. The reason for this is that if the members of the team brainstorming an item know they may be required to evaluate the feasibility and cost of the idea, they may be reluctant to add too many to the list or to propose an idea that is difficult to evaluate.

This phase begins with a summary review of the ideas from the creative phase to reduce the initial list to two to five ideas that are worthy of detailed evaluation.

With each remaining idea the need for each function is validated and each option for performing a function is examined individually as to feasibility, cost, and value received. Then various combinations of these ideas are tested as a system against the governing criteria for the facility. The most feasible combinations are isolated and given more detailed study. The combination with the most favorable balance of cost and value is further refined through more detailed study. *Which combination meets requirements and provides the greatest value or is most cost effective on a life cycle basis?*

Although total life cycle cost is often the only criteria used to rank the options, this is not always true. Another option, when everything cannot be reduced to monetary terms, is to list various criteria, weigh those criteria, and then grade each option against those criteria. Section 9.24 will provide an example of this.

**5. PROPOSAL PHASE.** In this phase the team's recommendations are summarized for presentation to the decision makers. This presentation must have a written version that details the basis for the recommendation. There also may be an oral presentation. *This is what we propose.*

This phase may be the weakest phase if the preparer assumes that the reader knows as much as the preparer and fails to provide the information needed to validate a recommendation or the proposal is so voluminous that the executives with decision-making authority would have difficulty finding time to read and digest it. When preparing a proposal, assume that the reader has no background about this idea and logically build the proposal one step at a time. If there is a lot of information involved in the proposal, prepare an executive summary and put backup detail in appendices so that the reader can get the flavor of the proposal very quickly and still have detail available if needed. Ensure that your proposal answers these questions:

What criteria were being satisfied?
What solution are you proposing?
What does it cost?
What value is anticipated?
What other options were seriously considered?

**6. FOLLOW-UP PHASE.** The cost performance of the selected option is tracked for purposes of comparing actual to planned and developing historical data for use in future studies. *Let's see how well we thought this out.*

Even though a design concept has been established through formal VE in the conceptual design phase, the principles of VE can continue during detailed design. A typical area for VE effort during this stage is in the area of specifications. All too often it is found convenient to utilize specifications directly from a computer database without questioning the need for them as written. Each should be questioned as to its need as written and the most cost-effective specification derived. During detailed design various design simplifications may also result from continued VE.

Some owners include "value engineering" clauses in their contracts for construction. These clauses usually set up a procedure whereby a contractor, after review of the construction documents, can submit proposals for more cost-effective designs than those shown on the contract drawings. If the VE proposal is accepted, the savings are shared with the contractor. As mentioned previously, such studies are more properly called "value analysis" studies since they come after completion of design.

## 9.24   APPLICATIONS

An example of a succinct application is the work of a VE team at Picatinny Arsenal, New York, which improved the design for the top charge of a blasting cap, eliminating ferruling in the assembly, as reported by Dobrow [11] and shown in Fig. 9.6. Loading was simplified and safety improved at no reduction in performance. Although unit savings were only $0.08, the yearly savings were $346,000.

Before

Ferrule required to
retain charge

Top charge

Intermediate charge

Base charge

After

Lower charges now retained by top charge
Ferrule eliminated
Loading procedures simplified
Improved safety of assembly operation

Unit cost reduced by $0.08
Validated savings: $346,900

**FIGURE 9.6**
Blast cap before and after modifications of the top charge. (*Chemical Engineering*, McGraw-Hill, Inc., New York, Aug. 21, 1972, pp. 122–126)

Value engineering (VE) can be applied to any of the following:

Hardware
Software
Methods and procedures
Services

It is not limited by the size of the study or the type of business.

A checklist of areas in which VE can be efficiently and profitably used to increase productivity is given in Table 9.10.

Section 9.23 outlined the principles and general approach to value studies. It was pointed out that *value engineering* is the proper term to be used for value studies in the design development stage, whereas *value analysis* is the term applicable to review of an existing design. Both have the objective of providing greatest value and the principles involved are the same. Since presentation of a complete VE study would be far too voluminous for reporting here and we can learn principles as well from a value analysis study, the results of such a study is illustrated by Example 9.12. This example is an actual case study.

**TABLE 9.10**
**Checklist for areas to apply value engineering**

1. Engineering
   a. Design
   b. Development
   c. Product improvement
2. Manufacturing
   a. Methods, material handling, and production
   b. Tool design
   c. Packaging
3. Purchasing
   a. Vendors' alternative proposals
   b. New product information and price against function evaluation
4. Sales
   a. Slow-selling projects
   b. VE service to customers
5. Systems and procedures
   a. Paperwork
   b. Forms processing
   c. Reproduction services
6. Maintenance
   a. Procedures
   b. Materials
   c. Mechanization
   d. Work scheduling

**Example 9.12.** A value analysis study was performed on the Veterans Administration hospital in Gainesville, Florida. Specifically reviewed was the fixed glass to be installed in a three-story, 50,000 square feet, 120-bed addition to that hospital.

The original design called for $\frac{1}{4}''$, single-glaze, tinted glass to be installed in fixed frames at various window locations. The windows were not designed to be opened and there was no requirement for them to be opened. Their basic function was to transmit natural light from outside to inside. A secondary function could be said to provide insulation.

The basic question was, "Is this the most cost-effective design on a life cycle basis?" The creative phase brainstorming session produced nine ideas as follows:

1. Visqueen
2. Glass, clear, single-glazed*
3. Glass, tinted, single-glazed (original design)*
4. Glass, tinted, double-glazed*
5. Saran wrap
6. Reflective glass*
7. Plastic
8. Clear glass with reflective 3M film*
9. Sun screens

As can be seen, some ideas were in the "ridiculous" category but they were listed nonetheless.

During the analytical phase the list of nine was initially reduced to the five shown with an asterisk. Each of these was then evaluated using the Weighted Constraint Chart shown in Fig. 9.7. The constraints or criteria used for this evaluation and their relative weights represent the consensus of the evaluators of those items representing value elements. Note that not all of them are directly translatable into a cost. Value items such as aesthetics and strength have been given a value in relation to those items for which value can be expressed in terms of money.

| Weights | | 5 | 6 | 3 | 4 | 1 | 2 | | | |
|---|---|---|---|---|---|---|---|---|---|---|
| CONSTRAINTS IDEAS | | Initial Cost | Annual Savings | Aesthetics | Ease of Installation | Maintenance | Strength | | Score | Rank |
| Clear Glass | | 4 | 0 | 2 | 3 | 4 | 4 | | 50 | 5 |
| Tinted Glass, Single Glazed | | 3 | 2 | 4 | 3 | 4 | 4 | | 63 | 3 |
| Tinted Glass, Double Glazed | | 2 | 4 | 4 | 3 | 4 | 4 | | 70 | 1 |
| Reflective Glass | | 3 | 2 | 3 | 3 | 4 | 4 | | 60 | 2 |
| Glass Plus 3M Film | | 3 | 3 | 3 | 2 | 3 | 4 | | 61 | 4 |

Scale: Excellent = 4  Good = 3  Fair = 2  Poor = 1  None = 0

**FIGURE 9.7**
Weighted constraint chart for Example 9.12.

The double-glazed tinted glass was the best among the five options evaluated. Detailed calculations contrasting the original design to the double-glazed tinted glass revealed that the latter's thermal efficiency was such that a smaller heating, ventilating, and air conditioning (HVAC) unit could be used. This yielded an initial savings of $28,000. Of course, this was offset by the $55,000 higher price tag for the double-glazed units.

The big savings comes in the form of reduced utility bills because of the double-glazed window's thermal efficiency. This amounts to $15,200 per year (based on rates existing at time of study). A 20-year life for the facility was used to project the total savings. Assuming 7 percent interest and assuming utility costs would increase 5 percent per year, the resultant savings over the 20 years amounted to $249,000 (present value).

Table 9.11 is a Value Analysis Report summarizing the results of this study.

# PRODUCTIVITY

## 9.25   DEFINING THE PROBLEM

Productivity has become an important topic in manufacturing and business. The Germans and the Japanese were the first to establish national centers to analyze and increase productivity. It was not until 1977 that the American Productivity Center was established in Houston, Texas, as a private undertaking.

Productivity is basically an engineering topic. The technological elements of productivity are not part of this discussion, but the subject has ancillary aspects that should be recognized by all cost engineers.

It is important that cost engineers properly use the terms *production* and *productivity*. *Production* refers to total quantities produced in a period of time, whereas *productivity* refers to quantities produced per employee hour of effort (hereinafter called work-hour). While production is of importance in terms of satisfying a requirement within a given period, this production may be achieved with varying productivity. Poor productivity results in higher unit costs; good productivity is the goal and is the subject of this section.

Productivity can be defined by any of the equations in Fig. 9.8. In construction, *total output,* as used in the last equation, is the complete project. To improve total productivity, the total work-hours must be reduced.

Table 9.12 shows the results on three projects as estimated by a public utility. *Prime productivity,* the direct productive work, as given in the tabulation is 34.7 percent, but this simple conclusion can be challenged. Some classifications not included as productive work, such as obtaining tools and material, may be essential in the opinion of others. Prime productivity is a concept that should be measurable by engineering standards, but the first observation is that even this simple concept is subject to human interpretation.

For purposes of discussion it is convenient to break out the total work-hours in a project as shown in Fig. 9.9. Next, to provide a point of reference for how these

**TABLE 9.11**
**Value analysis report for Example 9.12**

**PROJECT:** Gainesville VA Hospital     **DATE:** _____

**ITEM:** Glazing     **PROJECT NO.** _____

**Summary of Change (Description)**

**Before $:** 0/year     **After $:** $15.200/year

The existing design criteria required 1/4" single-glaze, tinted glass to be installed at the various window locations. Due to the geographical location, the building orientation, and sizeable area of glass in the building it was believed that other options might prove more cost effective. After value analysis of five options it was concluded that a tinted, double-glazed window unit would provide greatest value for the owner. The double-glazed unit has a much higher thermal efficiency and would result in annual savings in utility bills. Also, reduction in the size of the building's HVAC system can be achieved due to lower demand on the system.

**Estimated Cost Summary**

Initial Savings (Loss)

    Smaller HVAC equipment     $ 28,000

    Difference in cost of double-glazed tinted glass
    over original design     ($ 55,000)

Net Initial Savings (Loss)     ($ 27,000)

Life-Cycle Cost Savings

    Present value of 20 years of $15,200 annual
    savings at 7% interest and utility bills increasing
    5% per year     $ 249,000

Net Savings over 20 years, Present Value     $ 222,000

239

Productivity  = Output ÷ Input

= Units ÷ work-hours

= (Total output) ÷ (Total work-hours)

**FIGURE 9.8**
Productivity equations.

**TABLE 9.12**
**Activity of workers in public utility construction projects**

| Activity measured | Time, in % |
|---|---|
| Late starting, early quitting | 3.2 |
| Personal | 2.3 |
| Breaks | 2.1 |
| Waiting and unexplained idle time | 31.9 |
| Receiving instructions and planning work | 4.4 |
| Reading drawings | 0.7 |
| Traveling | 12.8 |
| Transport | 4.6 |
| Obtaining tools and materials | 3.3 |
| Direct productive work | 34.7 |
| | 100.0 |

*Source:* H. M. Hohns, "Productivity: An Updated Concept of Management," *Transactions of the American Association of Cost Engineers,* p. I.1.1, 1979.

Total work-hours $= X_T = X_1 + X_2 + X_3 + X_4 + X_5$

where:  $X_1$ = Direct hours

$X_2$ = Support hours

$X_3$ = Idle/nonproductive hours

$X_4$ = Rework hours, engineering error

$X_5$ = Rework hours, construction error

**FIGURE 9.9**
Total project work-hours equation.

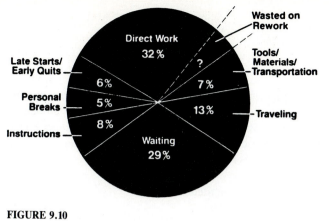

**FIGURE 9.10**
Typical worker's hour.

hours break down on a project, Fig. 9.10 provides in pie-chart form, the results of a study conducted on a number of U.S. power plant projects in the mid-1970s [32]. This study did not catalog hours by the five categories noted in Fig. 9.9, but the results help to make a point. About one-third the hours were spent in direct work, about 28 percent in support tasks (instructions, traveling, tools), and about 40 percent was wasted/idle time. Note that *rework* was not included as a category; therefore, some of the direct work must have been rework and this means that the percentage of usable direct time was really less than the 32 percent. Even if one assumes that one-third the time is a reasonable approximation of direct time, this is only 20 min out of an hour. Furthermore, it means that, to increase productivity 10 percent, one need only convert 2 min of nondirect work to direct each hour! A competent manager would never admit that could not be done. The next step is to derive a program for doing so.

## 9.26 PROGRAM BASIS

Referring back to the five categories of work included in a project's total, let us examine each in detail to determine what factors influence them.

Direct hours are the hours managers seek to maximize. This can be done in two ways: (1) creating more direct work minutes per clock hour by decreasing nondirect time, and (2) increasing the efficiency of work within each direct work minute. Decreasing nondirect time is a subject of later paragraphs. To increase efficiency, efforts must be targeted on these items which influence efficiency:

Individual worker skills

Work-force sociological characteristics

Crew stability

Capacity/condition of tools/equipment

Supervisor competence

Worker attitude

Project demographics

Technology employed

Point on learning curve

Management competence

Work area environment                   Relationship to other work
Project location                        Weather
Constructability of design              Use of crash scheduling
Work rules                              Layout of site

Support hours are next. With these, the objective is to minimize the total number by increasing efficiency of this work. Since support activity efficiency is influenced by the same factors as direct hour activity, the listing is the same as for direct hours.

Idle/nonproductive hours comprise another category of hours managers want to minimize. Factors influencing the number of these hours are:

Crew balance                     Worker attitudes
Crew stability                   Work rules
Weather                          Materials availability
Support equipment availability   Tool availability
Supervisor competence            Management competence
Job layout                       Holds for quality checks
Safety program

Rework hours due to engineering change, comprise the fourth category. The objective is complete elimination of these. The capability to do so will be influenced by:

Owner scope control                     Design engineer competence
Construction mode (fast track, etc.)    Constructability of design

Rework hours due to error is the last category. Again, the objective is total elimination. These factors affect the ability to achieve that objective:

Individual worker skills     Worker attitude
Crew stability               First-level quality control
Point on learning curve      Constructability of design
Supervisor competence

Having identified the many factors which influence productivity, the next step is to catalog them in terms of contractor capability to control. The first listing is of those over which the contractor has little or no control. This list includes:

Project location (owner can control during planning)
Work-force sociological characteristics (relates to location so owner can control)
Weather
Design engineer competence (owner can control in selection process)
Mode of construction contracting method (owner controls)
Project demographics (generally a function of project size)

Next is a listing of those over which the contractor may have some control:

| | |
|---|---|
| Individual worker skills | Point on learning curve |
| Relationship to other work | Work rules |
| Job layout | Constructability of design |
| Holds for quality checks | Work area environment |

Finally is a listing of those over which the contractor has significant control:

| | |
|---|---|
| Worker attitude | Crew stability |
| Technology employed | Capacity/condition of tools/equipment |
| Supervisor competence | Management competence |
| Use of crash scheduling | Crew balance |
| Materials availability | Support equipment availability |
| Tool availability | Job safety |
| First-level quality control | |

## 9.27 PROGRAMS FOR PRODUCTIVITY IMPROVEMENT

The factors identified as having potential for at least some control now become targets for improving productivity in a formal Improved Cost Effectiveness Program (ICEP). This leads to the creation of specific ICEP program elements. Major elements are summarized in the following paragraphs.

**MANAGEMENT TRAINING.** Competent management is essential to the success of an ICEP program. Key personnel must be proficient in motivational management, effective communication, construction planning, project controls, contract administration, construction methods and equipment, labor law, safety, and quality control. This proficiency is obtained through both structured, on-the-job training and formal instruction.

**SUPERINTENDENT AND FOREMAN TRAINING.** These first- and second-level supervisors need proficiency in leadership, effective communication, work planning, construction methods and equipment, project controls, safety, and quality control. Again, either structured, on-the-job development programs or formal instruction may be utilized.

**CRAFT TRAINING.** In addition to training in craft skills, craftsmen need training in safety and first-level quality control.

**SAFETY (LOSS CONTROL) PROGRAM.** A good loss control program not only prevents costly human, material, and time losses but is a morale booster for all workers — it shows that management cares. All of this contributes to improved productivity.

**FIRST LEVEL QUALITY CONTROL.** A quality control program oriented on *doing it right the first time* saves on costly rework and in that way contributes to overall higher productivity.

**MODERN MANAGEMENT SYSTEMS.** A comprehensive project controls system (cost, schedule, and materials) is needed to guide work through both planning and execution. To be truly effective, this system needs to integrate cost, schedule, and materials control. Additionally, there must be effective tools and equipment management programs. These systems are the tools of planners and controllers which help assure that personnel, tools, materials, and equipment are available in the right quantity when needed. Delays in the furnishing of these resources are major detractors of productivity.

**PARTICIPATIVE MANAGEMENT.** It has been well proven that a worker who contributes to the decision-making process will feel more committed to the work and be more productive. Also, worker ideas are often better than those of management because workers are closest to the work. Worker input can be solicited through quality circles, craftsmen surveys, and suggestion programs. Another effective technique is a frequent walk of the job by individual managers for the primary purpose of visiting with the workers and soliciting their ideas.

**COMMUNICATION.** Effective communication among all project participants can stimulate cooperation and productivity. Formal communication programs include pre-employment and follow-up orientations of all personnel, project newsletters, and bulletin boards. These programs are in addition to the interpersonal communications that are encouraged as part of effective motivational management.

**WORKER MOTIVATION.** Each of the previous programs contributes to worker motivation, but there are others specifically for that purpose. Examples are programs which *make the site a good place to work* (good worker facilities, well-organized site, etc.), motivational signage, a recognition and awards program, family days at the project, etc.

**ELIMINATION OF DISTURBANCES.** While it is obvious to everyone, one of the greatest detractors from productive work gets far too little attention. That is the effect of disturbances on productivity. When a worker is interrupted while doing a task the momentum is disrupted, time is lost during the interruption, and more time is lost as the worker attempts to reorganize and regain the momentum upon return to work. The lessons to be learned are twofold: (1) Provide all instructions and conduct all discussions *before* work begins. (2) Maintain policies which discourage disturbances between breaks by fellow workers or supervisors.

**CHANGE MANAGEMENT.** Changes are extremely disruptive to a project and lower its productivity by interrupting project momentum, causing costly rework, and interfering with efficient scheduling. A structured change management program should be employed within both engineering and construction phases.

**CONSTRUCTABILITY.** During both design and construction phases it is appropriate to utilize formal constructability review programs. These programs seek to select the most cost-effective methods of performing work without compromising quality or function.

## 9.28 WORK SAMPLING

Work sampling is often confused with productivity tracking, whereas it is only suitable for cataloging the way personnel are spending their time. There are two common work sampling techniques:

*Time-lapse photography* utilizes a movie camera that can be programmed to shoot a film one frame at a time with a predetermined time lapse between frames. Thus a single film reel can cover many hours or days of a project. When the film is developed, it can be analyzed for the purpose of identifying improper work practices, work inefficiencies, crew imbalances, safety problems, and other subjects that may be captured on film. It also may be used to provide a visual historical record of project progress. This system has become obsolete due to the replacement of 8mm photography by video.

*Human observation sampling* utilizes a matrix-type worksheet for logging the activity of a crew. Figure 9.11 is an example of such a worksheet. The observer randomly preselects a number of times that the crew is to be observed during a given shift. These times are selected before the shift and noted on the form as shown. At each designated time the observer observes the crew and puts in a check mark to indicate the activity each crew member is engaged in. In effect this method is a "snapshot" of crew activity at a point in time. After all observations have been made, the checks are summarized and data developed to show percentages of time for each category of activity. A video camera can be used to capture the activity, with logging handled in the office.

Unfortunately, work sampling does not distinguish among good work, improper work, and rework. Nor does it provide any indication of the rate work is being performed.

## NOMENCLATURE

| | |
|---|---|
| $B$ | Book value, $ |
| $C_{ex}$ | Irregular cost at the end of the $x$th year, $ |
| $C_i$ | Initial cost, $ |
| $C_{sal}$ | Salvage value, $ |
| $D_x$ | Fractional depreciation for the $x$th year, decimal |

**FIGURE 9.11**
Human observation work sampling form.

| | |
|---|---|
| $E_N$ | Effort required to produce the $N$th unit, e.g., work-hours per unit |
| $E_T$ | Cumulative effort to produce units 1 through $N$, e.g., work-hours |
| $E_T/N$ | Cumulative average unit effort, e.g., work-hours per unit |
| $F_{GP}$ | Factor to convert $G$ to $P$; uniform-gradient-series present-value factor, years |
| $F_{PK}$ | Capitalized cost factor, dimensionless |
| $F_{PR}$ | Capital recovery factor, year$^{-1}$ |
| $G$ | Uniform-gradient amount, \$/year |
| $i$ | Rate of return before taxes, decimal per year |
| $K$ | Capitalized cost, \$ |
| $K$ | Constant in Eq. (9.10); theoretical effort required for the first unit, e.g., work-hours |
| $L_D$ | Decimal learning ratio, dimensionless |
| $L_P$ | Percentage learning ratio, dimensionless |
| $n$ | A number, such as number of years |
| $N$ | Cumulative number of units produced, dimensionless |
| $p$ | Technological advancement rate, decimal per year |
| $P$ | Present value, \$ |
| $r$ | Rate of return after taxes, decimal per year |
| $R$ | Unacost, uniform end-of-year annual amount, \$/year |
| $R_p$ | Yearly operating expense subject to improvement by technological advancement, \$/year |
| $s$ | Slope constant, decimal, dimensionless |
| $t$ | Tax rate, decimal |
| VA | Value analysis |
| VE | Value engineering |
| $\Sigma E_N$ | Same as $E_T$ |
| $\dfrac{\Sigma E_N}{N}$ | Same as $E_T/N$ |
| $\psi$ | Present value of \$1 of depreciation, decimal |

## PROBLEMS

**9.1.** The book value of a machine is \$4000, with 4 years of life remaining at straight-line depreciation. Assume that the book value and salvage value are identical at all times. The operating cost for the machine is \$10,000/year as of the end of the year. A new machine costs \$30,000 and has an 8-year life and an operating cost of \$4000/year as of the end of the year. Neglect salvage value and taxes. If money is worth 10 percent per year, does it pay to install the new machine?

**9.2.** What is the actual rate of return realized with the new machine in the preceding problem?

**9.3.** Repeat Prob. 9.1 on the basis of a 34 percent tax rate and sum-of-the-years-digits depreciation for the new machine. The rate of return is 10 percent after taxes. Tax benefits can be taken as costs occur.

**9.4.** Assume that 5 years ago a machine cost \$35,000 and has been depreciated on a 10-year life with straight-line depreciation and \$1000 salvage value. If sold now, it will bring

$6000. The machine is part of a group, and depreciation continues at all times, just as if it were in service. If sold at the end of the year, it will bring $5000.

A new machine costs $75,000 with a 10-year life and has $4000 salvage value and an annual end-of-year savings of $12,000. Use sum-of-the-years-digits depreciation. If money is worth 12 percent per year after a 34 percent tax, should the new machine be bought?

**9.5.** Repeat Prob. 9.4 assuming that the machine is not part of a group and that any gain or loss on disposal can be taken as an immediate tax consideration.

**9.6.** A machine costs $20,000 and lasts 6 years, with $1000 salvage value at all times. Past records give the following data, where costs are as of the beginning of the year and include operating and repair costs:

| Year | Cost, $ |
| --- | --- |
| 1 | 3,000 |
| 2 | 3,500 |
| 3 | 8,000 |
| 4 | 15,000 |
| 5 | 5,000 |
| 6 | 15,000 |

If money is worth 10 percent per year and there is no tax, how long should the machine be kept?

**9.7.** If money is worth 12 percent per year after taxes, find the unacost and capitalized cost for the following after-tax expenses:

| End of year | Expense, $/year |
| --- | --- |
| 1 | 1000 |
| 2 | 1200 |
| 3 | 1400 |
| 4 | 1600 |
| ... | ... |
| 20 | 4800 |

**9.8.** In what respect is sum-of-the-years-digits depreciation related to a uniform-gradient series?

**9.9.** A company has produced machine parts as a subcontractor for a large military order for a number of years during which production was consistent with an 89 percent learning ratio. A commercial subcontract for making similar machine parts, but with a rigid schedule, is under consideration. Should you use the same percentage learning ratio for a schedule that is twice as fast as the military schedule? How would you develop the learning curve anticipated for this contract?

**9.10.** Use Eq. (9.9) and find the optimum value for $n$, the economic life under the following conditions: technological advancement rate 6 percent per year, 20 percent annual expense subject to technological advancement, 15 percent per year rate of return after a 50 percent tax, and sum-of-the-years-digits depreciation.

**9.11.** If 1000 light bulbs fail statistically according to the following pattern:

| Period | Failures |
|--------|----------|
| 1 | 50 |
| 2 | 100 |
| 3 | 250 |
| 4 | 500 |
| 5 | 75 |
| 6 | 25 |

and replacements are made at the beginning of each period if necessary, make a tabulation to find the steady-state number of replacements for an installation of 1000 bulbs and check this against the calculated value.

**9.12.** In Prob. 9.11 replacement costs are $1 per bulb but $0.50 per bulb as a group of 1000. Find the program for minimum cost.

**9.13.** Find the first unit value when the unit value of unit 100 is 30 h with an 80 percent learning ratio.

**9.14.** Find the value for unit 300 when the value for unit 3 is 1000 h with a 70 percent learning ratio.

**9.15.** Find the unit average hours for units 21 to 40 when unit 1 is 250 h with a learning ratio of (*a*) 90 percent and (*b*) 83 percent.

**9.16.** Units 1 through 200 have accumulated 45,000 h. Find the unit time for unit 100 with a learning ratio of (*a*) 70 percent and (*b*) 75 percent.

**9.17.** Using the rule-of-thumb lot-midpoint method, the geometric-mean lot-midpoint method, and the true lot-midpoint method for an 80 percent learning ratio, find the lot midpoints or lot equivalent points for:

Lot 1, units 1 through 14
Lot 2, units 15 through 40
Lot 3, units 41 through 80
Lot 4, units 81 through 100

**9.18.** Additional information for Prob. 9.17 is:

Lot 1, 15,000 h total time
Lot 2, 20,000 h total time
Lot 3, 25,000 h total time
Lot 4, 10,000 h total time

Continuing the problem, develop the slope constant by plotting on log-log graph paper. Using the method of least squares for the rule-of-thumb lot-midpoint method, establish the learning-curve function.

**9.19.** Analyze the answers to Prob. 9.18. What can be said about the slope of the curve?

# REFERENCES

**Learning curve**

1. Behrens, H. J.: "Improvement Curve, Real or Imaginary?" *AACE Bulletin*, vol. 8, June 1966, p. 75.
2. Fabrycky, W. J., P. M. Ghare, and P. E. Torgersen: *Industrial Operations Research*, 2d ed., Prentice-Hall, Inc., Englewood Cliffs, N.J., 1972.
3. Hirschmann, W. B.: "Profit from the Learning Curve," *Harvard Business Review*, January–February 1964, pp. 125–139.
4. Jordan, R.: *How to Use the Learning Curve*, Material Management Institute, Boston, 1965.
5. Kottler, J. L.: "The Learning Curve: A Case History in Its Application," *Journal of Industrial Engineering*, vol. 15, July–August 1964, pp. 176–180.
6. Noah, L. W., and R. W. Smith: *U.S. Air Force Project Rand Research Memorandum* MR-2786-Pr., January 1962.
7. Paterson, S. F.: "Predicting Learning Time," *Production*, July 1963.

**Value engineering**

8. Brown, B. H., Jr.: "Applied Value Engineering," *Skills and Knowledge of Cost Engineering*, 2d ed., American Association of Cost Engineers, Morgantown, W.Va., 1988, pp. 143–145.
9. Brown, B. H., Jr.: "Basic Value Engineering," *Skills and Knowledge of Cost Engineering*, 2d ed., American Association of Cost Engineers, Morgantown, W.Va., 1988, pp. 137–142.
10. Dell'Isola, A. J.: *Value Engineering in the Construction Industry*, 2d ed., Construction Publishing Company, New York, 1974.
11. Dobrow, P. V.: "Value Engineering—A Money Saving Tool," *Chemical Engineering*, vol. 79, August 21, 1972, pp. 122–126.
12. EIA Value Management Committee: *Value Engineering Support to Program Management*, Value Engineering Management Bulletin No. 8, Electronic Industries Association, Washington, D.C., 1979.
13. Falcon, W. D., ed.: *Value Analysis, Value Engineering*, American Management Association, New York, 1974.
14. Fallon, C.: *Value Analysis*, 2d rev. ed. Triangle Press, Southport, N.C., 1980.
15. Greve, J. W., ed.: *Value Engineering for Manufacturing*, Prentice-Hall, Inc., Englewood Cliffs, N.J., 1967.
16. Heller, E. D.: *Value Management, Value Engineering and Cost Reduction*, Addison-Wesley Publishing Company, Inc., Reading, Mass., 1971.
17. Miles, L. D.: *Techniques of Value Engineering and Analysis*, 2d ed., McGraw-Hill Book Company, New York, 1971.
18. Mudge, A. E.: *Value Engineering*, McGraw-Hill Book Company, New York, 1971.
19. Park, R. J.: *Practical Applications of Function Analysis System Technique "FAST,"* R. J. Park and Associates, Detroit, Mich., 1975.
20. Parker, D. E.: *Value Engineering Theory*, Value Foundation, Washington, D.C., 1977.
21. Ridge, W. J.: *Value Analysis for Better Management*, American Management Association, New York, 1969.
22. *SAVE Encyclopedia of Value*, Society of American Value Engineers, Southfield, Mich., 1977.
23. *Value World*, Society of American Value Engineers, Chicago, Ill.
24. Zimmerman, L. W., and G. D. Hart: *Value Engineering—A Practical Approach for Owners, Designers and Contractors*, Van Nostrand Reinhold, New York, 1982.

**Productivity**

25. *Fortune*: "The Disaster in Productivity," December 1974, p. 24.
26. Grayson, C. J., Jr.: "Eight Ways to Raise Productivity and Profits," *Nation's Business*, November 1972, pp. 30–36.
27. Hohns, H. M.: "Productivity: An Updated Concept of Management," *Transactions of the American Association of Cost Engineers*, 1979, pp. I.1.1–I.1.3.

28. Jelen, F. C., and M. S. Cole: "Can New Equipment Cost Less?" *Hydrocarbon Processing,* July 1971, pp. 97–100.
29. Jelen, F. C., and M. S. Cole: "Methods for Economic Analysis," *Hydrocarbon Processing,* Part 1, July 1974, pp. 133–139; Part 2, September 1974, pp. 227–233; Part 3, October 1974, pp. 161–163.
30. Neil, J. M.: "Productivity Management," *Skills and Knowledge of Cost Engineering,* 2d ed., American Association of Cost Engineers, Morgantown, W.Va., 1988, pp. 125–130.
31. Strandell, M.: "Productivity in the Construction Industry," *Cost Engineering,* March–April 1978, pp. 57–61. (Lists over 100 references.)
32. Strandell, M.: "Productivity in Power Plant Construction," *Transactions of the American Association of Cost Engineers,* 1976, pp. 334–337.
33. *Time*: "Productivity, Seeking That Old Magic," August 2, 1971, pp. 56–57.

# CHAPTER
# 10

# OPTIMIZATION

*F. C. Jelen*

**ONE VARIABLE**

## 10.1  OPTIMIZATION: A HUMAN TRAIT

*Optimization,* or the urge for efficiency, has a psychological origin. The human mind can confront a task or problem and recognize more than one course of action. This recognition is followed by a second phase, the selection of what is considered the best action. The second phase is the decision step. The two steps taken together, recognition of alternatives and decision, constitute optimization. Optimization can be qualitative (judged by human preference) or quantitative (detected by exact mathematical means). Optimization permeates society and technology to a far greater degree than is realized. The musical composer struggling for just the right combination of notes, the mathematician seeking an expression for a mathematical or physical principle, and the designer of a bridge are all involved in optimization. In a more prosaic sense, the golfer striving for a minimum score, the student hoping for a high IQ rating, and the newspaper headline announcing the greatest airplane fatality of all time likewise are manifestations of the allure of optimization and extreme values. Optimization is as old as antiquity, and even the ancients suspected that a circle has the greatest ratio of area to perimeter.

252

As engineering becomes more advanced and the business and industrial world more competitive, the methods for optimizing become more exact and the rewards increasingly great. Detailed consideration of optimization now pervades all successful human endeavor. There are basically three categories of optimization. The first is preferential, as exemplified by the musical composer struggling for just the right combination of notes, a category subject to preference and taste only. The second category is mathematical or physical, such as the demonstration that a sphere contains the maximum volume for a given surface, a fact not subject to human preference or fancy. The third category is economic optimization, which is a combination of these two categories but with the human preference elements introduced in a quantitative way. Economic optimization or engineering optimization, which is the subject of this chapter, has to consider human values, but they must be expressible in quantitative form. Thus the designer of a bridge is concerned with more than an engineering design for maximum strength-to-weight basis. He or she must consider myriad quantitative factors, such as the cost of on-site welding relative to the cost of prefabricated sections, extra construction costs for the safety of the workers, overdesign to meet future demand, etc.

## 10.2   NATURE OF OPTIMIZATION

Optimization in a general sense involves the determination of a highest or lowest value over some range. Which is desired is a matter of statement. Thus a problem can be maximized for profit or minimized for loss. *Optimization* will be used as a general term for either case. To the mathematician, maximum and minimum values have special qualities, and it is assumed that the analyst is familiar with these from a study of the calculus. There may be more than one mathematical maximum or minimum in a region, and the highest and lowest values in the region may or may not correspond to mathematical maximum and minimum values. Mathematical optima occur in the case of one independent variable when the slope of the curve is horizontal or zero. This is a necessary but not sufficient condition, as the point may be an inflection point. Certain other conditions must be met; e.g., the function must be continuous at the point in question. Figure 10.1 is a review of the circumstances. Points A and D are mathematical maxima, but the highest value is point F, which is not a mathematical maximum. Point C is a mathematical minimum and also the lowest value. Point B meets the condition that the slope is zero and the curve is horizontal, but B is neither a mathematical maximum nor a mathematical minimum; instead it is a point of inflection. Point E is not a mathematical minimum because the slope or first derivative is not continuous at point E. Maximum and minimum points which are not the highest and lowest values are referred to as *local optimum points*. Thus point A is a local maximum. Although the highest and lowest values in a region may not correspond to mathematical optima, a study of mathematical optima is fundamental to a study of optimization.

In working optimization problems, the analyst must formulate the problem in such a way as to optimize for the proper goal. This is elementary but frequently a

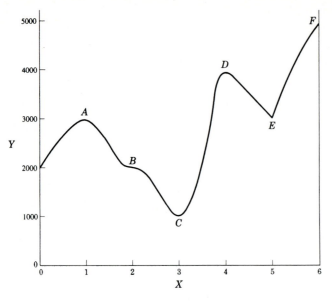

**FIGURE 10.1**
Various types of extreme values.

source of error because the result depends on the goal sought. Having obtained a result, the analyst should answer the following questions:

1. Is the answer feasible or realizable?
2. Is the answer optimal?
3. How sensitive is the optimum value to the parameters and how responsive is it to a variation of the independent variables from their optimum values?

1. The answer may not be feasible for several reasons. A physical barrier may exist. Thus mathematics may find that for optimum conditions, water at $-20°C$ should be pumped into the tubes of a heat exchanger, but water at $-20°C$ is solid. A computer can print out an optimum answer that requires an undisclosed mixture of benzene and toluene containing 120 percent benzene and $-20$ percent toluene. The two add up to 100 percent, but the mixture is not realizable. An answer may not be feasible because there are not enough days in the year to reach the yearly production requirement.

2. Is the answer optimum? It may be a minimum instead of a required maximum, or it may be a saddle point, such as point $B$ in Fig. 10.1, or it may be only a local maximum, such as point $A$.

3. Knowledge of the sensitivity of the optimum value to the parameters is essential. The analyst should know which components are most important and require the most study, and which are insignificant and can be ignored.

## 10.3 OPTIMIZATION METHODS

Many methods have been introduced to determine optimum procedures or policies. This chapter is fundamental and considers three powerful general methods:

1. Analytical
2. Graphical or tabular
3. Incremental

A large part of this book is devoted to optimization, and other chapters describe specialized methods. The search for or calculation of a highest or lowest value in a region can become exceedingly complicated in practice, and the analyst must be capable of originality in setting up solutions. Problems involving manufacturing or processing operations are referred to as *economic balance*. Usually cost data are involved, e.g., operation for maximum profit, but not necessarily so, e.g., operation for maximum output with cost disregarded. The problem may involve a single unit, such as a welding machine, or a combination of units, such as an assembly line. For multiunit systems, optimization is on an overall basis, and it does not follow that optimization of each unit on an individual basis will result in overall optimization.

In practice, many data will be empirical and economic in origin, so that optimization will be based on parameters which are not rigorously fixed or exact numbers. Also, the more variables and parameters involved, the flatter the optimum conditions generally become. Nevertheless, optimization as a goal is essential for successful operation even though there are some uncertainties.

Problems reducible to one independent variable will be discussed first, followed by an extension to more than one variable. The literature is replete with special problems leading to mathematical relationships which require only that analysts substitute numerical values. In doing so, they assume all the limitations, assumed values, and other fixed data in the model. Problems are so varied that analysts must be prepared to derive their own relationships.

## 10.4 ANALYTICAL METHOD FOR OPTIMIZATION

Let the problem be to optimize $Y$, which is a function of the single independent variable $X$. That is,

$$Y = F(X) \tag{10.1}$$

Generally an optimum value of $Y$ will be a mathematical maximum or minimum. For this condition, as given by the calculus,

$$\frac{dY}{dX} = \frac{dF(X)}{dX} = 0 \tag{10.2}$$

If the problem can be formulated so that $Y$ can be expressed in terms of $X$, the analytical method can be used. Equation (10.1) may represent an exact physical or mathematical relationship, an empirical relationship, or a combination. The means of establishing the relationship is not important; only its existence is necessary.

If $Y$ is not a continuous function of $X$, it can be treated as a series of regions in which it is continuous. Differentiation at a point having a discontinuity is not permissible since the value of Eq. (10.2), the derivative at the point, depends on from which side of the point $\Delta Y/\Delta X$ is taken. At times, too, $X$ can assume only integral values, e.g., the number of stories in an office building. Only continuous functions can be differentiated rigorously. Discrete variables should be treated by the calculus of finite differences or subjected to a tabular analysis. In many cases a function of discrete variables can be differentiated just as if it were a continuous function, but solutions so obtained should be investigated carefully.

If Eq. (10.1) can be set up, it is differentiated to satisfy Eq. (10.2), leading to a value or $X$ that is perhaps optimum. The value of $X$ so obtained can be substituted in Eq. (10.1) to give an optimum $X, Y$. The differentiation of Eq. (10.1) may be difficult and the resulting Eq. (10.2) so complicated that solution by trial and error may be required, but such considerations merely make the solution more tedious.

The solution to Eq. (10.2) will give perhaps several values of $X$, say, $X_1$, $X_2, \ldots, X_n$. Each value of $X$, say, $X_1$, can be tested analytically to determine whether it represents a maximum, a minimum, or a point of inflection, which is neither. For maximum and minimum, if

$$\frac{dY}{dX}\bigg|_{X=X_1} = 0 \quad \text{and} \quad \frac{d^2Y}{dX^2}\bigg|_{X=X_1} = +$$

then $\qquad\qquad Y$ at $X = X_1$ is a minimum

If

$$\frac{dY}{dX}\bigg|_{X=X_1} = 0 \quad \text{and} \quad \frac{d^2Y}{dX^2}\bigg|_{X=X_1} = -$$

then $\qquad\qquad Y$ at $X = X_1$ is a maximum

The sign of the second derivative indicates whether the point is a maximum or a minimum. The second derivative may be zero at the point, however, in which case the test is inconclusive. Further testing is necessary, which requires formulating higher-order derivatives. Find the lowest-order derivative that becomes nonzero for $X = X_1$. If this is an odd derivative, then $X_1$ is an inflection point. If this is an even derivative and positive in value for $X = X_1$, then $X_1$ is a minimum point, but if it is negative in value, $X_1$ is a maximum point.

In practice, it is rarely necessary to test beyond the second derivative. Another test is to calculate $Y$ in the neighborhood of $X = X_1$ and from the tabulation or plot to observe the behavior of $Y$ at $X_1$.

## 10.5  GRAPHICAL METHOD FOR OPTIMIZATION

If $\qquad\qquad\qquad\qquad Y = F(X) \qquad\qquad\qquad\qquad$ (10.3)

then a tabulation and plot of $Y$ against $X$ over the feasible range of $X$ will yield all the information for the optimization of $Y$. Figure 10.1 shows such a graph.

A plot or tabulation has the advantage of distinguishing among mathematical maxima and minima, inflection points, and highest and lowest values. Also, it discloses how responsive the optimum value is to a variation in the independent variable. The graphical method is widely applicable, and the data and information can be in analytical, tabular, or graphical form.

For one trial in one independent variable the analytical method has little advantage over the graphical or tabular method, although the analytical method will give the mathematical optima without a search procedure. Even so, the analytical method should be supplemented with some graphical or tabular data in a thorough analysis. The advantage of having an analytical expression for the optimum condition is very real, however, and permits a representation of the problem as mathematical relationships that are particularly valuable for theoretical studies or as expressions in compact, directly usable form.

## 10.6 INCREMENTAL METHOD FOR OPTIMIZATION

The incremental method is instinctive and basic. In essence, it amounts to operating at one value of the independent variable, changing the variable somewhat, and determining by calculation or observation whether the change is advantageous. Referring to Fig. 10.1, operation at $X = 3.5$ gives $Y = 2000$, and incrementing to $X = 3.6$ gives $Y = 2500$, an improvement. In this fashion small increments will lead to the local optimum $D$. Some precautions to be observed are evident from Fig. 10.1. It may be necessary to enter at several values of $X$ to avoid determining a local optimum only.

The incremental method is suitable for trial by experiment, including uncertainty in individual observations. With uncertainty, trials are repeated to establish probable values by means of statistical analysis. The incremental method is applicable to analytical or tabular data and can be used for continuous or discrete variables. It is analysis about a point.

## 10.7 A CYCLIC PROCESS

An example in optimization will be developed in detail to illustrate the principles. It is typical of a variety of problems in which at some point in a process it becomes economical to pay a penalty and start over again.

**Example 10.1.** In drilling wells in a certain type of terrain, an oil company finds that the depth to which it can drill before changing drill bits is given by the relationship

$$Q_c = 300T_d^{0.5}$$

where $Q_c$ = cumulative meters
$T_d$ = drilling time, days

Drilling costs are \$4250/day. It requires 18 h, or 0.75 day, to install a new drill bit at a total cost of \$10,500.

(a) Find the optimum cycle time for the minimum monthly cost at a drilling rate of 2000 m/month.

(b) Find the optimum cycle time for the minimum monthly cost at a drilling rate of 4750 m/month.

(c) Find the optimum cycle time for the maximum drilling depth per month without regard to cost.

(d) Find the optimum cycle time for the maximum profit if each meter drilled has a return value to the company of $60/m.

**Solution**

(a) The basis is 2000 m/month. Let $N$ = cycles per month. Then

$$NQ_c = 2000$$

$$N = \frac{2000}{Q_c} = \frac{2000}{300T_d^{0.5}} = \frac{6.667}{T_d^{0.5}} \qquad (10.4)$$

Set up an expression for cost in dollars per month for $N$ cycles per month:

$$\$/\text{month} = N\frac{\text{cycles}}{\text{month}}T_d\frac{\text{days}}{\text{cycle}}4250\frac{\$}{\text{day}} + 10,500\frac{\$}{\text{cycle}}N\frac{\text{cycles}}{\text{month}} \qquad (10.5)$$

$$\underset{\substack{\text{drilling}\\\text{expense}}}{} \qquad \underset{\substack{\text{bit}\\\text{expense}}}{}$$

Eliminate $N$ by Eq. (10.4):

$$\$/\text{month} = 28,333T_d^{0.5} + 70,000T_d^{-0.5} \qquad (10.6)$$

Differentiate and set equal to zero:

$$\frac{d(\$/\text{month})}{dT_d} = 0 = 0.5(28,333)T_d^{-0.5} - 0.5(70,000)T_d^{-1.5}$$

$$T_d = 2.471 \text{ drilling days per cycle}$$

Check for feasibility:

$$N = \frac{6.667}{T_d^{0.5}} = \frac{6.667}{2.471^{0.5}} = 4.241 \text{ cycles/month}$$

$$\text{Total days/cycle} = \underset{\text{drilling}}{2.471} + \underset{\text{changing}}{0.75} = 3.221$$

$$3.221\frac{\text{days}}{\text{cycle}}4.241\frac{\text{cycles}}{\text{month}} = 13.66 \text{ days/month}$$

which is feasible inasmuch as $13.66 < 30$:

$$\frac{\$}{\text{m}} = \frac{\$/\text{month}}{\text{m/month}} = \frac{4250NT_d + 10,500N}{2000}$$

and for $N = 4.241$ and $T_d = 2.471$, $\$/\text{m} = 44.53$. Check for the optimum by the second derivative. Differentiate Eq. (10.6):

$$\frac{d^2(\$/\text{month})}{dT_d^2} = -0.25(28,333)T_d^{-1.5} + 0.75(70,000)T_d^{-2.5}$$

which is positive for $T_d = 2.471$, and hence is a minimum cost.

*Graphical solution.* Dollars per month can be calculated from Eq. (10.6). Dollars per meter is obtained from dollars per month by dividing by 2000 m/month:

| $T_d$, days | $/month | $/m |
|---|---|---|
| 1.00 | 98,333 | 49.17 |
| 1.50 | 91,851 | 45.93 |
| 2.00 | 89,568 | 44.78 |
| 2.25 | 89,167 | 44.58 |
| 2.471 | 89,069 | 44.53 |
| 2.75 | 89,198 | 44.60 |
| 3.00 | 89,489 | 44.74 |
| 4.00 | 91,666 | 45.83 |

A plot of the results is given in Fig. 10.2, confirming that $T_d$ is both a mathematical minimum and a lowest value. It also shows how sensitive the monthly cost is to $T_d$.
(*b*) The basis is 4750 m/month:

$$NQ_c = 4750$$

$$N = \frac{4750}{300T_d^{0.5}} = 15.833T_d^{-0.5} \tag{10.7}$$

$$\$/\text{month} = NT_d(4250) + 10,500N$$

Eliminate $N$ by Eq. (10.7):

$$\$/\text{month} = 67,290T_d^{0.5} + 166,250T_d^{-0.5}$$

$$\frac{d(\$/\text{month})}{dT_d} = 0 = 33,645T_d^{-0.5} - 83,125T_d^{-1.5}$$

$$T_d = 2.471 \text{ drilling days per cycle}$$

Check for feasibility:

$$N = 15.833T_d^{-0.5} = 15.833(2.471^{-0.5}) = 10.07 \text{ cycles/month}$$

Total days/cycle = 2.471 + 0.75 = 3.221:

$$3.221\frac{\text{days}}{\text{cycle}} 10.07\frac{\text{cycles}}{\text{month}} = 32.44 \text{ days/month}$$

which is not feasible as 32.44 > 30. Collecting results, we obtain

$$N = 15.833T_d^{-0.5} \text{ cycles/month necessary}$$

$$\frac{30}{T_d + 0.75} \text{ cycles/month available}$$

and setting these equal gives

$$15.833T_d^{-0.5} = \frac{30}{T_d + 0.75}$$

**FIGURE 10.2**
Optimum drilling time per cycle for 2000 m/month.

or
$$T_d - 1.895T_d^{0.5} + 0.75 = 0$$

which is a quadratic equation in $T_d^{0.5}$. Solve this by the quadratic formula giving $T_d^{0.5}$ and then square to get $T_d$:

$$T_d = 0.317 \quad \text{or} \quad 1.775 \text{ days}$$

The feasible range for $T_d$ is 0.317 to 1.775 days. The following tabulation shows that the minimum cost is at $T_d = 1.775$ days:

| $T_d$, days | Cycles/month necessary | Cycles/month available | $/month | $/m |
|---|---|---|---|---|
| 0.250 | 31.67 | 30.00 | | |
| 0.317 | 28.12 | 28.12 | 333,160 | 70.14 |
| 1.000 | 15.83 | 17.14 | 233,510 | 49.16 |
| 1.500 | 12.92 | 13.33 | 218,120 | 45.92 |
| 1.775 | 11.89 | 11.89 | 214,510 | 45.16 |
| 2.000 | 11.20 | 10.91 | | |

Comparing parts (a) and (b), it should be understood that 2.471 days of drilling per cycle is optimum if production can be met. If not, it becomes necessary to change the drill bit more often. That $T_d = 1.775$ days in part (b) should be evident without the tabulation, since the closer the value is to 2.471, the less is the cost per meter drilled. Figure 10.2 gives the same conclusions. For $T_d$ in the range 0.317 to 1.775 the minimum cost per meter drilled is at $T_d = 1.775$. The mathematical minimum at $T_d = 2.471$ is outside the range and is not feasible.

(c) The meters per month drilled is maximized. Set up an expression for meter per month drilled:

$$\text{m/month} = \frac{30 \text{ days/month}}{(T_d + 0.75) \text{ days/cycle}} 300 T_d^{0.5} \frac{\text{m}}{\text{cycle}} = 9000 \frac{T_d^{0.5}}{T_d + 0.75}$$

For the maximum

$$\frac{d\text{m/month}}{dT_d} = 0 = 9000 \frac{(T_d + 0.75)(0.5) T_d^{-0.5} - T_d^{0.5}}{(T_d + 0.75)^2}$$

$$T_d = 0.75 \text{ drilling day per cycle}$$

$$T_{\text{tot}} = 0.75 + 0.75 = 1.50 \text{ days/cycle}$$

$$\text{m/cycle} = 300 T_d^{0.5} = 300(0.75^{0.5}) = 260$$

$$\text{\$/cycle} = 4250 T_d + 10,500 = 4250(0.75) + 10,500 = \$13,688$$

$$\text{\$/m} = \frac{13,688 \text{ \$/cycle}}{260 \text{ m/cycle}} = \$52.65$$

$$\text{\$/month} = 13,688 \frac{\text{\$}}{\text{cycle}} \frac{30}{0.75 + 0.75} \frac{\text{days/month}}{\text{days/cycle}} = \$273,760$$

(d) Maximum profit per month. Set up an expression for profit with

$$\text{Profit} = \text{income} - \text{expenses}$$

$$\frac{\text{Profit \$}}{\text{Cycle}} = \underbrace{300 T_d^{0.5} \frac{\text{m}}{\text{cycle}} 60 \frac{\text{\$}}{\text{m}}}_{\text{income}} - \underbrace{T_d \frac{\text{days}}{\text{cycle}} 4250 \frac{\text{\$}}{\text{day}}}_{\text{drilling}} - \underbrace{10,500 \frac{\text{\$}}{\text{cycle}}}_{\text{bits}}$$

$$\frac{\text{Profit \$}}{\text{Cycle}} = -4250 T_d + 18,000 T_d^{0.5} - 10,500$$

$$\frac{\text{Profit \$}}{\text{Month}} = \frac{\text{profit \$}}{\text{cycle}} \frac{\text{cycles}}{\text{month}}$$

$$\frac{\text{Profit \$}}{\text{Month}} = (-4250 T_d + 18,000 T_d^{0.5} - 10,500) \frac{30}{T_d + 0.75} \qquad (10.8)$$

For the maximum

$$\frac{d(\text{\$/month})}{dT_d} = 0$$

$$= 30 \frac{(T_d + 0.75)(-4250 + 9000 T_d^{-0.5}) - (-4250 + 18,000 T_d^{0.5} - 10,500)}{(T_d + 0.75)^2}$$

which is

$$-9000 T_d^{0.5} + 6750 T_d^{-0.5} + 7313 = 0$$

By trial and error or by the quadratic formula,

$$T_d = 1.86 \text{ drilling days per cycle}$$

$$\text{m/month} = 9000 \frac{T_d^{0.5}}{T_d + 0.75} = 9000 \frac{1.86^{0.5}}{1.86 + 0.75} = 4703$$

Income ($/month) = 60(4703) = $282,180

$$\text{Expenses (\$/month)} = [4250(1.86) + 10,500]\frac{\$}{\text{cycle}}\frac{30}{1.86 + 0.75}\frac{\text{cycles}}{\text{month}}$$

$$= \$211,550$$

Profit ($/month) = 282,180 − 211,550 = $70,630

$$\text{Expenses (\$/m drilled)} = \frac{211,550 \text{ \$/month}}{4703 \text{ m/month}} = \$44.98$$

A summary of the results for this example is given in Table 10.1. The optimum policy is different for each of the four conditions specified.

It is instructive to find the maximum profit by the *incremental method*. The profit per month is given by Eq. (10.8). Starting at $T_d = 1.6$ and adding increments to $T_d$ gives the following:

| $T_d$, days | Profit, $/month | Δ Profit, $/month | $\Delta T_d$ | Δ Profit/Δ $T_d$ |
|---|---|---|---|---|
| 1.60 | 69,810 | | | |
| | | 770 | 0.20 | 3850 |
| 1.80 | 70,580 | | | |
| | | 20 | 0.06 | 333 |
| 1.86 | 70,600 | | | |
| | | 0 | 0.04 | 0 |
| 1.90 | 70,600 | | | |
| | | −170 | 0.10 | −1700 |
| 2.00 | 70,430 | | | |

As $T_d$ is incremented, Δ profit/$\Delta T_d$ is positive until at $T_d = 1.86$ there is no further profit as $T_d$ is increased; thus Δ profit/$\Delta T_d$ is zero. As $T_d$ is increased beyond 1.86, the incremental profit is negative, a loss.

**TABLE 10.1**
## Summary of Example 10.1

| Condition optimized | $T_d$ optimum drilling, days/cycle | m/month | Cost, $/m | Cost, $/month | Profit, $/month |
|---|---|---|---|---|---|
| Drilling expense at 2000 m/month | 2.471 | 2000 | 44.53 | 89,069 | 30,931 |
| Drilling expense at 4750 m/month | 1.775 | 4750 | 45.16 | 214,510 | 70,490 |
| Meters drilled per month | 0.750 | 5198 | 52.65 | 273,760 | 37,640 |
| Profit per month at $60/m drilled | 1.86 | 4703 | 44.98 | 211,550 | 70,630 |

# 10.8 AN EXAMPLE INVOLVING RATE OF RETURN

The time value of money can be included. Thus if the time period is made 1 year and the end of the year is taken as the vantage point, all costs and receipts can be reduced to unacosts and the analysis made on that basis. The following example is instructive because rate of return is used as a criterion.

**Example 10.2.** An office building can be built 5, 10, 15, or 20 stories high. Investment and revenue representing receipts less expenses as a uniform end-of-year amount are:

|  | Stories | | | |
|---|---|---|---|---|
|  | 5 | 10 | 15 | 20 |
| Investment, $ | 5,000,000 | 7,500,000 | 11,000,000 | 15,000,000 |
| Revenue, $/year | 662,500 | 1,450,000 | 1,782,000 | 2,209,000 |

The figures are before a 34 percent tax rate. The life of the building is 31.5 years for both economic and tax purposes. MACRS depreciation with no salvage value will be used. Each dollar invested in the building must earn 10 percent per year or more after taxes. Find the maximum number of stories that should be constructed under these circumstances.

*Solution.* If $C_d$ represents the fully depreciable first cost MACRS depreciation is straight line. The unacost by item 1, Table 3.5, is

$$R = C_d(1 - 0.34F_{SLP,r,31.5})F_{PR,r,31.5}$$

Interpolating numerical values from Table A1.1 and designating $(1 - 0.34F_{SLP,r,31.5})$ $F_{PR,r,31.5}$ by $Y$, a tabulation is

| r, % | Y | r, % | Y |
|---|---|---|---|
| 4 | 0.0457 | 12 | 0.1127 |
| 6 | 0.0630 | 14 | 0.1315 |
| 8 | 0.0771 | 16 | 0.1507 |
| 10 | 0.0945 | 20 | 0.1899 |

If $R_v$ represents the end-of-year revenue, the unacost after taxes by item 2, Table 3.6, is $R = R_v(1 - t) = 0.66R_v$.

For the five-story building with expenses and receipts equal on the basis of unacost

$$C_d Y = 0.34R_v$$

$$5,000,000Y = 0.66(662,500)$$

$$Y = 0.0875$$

and from the tabulation,

$$r = 9.2 \text{ percent by interpolation}$$

So a five-story building cannot be considered. For a 10-story building

$$7,500,000Y = 0.66(1,450,000)$$

$$Y = 0.1276$$

$$r = 13.6 \text{ percent by interpolation}$$

So a 10-story building will pay. For a 15-story building

$$11,000,000Y = 0.66(1,782,000)$$

$$Y = 0.10692$$

$$r = 11.4 \text{ percent by interpolation}$$

but this is overall and includes the rate of return on the first 10 stories. Based on the incremental upper five stories

$$(11,000,000 - 7,500,000)Y = 0.66(1,782,000 - 1,450,000)$$

$$Y = 0.0626$$

$$r = 6.0 \text{ percent approx. by interpolation}$$

Hence a 15-story building is not justified because the investment in the upper five stories does not earn 10 percent per year. The calculation is extended to a 20-story building, and the results are given in Table 10.2. The only investment justified is the 10-story building if the investment, including the last dollar invested, is to earn 10 percent per year or more after taxes.

**TABLE 10.2**
**Summary of Example 10.2**

| | All stories | | | Incremental stories | | |
|---|---|---|---|---|---|---|
| Stories | Initial investment, $ | Yearly revenue, $ | Rate of return, % | Initial investment, $ | Yearly revenue, $ | Rate of return, % |
| 5 | 5,000,000 | 662,500 | 9.2 | | | |
| | | | | 2,500,000 | 787,500 | >20 |
| 10 | 7,500,000 | 1,450,000 | 13.6 | | | |
| | | | | 3,500,000 | 332,000 | 6.0 |
| 15 | 11,000,000 | 1,782,000 | 11.4 | | | |
| | | | | 4,000,000 | 427,000 | 7.1 |
| 20 | 15,000,000 | 2,209,000 | 10.3 | | | |

# 10.9  A TWO-STEP EXAMPLE WITH RECYCLE

The example developed in this section pertains to a two-step process. The requirement is not to optimize each step on an individual basis but to optimize the two steps together on an overall basis. In addition, the example includes feedback of material to an earlier step, a feature common in the process industries. In this example there is an economic balance between the two steps, the reactor and the separator. If the reactor is made large, the conversion is high and the separation costs are low; but there is an optimum size reactor which results in a lowest overall cost.

**Example 10.3.** The following homogeneous reaction

$$2A + B = D \qquad \text{(kilogram-moles basis)}$$

is carried out in a liquid-phase reactor in which the conversion of $B$ to $D$ depends on the residence time according to the relationship

$$r = 0.25T^{0.33} \qquad 0.5 < T < 10 \qquad (10.9)$$

where $r$ = mole $D$ formed per mole $B$ fed

$T$ = residence time in reactor, in hours

Fresh feed and recycle are always fed to the reactor in the proportion of $2A$ to $1B$. The density of the feed liquor is 9.6 kg mol/m$^3$.

Operation will be 4000 h/year to produce 3 kg mol of $D$ per onstream hour. Reactors cost $90,000/m$^3$ of contained liquid and have a 4-year life with no salvage value for economic and tax purposes. Use sum-of-the-years-digits depreciation and a 10 percent per year rate of return after a 46 percent combined federal, state, and local tax. The output from the reactor is run to a separator and is treated at a cost of $3.65/kg mol of unreacted $A$ and $B$. The unreacted $A$ and $B$ are recycled back to the reactor.

Find the size reactor in cubic meters that will give the minimum overall cost per year.

**Solution.** First, construct the flow sheet with a basis of 1 h and start with a basis of $2A$ and $1B$ fed to the reactor. If $T$ is the residence time in the reactor, then by Eq. (10.9), materials out of the reactor are as follows:

$0.25T^{0.33}$  kilogram-moles of $D$ or kilogram-moles of $B$ reacted per hour

$2 - 2(0.25)T^{0.33}$  kilogram-moles of unreacted $A$ per hour

$1 - 0.25T^{0.33}$  kilogram-moles of unreacted $B$ per hour

Fresh feed, which must be added to maintain $2A + 1B$ to the reactor, is

$0.50T^{0.33}$  kilogram-moles of $A$ per hour

$0.25T^{0.33}$  kilogram-moles of $B$ per hour

The flow sheet is given by Fig. 10.3.

**FIGURE 10.3**

Flow rates, in kilogram-moles per hour, basis $2A + 1B$, to a reactor per hour.

If every flow is multiplied by $3/0.25T^{0.33}$, the kilogram-moles per hour of $D$ produced becomes 3, the desired rate. Figure 10.4, so prepared, becomes the flow sheet for the process.

The volume of feed to the reactor in cubic meters per hour is

$$\left(\frac{6}{0.25T^{0.33}} + \frac{3}{0.25T^{0.33}}\right)\left(\frac{\text{kg mol}}{\text{h}} \quad \frac{1}{9.6 \text{ kg mol/m}^3}\right) = \frac{3.75}{T^{0.33}} \text{ m}^3/\text{h}$$

The volume $V$ of the reactor in cubic meters and the residence time are related by

$$T = \frac{V, \text{ m}^3}{\text{feed, m}^3/\text{h}}$$

so that, by the last two relationships,

$$V = \frac{3.75 \text{ m}^3}{T^{0.33} \text{ h}} T \frac{\text{h}}{1} = 3.75T^{0.67}$$

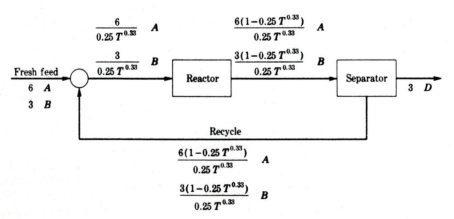

**FIGURE 10.4**

Flow rates, in kilogram-moles per hour, basis 3 kg mol/h, for product $D$.

The cost of reactors, basis unacost after taxes and item 1, Table 3.5, is

$$R = C_d(1 - tF_{SDP, 10\%, 4})F_{PR, 10\%, 4}$$

$$= 90,000(3.75T^{0.67})[1 - 0.46(0.83013)]0.31547$$

$$= 65,814T^{0.67} \$/year$$

The amount of recycle per hour is

$$\frac{6(1 - 0.25T^{0.33})}{0.25T^{0.33}} + \frac{3(1 - 0.25T^{0.33})}{0.25T^{0.33}} = 36T^{-0.33} - 9 \text{ kg mol/h}$$

The cost of separation, basis unacost after taxes and item 2, Table 3.5, assuming an end-of-year cost, is

$$R(1 - t) = (36T^{-0.33} - 9)\frac{\text{kg mol}}{\text{h}}4000\frac{\text{h}}{\text{year}}3.65\frac{\$}{\text{kg mol}}(1 - 0.46)$$

$$= 283,820T^{-0.33} - 70,960 \$/year$$

The total cost after taxes is

$$65,814T^{0.67} + 283,820T^{-0.33} - 70,960 \$/year \qquad (10.10)$$

For the minimum

$$\frac{d \$/year}{dT} = 0 = 44,095T^{-0.33} - 93,660T^{-1.33}$$

$$T = 2.124 \text{ h} \qquad \text{optimum retention time}$$

$$V = 3.75T^{0.67} = 6.21 \text{ m}^3 \qquad \text{optimum reactor volume}$$

A plot of Eq. (10.10) will verify that a minimum value for dollars per year is actually attained at $T = 2.124$ h.

Although this process involves two steps, it can be resolved in terms of a single variable. Eq. (10.10), of course, can be superseded by a relationship in terms of $V$ as the independent variable. The example illustrates that problems in economic balance demand a good model and care in setting up the expression for the condition to be optimized. In particular, the analyst must guard against errors in units and dimensions.

## 10.10 VARIOCYCLIC PROCESSES

Many processes are cyclic or can be treated as such for convenience. In Example 10.1, concerned with drilling an oil well, the underlying physical relationship was

$$Q_c = 300T_d^{0.5}$$

where $Q_c$ was cumulative meters drilled for a period of $T_d$ days after installing a new drill bit. It was inherent in the problem that each cycle duplicated the others. In some cases, however, there are cycles within cycles. Repetitions within the cycle will be

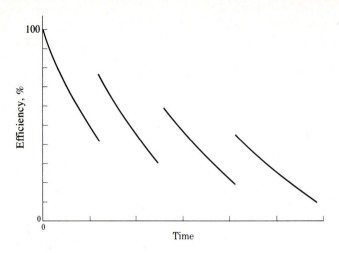

**FIGURE 10.5**
Efficiency in a variocyclic process.

referred to as *rounds*. A cycle comprises a number of rounds. The rounds within a cycle show differences, but each cycle repeats the previous cycle. Such cycles are referred to as *variocyclic* processes to denote the variation of the rounds with a cycle.

Catalytic processes offer some excellent examples of a variocyclic process. Frequently, a catalyst can be rejuvenated but becomes less efficient with each rejuvenation, until finally it is discarded and the cycle is repeated with new catalyst. Figure 10.5 is a graph of the situation, showing a series of rounds to complete one cycle. There is a processive degradation from round to round.

In practice, the situation can be very complicated, for more than one variable is usually involved. The temperature of the catalyst can have a pronounced effect, so that the optimization for the time-temperature relationship for all the rounds can become exceedingly intricate. The problem is discussed here in only an elementary and simplified way. In variocyclic processes it is important that the analyst understand the significance and use of such terms as *instantaneous rate, instantaneous efficiency, cumulative throughput, cumulative efficiency, average rate,* and *average efficiency.* An average depends on how the average is computed; e.g., an average with respect to time can be different from an average based on input.

If there is some systematic relationship between the rounds, an analytic solution is generally possible. Frequently, tabular solutions are convenient, and they may be necessary if the data are in tabular form or if the variables are discrete. Dynamic programming can be used successfully in many cases.

> **Example 10.4.** An absorbent clay is used to clarify a liquor. The clay costs $50/1000 kg and $25 for each activation, including the first round. One thousand kilograms of clay will treat 10,000 kg of liquor in the first round, diminishing to 85 percent of the previous round for subsequent rounds. How many rounds should the clay be used?

**TABLE 10.3**
**Optimum use of clay**

| Round | Treated, kg | Cumulative amount treated, kg | Catalyst cost, $ | Activation cost, $ | Total cost, $ | Cumulative cost, $ | Cumulative average, $/kg |
|-------|-------------|-------------------------------|------------------|--------------------|---------------|--------------------|--------------------------|
| 1 | 10,000 | 10,000 | 50 | 25 | 75 | 75 | 0.00750 |
| 2 | 8,500 | 18,500 | | 25 | 25 | 100 | 0.00541 |
| 3 | 7,225 | 25,725 | | 25 | 25 | 125 | 0.00486 |
| 4 | 6,141 | 31,866 | | 25 | 25 | 150 | 0.00471 |
| 5 | 5,220 | 37,086 | | 25 | 25 | 175 | 0.00472 |
| 6 | 4,436 | 41,523 | | 25 | 25 | 200 | 0.00481 |

*Solution.* The calculation is shown in Table 10.3. If the clay is used for four rounds, a minimum cost of \$0.00471/kg treated is obtained.

## MULTIVARIABLE OPTIMIZATION

### 10.11 ANALYTICAL METHOD

The same principles and methods that are useful for one independent variable apply to more than one independent variable but become much more complicated as the number of variables increases. Consider first the analytical method for some criterion $W$, a function of $X$ and $Y$, which is to be optimized as follows:

$$W = F(X, Y) \tag{10.11}$$

For a mathematical optimum two conditions must be met:

$$\frac{\partial W}{\partial X} = 0 \tag{10.12}$$

$$\frac{\partial W}{\partial Y} = 0 \tag{10.13}$$

Generally each of these relationships involves $X$ and $Y$; but there will be two equations with two unknowns, and sets of $X, Y$ can be obtained. Each set is a possible optimum for $W$ as given by Eq. (10.11).

An analytical test is available to establish whether a maximum or minimum exists. A maximum for $W$ exists if the three following conditions are satisfied for the $X, Y$ pair considered:

$$\frac{\partial^2 W}{\partial X^2} \text{ negative} \qquad \frac{\partial^2 W}{\partial Y^2} \text{ negative} \qquad \frac{\partial^2 W}{\partial X^2} \frac{\partial^2 W}{\partial Y^2} > \left(\frac{\partial^2 W}{\partial X \, \partial Y}\right)^2$$

Similarly, there will be a minimum if

$$\frac{\partial^2 W}{\partial X^2} \text{ positive} \qquad \frac{\partial^2 W}{\partial Y^2} \text{ positive} \qquad \frac{\partial^2 W}{\partial X^2}\frac{\partial^2 W}{\partial Y^2} > \left(\frac{\partial^2 W}{\partial X \partial Y}\right)^2$$

In either case if

$$\frac{\partial^2 W}{\partial X^2}\frac{\partial^2 W}{\partial Y^2} < \left(\frac{\partial^2 W}{\partial X \partial Y}\right)^2$$

$W$ is neither maximum nor minimum but a saddle point. If the last relationship is an identity, the nature of the point is inconclusive and further testing of higher-order derivatives is necessary.

If $W$ is a function of three variables, $X$, $Y$, and $Z$,

$$W = F(X, Y, Z)$$

the three following conditions are necessary, but not sufficient, for mathematical optima:

$$\frac{\partial W}{\partial X} = 0 \qquad \frac{\partial W}{\partial Y} = 0 \qquad \frac{\partial W}{\partial Z} = 0$$

In general these will each contain $X, Y, Z$, but the three equations in three unknowns can be solved for sets of $X, Y, Z$. The extension to four or more variables is obvious.

## 10.12  GRAPHICAL METHOD

A graphical method in the form of a three-dimensional space model can be used for two independent variables. For

$$W = F(X, Y) \tag{10.14}$$

$X$ and $Y$ can be measured in a horizontal plane and $W$ measured by the perpendicular distance from the $XY$ plane. In a complicated case the space model will appear like a mountain range. Inasmuch as space is limited to three dimensions, it is not possible to treat higher-order systems by this method.

There are techniques for simplifying multivariable systems which are explained here for the two-variable system of Eq. (10.14). Let $Y$ be given a series of successive fixed values and replace the single relationship by a series of relationships:

$$W_1 = F(X, Y_1)$$
$$W_2 = F(X, Y_2)$$
$$\cdots\cdots\cdots\cdots$$
$$W_N = F(X, Y_N)$$

which reduces the two-variable problem to a series of one-variable problems. If the range taken for $Y$ is wide enough, it becomes possible to find the optimum $W$ from a series of tabulations, augmented with a plot if desirable.

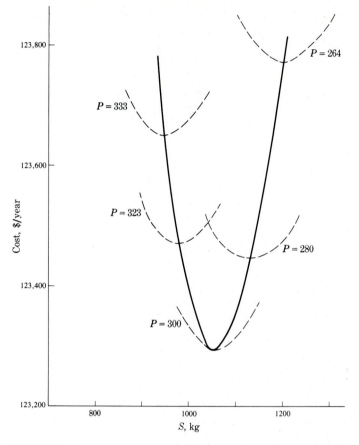

**FIGURE 10.6**
Optimum size $S$ for various powers to stirrer $P$.

Such a plot is shown in Fig. 10.6, where a cost in dollars per year is plotted for two variables, $S$ and $P$. $P$ is given successive fixed values, and the optimization for each is shown as a series of dotted curves. Connecting these incidental minima leads to the true minimum at $S = 1056$, $P = 300$.

Another method involves calculating $W$ for various points $X, Y$ using only a two-dimensional plot and connecting points corresponding to equal values of $W$ by contour curves. Such a plot resembles a typographical map. Details are given in Chap. 13.

## 10.13 INCREMENTAL METHOD

The incremental method involves changing the independent variables singly or in groups. As the number of variables increases, considerable precaution must be exercised to avoid false conclusions. An increase in $X$ alone may lower the value of $W$, the function to be optimized, yet the optimum may lie in the direction of increasing $X$ and $Y$ simultaneously. Chapter 13 should be consulted for details.

The incremental method with multivariables is suitable for trial by experiment, including uncertainty as mentioned in Sec. 10.6 for one variable. For further study see Chap. 13.

## 10.14 A TWO-VARIABLE OPTIMIZATION PROBLEM

**Example 10.5.** A manufacturer prepares a plastic in a special batch reactor-mixer. The time required to process the mixture varies directly as the square root of the capacity of the mixer and inversely as the square of the power to the stirrer. Quantitatively,

$$T = 1000 \frac{S^{0.5}}{P^2} \tag{10.15}$$

where $T$ = total time required, h/batch
$\quad\quad S$ = capacity of reactor-mixer, kg
$\quad\quad P$ = power to stirrer, kW

The cost for the reactor-mixer is $1265S^{0.5}$ \$/year. Electricity costs \$0.04/kWh, and there is an overhead or indirect cost expressible in terms of the power requirement as

$$\text{Overhead} = 137.2P \ \$/\text{year}$$

The full year is available, and $10^7$ kg/year must be processed. The time for loading and unloading the reactor-mixer is negligible. Find the optimum size and optimum stirrer power.

**Solution.** Set up an expression for the cost, basis 1 year of operation. Let $N$ be the number of batches per year and $S$ be the number of kilograms per batch:

$$SN = 10^7 \ \text{kg/year}$$

$$N = \frac{10^7}{S} \tag{10.16}$$

$$\$/\text{year} = 1265S^{0.5} + \left(T \frac{\text{h}}{\text{batch}}\right)\left(N \frac{\text{batch}}{\text{year}}\right)(P \ \text{kW})\left(0.04 \frac{\$}{\text{kWh}}\right) + 137.2P$$

Replace $T$ by Eq. (10.15) and $N$ by Eq. (10.16), yielding

$$\$/\text{year} = 1265S^{0.5} + 4 \times 10^8 P^{-1}S^{-0.5} + 137.2P \tag{10.17}$$

For the minimum cost, set the partial derivatives of the preceding expression equal to zero:

$$\frac{\partial, \ \$/\text{year}}{\partial S} = 0 = 0.5(1265)S^{-0.5} - 0.5(4 \times 10^8)P^{-1}S^{-1.5}$$

$$PS = 3.162 \times 10^5 \tag{10.18}$$

$$\frac{\partial, \ \$/\text{year}}{\partial P} = 0 = -4 \times 10^8 P^{-2}S^{-0.5} + 137.2$$

$$P^2S^{0.5} = 2.915 \times 10^6 \tag{10.19}$$

**TABLE 10.4**
**Yearly cost for example 10.5**

| S, kg | P, kW | Cost, $/year | Δ cost, $/year |
|-------|-------|--------------|----------------|
| 1056 | 300 | 123,298 | |
| 1056 | 315 | 123,402 | 104 |
| 1056 | 285 | 123,400 | 102 |
| 1003 | 300 | 123,323 | 25 |
| 1003 | 315 | 123,377 | 79 |
| 1003 | 285 | 123,481 | 181 |
| 1109 | 300 | 123,325 | 27 |
| 1109 | 315 | 123,476 | 178 |
| 1109 | 285 | 123,374 | 76 |

Solving for Eqs. (10.18) and (10.19) gives

$$S = 1056\text{-kg capacity}$$

$$P = 300\text{-kW power to stirrer}$$

Check for feasibility:

$$T = 1000\frac{S^{0.5}}{P^2} = 1000\frac{1056^{0.5}}{300^2} = 0.361 \text{ h/batch}$$

Using all the hours in a year, 8760, we obtain

$$1056\frac{\text{kg}}{\text{batch}}\frac{8760 \text{ h/year}}{0.361 \text{ h/batch}} = 2.56 \times 10^7 \text{ kg/year possible}$$

so that $10^7$ kg/year is feasible. A check is given in Table 10.4, where increments of ±5 percent in $S$ or $P$, or both simultaneously, are considered and the yearly cost is calculated from Eq. (10.17). Regardless of how $S$ and $P$ are varied, the cost increases over that for $S = 1056$ and $P = 300$. The cost is not very sensitive to $S$ and $P$, however. and a 5 percent variation in each simultaneously adds only $181/year to $123,298/year. Note further that the cost is more responsive to a 5 percent change in $P$ than to a 5 percent change in $S$.

The example can be solved as a one-variable problem by placing successive fixed values for $P$ in Eq. (10.17) and plotting the resulting curves. Figure 10.6 is such a plot, confirming an overall minimum at $S = 1056$ and $P = 300$.

## 10.15 LAGRANGE MULTIPLIERS

A powerful method for finding mathematical optima in multivariable problems involving supplementary conditions that must be observed, called *restraints*, was introduced by the French mathematician Lagrange (1736–1813). Consider

$$W_1 = F_1(X, Y, Z) \tag{10.20}$$

subject to the two conditional relationships, or restraints,

$$F_2(X, Y, Z) = 0 \tag{10.21}$$

$$F_3(X, Y, Z) = 0 \tag{10.22}$$

In an ideal situation the last two relationships can be used to eliminate, say, $Y$ and $Z$ from Eq. (10.20), leaving a relationship for $W$ in terms of $X$ only. This may not be possible, however, in an actual case. Lagrange's method is applicable to the general case and may make an analytical solution possible.

The method in application requires forming a new function, called the *Lagrange expression*, made up by linear addition of the original relationship and each restraint relationship multiplied by a Lagrange multiplier. There are as many multipliers as there are restraints. The Lagrange expression for the problem at hand is

$$LE = F_1(X, Y, Z) + \lambda_1 F_2(X, Y, Z) + \lambda_2 F_3(X, Y, Z) \tag{10.23}$$

and the optima for Eq. (10.20) are the optima of the Lagrange expression. Equation (10.23) includes the three original variables, $X, Y, Z$, and the two Lagrange multipliers, $\lambda_1$ and $\lambda_2$. Five relationships must be satisfied to find the optima for the Lagrange expression in this case:

$$\frac{\partial LE}{\partial X} = 0 \qquad \frac{\partial LE}{\partial Y} = 0 \qquad \frac{\partial LE}{\partial Z} = 0 \qquad \frac{\partial LE}{\partial \lambda_1} = 0 \qquad \frac{\partial LE}{\partial \lambda_2} = 0$$

whereas only three are necessary to find the optima for Eq. (10.20):

$$\frac{\partial W_1}{\partial X} = 0 \qquad \frac{\partial W_1}{\partial Y} = 0 \qquad \frac{\partial W_1}{\partial Z} = 0$$

Nevertheless, the Lagrange expression, with its five relationships, has the constraints built in and may be much easier to solve than Eq. (10.20) subject to the restraint equations (10.21) and (10.22).

For an example, return to Example 10.1, part ($a$), which has already been optimized. The requirement is to minimize

$$\$/\text{month} = NT_d(4250) + 10,500N \tag{10.24}$$

subject to the condition or restraint

$$N = 6.667T_d^{-0.5} \tag{10.25}$$

Write the restraint function as

$$N - 6.667T_d^{-0.5} = 0$$

The Lagrange expression becomes

$$LE = NT_d(4250) + 10,500N + \lambda(N - 6.667T_d^{-0.5}) = 0$$

Set the three partial derivatives equal to zero:

$$\frac{\partial LE}{\partial N} = 0 = 4250T_d + 10,500 + \lambda = 0$$

$$\frac{\partial LE}{\partial T_d} = 0 = 4250N - \lambda(6.667)(-0.5)T_d^{-1.5} = 0$$

$$\frac{\partial LE}{\partial \lambda} = N - 6.667T_d^{-0.5} = 0$$

Solving the three equations for $N$, $T_d$, and $\lambda$ gives

$$N = 4.241 \text{ cycles/month}$$

$$T_d = 2.471 \text{ h of drilling per cycle}$$

$$\lambda = -21,000$$

The values for $N$ and $T_d$ check those previously found.

In this example the use of Lagrange multipliers was not so direct as using Eq. (10.25) in Eq. (10.24) to reduce the latter to an equation in one independent variable. Nevertheless, the use of Lagrange multipliers can lead to a solution that otherwise would not be obtained. On the other hand, the use of Lagrange multipliers does not always produce expressions that are solvable by analytical means.

## 10.16 SENSITIVITY AND RESPONSE ANALYSIS

In a detailed analysis it is desirable to know the optimum and also how sensitive the optimum is to changes in various parameters that make up the problem. For example, suppose $Y$ is expressed by the relationship

$$Y = AX + \frac{B}{X} \tag{10.26}$$

where $X$ is the independent variable and $A$ and $B$ are parameters. In a given problem $A$ and $B$ may be constant, but they may be subject to change in practice. For example, $A$ and $B$ may be estimates, and it becomes desirable to know how much $Y_{opt}$ is changed by changes in the estimates of $A$ and $B$. What is desired mathematically is

$$\frac{\partial Y_{opt}}{\partial A} \quad \text{and} \quad \frac{\partial Y_{opt}}{\partial B}$$

The determination of such values is called *sensitivity analysis*.

An analytical solution is easily obtained for the relationship Eq. (10.26). For $Y_{opt}$

$$\frac{dY}{dX} = 0 = A - \frac{B}{X^2} \quad \text{and} \quad X_{opt} = A^{-0.5}B^{0.5}$$

which, substituted in Eq. (10.26), gives

$$Y_{\text{opt}} = 2A^{0.5}B^{0.5}$$

and leads directly to the relationships

$$\frac{\partial Y_{\text{opt}}}{\partial A} = A^{-0.5}B^{0.5} \qquad \frac{\partial Y_{\text{opt}}}{\partial B} = A^{0.5}B^{-0.5}$$

For the specific case for $A = 10$, $B = 40$

$$\frac{\partial Y_{\text{opt}}}{\partial A} = \sqrt{\frac{40}{10}} = 2$$

The value can be obtained in another way which is particularly useful if the analytical method is too involved or not available. The procedure is to start with Eq. (10.26) or graphical or tabular data and for $A = 10$ and $B = 40$ to find $Y$ optimum. This is found to be

$$Y_{\text{opt}} = 40 \qquad A = 10$$
$$B = 40$$

Next, repeat the calculation for, say, $A = 10.1$, $B = 40$. $Y_{\text{opt}}$ is found to be

$$Y_{\text{opt}} = 40.1995 \qquad A = 10.1$$
$$B = 40$$

From the last two relationships,

$$\frac{\Delta Y_{\text{opt}}}{\Delta A} = \frac{40.1995 - 40.000}{10.1 - 10} = 1.995$$

Thus a sensitivity analysis can be made even if analytical expressions are unwieldy or unavailable. A sensitivity analysis is also made in the same manner if there is more than one independent variable.

The sensitivity analysis as such cannot be applied to the independent variable. Thus for the case under study

$$\frac{\partial Y_{\text{opt}}}{\partial X} = 0$$

That is, $Y_{\text{opt}}$ is determined by $A$ and $B$ and not by $X$. However, it is possible to evaluate the change in $Y$ as a consequence of not operating at the optimum $X$. This is

$$\frac{Y - Y_{\text{opt}}}{X - X_{\text{opt}}} = \frac{\Delta Y}{\Delta X}$$

and $X$ can be taken as, say, a 1 percent change. The procedure can be called a *response analysis* to avoid confusion with a true sensitivity analysis. The reader should refer to Table 10.4, which was essentially a response analysis for a reactor-mixer having independent variables $S$ and $P$.

## 10.17 SIMPLIFICATION OF MULTIVARIABLE PROBLEMS

Optimization in the fields of economic balance can become exceedingly complicated because the problems can be intricately involved with physical, engineering, and cost considerations. This chapter was concerned with general methods, but the analyst must acquire the facility of representing engineering systems in terms of real models, either mathematical or empirical.

As the number of variables mounts, the complexity and scope of the problem increase disproportionately; hence every effort should be made to reduce the number of variables. This can be done by taking fixed values for some variables, particularly if practical considerations will limit the range. If a single value is too restricted, the study might be limited to a few values. Complicated expressions can sometimes be replaced with simpler expressions which might be satisfactory. Thus a linear expression might replace a quadratic expression for which the curvature is small in the feasible range.

As the number of variables and parameters increases and the number of constraints and practical limitations grows larger, the optimum has a propensity to grow flatter, with less likelihood that the realizable optimum will be a mathematical optimum. The tendency toward flatness is real in practical large-scale problems. This does not signify that a problem can be slighted, but it does indicate that some optimization should be considered in deciding on how much detail a particular problem merits.

Finally, optimization is a continuous study, and old problems take on a fresh challenge as technology, economics, and society change.

## 10.18 FORMULATING AN OPTIMIZATION PROBLEM

An optimization problem is an exercise in mathematical modeling that requires great care in setting up the model. Four steps are involved:

1. Decide the exact objective to be optimized. As shown in Example 10.1, many different objectives are possible.
2. Set up the objective function using as many variables as are required. Try for accuracy rather than compactness. Make sure that all the terms have the same dimensional units.
3. Set up all the restraints and relationships between the variables.
4. If possible, reduce the objective function in step 2 to independent variables. Step 2 may contain both dependent and independent variables.
5. The objective function is now ready for solution. If it contains independent variables only, the differentials can be set equal to zero to optimize the expression, or alternatively, a tabulation can be made. If the objective function contains dependent variables in addition to independent variables, a Lagrange expression can be

tried. If this fails, the objective function and its restraints must be optimized, using the skill and ingenuity of the analyst.

In formulating problems, it is necessary to recognize the difference among instantaneous rates, cumulative amounts, and various averages that may arise. Careful detailed labeling is recommended.

**Example 10.6.** The instantaneous rate of production for a material in kilograms per day with $T$ in days is given by

$$\frac{dP}{dT} = 1000e^{-0.1T}$$

Find the production rate at the end of the tenth day, the total production for 10 days, the average production rate for 10 days, and the average production rate for the last 5 of the 10 days.

**Solution.** The instantaneous production rate at $T = 10$ is

$$1000e^{-0.1T} = 1000e^{-0.1(10)} = 367.88 \text{ kg/day}$$

The total production for 10 days is

$$\int_0^{10} 1000e^{-0.1T} \, dT = \frac{1000}{-0.1}(e^{-0.1(10)} - e^0) = 6321.2 \text{ kg}$$

The average production for 10 days is

$$\frac{6321.2}{10} = 632.12 \text{ kg/day}$$

The average production for the last 5 days is

$$\int_5^{10} \frac{1000e^{-0.1T} \, dT}{10 - 5} = 477.3 \text{ kg/day}$$

## NOMENCLATURE

| | |
|---|---|
| $A$ | Parameter, as in Eq. (10.26) |
| $B$ | Parameter, as in Eq. (10.26) |
| $C_d$ | Depreciable cost, $ |
| $F_{SDP}$ | Present value for $1 for sum-of-the-years-digits depreciation, decimal |
| $F_{PR}$ | Factor to convert a present value to unacost, year$^{-1}$ |
| LE | Lagrange expression |
| $N$ | Cycles per month in Example 10.1; batches per year in Example 10.5 |
| $P$ | Power to the stirrer in Example 10.5, kilowatts |
| $Q_c$ | Cumulative meters drilled from the start of the cycle, Example 10.1 |
| $r$ | Rate of return after taxes, decimal per year |

| | |
|---|---|
| $r$ | Moles of $D$ formed per mole of $B$ fed in Example 10.3; depends on the residence time |
| $R$ | Unacost, uniform end-of-year amount, \$/year |
| $R_v$ | Revenue, end of year, in Example 10.2, \$/year |
| $S$ | Capacity of the reactor-mixer in Example 10.5, kilograms |
| $t$ | Tax rate, decimal |
| $T$ | Residence time in the reactor in Example 10.3, hours |
| $T$ | Total time required for a batch in Example 10.5, hours |
| $T_d$ | Drilling time from the start of the cycle in Example 10.1, days |
| $V$ | Volume of reactor in Example 10.3, cubic meters |
| $W$ | Dependent variable |
| $Y$ | Independent variable, generally |
| $Y$ | $(1 - 0.48F_{SDP, r, 25})F_{PR, r, 25}$ in Example 10.2 |
| $\lambda$ | Lagrange multiplier |

## PROBLEMS

**10.1.** A causeway is to be built over a long swamp. If $X$ is the distance in meters between piers for the range $20 \le X \le 100$, the following cost data apply:

$$\$ \text{ per pier} = 5X^2$$

$$\$/\text{m, span between piers} = 5000X^{-0.5}$$

Find the optimum distance between piers for the minimum cost.

**10.2.** The cost $Y$ in dollars per hour to produce electrical energy in a unit rated at 20,000 kW, where $X$ is the load generated in megawatts, is given by

$$Y = 12 + 0.2X + 0.27X^2$$

Income in dollars per hour from the sale of electricity is

$$16X - 0.2X^2$$

Find the generated load that gives the maximum profit.

**10.3.** The fixed cost for a steam line per meter of pipe is

$$350X + 18 \qquad \$/\text{year}$$

and the cost for loss of heat from the pipe per meter is

$$\frac{1.6}{X} \qquad \$/\text{year}$$

where $X$ is the thickness of insulation in meters. Find the optimum thickness of the insulation.

**10.4.** A commercial location makes a study to predict added gross income in dollars per year for investment in additional parking space. The survey predicts

| Invested, in parking lot, k$ | Added to gross income, k$/year | Invested in parking lot, k$ | Added to gross income, k$/year |
|---|---|---|---|
| 18  | 48  | 153 | 153 |
| 36  | 69  | 189 | 165 |
| 54  | 90  | 225 | 174 |
| 75  | 111 | 270 | 183 |
| 99  | 126 | 315 | 192 |
| 126 | 141 |     |     |

Added net income can be taken as 33.3 percent of added gross income. Use the incremental method and find how much should be invested in additional parking-lot space if the return on investment is 17 percent or more.

**10.5.** An office building can be built 5, 10, 15, or 20 stories high. The building will be depreciated uniformly over a 22-year life, and taxes may be neglected. Analysis gives the following, with all amounts in megadollars:

| Stories | 5 | 10 | 15 | 20 |
|---|---|---|---|---|
| Investment, M$ | 20 | 32 | 44 | 58 |
| Rent, M$/year | 3.6 | 7 | 10.2 | 13.2 |
| Operation, M$/year | 1.5 | 2 | 2.8 | 4.0 |

The minimum acceptable rate of return is 10 percent per year. How high should the building be built?

**10.6.** Costs associated with an extension to a plant are tabulated below in megadollars. Four production capacities are being considered, and the fixed investment will be written off in 10 years. Neglect taxes. A minimum 13.5 percent return on the original investment is required. What size plant, if any, should be built?

|  | *A* | *B* | *C* | *D* |
|---|---|---|---|---|
| Mkg/year | 5 | 10 | 15 | 20 |
| Fixed cost, M$ | 12 | 20 | 32 | 35 |
| Sales price, $/kg | 0.75 | 0.73 | 0.63 | 0.61 |
| Utilities, M$/year | 0.21 | 0.34 | 0.42 | 0.50 |
| Raw materials, M$/year | 0.2 | 0.4 | 0.6 | 0.8 |
| Labor, M$/year | 1.5 | 1.5 | 2.4 | 2.4 |

**10.7.** A heat treating furnace must process 1 million parts per year. The furnace can process only one batch per 8-h shift for one shift per day for a maximum of 250 days/year. The cost of the furnace is

$$25X^{0.7} \ \$/\text{year}$$

where $X$ is the number of machine parts processed per batch. The operating cost for the furnace is $200 per shift. Find the optimum size for the furnace to be installed.

**10.8.** Repeat Prob. 10.7, letting the furnacing time be dependent on the furnace size:

$$0.04X^{0.5} \text{ h}$$

The operating cost is a fixed cost of $104 per batch and $12/h for labor. A shift may run over 8 h, but only one shift per day for 250 days/year is allowed.

**10.9.** The cost for a stamping machine in dollars per year can be expressed as

$$48,000T^{-0.5}$$

where $T$ is the time in hours required to form the stamping. It costs $64/h to operate the machine. Required are 150,000 stampings per year for 8 h/day, not to exceed 300 days/year. Find the size of the stamping machine to be bought and the cost per year for the stamping operation.

**10.10.** In certain operations it is necessary to provide one spare machine to ensure continuity of operations at full capacity when one machine is being overhauled. If the cost of machines varies as the 0.8 power of capacity, find the optimum number of machines, including the spare, that should be provided.

**10.11.** A pressure vessel is to be made from a cylinder of diameter $X$ meters capped with a hemisphere at each end. The cost for the cylindrical section is

$$1800X^{0.5} \text{ \$/m of length}$$

and the cost for one hemispherical end is

$$1400X^{2.4} \text{ \$}$$

The vessel is to hold 40 m$^3$. Find the optimum dimensions and cost.

**10.12.** A chemical is produced by a batch process. Chemicals $X$ and $Y$ are used to make $Z$, with the following relationship for the kilograms of Z produced and the kilograms of $X$ and $Y$ used:

$$Z = 1.5(1.1XZ + 1.3YZ - XY)^{0.5}$$

$X$ costs $0.18/kg and $Y$ $0.08/kg, and $Z$ sells for $1.60/kg. One-half of the selling price for $Z$ is due to costs other than raw materials. Only $Z$ is recovered from the process. Find the maximum profit obtainable per kilogram of $Z$.

**10.13.** A machine costs $100,000 and has expenses which increase continuously according to the relationship

$$R = 50,000e^{0.2T}$$

where $R$ is the instantaneous rate of expense in dollars per year and $T$ is the time in years. Use continuously compounded interest with money worth 10 percent per year expressed as the continuous rate. Neglect taxes and find the optimum time to keep the machine in service.

**10.14.** Suppose that 86,000 kg/day of water for 300 days/year is to be evaporated from a salt solution. One kilogram of steam will evaporate $0.7N$ kg water, where $N$ is the number of effects.

Steam costs $2.50/1000 kg. The cost of the first effect is $75,000 and each additional effect costs $60,000, to be written off over a 10-year period. Repairs, local taxes, and insurance per year cost 10 percent of the first cost. Other costs may be neglected since they are independent of the number of effects.

If the charge for the use of capital is 13.8 percent per year based on the initial investment, find the optimum number of effects to install, neglecting the income tax.

**10.15.** Water at $3.2 \times 10^6$ kg/year is to be obtained from a slurry containing 8 percent solids by filtering. The cake will contain 40 percent solids. The relationship is

$$Q = 13,000T^{0.5}$$

where $Q$ is the kilograms of filtrate water obtained during $T$ hours of filtering time. The cake is not washed. The dumping plus cleaning time is 3 h and costs \$100. The filtration cost is \$40/h. Find (a) the cycle time for the minimum cost, (b) the optimum amount of cake per cycle for the minimum cost, and (c) the cycle time for the maximum production.

**10.16.** The feed of material $A$ to a process varies with time and is given by

$$F = 1000 - 10T$$

where $F$ is kilograms per day and $T$ is time in days. Material $A$ converts to material $B$ with no change in weight. The instantaneous efficiency $E$ is given by

$$E = \frac{\text{kg } B \text{ formed}}{\text{kg } A \text{ fed}}$$

and $E$ varies with time according to

$$E = 1 - 0.05T \qquad T < 20$$

(a) Find the instantaneous rates of $A$ in, $A$ out, $B$ out, and the efficiency, all at $T = 10$ days.

(b) Find the cumulative amounts of $A$ in, $A$ out, and $B$ out for the first 10 days. Also, find their averages and the average efficiency.

(c) Repeat part (b) for the tenth day alone, i.e., from the end of day 9 to the end of day 10.

**10.17.** An evaporator shows the following variation of $U$, the overall heat-transfer coefficient, in J/(m² · s · K), with $T$ in seconds measured from the starting time with clean tubes:

$$\frac{1}{U^2} = 0.0002 \times 10^{-8}T + 8 \times 10^{-8}$$

A cleanout costs \$330 and causes 10 h downtime. The allocable operating cost is \$15/h; 500,000 kg/day for 250 days/year must be evaporated from a 280-m² area. The latent heat of vaporization is $2.3 \times 10^6$ J/kg, and the driving force is 28 K. Find the optimum time for operation before recleaning.

**10.18.** Jewels, along with recycled jewels, are polished in a special machine for a space project. The retention time in the polisher is inversely proportional to the number of jewels fed per hour and is 1 h when 1000 jewels per hour are fed.

The fraction of acceptable jewels produced is $1 - e^{-0.5T}$ where $T$ is the retention time in hours. Rejected jewels are returned in a recycle stream to the polisher where they perform like raw jewels.

It costs \$1000/h to operate the polisher. The cost of separating good jewels from rejects is proportional to the number of rejects per hour and is \$2/h for one reject per hour. Acceptable jewels are worth \$10 per jewel above the cost for raw jewels.

Find (a) the maximum profit per hour and (b) the number of acceptable jewels produced per hour under maximum profit-per-hour conditions.

**10.19.** A process uses a catalyst that must be regenerated periodically because of reduction in conversion efficiency. The cost data are

| | |
|---|---|
| One regeneration | = $800 |
| Feed | = $2.50/kg |
| Operation | = $300/day, 300 days/year |
| Value of product | = $14/kg |
| Overhead | = $100,000/year |

The feed rate is maintained constant at a rate of 150 kg/day.

The efficiency of the catalyst expressed as kilograms of product to kilograms of feed is

$$0.87 \text{ average} \qquad 0 \leq T = 1$$

$$\frac{0.87 \text{ instantaneous}}{T^{0.25}} \qquad T > 1$$

where $T$ is the time in days from the last regeneration.

If no other costs are involved, find the maximum annual profit that can be obtained.

**10.20.** Lubricating oil is to be treated continuously in a tower using a selective solvent to remove naphthenic material. The parameter $R$ is defined as

$$R = \frac{\text{bbl solvent used}}{\text{bbl oil treated}}$$

and the treatment cost, without recovery of the solvent, in dollars per year, is given by

$$18,600R^2 - 111,000R + 217,000$$

It costs $0.40 to separate each barrel of solvent circulated through the tower, and 300 bbl/day of oil will be processed for 320 days/year. Find the most economical value of $R$ to be used.

**10.21.** The savings in dollars per day obtained while operating a heat exchanger is given by the expression

$$(A - C)e^{-kT+C}$$

where $T$ is the time in days from the start of the cycle since cleaning. At the instant the cycle is started savings are $90/day, and after a long time they level off at $30/day. Also, during the first month, savings are $111/month more than the average savings per month for the first 4 months after cleaning.

It costs $1000 to clean the heat exchanger, and 3 days are lost with no savings. In this problem 1 month can be taken as 30 days.

Find the number of days between cleanings to yield the maximum savings.

**10.22.** Water is treated with an ion-exchange resin which must be regenerated. A large tank is available at no cost for storing softened water to smooth any variations in demand and to supply water while the resin is being regenerated at negligible time.

Demand averages 2.3 m³/h for all the hours in a year. The time $T$ in hours that the resin can be used before it must be regenerated is given by

$$T = \frac{160V}{F}$$

where $V$ is the volume of the resin in cubic meters and $F$ is the flow rate of the water in cubic meters per hour.

Fixed costs are $9000/m³ for the resin and equipment for a 5-year economic life. Regeneration costs vary as the 0.8 power of the volume of the resin and are $55 for a 1-m³ unit.

Find the optimum size unit to install and the cost per 1000 m³ of softened water.

**10.23.** When 100 kg of $A$ reacts, it forms 75 kg of $B$ and 25 kg of $C$. A batch-processing unit is available for processing 1000 kg of $A$. The fraction of $A$ that reacts is $\alpha$:

$$\alpha = \frac{T}{T + 1}$$

where $T$ is the time in hours that a batch is processed. Material $A$ is worth $1/kg and $B$ $10/kg. Unreacted $A$ and $C$ formed are not recovered, but $B$ is. The setup time between batches is 2 h.

(*a*) Find the maximum production rate for $B$ in kilograms per day.
(*b*) Find the maximum profit in dollars per day.

**10.24.** An ion-exchange resin is used to make ultrapure water. It is necessary to slow down the process with time according to the relationship

$$q = 75e^{-0.07T}$$

where $q$ is the rate in cubic meters per hour and $T$ is the time in hours since the start of a cycle. Demand is steady at 40 m³/h, and when the softening unit is delivering at a rate higher than that, the excess water is given a credit of $2.50/m³. When the rate is less than 40 m³/h, the deficit below that rate is charged at $3/m³.

The cost for rejuvenating the resin and setting up the equipment is $2500, which includes the charge for no water being produced in the period. The downtime is 5 h. The operating charge while producing water is $12/h. Find the optimum time between rejuvenations.

**10.25.** A waste material contains a product, $P$, which can be leached out with water. The water solution is concentrated by evaporation, at a cost, to yield an extract having value.

A batch of waste is leached using 28 m³, 28,000 kg, of water per hour at a constant rate. The concentration of $P$ in the exit water decreases with time and is given by the rate expression

$$\frac{dc}{dT} = -11.8c$$

where $c$ is the concentration of $P$ in kilograms per cubic meter, and at the start is 4.8.

The final solution to be prepared after evaporation is to be 20 percent $P$. Each 1000 kg of water evaporated is charged at $3, and each 1 kg of $P$ recovered in the 20 percent concentrate receives a credit of $2.

Find how long a batch should be leached to obtain the maximum profit from each batch.

**10.26.** A process is to be designed to process $5 \times 10^6$ kg/year. Processing is by batches, and the processing time is proportional to the square root of the batch size; the processing time is 4 h for a batch of 10,000 kg. The cost of processing units is given by

$$425X^{0.7} \text{ \$}$$

where $X$ is the size of a batch in kilograms, and the cost will be written off in 5 years. Other costs are a \$250 setup cost for each batch, with negligible setup time, and \$22/h for labor while the processing unit is in operation. A batch must be finished in 8 h within a shift of 8 h, and only one shift per day for 200 days/year is allowed. Find the optimum batch size.

**10.27.** One kilogram of $A$ is processed to produce 1 kg of $B$ in a primary converter. At the end of a processing cycle, 1 day is lost in cleaning the converter. The feed rate to the converter is decreased uniformly with time and is 1000 kg/day of $A$ at zero time and would be 500 kg/day at the end of 30 days.

Efficiency in operating the converter is measured by the ratio of kilograms of $B$ produced to kilograms of $A$ fed. The efficiency decreases linearly with time in proportion to the *cumulative* amount of $A$ fed to the converter. The efficiency is 0.9 at the start and 0.75 at the end of 3 days.

Unreacted $A$ from the primary converter can be converted in part to $B$ in a secondary converter, which recovers 20 percent of the unreacted $A$ by converting it to $B$ by a kilogram-to-kilogram conversion. The amount of $B$ so formed is added to that obtained in the primary converter.

Determine how long the primary converter should be run before being cleaned to maximize the total production of $B$.

**10.28.** It costs \$150,000 to overhaul an auxiliary electrical power unit. Immediately after the overhaul, there is a savings of \$1000/day by avoiding purchased electricity. The savings per day decreases with time because of scaling, outages, etc. as it becomes necessary to purchase electricity from outside. The purchased electricity increases linearly with time, and at the end of 1 year it would be purchased at a rate amounting to 25 percent less than the savings rate immediately after the overhaul.

Penalty charges arise with time, and they increase continuously at a rate proportional to the square of the time since the overhaul. For 3 years they would total \$300,000.

The time required for the overhaul, during which extra power must be purchased, the time value of money, and taxes may be neglected. Find the optimum time between overhauls.

**10.29.** A vehicular tunnel is to be built from point $A$ on one side of a river to point $C$ on the other side. The river is 800 m wide at $AB$, but the tunnel will be dug in a straight line

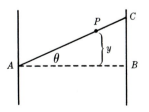

at some angle $\theta$ with $AB$. The ease of drilling depends on the distance from line $AB$. A factor, $g$, is given by

$$g = \frac{2 \times 10^5}{10^5 + y^2}$$

where $y$ is expressed in meters. The cost for drilling 1 m is proportional to $g$ and would be $33,000/m along line $AB$.

All other costs, such as those for casing, tile, and ventilation, amount to $25,000/m. Find the optimum angle with respect to line $AB$ at which the tunnel should be built.

**10.30.** Show how you would obtain the average efficiency in a variocyclic process, such as from the graph of Fig. 10.5. Does it matter if the input to the equipment is not constant?

**10.31.** The number of items $Q$ that can be made in unit time depends on the mixture of the number of units of labor and capital and is given by

$$Q = L^{0.6}C^{0.7}$$

where $L$ is the number of units of labor supplied and $C$ is the number of units of capital supplied.

The cost per unit of time for $L$ units of labor is $aL$, and the cost per unit of time for $C$ units of capital is $bC$. The sales value of $Q$ items is $sQ$. Find the optimum mix of labor and capital so that profit is maximized per unit time.

**10.32.** An evaporator's scales and tests show that the steam condensed in the heating tubes varies with time according to the relationship

$$S = A(50 - 0.05T) \qquad T < 1000$$

where $S$ is the condensation rate in kilograms per hour, $A$ is the heating area in square meters, and $T$ is the evaporating time in hours since the last cleanout. There will be 0.9 kg of water evaporated per kilogram of steam condensed. The cost for the evaporators is $2100/m^2$ heating area and is written off over a 10-year period. The allocable labor cost is $15/h of operation. Each cleaning costs $2500. The water to be evaporated per year is $4.5 \times 10^6$ kg, and the operating time is not to exceed 6000 h/year.

Find ($a$) the optimum evaporating time per cycle before recleaning the tubes and ($b$) the optimum heating area to be installed.

**10.33.** A processing unit requires cooling supplied by a chilling liquid. The cost to cool the chilling liquid depends on $\Delta T$ in °C that it is cooled and the flow rate $F$ in kilograms per second at which it is pumped.

The pumping cost is proportional to the cube of the pumping rate, and for a pumping rate of 100 kg/sec it is $12.60/h. The cost for chilling the chilling liquid is proportional to the square of $\Delta T$, and for a drop of 50°C it is $17.5 \times 10^{-5}$ for 1 kg of liquid. There is a fixed cost of $54/h.

The cooling accomplished in the processing unit depends on both the flow rate and the $\Delta T$ of the chilling liquid and is given by

$$\text{MJ/h removed} = 10.8F(\Delta T)$$

where $F$ is in kilograms per second.

Find the flow rate $F$ and the $\Delta T$ for the chilling liquid that will give the minimum cost per unit time per joule per unit time removed in the process.

**10.34.** A company owns a section of land which is shaped like a right triangle, with legs $A$ and $B$ and hypotenuse $C$. The company desires to build on the largest rectangular area that can be obtained from the lot. Find the area of that lot ($a$) by using ordinary calculus and ($b$) by using a Lagrange multiplier.

**10.35.** Maximize the product $XY$ subject to the restraint that $X$ and $Y$ lie on a circle with a radius of 5; i.e., $X^2 + Y^2 = 5^2$. ($a$) Use ordinary calculus and ($b$) use a Lagrange multiplier.

**10.36.** From the Lagrange expression, find the values of $X$, $Y$, and $Z$ that minimize the function $X + 2Y^2 + Z^2$, subject to the constraint that $X + Y + Z = 1$.

**10.37.** For the expression

$$Y = AX^2 - ABX + \frac{BC}{X^2}$$

verify that for $A = B = C = 2$, $Y$ has a minimum value at $X = 1.54$. Find which parameter, $A$, $B$, or $C$, has the greatest effect on the minimum value of $Y$ for the conditions given.

# REFERENCES

1. Beightler, C. S., D. T. Phillips, and D. J. Wilde: *Foundations of Optimization*, 2d ed., Prentice-Hall, Inc., Englewood Cliffs, N.J., 1979.
2. Boas, A. H.: "How to Use Lagrange Multipliers," *Chemical Engineering*, January 7, 1963, pp. 95–98.
3. Conley, W.: "Production Planning with Nonlinear Cost Considerations," *Cost Engineering*, vol. 28, No. 6, 1986, pp. 16–20.
4. Denn, M. H.: *Optimization by Variational Methods*, McGraw-Hill Book Company, New York, 1969.
5. Fan, L. T.: *The Continuous Maximum Principle*, John Wiley & Sons, Inc., New York, 1966.
6. Hestenes, M. R.: *Optimization Theory—The Finite Dimensional Case*, John Wiley & Sons, Inc., New York, 1975.
7. Iscol, L.: "How to Solve Optimization Problems," *Chemical Engineering*, February 19, 1962, pp. 107–116.
8. Miller, R. E.: *Modern Mathematical Methods for Economics and Business*, Robert E. Krieger Publishing Company, Inc., Huntington, N.Y., reprinted 1978.
9. Peters, M. S., and K. D. Timmerhaus: *Plant Design and Economics for Chemical Engineers*, 3d ed., McGraw-Hill Book Company, New York, 1980.
10. Tao, L. C.: "Process Optimization," *Chemical Engineering*, April 26, 1965, pp. 143–148.

# chapter 11

# SUBSETS IN
# OPTIMIZATION

*F. C. Jelen*

**INVENTORY PROBLEMS**

## 11.1 INTRODUCTION

This chapter discusses two types of problems — inventory and queuing — which are treated in operations research but are also important in cost engineering. Inventories are maintained at a cost to gain advantages having monetary value, such as avoiding a shutdown due to a temporary absence of supplies or permitting uniform production for a seasonable supply or demand. There is also an economic balance leading to some optimum policy for minimum cost.

*Inventory* is simply a stock of physical assets having value, which can be material, money, or labor. Material inventory can be raw material, tools and accessories including spare parts used in production, unfinished or in-process inventory, and finished products. Figure 11.1 illustrates the classes of inventory in the manufacture of steel.

## 11.2 GENERAL INVENTORY MODEL

Mathematical models employ the language and symbols of mathematics and may be deterministic, for which all information and values are treated as being definite, or probabilistic (stochastic), for which uncertainty in some of the values is recognized and considered.

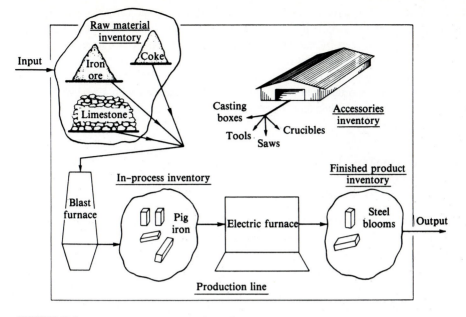

**FIGURE 11.1**

Classification of a materials inventory in the manufacture of steel.

The general inventory system is shown in Fig. 11.2. Ordinates represent the number of units available in stock against time as the abscissa. Consider the quantity $Q$ units received at the beginning of the period $T_1$. The use during this period $U_1$ is subtracted from the stock to give a lower stock for the second period, etc. At the end of the fourth period $T_4$ the level of inventory reaches the reorder level $R$. At that time an order equal to $Q$ units is placed, which is eventually received at the beginning of

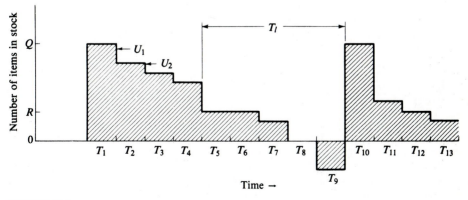

**FIGURE 11.2**

Analysis of inventory against time.

the tenth period. The procurement time, or lead time $T_l$, is the time required in periods to receive an order after it is placed. During the procurement time the level of stock may go below zero, creating a shortage for which a penalty can arise.

The object is to minimize the total cost of the system, which is the minimum cost per period, not per cycle. Some costs are connected with the cycle, such as placing and receiving an order, whereas other costs are connected with a period, such as storage costs. Confusion between a period and a cycle is the most common source of error for a beginner.

## 11.3   ECONOMIC LOT SIZE

The simplest case makes the following assumptions:

1. All items are delivered simultaneously at the time the stock becomes zero.
2. The cost of placing and receiving an order is constant and is independent of the lot size.
3. Usage is at a constant rate.
4. No safety stock is provided.
5. No shortage is allowed.
6. The warehousing cost is calculated on a continuous basis, and at any instant of time is proportional to the number of items in the inventory.

The schematic diagram is illustrated in Fig. 11.3. Let $C_o$ be the cost for placing and receiving an order, $C_h$ the storage or holding cost *per item per period*, $M = Q$ the maximum inventory where $Q$ is the lot order, and $U$ the rate of use in items per period. Consider a time element $T$ units of time from zero time lasting $dT$ units of time. The inventory at any time is a straight-line function of time and is $Q$ at $T = 0$, and 0 at $T = N$ where $N$ is the duration of the cycle. Then

$$\text{Inventory} = -\frac{Q}{N}T + Q$$

**FIGURE 11.3**
Diagram for a simple inventory model.

The number of periods of time $N$ that the lot $Q$ will last is given by

$$N = \frac{\text{items}}{\text{items/period}} = \frac{Q}{U}$$

so that the inventory at any time is

$$\text{Inventory} = -UT + Q$$

The total cycle cost is

$$C_{\text{cycle}} = C_o + \int_0^{N=Q/U} C_h(-UT + Q)\,dT = C_o + C_h\left(-\frac{U}{2}\frac{Q^2}{U^2} + Q\frac{Q}{U}\right)$$

$$= C_o + \frac{C_h Q^2}{2U}$$

The function to be optimized is the cost per period $C_t$ and with

$$\text{Cycles/period} = \frac{\text{item/period}}{\text{item/cycle}} = \frac{U}{Q}$$

the cost per period becomes

$$C_t = \frac{\$}{\text{cycle}}\frac{\text{cycle}}{\text{period}} = \left(C_o + \frac{C_h Q^2}{2U}\right)\frac{U}{Q} = C_o\frac{U}{Q} + \frac{C_h Q}{2} \tag{11.1}$$

Differentiating and setting equal to zero, we obtain

$$\frac{dC_t}{dQ} = 0 = -C_o\frac{U}{Q^2} + \frac{C_h}{2}$$

Finally,

$$Q_{\text{opt}} = \sqrt{\frac{2C_o U}{C_h}} \tag{11.2}$$

The lead time $T_l$ multiplied by the rate of use per period $U$ gives the reorder level $R$:

$$R = T_l U \tag{11.3}$$

By placing $Q_{\text{opt}}$ in Eq. (11.1), we get for the optimum period cost

$$C_t = C_o\frac{U}{\sqrt{2C_o U/C_h}} + C_h\sqrt{\frac{C_o U}{C_h}} = \sqrt{2C_h C_o U} \tag{11.4}$$

**Example 11.1.** The demand for an item is 100/day. The cost of placing and receiving an order is $50 with an infinite delivery rate. The storage cost is $10 per item per year based on the actual inventory at any time. No shortages are allowed. The reorder level is for an 8-day delivery time.

The storage charge per item per day is $10/365 = 0.0274$. Equation (11.2) is

$$Q_{opt} = \sqrt{\frac{2(50)(100)}{0.0274}} = 604 \text{ items}$$

The minimum cost, by Eq. (11.4), is

$$C_t = \sqrt{2 \times 0.0274 \times 50 \times 100} = \$16.55/\text{day}$$

The reorder level, by Eq. (11.3), is

$$R = 8(100) = 800 \text{ items in stock}$$

## 11.4 SAFETY STOCK

In some situations the decision maker wants to be sure that a shortage in an item will not occur, and so adds more items to the inventory called *safety stock*. See Fig. 11.4, where the safety stock is represented by $S$. The model has the same assumptions as in the preceding section. The procedure is the same, and the inventory at any time $T$ is

$$\text{Inventory} = -UT + Q + S$$

Individual steps are not shown, but the cost per cycle will be

$$C_{cycle} = C_o + C_h \left( \frac{1}{2} \frac{Q^2}{U} + S \frac{Q}{U} \right)$$

The cost per period will be

$$C_t = C_o \frac{U}{Q} + C_h \frac{Q}{2} + SC_h$$

The optimum order for the minimum cost will be

$$Q_{opt} = \sqrt{\frac{2C_o U}{C_h}} \tag{11.5}$$

**FIGURE 11.4**
Diagram for a simple inventory model with safety stock.

and the optimum period cost is

$$C_t = \sqrt{2C_h C_o U} + SC_h \qquad (11.6)$$

Equation (11.5) is the same as Eq. (11.2). When $S = 0$, Eq. (11.6) reduces to Eq. (11.4).

## 11.5  FINITE RATE DELIVERY

The model has important applications since a finite delivery rate is associated with manufacturing. The schematic diagram is shown in Fig. 11.5. Here $C_o$ is a setup cost, and $D$ items per unit of time is a manufacturing rate. The use rate in items per unit of time is $U$. The value for $Q$ as shown is the total number of items produced and used in the cycle of $(T_1 + T_2)$ periods. The safety stock is $S$ items, and $M$ is the maximum inventory. In time $T_1$ the inventory builds up by an amount $(M - S)$ at a rate $(D - U)$. Thus

$$M - S = T_1(D - U)$$

Also, in time $T_1$ the amount $Q$ is made at a rate $D$:

$$Q = T_1 D$$

Eliminating $T_1$ gives

$$M - S = Q\frac{D - U}{D} \qquad (11.7)$$

**FIGURE 11.5**
Diagram for a model with a finite delivery rate and safety stock.

but $Q$ is used in $N$ periods at a rate $U$:

$$Q = UN \tag{11.8}$$

Placing this value for $Q$ in Eq. (11.7) gives

$$M - S = UN\frac{D - U}{D} \tag{11.9}$$

In this model $(M - S)$ is linear with time on both sides of the maximum inventory. The average inventory for the total $N$ periods is one-half $(M - S)$ plus the safety stock:

$$\text{Average inventory} = \frac{UN}{2}\frac{D - U}{D} + S \tag{11.10}$$

The average inventory is held at a cost $C_h$ per item per period. The cost per period is

$$C_t = \left[C_o + \left(\frac{UN}{2}\frac{D - U}{D} + S\right)C_hN\right]\frac{\$}{\text{cycle}}\frac{1}{N}\frac{\text{cycle}}{\text{period}}$$

$$= \frac{C_o}{N} + \frac{UN}{2}\frac{D - U}{D}C_h + C_hS \tag{11.11}$$

For a minimum

$$\frac{dC_t}{dN} = 0 = -\frac{C_o}{N^2} + \frac{C_hU}{2}\frac{D - U}{D}$$

$$N = \sqrt{\frac{2C_oD}{C_hU(D - U)}} \tag{11.12}$$

Placing $N$ in Eq. (11.8) gives

$$Q_{\text{opt}} = \sqrt{\frac{2C_oUD}{C_h(D - U)}} \tag{11.13}$$

and placing $N$ in Eq. (11.11) yields

$$C_t = \sqrt{\frac{2C_hC_oU(D - U)}{D}} + C_hS \tag{11.14}$$

**Example 11.2.** The setup cost for a small operation is $600, and the storage cost is $0.40 per item per day. Usage is uniform at 300 items per day, and the production rate is 500 times per day. A safety stock of 100 units is required. Find the optimum production lot size that will give the minimum cost.

*Solution.* By Eq. (11.13), the optimum lot size is

$$Q_{\text{opt}} = \sqrt{\frac{2(600)(300)(500)}{(0.40)(500 - 300)}} = 1500 \text{ items}$$

The period cost, by Eq. (11.14), is

$$C_t = \sqrt{\frac{2(600)\,(0.40)\,(300)\,(500-300)}{500}} + 0.40(100) = \$280/\text{day}$$

The production time is $T_1$, given by

$$T_1 = \frac{Q}{D} = \frac{1500}{500} = 3 \text{ days}$$

The total cycle time is $N$ days, given by

$$N = \frac{Q}{U} = \frac{1500}{300} = 5 \text{ days}$$

## 11.6 SETTING UP INVENTORY PROBLEMS

The number of models used to represent inventory problems is infinite, and analysts must learn how to set up their own mathematical models and optimize them by analytical or graphical methods. In some models the inventory may reduce to zero, and unfulfilled demand may be assessed a penalty. The cost of placing or receiving an order may depend on the size. The variations are endless. The general method of setting up a model is illustrated in the following example.

**Example 11.3.** The demand for an item is uniform at 1000 items per day. Delivery when placing an order is immediate. The cost for placing and receiving an order depends on the order size and is given by

$$C_o = 13{,}000 + 5Q^{0.5}$$

It costs \$0.002/day for each item in storage.

In addition, there is an antispoilage cost while in storage, which is proportional to the age of the inventory multiplied by the number of units in the inventory at the time. If an inventory is 20 days old and there are 8000 items in stock, the antispoilage cost would be

$$(0.000015)\,(20)\,(8000) \ \$/\text{day}$$

where 0.00015 is a constant. Find the optimum lot order size.

*Solution.* Let a cycle last $N$ days. The lot size will then be

$$Q = 1000N$$

for which

$$C_o = 13{,}000 + 5(1000N)^{0.5} = 13{,}000 + 158.11N^{0.5}$$

The inventory declines linearly from $1000N$ to 0 in $N$ days. The inventory at any time is

$$\text{Inventory} = 1000(N-T)$$

where $T$ is time in days. The cost for antispoilage at that time is

$$(0.000015)T[1000(N-T)] = 0.015(NT - T^2) \ \$/\text{day}$$

The cost for one cycle is

$$\$/\text{cycle} = 13{,}000 + 158.11N^{0.5} + \int_0^N 0.002[1000(N - T)]\,dT$$

$$+ \int_0^N 0.015(NT - T^2)\,dT$$

which after integration reduces to

$$\$/\text{cycle} = 13{,}000 + 158.11N^{0.5} + N^2 + 0.0025N^3$$

Multiplying by $1/N$ cycles per day gives the dollars per day:

$$\$/\text{day} = 13{,}000N^{-1} + 158.11N^{-0.5} + N + 0.0025N^2$$

By repeated trials, the minimum occurs at $N = 96.4$, for which

$$Q_{\text{opt}} = 1000N = 96{,}400 \text{ items per lot}$$

# PROBABILISTIC MODELS

## 11.7 OPTIMUM STOCK BY INCREMENTAL ANALYSIS

When uncertainty is involved in any of the factors, the problem can become complicated. This section and the two following illustrate a few examples to acquaint the analyst with some methods.

**Example 11.4.** The cumulative probability for the sales of a certain item in season is

| Cumulative sales | Cumulative probability of this many sales or less | Cumulative probability of this many sales or more |
|---|---|---|
| 0 | 0.00 | 1.00 |
| 1,000 | 0.03 | 0.97 |
| 2,000 | 0.09 | 0.91 |
| 3,000 | 0.18 | 0.82 |
| 4,000 | 0.29 | 0.71 |
| 5,000 | 0.44 | 0.56 |
| 6,000 | 0.60 | 0.40 |
| 7,000 | 0.80 | 0.20 |
| 8,000 | 0.94 | 0.06 |
| 9,000 | 0.98 | 0.02 |
| 10,000 | 1.00 | 0.00 |

An item costs $3, sells for $4 in season, and is disposed of for $1 if not sold in season. There is no storage charge. Determine the optimum number to stock at the start of the season.

*Solution.* The number to stock is such that the last item stocked has an expectation equal to its cost. If another were stocked, its expectation would be less than the cost. Recall that

$$\text{Expectation} = \sum (\text{value})(\text{probability})$$

and let $P(X)$ be the probability of selling the last unit and $1 - P(X)$ be the probability of not selling the last unit. If bought, the last unit costs \$3 and it brings either \$4 or \$1. The mathematical relationship for break-even is

$$3.00 = 4.00P(X) + 1.00[1 - P(X)]$$

$$P(X) = 0.67$$

The probability of selling the last unit must be 0.67 and of not selling it $1 - 0.67$, or 0.33. By interpolation in the tabulation, or from a plot, a stock of 4300 meets the requirements. If a stock of 4300 is ordered, the probability of selling the last unit is the probability of selling 4300 or more, i.e., 0.67. Analysts must be wary of incorrect reasoning and can assist themselves by recognizing the two complementary cumulative probabilities, as was done in the tabulation of data for the example.

# 11.8   OPTIMUM INVENTORY BASED ON EXPECTATION

The expectation of a variable can be used in many ways to reduce a problem to a tabular or graphical solution. The following example is an illustration.

**Example 11.5.** The average demand for an item is four per week and follows a Poisson distribution. Find the number of units to be stocked at the beginning of each week so that the number of customers turned away on the average will not exceed 10 percent of demand.

*Solution.* For a Poisson distribution, with an average demand of four the probability of a demand for exactly $X$, by Eq. (8.11),

$$P(X) = \frac{4^x e^{-4}}{X!}$$

and by calculation or from tables,

| | | |
|---|---|---|
| $P(0) = 0.0183$ | $P(1) = 0.0732$ | $P(2) = 0.1464$ |
| $P(3) = 0.1950$ | $P(4) = 0.1950$ | $P(5) = 0.1560$ |
| $P(6) = 0.1040$ | $P(7) = 0.0594$ | $P(8) = 0.0296$ |
| $P(9) = 0.0132$ | $P(10) = 0.0053$ | $P(11) = 0.0019$ |
| $P(12) = 0.0006$ | $P(13) = 0.0002$ | $P(14) = 0.0001$ |
| $P(15) = 0.0000$ | | |

**TABLE 11.1**
**Example 11.5 with an initial stock of 6**

| Demand X | $P(X)$ | Lost sales $(X - 6)$ | $(X - 6)P(X)$ |
|---|---|---|---|
| 7 | 0.0594 | 1 | 0.0594 |
| 8 | 0.0296 | 2 | 0.0592 |
| 9 | 0.0132 | 3 | 0.0396 |
| 10 | 0.0053 | 4 | 0.0212 |
| 11 | 0.0019 | 5 | 0.0095 |
| 12 | 0.0006 | 6 | 0.0036 |
| 13 | 0.0002 | 7 | 0.0014 |
| 14 | 0.0001 | 8 | 0.0008 |
| 15 | 0.0000 | 9 | 0.0000 |
| | | | $\Sigma = 0.195$ expected lost sales per week |

Let the starting stock be six. Sales will be lost only when the demand is seven or greater. Table 11.1 shows how the expected lost sales are computed using the following underlying relationship:

$$\text{Expected lost sales} = \sum (\text{lost sales}) (\text{probability})$$

If the initial stock is six, the expected lost sales is 0.195, and the average demand is four. Lost sales as a fraction of demand is

$$\frac{0.195}{4} = 0.05$$

Inasmuch as the ratio may be 0.10, a stock of six is more than adequate.

If the calculation is repeated on the basis of a starting stock of five, the expected lost sales is found to be 0.409, which as a fraction of demand is $0.409/4 = 0.102$. The conditions of the example can be very nearly met by stocking five units.

# 11.9  MONTE CARLO SIMULATION IN INVENTORY

Monte Carlo simulation can be used very effectively to solve probabilistic inventory problems. In this method variables subject to uncertainty are generated by the use of random numbers in such a way that real frequencies of distribution are obeyed on the average by the generated values. A number of cycles are simulated to establish the most probable values. Machine computation makes the method very versatile. Not only can average and expected values be predicted, but frequency distributions are easily established.

The method will be illustrated in a simple form in this section with an inventory problem in which the lead time and demand each follow known probabilistic distributions. The Monte Carlo method itself was described in Chap. 8.

**Example 11.6.** The mean demand for an article $U_m$ is 4.7 units per day. The distribution of this demand is as follows, with the last column representing the assignment of random rectangular variates:

| Demand | $P(U)$ | Cumulative relative frequency | Random numbers assigned |
|---|---|---|---|
| 2 | 0.05 | 0.05 | 01–05 |
| 3 | 0.15 | 0.20 | 06–20 |
| 4 | 0.20 | 0.40 | 21–40 |
| 5 | 0.30 | 0.70 | 41–70 |
| 6 | 0.20 | 0.90 | 71–90 |
| 7 | 0.10 | 1.00 | 91–00 |

The mean lead time $T_{lm}$ is 4 days. The distribution of this time is as follows, with the last column representing the assignment of random rectangular variates:

| Lead time $T_l$ | $P(T_l)$ | Cumulative relative frequency | Random numbers assigned |
|---|---|---|---|
| 3 | 0.25 | 0.25 | 01–25 |
| 4 | 0.50 | 0.75 | 26–75 |
| 5 | 0.25 | 1.00 | 76–00 |

It costs $15 to place an order, inventory cost is $1 per unit per day, and shortage cost is $2 per unit per day. Assume that orders are placed and arrive at the end of a day and that charges are based on the end-of-day inventory but before arrival of a new lot.

Find the optimum lot size and reorder level.

*Solution.* Make a Monte Carlo simulation assuming a lot size $Q$ of 30 and a reorder level $R$ of 20. Start the first cycle with 20 units on hand. This is arbitrary, but the effect will be washed out when a great number of cycles is considered.

The tabulation is made in Table 11.2. All random numbers were selected in an unbiased fashion from a table of random rectangular variates. A cycle is assumed to start the day after a new order is placed, which occurs whenever inventory falls to 20 or less at the end of a day.

At zero time the random number 79 indicates a lead time of 5 days for the order placed at that time. During the first period the demand is 3, and the inventory drops to 17 at the end of the first day. This continues, but between the end of the fifth day and the start of the sixth day, a new lot of 30 arrives and is available for the sixth day. At the end of the sixth day inventory is 25, being the 30 received less the 2 short the previous day less the 3 used during the sixth day. This day is also charged with the lot order cost. The cumulative cost divided by the cumulative days or periods is the average cost per day for $Q = 30$ and $R = 20$. If enough cycles are taken, the simulation will be exceedingly good. Other combinations are tried to find the optimum $Q$ and $R$.

# TABLE 11.2
## Monte Carlo method for Example 11.6 with $Q = 30$, $R = 20$

| Cycle | Period cumulative | Random no. | Demand | Random no. | Lead time | Stock, end of day | Shortage cost, $ | Inventory cost, $ | Lot order cost, $ | Cumulative cost, $ | Average cost per period, $ |
|---|---|---|---|---|---|---|---|---|---|---|---|
|  | 0 |  |  | 79 | 5 | 20 |  |  |  |  |  |
| 1 | 1 | 13 | 3 |  |  | 17 | 0 | 17 | 0 | 17 | 17 |
|  | 2 | 28 | 4 |  |  | 13 | 0 | 13 | 0 | 30 | 15 |
|  | 3 | 77 | 6 |  |  | 7 | 0 | 7 | 0 | 37 | 12 |
|  | 4 | 60 | 5 |  |  | 2 | 0 | 2 | 0 | 39 | 10 |
|  | 5 | 29 | 4 |  |  | -2 | 4 | 0 | 0 | 43 | 9 |
|  | 6 | 16 | 3 | 68 |  | $30 - 2 - 3 = 25$ | 0 | 25 | 15 | 83 | 14 |
|  | 7 | 70 | 5 |  | 4 | 20 | 0 | 20 | 0 | 103 | 15 |
| 2 | 8 | 09 | 3 |  |  | 17 | 0 | 17 | 0 | 120 | 15 |
|  | 9 | 24 | 4 |  |  | 13 | 0 | 13 | 0 | 133 | 14 |
|  | 10 | 16 | 3 |  |  | 10 | 0 | 10 | 0 | 143 | 14 |
|  | 11 | 19 | 3 |  |  | 7 | 0 | 7 | 0 | 150 | 14 |
|  | 12 | 14 | 3 |  |  | $30 + 7 - 3 = 34$ | 0 | 34 | 15 | 199 | 17 |
|  | 13 | 29 | 4 |  |  | 30 | 0 | 30 | 0 | 229 | 18 |
|  | 14 | 49 | 5 |  |  | 25 | 0 | 25 | 0 | 254 | 18 |
|  | 15 | 08 | 3 |  |  | 22 | 0 | 22 | 0 | 276 | 18 |
|  | 16 | 80 | 6 | 45 | 4 | 17 | 0 | 17 | 0 | 293 | 18 |

300

The method can be used to get averages and distributions of the various variables. Thus by the proper tabulation it is possible to determine the average shortage and distribution of shortage per cycle.

# QUEUING PROBLEMS

## 11.10 PREVALENCE

A queue is a waiting line, but *queuing* is a term now widely used in a broad sense to cover a wide variety of problems, usually for economic balance and optimization, involving waiting and delay in serving people or servicing machines and equipment. Queuing also covers all facets of the problem, such as the optimum number of long-distance lines required between two cities or the optimum number of repairers and their equipment required to keep an assembly line in operation.

Modern society, with its large population and interdependent functions, has focused considerable attention on queuing problems. There is now strong competition among service industries to gain and retain the customer, and all companies must consider internal queuing for the economic balance of manufacturing and operational efficiency.

## 11.11 DESCRIPTION OF A QUEUING SYSTEM

A queuing system is illustrated in Fig. 11.6. The term *machine* will be used in a general way for the object to be serviced. The main elements in the system are the input

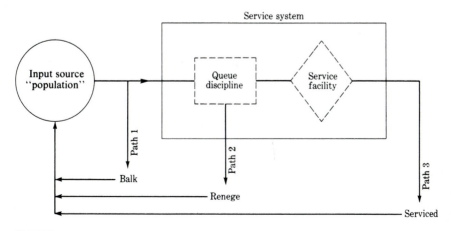

**FIGURE 11.6**
Chart of a queuing system.

source and the service system. The latter is characterized by queue discipline and service facility.

A machine needing servicing can follow three possible paths:

**1.** Return to the population without servicing, called *balking,* path 1.
**2.** Join the queue but return to population without servicing, called *reneging,* path 2.
**3.** Return to the population after being serviced, path 3.

## 11.12  CLASSIFICATION OF QUEUING SYSTEMS

A queuing system can be classified with respect to input source, queue, and service facility.

The input source, or population, is classified as *infinite* or *finite,* which is chiefly the size of the population relative to the number in the queue and being serviced. If the characteristics of the input source (Fig. 11.6) are changed by the number of withdrawals, the population is considered finite and the problem is solved with the changes in the population taken into account.

The queue itself can be classified as *infinite* if it is allowed to grow to any size or *finite* if characterized by a maximum permissible size. Queues can be classified further as *single* or *multiple.* A one-chair barbershop is a good example of a single queue. The multiple-queue case is exemplified by a work center having two input queues from different operations.

## 11.13  THE SIMPLEST MODEL

The simplest model is a single-channel, single-phase, single queue with an infinite population. Arrivals come at equally spaced intervals and require a constant servicing time. The time periods between arrivals $T_a$ must be greater than the servicing time $T_s$; otherwise an infinite queue would form. The waiting time or lost time for each item is the servicing time only, since no queue forms.

Consider an interval of time $T_a$ during which one item will enter the service channel and remain there for $T_s$ time periods. If $C_w$ is the cost for one item waiting one time period, the cost for the waiting time is

$$C_w T_s \tag{11.15}$$

In addition, there is the cost of servicing. The size of the servicing channel is optional. It will be assumed that the time for servicing an item $T_s$ is inversely proportional to the size of the channel, and that the period cost of a servicing channel is proportional to its size. Let a service channel which can service one item in one time period cost $C_f$ dollars to operate for one time period. If the servicing time to service one unit is to be $T_s$ and not a unity time period, then a servicing channel of a different size will be required, and its cost for one time period will be

$$\frac{C_f}{T_s} \tag{11.16}$$

The cost for a time interval $T_a$ will be the sum of the cost for the lost time or waiting time for the unit being serviced plus the cost of operating the servicing channel for $T_a$ time periods, which is

$$C_w T_s + \frac{C_f}{T_s} T_a \tag{11.17}$$

Obtain the total cost for one time period $C_t$ by dividing the preceding by $T_a$, giving

$$C_t = C_w \frac{T_s}{T_a} + \frac{C_f}{T_s} \tag{11.18}$$

The minimum cost per period is obtained by differentiation with respect to $T_s$ and setting equal to zero:

$$\frac{dC_t}{dT_s} = 0 = \frac{C_w}{T_a} - \frac{C_f}{T_s^2}$$

$$T_{s,\,\text{opt}} = \sqrt{\frac{C_f T_a}{C_w}} \tag{11.19}$$

and placing this value in Eq. (11.18) gives

$$C_{t,\,\text{opt}} = 2\sqrt{\frac{C_f C_w}{T_a}} \tag{11.20}$$

It is important to note that $C_f$ is the cost for one time period for a servicing channel which when working full time would service one item.

If there are $L$ channels or service facilities instead of one, as before, then Eq. (11.18) becomes

$$C_t = C_w \frac{T_s}{T_s} + \frac{C_f L}{T_s} \tag{11.21}$$

and Eqs. (11.19) and (11.20) become

$$T_{s,\,\text{opt}} = \sqrt{\frac{C_f T_a L}{C_w}} \tag{11.22}$$

$$C_{t,\,\text{opt}} = 2\sqrt{\frac{C_f C_w L}{T_a}} \tag{11.23}$$

**Example 11.7.** The arrival rate is constant at three items per hour. The cost of providing and maintaining a service facility is \$25/h, and it can service four items per hour working full time. If an item waits 1 day, it represents a cost of \$2400. Find the optimum time to service one item and the minimum variable cost per item.

***Solution.*** Take 1 h as one period:

$$C_f = \frac{25}{4} = \$6.25$$

$$T_a = \frac{1}{3} \text{h between arrivals}$$

$$C_w = 2400/24 = \$100/\text{h} \cdot \text{item}$$

By Eq. (11.19),

$$T_{s,\text{opt}} = \sqrt{\frac{6.25(1/3)}{100}} = 0.1443 \text{ h, optimum servicing time}$$

The total cost per period, by Eq. (11.20), is

$$C_{t,\text{opt}} = 2\sqrt{\frac{6.25 \times 100}{\frac{1}{3}}} = \$86.60$$

but three items enter the system per period. The total cost for servicing one item is $86.60/3 = \$28.87$.

**Example 11.8.** Redo the previous example. Here the cost of a service facility is proportional to the 0.6 power of its size.

*Solution.* Let $S$ be the size of the service station. Then $S$ is inversely proportional to the time required to service one item:

$$S = \frac{k}{T_s}$$

and the cost of the service facility for 1 h is

$$K'\left(\frac{k}{T_s}\right)^{0.6} = K\left(\frac{1}{T_s}\right)^{0.6}$$

When $T_s = 1/4$, the cost is $25:

$$25 = K\left(\frac{1}{T_s}\right)^{0.6} \qquad K = 10.88$$

and the cost for a service facility in dollars per hour is

$$10.88\left(\frac{1}{T_s}\right)^{0.6} = 10.88T_s^{-0.6} \tag{11.24}$$

For a time interval $T_a$ when one item is serviced, the cost for waiting plus the cost for the service facility is

$$C_w T_s + 10.88T_s^{-0.6}T_a \tag{11.25}$$

Dividing by the time $T_a$ gives the cost per hour:

$$C_t = C_w\frac{T_s}{T_a} + 10.88T_s^{-0.6} \tag{11.26}$$

Differentiate with respect to $T_s$ and set equal to zero:

$$0 = \frac{C_w}{T_a} - 0.6(10.88)T_s^{-1.6}$$

This yields for the optimum servicing time

$$T_s = \left(\frac{6.528T_a}{C_w}\right)^{1/1.6} = \left(\frac{6.528 \times 1/3}{100}\right)^{1/1.6} = 0.0914 \text{ h}$$

The total cost per period, from Eq. (11.26), is

$$C_{t,\,opt} = 100\frac{0.0914}{\frac{1}{3}} + 10.88(0.0914)^{-0.6} = \$73.14/\text{h}$$

## 11.14 MODELS INVOLVING PROBABILITY

In practice, the arrival rate and servicing time will not be constant. The mathematics for these models can become quite complicated, and only the results for some simple cases are given here.

If the number of arrivals per period follows a Poisson distribution with $\lambda$ average arrivals per period, and the number of items serviced per period follows a Poisson distribution with $(\mu > \lambda)$, then minimum cost per period is obtained at

$$C_{t,\,opt} = 3C_w \tag{11.27}$$

for which

$$\lambda'_{opt} = \frac{C_w}{C_f} \tag{11.28}$$

$$\mu'_{opt} = 2\frac{C_w}{C_f} \tag{11.29}$$

where $\lambda'_{opt}$ and $\mu'_{opt}$ are specific values of the independent variables $\lambda'$ and $\mu'$ at which the lowest cost of the system occurs.

**Example 11.9.** The cost of waiting per period is \$4, and the cost per hour for servicing an item in a service center which can handle one item in 1 h is \$2. If arrivals and servicing rates follow Poisson distributions, find the lowest cost policy.

*Solution.* By Eq. (11.27),

$$C_{t,\,opt} = 3C_w = 3(4) = \$12 \text{ per period}$$

for which, by Eqs. (11.28) and (11.29),

$$\lambda'_{opt} = \frac{C_w}{C_f} = \frac{4}{2} = 2 \text{ items per period}$$

$$\mu'_{opt} = 2\frac{C_w}{C_f} = 2\frac{4}{2} = 4 \text{ items per period}$$

## 11.15 FINITE QUEUING

In practice, the number of items being serviced might have a significant effect on the number of items still in use. The situation becomes one of finite queuing. Analytical solutions for finite queuing problems can become quite complicated and are not considered here. The problem can be treated by the general methods described subsequently, noting that probabilities will be affected by the number of items requiring servicing.

## 11.16 GENERAL METHOD WITH CERTAINTY

In many cases an analytical solution is difficult or perhaps impossible. Such cases must be reduced to a tabulation, and considerable ingenuity may be required to represent the problem on paper. The following example is straightforward but would be quite complicated if a reduction to mathematical equations were attempted.

**Example 11.10.** Five units of a product must be processed at three different stations, $A$, $B$, and $C$. The processes are independent. The process time in days at each station is:

| | |
|---|---|
| $A$ | 5 days |
| $B$ | 7 days |
| $C$ | 9 days |

The five products must have the same operating sequence. The handling time and cost between stations are negligible.

Stations are all started at zero time and are shut individually when the fifth unit has gone through the station. Idle time for a product is counted whenever a product arrives at a station and must wait for entry. Idle time for a station is counted whenever a station is waiting for work.

Find the most economical sequence of work centers for the following conditions, where $C_w$ is the waiting cost in dollars per part per day and $C_e$ is the cost of idle time in dollars per station per day:

| | $C_w$ | $C_e$ |
|---|---|---|
| (a) | 10 | 30 |
| (b) | 20 | 20 |
| (c) | 30 | 10 |

**Solution.** Table 11.3 shows the waiting time of products and the idle time of machines for all possible sequences of work centers. The total time required to produce five parts is 57 days for all the sequences; but the waiting and idle times vary, and in summary are

| Sequence | Part waiting | Station idle |
|---|---|---|
| ABC | 40 | 17 |
| ACB | 40 | 27 |
| BAC | 20 | 27 |
| BCA | 20 | 39 |
| CAB | 0 | 47 |
| CBA | 0 | 49 |

**TABLE 11.3**
**Time-analysis tabulation for Example 11.10**

| Sequence | Part no. | Station 1 In | Station 1 Out | Station 2 In | Station 2 Out | Station 2 Part waiting | Station 2 Station idle | Station 3 In | Station 3 Out | Station 3 Part waiting | Station 3 Station idle |
|---|---|---|---|---|---|---|---|---|---|---|---|
| ABC | 1 | 0 | 5 | 5 | 12 | 0 | 5 | 12 | 21 | 0 | 12 |
|  | 2 | 5 | 10 | 12 | 19 | 2 | 0 | 21 | 30 | 2 | 0 |
|  | 3 | 10 | 15 | 19 | 26 | 4 | 0 | 30 | 39 | 4 | 0 |
|  | 4 | 15 | 20 | 26 | 33 | 6 | 0 | 39 | 48 | 6 | 0 |
|  | 5 | 20 | 25 | 33 | 40 | 8 | 0 | 48 | 57 | 8 | 0 |
| ACB | 1 | 0 | 5 | 5 | 14 | 0 | 5 | 14 | 21 | 0 | 14 |
|  | 2 | 5 | 10 | 14 | 23 | 4 | 0 | 23 | 30 | 0 | 2 |
|  | 3 | 10 | 15 | 23 | 32 | 8 | 0 | 32 | 39 | 0 | 2 |
|  | 4 | 15 | 20 | 32 | 41 | 12 | 0 | 41 | 48 | 0 | 2 |
|  | 5 | 20 | 25 | 41 | 50 | 16 | 0 | 50 | 57 | 0 | 2 |
| BAC | 1 | 0 | 7 | 7 | 12 | 0 | 7 | 12 | 21 | 0 | 12 |
|  | 2 | 7 | 14 | 14 | 19 | 0 | 2 | 21 | 30 | 2 | 0 |
|  | 3 | 14 | 21 | 21 | 26 | 0 | 2 | 30 | 39 | 4 | 0 |
|  | 4 | 21 | 28 | 28 | 33 | 0 | 2 | 39 | 48 | 6 | 0 |
|  | 5 | 28 | 35 | 35 | 40 | 0 | 2 | 48 | 57 | 8 | 0 |
| BCA | 1 | 0 | 7 | 7 | 16 | 0 | 7 | 16 | 21 | 0 | 16 |
|  | 2 | 7 | 14 | 16 | 25 | 2 | 0 | 25 | 30 | 0 | 4 |
|  | 3 | 14 | 21 | 25 | 34 | 4 | 0 | 34 | 39 | 0 | 4 |
|  | 4 | 21 | 28 | 34 | 43 | 6 | 0 | 43 | 48 | 0 | 4 |
|  | 5 | 28 | 35 | 43 | 52 | 8 | 0 | 52 | 57 | 0 | 4 |
| CAB | 1 | 0 | 9 | 9 | 14 | 0 | 9 | 14 | 21 | 0 | 14 |
|  | 2 | 9 | 18 | 18 | 23 | 0 | 4 | 23 | 30 | 0 | 2 |
|  | 3 | 18 | 27 | 27 | 32 | 0 | 4 | 32 | 39 | 0 | 2 |
|  | 4 | 27 | 36 | 36 | 41 | 0 | 4 | 41 | 48 | 0 | 2 |
|  | 5 | 36 | 45 | 45 | 50 | 0 | 4 | 50 | 57 | 0 | 2 |
| CBA | 1 | 0 | 9 | 9 | 16 | 0 | 9 | 16 | 21 | 0 | 16 |
|  | 2 | 9 | 18 | 18 | 25 | 0 | 2 | 25 | 30 | 0 | 4 |
|  | 3 | 18 | 27 | 27 | 34 | 0 | 2 | 34 | 39 | 0 | 4 |
|  | 4 | 27 | 36 | 36 | 43 | 0 | 2 | 43 | 48 | 0 | 4 |
|  | 5 | 36 | 45 | 45 | 52 | 0 | 2 | 52 | 57 | 0 | 4 |

**TABLE 11.4**
**Cost tabulation for Example 11.10**

| Sequence | Total product waiting, product-days | Cost of waiting, $ (a) | (b) | (c) | Total station idle, station-days | Cost of idle stations, $ (a) | (b) | (c) | Total variable cost per period, $ (a) | (b) | (c) |
|---|---|---|---|---|---|---|---|---|---|---|---|
| ABC | 40 | 400 | 800 | 1200 | 17 | 510 | 340 | 170 | 910 | 1140 | 1370 |
| ACB | 40 | 400 | 800 | 1200 | 27 | 810 | 540 | 270 | 1210 | 1340 | 1470 |
| BAC | 20 | 200 | 400 | 600 | 27 | 810 | 540 | 270 | 1010 | 940 | 870 |
| BCA | 20 | 200 | 400 | 600 | 39 | 1170 | 780 | 390 | 1370 | 1180 | 990 |
| CAB | 0 | 0 | 0 | 0 | 47 | 1410 | 940 | 470 | 1410 | 940 | 470 |
| CBA | 0 | 0 | 0 | 0 | 49 | 1470 | 980 | 490 | 1470 | 980 | 490 |

Table 11.4 is the calculation for costs based on the following relationship:

$$C_t = C_w \left( \begin{array}{c} \text{total waiting time} \\ \text{for product} \end{array} \right) + C_e \left( \begin{array}{c} \text{total idle time} \\ \text{for stations} \end{array} \right)$$

The minimum cost for (a) is $910 per period for sequence ABC, for (b) $940 per period for BAC or CAB, and for (c) $470 per period for CAB.

# 11.17 GENERAL METHOD WITH UNCERTAINTY

The analysis of queuing problems with uncertainty in arrival or servicing time or both can become exceedingly complicated. The most general method is to subject the data to a Monte Carlo computation, which is a simulation process. The distributions are simulated by the use of random numbers, and a large number of periods are taken to get a good average and a good summary of final distributions and costs. The Monte Carlo method was discussed in Chap. 8 and used in Sec. 11.9 for an inventory problem. The following example is an application to a queuing problem. The example is very simple and can actually be evaluated without the Monte Carlo technique.

**Example 11.11.** A company has a large number of switches. The number of switches out of service follows a Poisson distribution, with an expectancy of 0.5 switches out of service for a 1-h period. A repair crew makes a round of all the switches and can restore all switches to service in a very short time at negligible cost. A penalty is charged for switches out of service based on the *average* number out of service at a rate of $500/h per switch. The cost for a repair crew depends on how fast they can make a round, and for an interval of 10 to 60 min it can be expressed as

$$-50\sqrt{T} + 550 \qquad \text{\$/h}$$

where $T$ is the time in minutes to make a round.

Assume that a crew makes a round in 36 min, or 0.6 h. For the Poisson distribution of 1 h the expectancy is 0.5 switches down, and for 0.6 h it will be $0.6 \times 0.5 = 0.3$ expected switches down following the Poisson distribution:

$$P(N) = \frac{e^{-0.3}(0.3)^N}{N!}$$

with a cost in dollars per hour for the crew equal to

$$-50\sqrt{36} + 550 = \$250/h$$

Assignment of random numbers is

| N | P(N) | Cumulative P(N) | Random numbers |
|---|------|------------------|----------------|
| 0 | 0.74082 | 0.74082 | 0001–7408 |
| 1 | 0.22225 | 0.96307 | 7409–9631 |
| 2 | 0.03334 | 0.99641 | 9632–9964 |
| 3 | 0.00333 | 0.99974 | 9965–9997 |
| 4 | 0.00025 | 0.99999 | 9998–0000 |
| 5 | 0.00002 | 1.00000 | |

Random numbers are generated with the following tabulation:

| Random no. | N switch down | Cost, $/h | Crew, $/h | Total, $/h | Cumulative average, $/h |
|---|---|---|---|---|---|
| 9804 | 2 | 1000 | 250 | 1250 | 1250 |
| 0472 | 0 | 0 | 250 | 250 | 750 |
| 3167 | 0 | 0 | 250 | 250 | 583.33 |
| 8012 | 1 | 500 | 250 | 750 | 625 |

A large number of trials gave \$400.80/h as the cost for a crew time of 36 min, while the exact cost is $500(0.3) + 250 = \$400$, but generally an exact solution will not be possible. Other crew times can be tried to obtain the dollars per hour against crew time. Some results are

| Crew time, min | $/h |
|---|---|
| 12 | 427 |
| 36 | 400 |
| 60 | 413 |

The minimum cost actually occurs for a crew time of 36 min.

# NOMENCLATURE

$C_f$      Cost per period for a service facility which can service one item per period when working full time

$C_h$      Storage or holding cost per item per period, $/(item)(period)

$C_o$      Cost of placing and receiving an order, or setup cost, $

$C_t$      Total variable cost per period, $/period

$C_T$      Total cost, $

$C_w$      Cost for one item waiting one period, $/(item)(period)

$D$      Production rate, items per period

$L$      Number of channels in the service facility, dimensionless

$M$      Maximum inventory, number of items

$N$      Number of periods per cycle, dimensionless

$P(X)$      Probability of $X$

$Q$      Lot size, items

$R$      Reorder level, items

$T$      Time, periods

$T_a$      Time between two consecutive arrivals, periods per item

$T_l$      Procurement or lead time, periods

$T_s$      Time in periods to service one item, periods per item

$U$      Usage rate, items per period

# PROBLEMS

**11.1.** The rate of use of an item is 20 units per year. The cost of placing and receiving an order is $40. The cost of storing one item for 1 year is $0.16, and it depends on the average stock. Find (*a*) the economic lot size, (*b*) the minimum total variable cost per period, and (*c*) the reorder level if the lead time is 3 months.

**11.2.** Repeat Prob. 11.1 with a safety stock of four items.

**11.3.** For this problem the storage cost is based on maximum inventory rather than average inventory. Derive expressions for $Q_{opt}$ and $C_{t, opt}$ corresponding to Eqs. (11.2) and (11.4).

**11.4.** A store has to satisfy a daily demand of 10 items of a product. The holding cost is $0.10 per item per day based on average inventory. The cost of procuring an order is $20, and it takes 3 days for delivery. The minimum number of items per order is 100 and the maximum is 500 with increments of 5 items. The replenishment rate is infinite. Find the optimum lot size.

**11.5.** An appliance manufacturer can sell 400 per day of an item at $5.99 each with a manufacturing cost of $4.92 each. He can produce them at a rate of 1000 units per day with a setup cost of $1200. The inventory cost is based on carrying the inventory at his cost at 20 percent per year. Find the economic production run and how long the lot will last.

**11.6.** For this problem the storage cost is based on maximum inventory rather than average inventory. Derive expressions for $Q_{opt}$ and $C_{t, opt}$ corresponding to Eqs. (11.13) and (11.14).

**11.7.** The cumulative probability curve for the number of items sold in season can be plotted from the following information:

| Cumulative probability of selling this many items or less in season | Number of items |
|---|---|
| 0 | 0 |
| 0.15 | 2,000 |
| 0.52 | 4,000 |
| 0.82 | 6,000 |
| 0.98 | 8,000 |
| 1.00 | 10,000 |

Inventory is built up at the start of a season and cannot be replenished during a season. An item costs $3 and can be sold for $6 in season, but if disposed of after the season it will bring only $1. Find the optimum number with which to start the season. The last one ordered should have a sales expectation based on in-season and out-of-season value equal to its cost.

**11.8.** Redo Example 11.5 using a binomial distribution. Assume that the total demand $N$ might be 40 per week and that with an average of four per week, the probability for demand is $p = 4/40 = 0.1$ in a population of $N = 40$.

**11.9.** A pharmaceutical company produces 10,000 m³/day of an antibiotic broth. The broth is collected for $T$ days, and then the entire collection is converted to a finished product.

The broth spoils in storage such that the spoilage for each addition is proportional to the number of days that the addition has been held in storage prior to conversion. If a fixed amount of broth is held in storage for 10 days, it loses 0.725 percent of its biological effectiveness.

Broth is held in inventory at $2/m^3$ at all times. The fractional holding charge for inventory is $0.10/year. Spoilage is taken at the inventory cost. Broth is produced 365 days/year. It costs $1000 to set up the conversion process of broth to the finished product.

How often should the broth be converted to the finished product?

**11.10.** An item costs $2 and is used at a uniform rate of 10 items per day. No shortage is allowed, and the delivery rate is infinite. It costs $10 to process an order. Inventory charge is $0.02/day for each item in inventory. In addition, there is a flat charge of $3/year per item based on the maximum number of items in inventory. Find the optimum procurement quantity.

**11.11.** In an emergency repair shop machinists cost the company $27/h. If a machinist is working on an emergency job, he or she is worth $60/h; otherwise the value is $12/h. The demand for machinists for emergency work is

| Percent of total time | Machinists on emergency work | Percent of total time | Machinists on emergency work |
|---|---|---|---|
| 5 | 1 or less | 85 | 6 or less |
| 15 | 2 or less | 90 | 7 or less |
| 30 | 3 or less | 95 | 8 or less |
| 50 | 4 or less | 98 | 9 or less |
| 70 | 5 or less | 100 | 10 or less |

How many machinists should be kept on hand for emergencies?

**11.12.** The average demand for a certain item in inventory is 0.7 per week, with the probability for a demand of exactly $N$ items in a week being $P(N)$ and

$$P(0) = 0.500$$

$$P(1) = 0.300$$

$$P(2) = 0.200$$

The time in weeks required to get a new item if it is not in stock averages 1.85 weeks, with the distribution as follows:

$$P(1) = 0.400$$

$$P(2) = 0.350$$

$$P(3) = 0.250$$

If an order is placed when the level on hand is one, find the probability of going short before the order is received.

**11.13.** A food company makes a flavoring material at the rate of 100 kg/day. It is made batchwise at a cost of $3000 per batch.

Inventory cost is $2/kg/day since very expensive methods are required to keep the material from spoiling. In addition, there is an extra cost which is proportional to the batch size multiplied by the square of the time that the inventory is held before it is exhausted. If an initial inventory of 1000 kg lasting 10 days is made, this extra cost would be $10,900. Find the optimum batch size.

**11.14.** Coal is stored and will form a conical pile at all times, with a slope of 45 degrees with the horizontal. The packed coal occupies 1 m³/1600 kg.

The demand for coal is 455,000 kg/day. The cost for ordering and procuring is $1500 per order. The coal is valued at $27.50/1000 kg. Inventory cost is at a rate of 20 percent per year of the value of the coal in the pile at any time. The delivery rate is infinite. There is also a charge of $5.38/m³/year for the maximum ground area that the coal occupies.

Find the optimum amount of coal to be purchased per order.

**11.15.** It costs $50 to place an order. The cost of putting the order into the warehouse is proportional to the square root of the size of the order and is $25 when the order is for 25 items. A safety stock of 15 items is to be maintained. The cost of maintaining the inventory in storage is proportional to the 1.5 power of the number in storage at any instant and would be $12.50/week for an inventory of 25 items. Usage is continuous at a rate of five items per week, and delivery of an order may be considered instantaneous after it is placed. Find the optimum order size.

**11.16.** Redo Prob. 11.12 by the Monte Carlo method.

**11.17.** The demand for a spare part averages two per week and follows a Poisson distribution. The cost for stocking a part is $5/week. If the demand for the part cannot be met due to a shortage, the penalty for each item short is $30. Find the optimum number of parts to stock at the beginning of the week.

**11.18.** A gasoline station receives an automobile every 10 min. The cost of waiting in the station is $0.10 per automobile per minute. A service station that can service one car in 1 min costs $0.16/min.
(a) Plot the service cost as a function of the service time in minutes.
(b) Plot the waiting cost as a function of the service time in minutes.
(c) Plot the total variable cost as a function of the service time in minutes.
(d) Find the service time for the lowest cost per automobile.

**11.19.** If in Prob. 11.18 the station has two or three pumps, find the lowest cost situation.

**11.20.** The demand for a test is four per hour on the average and follows a Poisson distribution. A testing machine can perform one test per hour. The company now has three testing machines. If the demand for a test cannot be met at once, it costs the company $100. How much can the company afford per hour for three additional testing machines?

**11.21.** An attendant arrives at an inspection point in a plant every 5 h to pick up rejected assemblies. Her pickup truck can accommodate only three assemblies at the most, and the probabilities for vacant space on the truck are

| No. of vacant spaces | Probability |
| --- | --- |
| 0 | 0.05 |
| 1 | 0.10 |
| 2 | 0.50 |
| 3 | 0.35 |

The number of rejected assemblies arriving at the inspection point during the 5 h has the following probabilities:

| Number | Probability |
|--------|-------------|
| 0 | 0.05 |
| 1 | 0.50 |
| 2 | 0.30 |
| 3 | 0.15 |

Any assemblies that cannot be picked up by the attendant must be picked up by an emergency vehicle. What fraction of the rejected assemblies must be accommodated by the emergency vehicle?

**11.22.** Redo Prob. 11.21 by the Monte Carlo method.

**11.23.** A testing station runs tests on motors. Twenty percent of the motors are class $A$, 30 percent are class $B$, and 50 percent are class $C$. A motor arrives every 10 h with negligible variation. While a motor is being serviced the cost to the company for waiting in dollars per hour is $100 for class $A$ motors, $150 for class $B$, and $180 for class $C$.

Any size inspection station can be installed, and the inspection time per motor for one test will be inversely proportional to the size of the station. A station that can make 200 tests per hour will cost $47.60/h.

Class $A$ motors require 100 tests, class $B$ 200, and class $C$ 300. Find the optimum size inspection station.

**11.24.** A company has automatic machines which require overhauls on the average of one per week with a Poisson distribution. A single crew can overhaul a machine in 4 days at a cost of $1000/week, two crews can overhaul a machine in 2 days at $2000/week, etc. Lost time is charged against a machine only while it is actually being overhauled, i.e., no waiting in line occurs, at a charge of $2000/day. Find the optimum number of crews to be hired and the cost per week.

**11.25.** The cost in dollars per day for an inspection crew is given by

$$150M^{0.9}$$

where $M$ is the number of persons in the crew. The time in days to complete an inspection is given by

$$\frac{2}{M^{0.8}}$$

An inspection is made every 3 days, and the loss for the operating item while it is being inspected is $1000/h. Find the optimum number of persons for the inspection crew.

**11.26.** Every 10 min a bus arrives at a stop with a distribution of empty seats as follows:

| No. of empty seats | Times occurring |
|--------------------|-----------------|
| 0 | 1 |
| 1 | 3 |
| 2 | 5 |
| 3 | 6 |
| 4 | 4 |
| 5 | 1 |
| 6 | 0 |

The number of persons arriving at the stop in a 10-min period has a distribution as follows:

| No. arriving | Times occurring |
|---|---|
| 0 | 5 |
| 1 | 4 |
| 2 | 3 |
| 3 | 3 |
| 4 | 1 |
| 5 | 2 |
| 6 | 2 |
| 7 | 0 |

No standees are allowed in the bus. What fraction of the total persons arriving will have to wait for the next bus? Use a Monte Carlo technique.

**11.27.** Ten items of product $X$ are to be processed on machines $A$, $B$, $C$, and $D$. Machine $A$ must be used first, but the other machines can be selected in any order. The process times in days are

| $A$ | $B$ | $C$ | $D$ |
|---|---|---|---|
| 3 | 2 | 7 | 5 |

The waiting costs are \$100 per item per day and \$150 per machine per day.

The same sequence of operations must be used for all the items. A machine must finish one item before starting another. The handling time between machines is negligible.

Find the optimum sequence of operations.

# REFERENCES

## Inventory problems

1. Arrow, K. J., S. Karlin, and H. Scarf: *Studies in the Mathematical Theory of Inventory and Production,* Stanford University Press, Stanford, Calif., 1958.
2. Banks, J., and W. J. Fabrycky: *Procurement and Inventory Systems Analysis,* Prentice-Hall, Inc., Englewood Cliffs, N. J., 1987.
3. Chikan, A., ed.: *Studies in Production and Engineering Economics: Inventory in Theory and Practice,* Elsevier Science Publishers, Amsterdam, 1986.
4. Fabrycky, W. J., P. M. Ghare, and P. E. Torgersen: *Industrial Operations Research,* rev. ed., Prentice-Hall, Inc., Englewood Cliffs, N. J., 1972.
5. Hadley, G., and T. M. Whitin: *Analysis of Inventory Systems,* Prentice-Hall, Inc., Englewood Cliffs, N. J., 1963.
6. Larson, S. E.: *Inventory Systems and Controls Handbook,* Prentice-Hall, Inc., Englewood Cliffs, N. J., 1976.
7. Lewis, C. D.: *Demand, Analysis, and Inventory Control,* Lexington Books, Lexington, Mass., 1975.
8. Love, S. F.: *Inventory Control,* McGraw-Hill Book Company, New York, 1979.
9. Starr, M. K., and D. W. Miller: *Inventory Control, Theory and Practice,* Prentice-Hall, Inc., Englewood Cliffs, N. J., 1962.
10. Welch, W. E.: *Scientific Inventory Control,* Management Publishing Company, Greenwich, Conn., 1956.

**Queuing problems**

11. Allen, A. O.: *Probability and Statistics, and Queuing Theory*, Academic Press, Inc., New York, 1978.
12. Cooper, R. B.: *Introduction to Queuing Theory*, 2d ed., North Holland, Inc., New York, 1981.
13. Fabrycky, W. J., P. M. Ghare, and P. E. Torgersen: *Industrial Operations Research*, rev. ed., Prentice-Hall, Inc., Englewood Cliffs, N. J., 1972.
14. Gross, D., and C. M. Harris: *Fundamentals of Queuing Theory*, John Wiley & Sons, Inc., New York, 1974.
15. Kleinrock, L.: *Queuing Systems,* vol. 1, *Theory*, John Wiley & Sons, Inc., New York, 1975.
16. Page, E.: *Queuing Theory in OR*, C. Russak and Company, Inc., New York, 1972.
17. Peck, L. G., and R. N. Hazelwood: *Finite Queuing Tables*, John Wiley & Sons, Inc., New York, 1958.
18. Prabhu, N. U.: *Queues and Inventories*, John Wiley & Sons, Inc., New York, 1965.
19. White, J. A., J. W. Schmidt, and G. K. Bennett: *Analysis of Queueing Systems*, Academic Press, Inc., New York, 1975.

# CHAPTER
# 12

# LINEAR, DYNAMIC, AND GEOMETRIC PROGRAMMING

*A.H. Boas*

## LINEAR PROGRAMMING

## 12.1 OPTIMIZATION OF AN OBJECTIVE FUNCTION

*Linear programming* (*LP*) is a mathematical tool for finding the optimum solution to a certain class of problems. The word *linear* implies that the relations involved will be linear, i.e., in all relationships every term can be expressed in only one variable raised to the first power. Thus

$$A_1 X_1 + A_2 X_2 + A_3 X_3 = B_1$$

is linear, while the following are not:

$$A_1 X_1 + A_2 X_2^2 + A_3 X_3 = B_1$$

$$A_1 \ln X_1 + A_2 X_2 + A_3 X_3 = B_1$$

$$A_1 X_1 + A_2 X_1 X_2 + A_3 X_3 = B_1$$

The term *programming* in this context means planning of activities.

In any type of optimization problem, linear or nonlinear, the problem can be represented as indicated in Fig. 12.1. The black box constitutes the mathematical

**317**

**FIGURE 12.1**
A mathematical model.

model, i.e., relations that describe some physical situation. The input to the black box is composed of two quantities, fixed and variable. The fixed quantities are those which are inherent in the problem; the formulator has no control over these. On the other hand, the variable inputs are the unknowns in the problem; the values of these variables determine the optimum solution to the problem. Hence the interest is in the values of the variables that will give the optimum value to a term called the *objective function*, which is the output from the black box.

## 12.2 DEVELOPMENT OF LINEAR-PROGRAMMING EQUATIONS

Define each of the variable inputs by $X_j$ and assume that there are $n$ of them. These variables are also called *activity levels*. In LP problems a restriction that is imposed on these variables is that they be nonnegative; i.e.,

$$X_j \geq 0 \tag{12.1}$$

The objective function, the output from the black box, will be called $Z$ and defined as

$$Z = C_1 X_1 + C_2 X_2 + \cdots + C_n X_n \tag{12.2}$$

where the $C_j$ values, $j = 1, 2, \ldots, n$, are unit costs. Associated with each variable $X_j$ is a cost coefficient $C_j$, which may be zero, positive, or negative. Some examples of $C_j$ values are

| $X_j$ | $C_j$ | $C_j X_j$ |
|---|---|---|
| kg/h | $/kg | $/h |
| Btu/h | $/Btu | $/h |
| work-h/h | $/work-h | $/h |

In each case the product of $X_j$ and $C_j$ represents a cost in dollars per hour. These may be added together to represent the total cost, in dollars per hour, and could therefore represent a meaningful objective function. Mathematically, the objective function may be represented as

$$Z = \sum_{j=1}^{n} C_j X_j \tag{12.3}$$

Note that this equation is linear, as required in LP models.

The mathematical model itself may be represented as

$$A_{11}X_1 + A_{12}X_2 + \cdots + A_{1j}X_j + \cdots + A_{1n}X_n = B_1 \qquad (12.4)$$

$$A_{21}X_1 + A_{22}X_2 + \cdots + A_{2j}X_j + \cdots + A_{2n}X_n = B_2 \qquad (12.5)$$

$$\cdots\cdots\cdots\cdots\cdots\cdots\cdots\cdots\cdots\cdots\cdots\cdots\cdots\cdots$$

$$A_{m1}X_1 + A_{m2}X_2 + \cdots + A_{mj}X_j + \cdots + A_{mn}X_n = B_m \qquad (12.6)$$

Each equation in an LP model takes this form and is called a *row* or *constraint*. The right-hand sides are the $B_i$ values. Each $A_{ij}$ is a coefficient with the usual mathematical convention of rows before columns; thus $i$ represents the row and $j$ the column. The coefficients may be zero, positive, or negative. Equations (12.4) to (12.6) can be expressed mathematically:

$$\sum_{j=1}^{n} A_{ij}X_j = B_i \qquad i = 1, 2, \ldots, m \qquad (12.7)$$

## 12.3 SLACK VARIABLES

Suppose that a process unit can handle a maximum of 100 kg/h. Mathematically,

$$X_1 \le 100 \qquad (12.8)$$

It is necessary to express this fact within the LP model. Any solution to the problem which violates this capacity limit is not a valid solution because the process unit cannot physically handle such a quantity of stream 1. Equation (12.8) is not in the form of a linear equation, and it becomes necessary to convert the inequality equation to an equality relationship. This is done through the introduction of a *slack variable*. Express Eq. (12.8) as

$$X_1 + X_4 = 100 \qquad (12.9)$$

The variable $X_4$, the slack variable, is treated like any other variable with the restriction of nonnegativity imposed on it. When $X_4$ assumes its minimum value of 0, then $X_1 = 100$, and for the maximum value $X_4 = 100$, $X_1 = 0$.

In contrast, suppose the following inequality appears:

$$X_2 \ge 10 \qquad (12.10)$$

This equation states that $X_2$ must equal *at least* 10 kg/h. In a manner similar to the foregoing illustration, a *negative* slack variable $X_5$ is introduced, so that Eq. (12.10) becomes

$$X_2 - X_5 = 10 \qquad (12.11)$$

The slack variable $X_5$ must meet the nonnegativity restriction, of course. Thus a method is available for converting inequalities to equalities and satisfying the form of the standard LP constraint equations.

## 12.4 QUALITY CONSTRAINTS

Another type of constraint encountered in LP work is the *quality* constraint. Normally some type of specification or quality of product must be met. As an example, let us consider the blending of gasoline components to meet a certain octane number. The assumption of linear blending will be made so that the following relation holds:

$$Q = \frac{X_1 Q_1 + X_2 Q_2 + X_3 Q_3}{X_1 + X_2 + X_3} \tag{12.12}$$

where $X_i$ = quantity of blending stock $i$, bbl

$Q_i$ = quality of $i$th blending stock, octane numbers/bbl

Equation (12.12) then represents the octane number of a three-component mixture. In general

$$Q = \frac{\sum X_i Q_i}{\sum X_i} \tag{12.13}$$

**Example 12.1.** Three components with octane numbers 108, 93, and 100 are to be linearly blended such that the final octane number of the mixture is at least 98. Express this condition as a standard equation in an LP model.

*Solution.* The final octane number of the mixture can be obtained from Eq. (12.12), and it must be a minimum of 98. That is,

$$\frac{108X_1 + 93X_2 + 100X_3}{X_1 + X_2 + X_3} \geq 98$$

or
$$108X_1 + 93X_2 + 100X_3 \geq 98X_1 + 98X_2 + 98X_3$$

or
$$10X_1 - 5X_2 + 2X_3 \geq 0$$

Introducing the slack variable $X_4$, the required relationship is

$$10X_1 - 5X_2 + 2X_3 - X_4 = 0$$

which is in standard LP form and can be used as one of the equations in the model to represent the quality constraint on octane blending.

## 12.5 METHOD OF SOLUTION

Refer to Fig. 12.2 and identify the various inputs to a standard LP problem. The fixed inputs would be all the constants in the problem, $A_{ij}$, $B_i$, and $C_j$; the variables would be the $X_j$ values. The model within the black box would be all the constraint equations, given by Eq. (12.7). The output from the black box is represented by the objective function, Eq. (12.3). Figure 12.2 is a representation of the standard LP problem.

The formulation of the problem, the variables, the constraint equations, and the objective function have all been discussed. The problem of finding the optimal solution remains. Examine the constraint Eqs. (12.7). Where the number of variables $n$ is

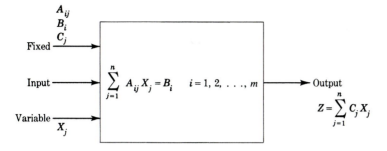

**FIGURE 12.2**
A linear-programming model.

equal to the number of equations $m$, one can solve this set of independent linear equations. One unique solution will be obtained. There is no problem in optimization here because there is only one solution. The values obtained for the variables from the constraint equations are substituted in the equation for the objective function, and the solution is obtained immediately. It is when there are more variables than equations, i.e., $n > m$, that a problem arises. Now there is not one unique solution but an infinite number of solutions. From all these solutions, one wishes to find the one which optimizes the objective function.

The method used is to set $n - m$ variables equal to zero, i.e., the excess number of variables. Then there will be $m$ variables left and $m$ equations to solve. For example, if $n = 23$ variables and $m = 10$ equations, set $23 - 10$, or 13, variables equal to zero and solve the 10 equations for the 10 unknowns. The object is to determine which variables to set equal to zero. By examining every possible way of selecting $n - m$ variables from $n$ variables, setting these equal to zero, and solving for the $m$ remaining variables, it can be shown that the optimum solution lies among these many solutions [3].

The $n - m$ variables set equal to zero are called *nonbasic variables*; the remaining $m$ variables are known as *basic variables,* and the collection of them is called a *basis*. The solution of the equations involving the basic variables is termed a *basic solution*. If the values of the basic variables are nonnegative and do not violate any of the constraint equations, the solution is called a *basic feasible solution*. The optimal solution is to be found among the basic feasible solutions. Therefore, once a basic feasible solution is obtained, the object is to find a more optimal solution. This process is continued until a point is reached at which no further improvement is possible.

**Example 12.2.** A production facility manufactures two products, $A$ and $B$. Each product must pass through a bank of type 1 machines and a bank of type 2 machines serially, as shown in Fig. 12.3. Each bank consists of many machines of its particular type. Product $A$ requires 2 h on type 1 machines and 1 h on a type 2 machine, while product $B$ requires 1 h on type 1 and 4 h on type 2 machines. The total time available per week on machines of type 1 is 6000 h, whereas the time available for type 2 machines is 10,000 h. The net profit is \$5 per unit of product $B$ and \$3.5 per unit of product $A$. Find the optimal solution, the production schedule which will maximize profit per week.

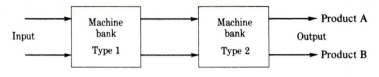

**FIGURE 12.3**
A model for Example 12.2.

***Solution.*** First present all the data in tabular form:

| Machine type | Time required, h | | Maximum |
| | Product A | Product B | h/week |
| --- | --- | --- | --- |
| 1 | 2 | 1 | 6,000 |
| 2 | 1 | 4 | 10,000 |
| | Profit, $ per unit | | |
| | 3.5 | 5.0 | |

The first task is to represent all the data in the form of standard LP equations. Let $X_1$ be the quantity of product A produced per week, and $X_2$ correspondingly will be the quantity of product B produced per week. For each unit of product A, $2X_1$ h will be required on type 1 machines; similarly, $X_2$ hours will be required on type 1 machines for product B. The total time required on type 1 machines is $2X_1 + X_2$, and the maximum weekly time is 6000 h; hence

$$2X_1 + X_2 \le 6000 \tag{12.14}$$

which, upon introduction of slack variable $X_3$, becomes

$$2X_1 + X_2 + X_3 = 6000 \tag{12.15}$$

Similarly, for type 2 machine with slack variable $X_4$

$$X_1 + 4X_2 + X_4 = 10,000 \tag{12.16}$$

Equations (12.15) and (12.16) are the constraint equations of the LP model. The variables are $X_1$, $X_2$, $X_3$, and $X_4$. It remains to define the objective function and solve the problem. Based on the unit profit figures, the objective function is

$$Z = 3.5X_1 + 5X_2 \tag{12.17}$$

At this point it might be instructive to see a geometric interpretation of the problem. Equation (12.14) is plotted in Fig. 12.4 as line *BD*. Any point $(X_1, X_2)$ that lies above this line violates the constraint on maximum available time on machine type 1. The solution must, therefore, lie in the region *OBD*. However, there is another restriction on the solution. Equation (12.16) is plotted as line *AC*. A feasible solution that does not violate the maximum available time on machine 2 must lie in area *OCA*. Thus, in order to satisfy both constraints of the problem, the solution must lie in area *OBEA*. Thanks to

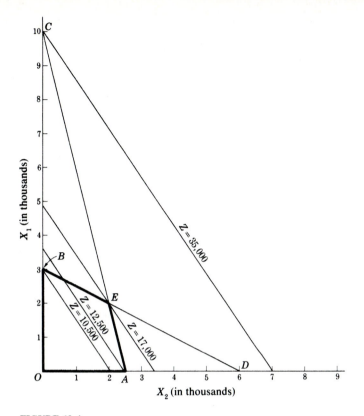

**FIGURE 12.4**
A graph for Example 12.2.

the nonnegativity restriction on the variables, it was necessary to consider only the first quadrant. The problem, therefore, is to seek a point $(X_1, X_2)$ that lies within the region *OBEA* and maximizes the profit function. Equation (12.17) represents the profit function and can be plotted on Fig. 12.4 for various values of $Z$. Four of these lines have been plotted for illustrative purposes. Any point along the line $Z = 10,500$ indicates a profit of \$10,500. There are an infinite number of these points along the line. However, the line for $Z = 12,500$ is a line of higher profit and thus indicates a more favorable direction. The line for $Z = 17,000$ has only one point within the feasible region, the intersection of the two constraint lines. Any line to the right of this line, such as $Z = 35,000$, contains points that yield higher profits, but the constraints of the problem are violated and hence these solutions are not feasible. By inspection, therefore, the optimal solution is

$$X_1 = 2000 \text{ units per week of product } A$$

$$X_2 = 2000 \text{ units per week of product } B$$

$$X_3 = 0$$

$$X_4 = 0$$

$$Z = \$17,000/\text{week net profit}$$

In describing the method of solution, we stated that the excess number of variables should be set equal to zero and the remaining variables solved for. If every possible way of doing this is investigated, we should be able to find the optimal solution. Let us now illustrate this. Equations (12.15) and (12.16) indicate that there are four variables and two equations. Two of these four variables must be set equal to zero and the other two solved for. For example, setting $X_1$ and $X_2$ equal to zero and solving for $X_3$ and $X_4$, we obtain $X_3 = 6000$ and $X_4 = 10,000$. How many ways are there to set two variables equal from a total of four? This is the well-known problem for the number of combinations of $P$ things taken $Q$ at a time, which is

$$\frac{P!}{Q!(P-Q)!} \tag{12.18}$$

which for $P = 4$ and $Q = 2$ gives

$$\frac{4!}{2!(4-2)!} = \frac{4(3)(2)(1)}{2(1)(2)(1)} = 6 \text{ combinations}$$

Tabulating all six solutions, we obtain

| $X_1$ | $X_2$ | $X_3$ | $X_4$ | $Z$ | Fig. 12.4 |
|---|---|---|---|---|---|
| 0 | 0 | 6,000 | 10,000 | 0 | $O$ |
| 0 | 6,000 | 0 | $-14,000$ | NF | |
| 0 | 2,500 | 3,500 | 0 | 12,500 | A |
| 3,000 | 0 | 0 | 7,000 | 10,500 | B |
| 10,000 | 0 | $-14,000$ | 0 | NF | |
| 2,000 | 2,000 | 0 | 0 | 17,000 | E |

where NF indicates a solution that is nonfeasible. It will be noted that the four feasible solutions correspond to $O$, $A$, $B$, and $E$ of Fig. 12.4. These points are called *extreme points* or *vertices*. In all LP problems these extreme points correspond to basic feasible solutions. In examining the six solutions, the nonfeasible ones were eliminated immediately; of the remaining basic feasible solutions, the optimal solution was found.

## 12.6 ALGEBRAIC METHOD

The technique used will follow four steps:

**1.** Select the basic variables.
**2.** Solve the constraint equations and the objective function for the basic variables in terms of the nonbasic variables.
**3.** Inspect the objective function equation. Decide which variable will leave the basis and which will enter in order to improve the value of the objective function.
**4.** Continue until no further improvement is possible.

**Example 12.3.** Repeat Example 12.2 using the algebraic technique.
**Step 1.** Consider a feasible mixture that violates no restraints. In this problem it is feasible to have a mixture of $X_1$ and $X_2$ only. It may be the worst mixture possible for a maximization, but it is a feasible mixture. As per Sec. 12.5, $X_1$ and $X_2$ can be set equal to zero, becoming nonbasic variables, and $X_3$ and $X_4$ become the basic variables.

**Step 2.** Solve the constraint Eqs. (12.15) and (12.16) in terms of the nonbasic variables $X_1$ and $X_2$:

$$X_3 = 6000 - 2X_1 - X_2 \qquad (12.19)$$

$$X_4 = 10,000 - X_1 - 4X_2 \qquad (12.20)$$

The objective function was given by Eq. (12.17):

$$Z = 3.5X_1 + 5X_2 \qquad (12.21)$$

For $X_1 = X_2 = 0$ the initial solution from the preceding is

$$X_1 = 0 \qquad X_2 = 0 \qquad X_3 = 6000 \qquad X_4 = 10,000 \qquad Z = 0$$

This basic feasible solution corresponds to point $O$ in Fig. 12.4.

**Step 3.** By inspecting Eq. (12.21), we note that it is better to introduce $X_2$ than $X_1$ since $X_2$ has the larger coefficient and will improve the objective function more than $X_1$. How much $X_2$ should be introduced? Obviously, as much as possible. Equation (12.20) indicates that when $X_2$ exceeds 2500, $X_4$ becomes negative. Therefore, the most that $X_2$ can be is 2500. The variable $X_2$ enters the *basis*; it becomes a *basic variable*, and $X_4$ leaves the basis and becomes a *nonbasic* variable. When one variable enters the basis, another must leave to make room for it. The variable that is driven to zero is the one that leaves the basis. Now repeat steps 1, 2, and 3 for the next iteration.

Solving the constraint equations and the objective-function equation for $X_2$ and $X_3$ in terms of $X_1$ and $X_4$ gives, from Eqs. (12.20) and (12.21), respectively,

$$X_2 = 2500 - \tfrac{1}{4}X_1 - \tfrac{1}{4}X_4 \qquad (12.22)$$

$$X_3 = 6000 - 2X_1 - (2500 - \tfrac{1}{4}X_1 - \tfrac{1}{4}X_4) = 3500 - \tfrac{7}{4}X_1 + \tfrac{1}{4}X_4 \qquad (12.23)$$

and $\qquad Z = 3.5X_1 + 5(2500 - \tfrac{1}{4}X_1 - \tfrac{1}{4}X_4) = 12,500 + \tfrac{9}{4}X_1 - \tfrac{5}{4}x_4 \qquad (12.24)$

For $X_1 = 0$ and $X_4 = 0$ the solution is

$$X_1 = 0 \qquad X_2 = 2500 \qquad X_3 = 3500 \qquad X_4 = 0 \qquad Z = 12,500$$

This basic feasible solution corresponds to point $A$ in Fig. 12.4. The improvement in the objective function has been from 0 to 12,500. Upon examination of Eq. (12.24), we see that $X_1$ should be introduced since it has a positive coefficient. Inspection of Eqs. (12.22) and (12.23) shows that Eq. (12.23) limits the value of $X_1$ to 2000, at which point $X_3$ goes to zero. Therefore, $X_1$ enters the basis and $x_3$ leaves. Repeating the technique of solving for the basic variables in terms of the nonbasic variables, we obtain

$$X_1 = 2000 - \tfrac{4}{7}X_3 + \tfrac{1}{7}X_4 \qquad (12.25)$$

$$X_2 = 2000 + \tfrac{1}{7}X_3 - \tfrac{2}{7}X_4 \qquad (12.26)$$

$$Z = 17,000 - \tfrac{9}{7}X_3 - \tfrac{13}{14}X_4 \qquad (12.27)$$

For $X_3 = 0$ and $X_4 = 0$ the solution is

$$X_1 = 2000 \qquad X_2 = 2000 \qquad X_3 = 0 \qquad X_4 = 0 \qquad Z = 17,000$$

This basic solution corresponds to point $E$ in Fig. 12.4. It is the optimal solution because both coefficients of the nonbasic variables in Eq. (12.27) are negative; introducing any of these nonbasic variables would only lower the value of $Z$ and lead to a less optimal solution.

## 12.7 SIMPLEX METHOD

The method of solution most widely used is the simplex method developed by Dantzig [2]. It will be seen that this method facilitates the actual calculations of the algebraic method described here. The development will make many references to the equations just derived in the algebraic method.

The first step requires a representation of the problem in matrix form. Start by listing the variables $X_1$, $X_2$, $X_3$, and $X_4$ with the respective cost coefficients directly above. Note that the slack variables $X_3$ and $X_4$ have zero cost coefficients because they do not appear in the objective function. The original Eqs. (12.15) and (12.16) are now entered, with the right-hand sides given under $B_i$:

| $j$ | 3.5 | 5 | 0 | 0 | |
|---|---|---|---|---|---|
| $i$ | $X_1$ | $X_2$ | $X_3$ | $X_4$ | $B_i$ |
| | 2 | 1 | 1 | 0 | 6,000 |
| | 1 | 4 | 0 | 1 | 10,000 |

At this point the basic variables must be selected. This requires a feasible mixture since the computation must start with a feasible mixture. As explained in Example 12.3, step 1, $X_3$ and $X_4$ become the basic variables and are entered in the first column of the tabulation, with the cost coefficients alongside, as shown in Table 12.1. The rest of the manipulation is the solution of the constraint equations and the objective function for the basic variables in terms of the nonbasic variables. The following calculations are required:

$$Z_j = \sum_{i=1}^{m} C_i A_{ij} \tag{12.28}$$

$$\Delta_j = Z_j - C_j \tag{12.29}$$

where $C_i$ refers to the cost coefficient of the $i$th basic variable. In the first iteration $X_3$ and $X_4$ have been chosen as basic variables. Therefore, $C_1 = C_2 = 0$ since these are

**TABLE 12.1**
**First tableau of the simplex method**

| | | 3.5 | 5 | 0 | 0 | | Corresponding | |
|---|---|---|---|---|---|---|---|---|
| | | $X_1$ | $X_2$ | $X_3$ | $X_4$ | $B_i$ | equation | |
| $X_3$ | 0 | 2 | 1 | 1 | 0 | 6,000 | (12.15) $\frac{6,000}{1} = 6,000$ | |
| $X_4$ | 0 | 1 | 4 | 0 | 1 | 10,000 | (12.16) $\frac{10,000}{4} = 2,500$ ⟵ Smallest positive |
| $Z_j$ | | 0 | 0 | 0 | 0 | 0 | | |
| $\Delta_j$ | | −3.5 | −5 | 0 | 0 | | | |

↑
Largest
negative

the values of the cost coefficients for $X_3$ and $X_4$ in the objective function. For the first basic variable $X_3$ use the first line of the matrix for $i = 1$ and read

$$A_{11} = 2 \qquad A_{12} = 1 \qquad A_{13} = 1 \qquad A_{14} = 0$$

For the second basic variable $X_4$ use the second line of the matrix for $i = 2$ and read

$$A_{21} = 1 \qquad A_{22} = 4 \qquad A_{23} = 0 \qquad A_{24} = 1$$

Therefore

$$Z_1 = C_1 A_{11} + C_2 A_{21} = 0(2) + 0(1) = 0$$

$$Z_2 = C_1 A_{12} + C_2 A_{22} = 0(1) + 0(4) = 0$$

$$Z_3 = C_1 A_{13} + C_2 A_{23} = 0(1) + 0(0) = 0$$

$$Z_4 = C_1 A_{14} + C_2 A_{24} = 0(0) + 0(1) = 0$$

The $\Delta_j$ values can now be calculated:

$$\Delta_1 = Z_1 - C_1 = 0 - 3.5 = -3.5$$

$$\Delta_2 = Z_2 - C_2 = 0 - 5.0 = -5.0$$

$$\Delta_3 = Z_3 - C_3 = 0 - 0 = 0$$

$$\Delta_4 = Z_4 - C_4 = 0 - 0 = 0$$

Note that the $\Delta_j$ values are the coefficients of the nonbasic variables with the sign reversed.

These values are entered in the appropriate spaces, giving the first tableau in Table 12.1. The value of $Z$, the objective function, is also listed in the lower right-hand corner. It is obtained from the relationship

$$Z = \sum_{i=1}^{m} C_i B_i \tag{12.30}$$

The bottom row is examined for the most negative coefficient. This is seen to be under column $X_2$, which indicates that it is desirable to introduce this variable into the basis. Each right-hand side is divided by the element in the $X_2$ column to determine which is limiting. It is that calculation which gives the smallest *positive* number. For this example

$$\frac{6000}{1} = 6000$$

$$\frac{10,000}{4} = 2500$$

The second row controls since $X_4$ cannot be larger than 2500 (lest $X_4$ become a negative number). Hence $X_2$ enters the basis and $X_4$ leaves. Element $A_{22} = 4$ is called the *pivot element*. The variable that enters the basis is on the *pivot column,* and the variable leaving lies on the *pivot row.* Their intersection is the pivot element.

The method now prepares a new tableau without violating any restraint. The procedure is based on a matrix manipulation in which each row in the first tableau is an algebraic equation, and algebraic manipulation is permissible.

The old pivot row is operated on to make the pivot element unity. In this case row 2 of tableau 1 is divided by 4, giving

$$
\begin{array}{ccccc}
X_1 & X_2 & X_3 & X_4 & B_i \\
\frac{1}{4} & 1 & 0 & \frac{1}{4} & 2500
\end{array}
\tag{12.31}
$$

which becomes row 2 of tableau 2.

Next, each of the other rows (there is only one other row here) is manipulated so as to make the element in the pivot column equal to zero. Proceed exactly as follows. Write the row. Below it, write Eq. (12.31) multiplied by a factor such that addition will give zero in the pivot column. Here, multiply Eq. (12.31) by minus unity, giving

$$
\begin{array}{ccccc}
X_1 & X_2 & X_3 & X_4 & B_i \\
2 & 1 & 1 & 0 & 6000 \\
-\frac{1}{4} & -1 & 0 & -\frac{1}{4} & -2500 \\
\hline
\frac{7}{4} & 0 & 1 & -\frac{1}{4} & 3500
\end{array}
$$

which becomes row 1 in the second tableau, Table 12.2. Note also that $X_2$ with a value of 5 has replaced $X_4$.

The $Z_j$ and $\Delta_j$ calculations are then carried out, with the results given in Table 12.2.

The final iteration is carried out in the same way. Briefly, in tableau 2 the first column is the pivot column and the first row is the pivot row, with $\frac{7}{4}$ the pivot element. Multiplying the first row by $\frac{4}{7}$ makes the pivot element unity and gives

$$
\begin{array}{ccccc}
X_1 & X_2 & X_3 & X_4 & B_i \\
1 & 0 & \frac{4}{7} & -\frac{1}{7} & 2000
\end{array}
\tag{12.32}
$$

which becomes row 1 of the new tableau. Write row 2 and below it Eq. (12.32) multiplied by $-\frac{1}{4}$, giving

$$
\begin{array}{ccccc}
X_1 & X_2 & X_3 & X_4 & B_i \\
\frac{1}{4} & 1 & 0 & \frac{1}{4} & 2500 \\
-\frac{1}{4} & 0 & -\frac{1}{7} & \frac{1}{28} & -500 \\
\hline
0 & 1 & -\frac{1}{7} & \frac{2}{7} & 2000
\end{array}
$$

**TABLE 12.2**
**Second tableau of the simplex method**

|       |   | 3.5 $X_1$ | 5 $X_2$ | 0 $X_3$ | 0 $X_4$ | $B_i$ | Corresponding equation |
|-------|---|-----------|---------|---------|---------|-------|------------------------|
| $X_3$ | 0 | $\frac{7}{4}$ | 0 | 1 | $-\frac{1}{4}$ | 3,500 | (12.23) |
| $X_2$ | 5 | $\frac{1}{4}$ | 1 | 0 | $\frac{1}{4}$ | 2,500 | (12.22) |
| $Z_j$ |   | $\frac{5}{4}$ | 5 | 0 | $\frac{5}{4}$ | 12,500 | |
| $\Delta_j$ |   | $-\frac{9}{4}$ | 0 | 0 | $\frac{5}{4}$ | | |

**TABLE 12.3**
**Third tableau of the simplex method**

|        |     | 3.5 | 5 | 0 | 0 | | Corresponding |
|--------|-----|-----|---|---|---|---|---------------|
|        |     | $X_1$ | $X_2$ | $X_3$ | $X_4$ | $B_i$ | equation |
| $X_1$  | 3.5 | 1 | 0 | $\frac{4}{7}$ | $-\frac{1}{7}$ | 2,000 | (12.25) |
| $X_2$  | 5   | 0 | 1 | $-\frac{1}{7}$ | $\frac{2}{7}$ | 2,000 | (12.26) |
| $Z_j$  |     | 3.5 | 5 | $\frac{9}{7}$ | $\frac{13}{14}$ | 17,000 | |
| $\Delta_j$ |  | 0 | 0 | $\frac{9}{7}$ | $\frac{13}{14}$ | | |

which is row 2 for the third tableau shown completed in Table 12.3. The bottom row has no negative coefficients, and hence no further improvement is possible. The best mixture is $X_1 = X_2 = 2000$ for $Z = \$17,000$.

## 12.8 APPLICATIONS

The previous development of the simplex method was based on maximization. However, it can be applied to a minimization problem by changing the sign of the objective function and treating the problem as one in maximization. (This is equivalent to changing the signs of all cost coefficients from positive to negative.)

Computer programs are available which can accommodate thousands of restraints and variables. In practice, an LP problem requires setting up all the restraint equations to represent the simulation realistically. Restraints may arise from limited resources (raw materials), equipment capacities, quality specifications on products, physical constraints (e.g., freezing of water at 0°C to prevent pumping), etc. An implied constraint of nonnegativity is assumed for all the variables so that all variables will have values of 0.0 or positive numbers.

Once all the restraint equations have been formulated, a mathematical model is available. The LP computer program can now be applied. It will find a feasible solution that will satisfy all the constraints. The procedure will then be to improve this initial solution to obtain the optimal solution using the objective function as the criterion of optimality. As a by-product of the final solution, a sensitivity analysis is also obtained.

## DYNAMIC PROGRAMMING

## 12.9 AN ALLOCATION EXAMPLE

*Dynamic programming (DP)* is an optimization technique that is especially applicable to the solution of multistage problems. *Programming* is used in the same context as in linear programming, i.e., the planning of activities. Bellman introduced the concept of dynamic programming through the principle of optimality. "An optimal policy has the property that whatever the initial state and initial decision are, the remaining

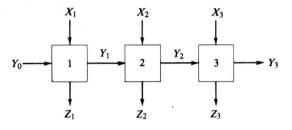

**FIGURE 12.5**
Model for a multistage problem.

decisions must constitute an optimal policy with regard to the state resulting from the first decision."*

Consider the following optimization problem, consisting of three black boxes, shown in Fig. 12.5. The output $Y_3$ from black box 3 is related to the input $Y_2$ and the *decision variable* $X_3$. The profit function $Z_3$ from black box 3 is also related to $Y_2$ and $X_3$. Once $Y_2$ is known, the optimization problem is to find the best value of $X_3$. The path taken to arrive at point $Y_2$ is no longer important. The object is to do the best possible from this point to $Y_3$. If the procedure is repeated for all possible values of $Y_2$, then the optimal values of $X_3$ corresponding to every possible value of $Y_2$ become known.

Now consider black box 2. Assume that $Y_1$ is known and the best value of $X_2$ must be chosen. By selecting a value of $X_2$, we can calculate $Y_2$ from the model for black box 2, as well as the profit function $Z_2$. Once $Y_2$ is calculated, $X_3$ has already been determined, as well as $Z_3$ from the study made on black box 3. It will be assumed that the profit functions are additive so that the total profit function at this point is $Z_2 + Z_3$. Various values of $X_2$ are tried for an assumed value of $Y_1$ until that value is found which optimizes the objective function $Z_2 + Z_3$. We, therefore, have found the best values of $X_2$ and $X_3$ for a given $Y_1$. This process is repeated for all possible values of $Y_1$. We are now in a position to specify the optimal policy for any $Y_1$. The quantities $Y_1$ and $Y_2$ are called *stage variables,* being the outputs from one stage and the inputs to another stage. Then for any stage variables we have determined the best decision variables. The process is continued until $Y_0$ is reached. This is the input variable, and it is assumed that this quantity is known. The optimal policy can actually be determined for any input. The technique of dynamic programming can best be illustrated by examples.

**Example 12.4.** A decision must be made with respect to the allocation of a fixed sum of money, $4 million. Three types of research and development are shown graphically in Figs. 12.6, 12.7, and 12.8, respectively, and in Table 12.4. In project $A$ returns do not materialize until appreciable money is invested; in project $B$ the law of diminishing returns applies; and in project $C$ the return is linear with respect to the investment.

---

*R. E. Bellman, *Dynamic Programming,* Princeton University Press, Princeton, N.J., 1957, p. 83.

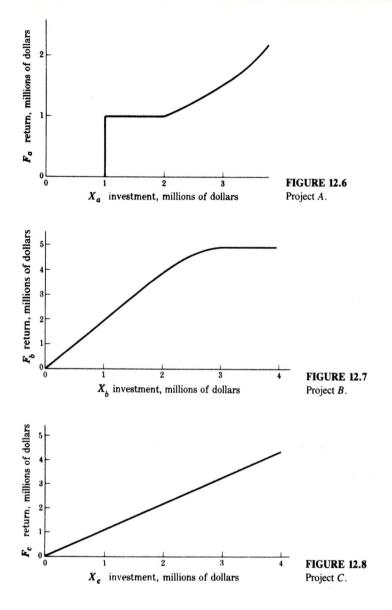

**FIGURE 12.6**
Project A.

**FIGURE 12.7**
Project B.

**FIGURE 12.8**
Project C.

If $4 million is available for the three projects, what is the optimal way to allocate these funds assuming integral million dollars investments in each project?

**Solution.** Consider projects $B$ and $C$ only. Table 12.5 gives all possible combinations for investing $4 million or less among the two projects. A summary of the optimal conditions for projects $B$ and $C$ together is given in Table 12.6. The table was prepared from the optimum conditions found in Table 12.5. Thus if $3 million is the total investment available for projects $A$ and $B$, the maximum profit, $5.05 million, is obtained by in-

**TABLE 12.4**
**Investments for Example 12.4**

| Project A | | Project B | | Project C | |
|---|---|---|---|---|---|
| Investment $X_a$, M\$ | Return $F_a$, M\$ | Investment $X_b$, M\$ | Return $F_b$, M\$ | Investment $X_c$, M\$ | Return $F_c$, M\$ |
| 0 | 0 | 0 | 0 | 0 | 0 |
| 1.0 | 1.0 | 1.0 | 2.0 | 1.0 | 1.05 |
| 2.0 | 1.0 | 2.0 | 4.0 | 2.0 | 2.10 |
| 3.0 | 1.5 | 3.0 | 5.0 | 3.0 | 3.15 |
| 4.0 | 2.5 | 4.0 | 5.0 | 4.0 | 4.20 |

**TABLE 12.5**
**Investment in projects B and C only**

| Total investment, M\$ | Invest project B, M\$ | Return $F_b$, from B, M\$ | Invest project C, M\$ | Return $F_c$, from C, M\$ | Return from B and C, M\$ |
|---|---|---|---|---|---|
| 4 | 4.0 | 5.0 | 0.0 | 0.0 | 5.0 |
|  | 3.0 | 5.0 | 1.0 | 1.05 | 6.05 |
|  | 2.0 | 4.0 | 2.0 | 2.10 | 6.10 |
|  | 1.0 | 2.0 | 3.0 | 3.15 | 5.15 |
|  | 0.0 | 0.0 | 4.0 | 4.20 | 4.20 |
| 3 | 3.0 | 5.0 | 0.0 | 0.0 | 5.0 |
|  | 2.0 | 4.0 | 1.0 | 1.05 | 5.05 |
|  | 1.0 | 2.0 | 2.0 | 2.10 | 4.10 |
|  | 0.0 | 0.0 | 3.0 | 3.15 | 3.15 |
| 2 | 2.0 | 4.0 | 0.0 | 0.0 | 4.0 |
|  | 1.0 | 2.0 | 1.0 | 1.05 | 3.05 |
|  | 0.0 | 0.0 | 2.0 | 2.10 | 2.10 |
| 1 | 1.0 | 2.0 | 0.0 | 0.0 | 2.0 |
|  | 0.0 | 0.0 | 1.0 | 1.05 | 1.05 |

**TABLE 12.6**
**Optimum profit from projects B and C only**

| Total investment in B and C, M\$ | Invest in project B, M\$ | Invest in project C, M\$ | Optimal profit from B + C, M\$ |
|---|---|---|---|
| 0 | 0 | 0 | 0 |
| 1.0 | 1.0 | 0 | 2.0 |
| 2.0 | 2.0 | 0 | 4.0 |
| 3.0 | 2.0 | 1.0 | 5.05 |
| 4.0 | 2.0 | 2.0 | 6.10 |

**TABLE 12.7**
**A 4 M\$ investment in projects *A*, *B*, and *C***

| Total investment, M\$ | Invest project *A*, M\$ | Return from *A*, $F_a$, M\$ | Invest *B* + *C*, M\$ | Optimal return *B* + *C*, M\$ | Return from *A* + *B* + *C*, M\$ |
|---|---|---|---|---|---|
| 4 | 0.0 | 0.0 | 4.0 | 6.10 | 6.10 |
| | 1.0 | 1.0 | 3.0 | 5.05 | 6.05 |
| | 2.0 | 1.0 | 2.0 | 4.0 | 5.0 |
| | 3.0 | 1.5 | 1.0 | 2.0 | 3.5 |
| | 4.0 | 2.5 | 0.0 | 0.0 | 2.5 |

vesting \$2 million in project *A* and \$1 million in project *C*. It should be noted that the 14 calculations of Table 12.5 can now be replaced by the five calculations of Table 12.6.

Projects *A*, *B*, and *C* can now be considered together, but this requires consideration only of project *A* in conjunction with the optimum policies for projects *B* and *C* together. A tabulation is given in Table 12.7. In the tabulation the investment in projects *A*, *B*, and *C* all together is \$4 million, inasmuch as there are no other projects to invest in.

The optimal policy is to invest \$0 million in project *A* and \$4 million in projects *B* and *C* together, which in turn should be \$2 million in project *B* and \$2 million in project *C* for a maximum total profit of \$6.10 million.

## 12.10   A TRANSPORTATION EXAMPLE

**Example 12.5.** A sales representative must travel from city *A* to city *B*. The territory is divided into three districts, and she is required to visit at least one city in each district. Travel costs between cities are indicated in Fig. 12.9. The object is to meet the objective of visiting at least one city in each district at the minimum total travel cost. Use the technique of dynamic programming to solve this problem.

*Solution.* The solution is given in Tables 12.8 to 12.13, obtained as follows. Consider destination city *B*. There are three routes from district 1 to city *B*: J–B, K–B, L–B. Hav-

**TABLE 12.8**
**District 1 to city *B***

| Route | Cost |
|---|---|
| J–B | 4 |
| K–B | 6 |
| L–B | 3 |

**TABLE 12.9**
**District 2 to city *B***

| Route | Cost | |
|---|---|---|
| G–J–B | 2 + 4 = | 6 |
| G–K–B | 5 + 6 = | 11 |
| G–L–B | 4 + 3 = | 7 |
| H–J–B | 5 + 4 = | 9 |
| H–K–B | 1 + 6 = | 7 |
| H–L–B | 5 + 3 = | 8 |
| I–J–B | 2 + 4 = | 6 |
| I–K–B | 3 + 6 = | 9 |
| I–L–B | 5 + 3 = | 8 |

**TABLE 12.10**
**Optimal routes, district 2 to city *B***

| Route | Cost |
|---|---|
| G–J–B | 6 |
| H–K–B | 7 |
| I–J–B | 6 |

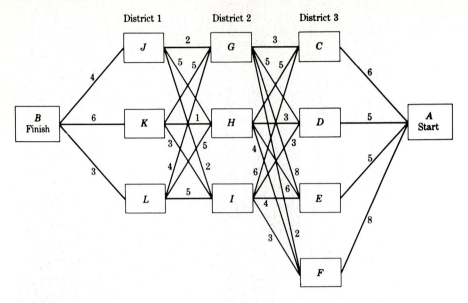

**FIGURE 12.9**
Travel costs between cities.

**TABLE 12.11**
**District 3 to city B**

| Route | Cost |
|---|---|
| C–(G–J–B) | 3 + 6 =  9 |
| C–(H–K–B) | 5 + 7 = 12 |
| C–(I–J–B) | 6 + 6 = 12 |
| D–(G–J–B) | 5 + 6 = 11 |
| D–(H–K–B) | 3 + 7 = 10 |
| D–(I–J–B) | 3 + 6 =  9 |
| E–(G–J–B) | 8 + 6 = 14 |
| E–(H–K–B) | 6 + 7 = 13 |
| E–(I–J–B) | 4 + 6 = 10 |
| F–(G–J–B) | 2 + 6 =  8 |
| F–(H–K–B) | 4 + 7 = 11 |
| F–(I–J–B) | 3 + 6 =  9 |

**TABLE 12.12**
**Optimal routes, district 3 to city B**

| Route | Cost |
|---|---|
| C–G–J–B | 9 |
| D–I–J–B | 9 |
| E–I–J–B | 10 |
| F–G–J–B | 8 |

**TABLE 12.13**
**City A to B**

| Route | Cost |
|---|---|
| A–(C–G–J–B) | 6 +  9 = 15 |
| A–(D–I–J–B) | 5 +  9 = 14 |
| A–(E–I–J–B) | 5 + 10 = 15 |
| A–(F–G–J–B) | 8 +  8 = 16 |

ing arrived in district 1, therefore, there is no decision to be made. Only one route is available from any city in district 1 to city B as given in Table 12.8.

Now go back one stage, one district, and examine the situation from district 2 to city B. All the combinations are given in Table 12.9 and the optimal routes in Table 12.10. Thus if the salesperson is in city H, her best route is H–K–B and she need not consider other routes originating from H.

Next, go back another stage to district 3. All the *pertinent* combinations are given in Table 12.11 prepared from Table 12.10 and the costs from district 3 to district 2. Table 12.12 gives the optimal routes from district 3 to city *B*.

Finally, Table 12.13 gives all the pertinent combinations from city *A* to city *B*. The optimal path is route *A–D–I–J–B* at a cost of 14.

There may of course be ties, so that more than one optimal path can exist. The problem is subject to many variations. If no path exists between two cities, it can be included by placing a very high cost on the route so that it will eliminate itself but not break up the systematic formulation of the solution.


# 12.11  PROS AND CONS OF DYNAMIC PROGRAMMING

In the previous dynamic programming examples it has been assumed tacitly that the manner in which a stage has been reached has no influence on the stages ahead. The existence and importance of this limitation should be appreciated. Thus in Example 12.5 the road ahead is not influenced by the road behind, and use of the optimality principle was very effective in reducing the number of calculations. However, the later stages can be influenced by the manner in which the earlier stages have occurred. For example, suppose in Example 12.5 the fatigue of the sales representative must be considered and her fatigue contributes to the cost of travel. That is, the more tired she becomes, the more she will require extra service, and the greater the travel cost will be. It then becomes necessary to consider the degree of fatigue of the salesperson at each city, which in turn depends on the manner in which the previous steps were executed.

Some of the pros and cons of dynamic programming are as follows:

1. At any stage the optimization technique to use is left up to the user. The most efficient method can, therefore, be used for any particular problem.
2. The problem of constraints usually hinders most optimization techniques. In dynamic programming, however, the constraints are actually helpful because they limit the range to be investigated.
3. Discontinuities in the objective function can be handled by dynamic programming because no analytical function need be used; tables and curves are adequate.
4. Dynamic programming greatly reduces the number of optimization problems to be examined. In general, where there are $N$ stages and $K$ decisions to be made at each stage, the overall optimization problem involves $K^N$ possible solutions. For a five-stage problem with three decisions at each stage this means $3^5$, or 243, possible combinations. By dynamic programming, only $3 \times 5$, or 15, one-stage optimization problems would have to be solved. The main disadvantage of dynamic programming is the problem of dimensionality (many variables), which arises in the optimization at each stage. For example, when many decisions are to be made at one time, the single-stage optimization itself may become quite complex.

# GEOMETRIC PROGRAMMING

## 12.12 THE FORMAT

*Geometric programming* (GP) is an optimization technique that can handle a certain form of functions with fractional and negative exponents in an efficient manner. The technique was introduced by Duffin, Paterson, and Zener [13] and reviewed by a number of others [15, 16, 17].

A polynomial of the form

$$g = u_1 + u_2 + \cdots + u_n \tag{12.33}$$

is known as a *posynomial* when each term on the right-hand side has the form

$$u_i = c_i X_1^{a_1} X_2^{a_2} X_3^{a_3} \ldots X_n^{a_n}$$

where $c_i$ must be a positive constant and the exponents can be positive, negative, or zero. The variables $X_1 \ldots X_n$ are also positive.

**Example 12.6.** Which of the following are permissible terms in a posynomial?
(a) $4X_1^2 X_2^{-1.5}$
(b) $-4X_1^2 X_2^3$
(c) $4X_1 \sqrt{2}$
(d) $4X_1 \log X_2$
(e) $12$

**Solution.** Only (a), (c), and (e) are permissible; (b) is negative, and (d) does not have the correct form.

## 12.13 A SIMPLE APPLICATION

If the function to be optimized is a posynomial, geometric programming is an efficient method for optimization. Only the application of the method is described here.

**Example 12.7.** Reconsider the two-variable optimization given by Eq. (10.17):

$$Z = 1265S^{0.5} + 4 \times 10^8 P^{-1} S^{-0.5} + 137.2P \tag{12.34}$$

Let $\alpha$ denote a weighting factor and consider the following expression for $v$, known as the *dual function*:

$$v = \left(\frac{1265S^{0.5}}{\alpha_1}\right)^{\alpha_1} \left(\frac{4 \times 10^8 P^{-1} S^{-0.5}}{\alpha_2}\right)^{\alpha_2} \left(\frac{137.2P}{\alpha_3}\right)^{\alpha_3} \tag{12.35}$$

Now find relationships between the $\alpha$ factors which will make the expression dimensionless.

*Solution.* This dimensionless expression is obtained when both $S$ and $P$ are reduced to the zero power. For the terms in $S$

$$S^{0.5\alpha_1} S^{-0.5\alpha_2} = S^0$$

$$0.5\alpha_1 - 0.5\alpha_2 = 0$$

and similarly, from the terms in $P$,

$$-\alpha_2 + \alpha_3 = 0$$

Since the $\alpha$ factors are only weighting factors, one $\alpha$ can be chosen arbitrarily, but it is most convenient to set their sum equal to unity. Thus there are three relationships:

$$\alpha_1 + \quad \alpha_2 + \alpha_3 = 1 \tag{12.36}$$

$$0.5\alpha_1 - 0.5\alpha_2 \quad\quad = 0 \tag{12.37}$$

$$- \quad \alpha_2 + \alpha_3 = 0 \tag{12.38}$$

The first is known as the *normality condition*; the others are called the *orthogonality conditions*.

The three equations are easily solved:

$$\alpha_1 = \alpha_2 = \alpha_3 = \tfrac{1}{3} \tag{12.39}$$

but in general the weighting factors will not be equal.

With $\alpha_1$, $\alpha_2$, and $\alpha_3$ now known and with the dual function dimensionless in $S$ and $P$, Eq. (12.35) becomes

$$v = \left(\frac{1265}{\alpha_1}\right)^{\alpha_1} \left(\frac{4 \times 10^8}{\alpha_2}\right)^{\alpha_2} \left(\frac{137.2}{\alpha_3}\right)^{\alpha_3} = \$123{,}298$$

Here $v$ is the optimum sought, and it checks the value given previously in Table 10.4. The optimum value has been obtained without knowing the individual values for $S$ and $P$ that lead to it.

The optimum values for $S$ and $P$ at the optimum are easily obtained from $v$, the dual function, and the weighting $\alpha$ factors:

$$1265S^{0.5} = \alpha_1 v = \tfrac{1}{3}(123{,}298) = 41{,}099$$

$$S = 1056$$

$$137.2P = \alpha_3 v = \tfrac{1}{3}(123{,}298) = 41{,}099$$

$$P = 300$$

which check the optimum values for $S$ and $P$ found previously in Example 10.5.

## 12.14  SOME COMMENTS ON GEOMETRIC PROGRAMMING

Equation (12.33) is known as the *primal function* in geometric programming. The basis of geometric programming work is the inequality which states that the arithmetic mean is equal to or greater than the geometric mean. The dual function, the optimum, can be calculated without finding the values of the problem variables, a unique feature of geometric programming.

The degree of difficulty $D$ in geometric programming is defined as the total number of terms $T$ in the posynomial minus one more than the number of independent variables $N$:

$$D = T - (N + 1)$$

In Example 12.7 there are three terms and two variables, and therefore zero degree of difficulty.

Geometric programming has been expanded to overcome the limitations of a simple problem such as that expressed by Eq. (12.34). The removal of each limitation is obtained at the expense of increased mathematical difficulty. Restraints can be included using Lagrange's method of undetermined multipliers, and even inequality constraints can be accommodated. Also, the severe requirement that all terms in the objective function be positive can be circumvented.

The advantages of geometric programming are that functions with fractional and negative exponents can be handled efficiently, the value of each variable need not be found, and constraints may be incorporated in the solution techniques.

The disadvantages are that if the degree of difficulty is more than zero, the problem formulated can be more complicated than the original problem, all variables must be continuous or at least treated as such, and it is not applicable to large, complex design problems since the equations will not necessarily be in the proper format.


## NOMENCLATURE

| | |
|---|---|
| $A_{ij}$ | Coefficient in a restraint equation $i$th row, $j$th column, LP |
| $a_n$ | Exponent in a GP term |
| $B_i$ | Right-hand side of the $i$th restraint equation, LP |
| $c_i$ | Constant in a GP term |
| $C_i$ | Unit cost associated with a basic variable in the $i$th row, LP |
| $C_j$ | Unit cost associated with $X_j$, LP |
| $D$ | Degree of difficulty, GP |
| DP | Dynamic programming |
| $F$ | Objective function |
| $F_a$ | Return from project $A$, DP |
| $g$ | A posynomial, GP |
| GP | Geometric programming |
| $i$ | The $i$th row |
| $j$ | The $j$th column |
| LP | Linear programming |
| $N$ | The $N$th stage, DP |
| $N$ | Number of variables, GP |
| $Q$ | Quality coefficient |
| $T$ | Number of terms in a posynomial, GP |
| $u$ | A term in a posynomial |
| $v$ | Dual function, GP |
| $X_a$ | Investment in project $A$, DP |

| | |
|---|---|
| $X_j$ | Variable input, LP |
| $X_N$ | Decision variable associated with the $N$th stage, DP |
| $Y_N$ | Output from the $N$th stage, input to the $(N + 1)$ stage, DP |
| $Z$ | Objective function |
| $Z_g$ | $\sum_{i=1}^{m} C_i A_{ij}$, LP |
| $\alpha$ | Weighting factor, GP |
| $\Delta_j$ | $Z_j - C_i$, LP |

## PROBLEMS

**12.1.** Verify all terms in the first and second rows of the second and third tableaus in Tables 12.2 and 12.3.

**12.2.** Verify the $Z_j$ and $\Delta_j$ values in Table 12.3.

**12.3.** Set up the equations for the following diet problem given by Garvin [3]. One unit of hog's liver contains 1 unit of carbohydrates, 3 units of vitamins, 3 units of proteins, and costs 50 units. Suppose 1 unit of castor oil contains 3, 4, and 1 units of these, respectively, and costs 25 units. Minimum daily requirements are 8 units of carbohydrates, 19 units of vitamins, and 7 units of proteins. Obtain the minimum-cost diet subject to the constraints of the problem.

**12.4.** Express Prob. 12.3 geometrically, indicating the feasible region.

**12.5.** Solve Prob. 12.3 by the enumeration of all solutions.

**12.6.** Solve Prob. 12.3 by the algebraic method.

**12.7.** Verify the equations in the algebraic method of Prob. 12.6 by solving the problem using the simplex method.

**12.8.** A student is given an examination and is to solve no more than 100 problems, with the problems divided into three groups.

First group—moderately difficult $(X_1)$, worth five points each, with an average time of 3 min per problem.

Second group—least difficult $(X_2)$, worth four points each, with an average time of 2 min per problem.

Third group—most difficult $(X_3)$, worth six points each, with an average time of 4 min per problem.

The total time available to the student is 3.5 h. The problems in groups 1 and 2 are the most tedious, and the student can take only 2.5 h of this type of work. How many problems of each type should the student solve to obtain a maximum grade on the examination?

**12.9.** An advertising budget is to be prepared which aims to reach 32 million potential customers, of whom 21 million are required to have incomes over \$18,000/year. The following data are given:

| Medium | Cost, \$/thousand | Audience, million per ad | Audience with income over \$18,000, millions |
|---|---|---|---|
| Magazine | 28 | 1 | 0.6 |
| Television | 400 | 9 | 2 |
| Radio | 20 | 0.8 | 0.7 |

Set up the model and find the optimal solution by the enumeration method.

**12.10.** A small machine shop has capabilities in turning, milling, drilling, as follows:

Turning,    140 h/day
Milling,     80 h/day
Drilling,    120 h/day

Two products, A and B, are made, yielding a net profit of $2.20 and $3.80 per item, respectively. Each will require the following amount of machine time in hours:

|  | A | B |
|---|---|---|
| Turning | 2.4 | 3.2 |
| Milling | 0.0 | 2.6 |
| Drilling | 4.1 | 0.0 |

Use the simplex method and find the policy that gives the maximum profit per day.

**12.11.** A customer needs 1000 kg of a mixture made up from three different raw materials with the following costs:

$X_1$,    2 \$/kg
$X_2$,    3 \$/kg
$X_3$,    4 \$/kg

The mixture must not contain more than 400 kg of $X_1$, at least 200 kg of $X_2$, and at least 100 kg of $X_3$.
      Use the simplex method and find the least-cost mixture that will satisfy all the requirements.

**12.12.** Crude A is limited to 15,000 bbl/day and if sent to process C gives a profit of $1.10/bbl. If sent to process D, it gives a profit of $1.20/bbl.
      Crude B is limited to 18,000 bbl/day and if sent to process C gives a profit of $1.60/bbl. If sent to process D, it gives a profit of $1.40/bbl.
      Process C has a limit of 12,000 bbl/day and process D a limit of 20,000 bbl/day. Find the policy that maximizes the profit per day.

**12.13.** Refer to Example 12.4. What is the optimal policy if 3 M\$ instead of 4 M\$ were available for all three projects?

**12.14.** Repeat Prob. 12.13 for (a) 2 M\$ and (b) 1 M\$. Plot the optimal return against available funds.

**12.15.** Repeat Example 12.5, traveling from city B to city A.

**12.16.** Describe some application of DP with which you are familiar, e.g., investment decisions, staged engineering calculations, or network problems.

**12.17.** Find the number of each of three items to include in a package so that the value of the package will be a maximum subject to the limitation that the mass must not exceed 6 kg:

| Item | Mass, kg | Value, \$ |
|---|---|---|
| 1 | 3 | 6 |
| 2 | 2 | 4 |
| 3 | 4 | 7 |

**12.18.** The following type of industrial operation is being evaluated:

$$\longrightarrow \text{Unit 1} \longrightarrow \text{unit 2} \longrightarrow \text{unit 3} \longrightarrow$$

Two types of units, $A$ and $B$, costing $10,000 and $20,000, respectively, are being considered for unit 1.

The operating costs for unit 1 are

| Efficiency, % | Operating cost, $/year | |
|---|---|---|
| | $A$ | $B$ |
| 60 | 3,000 | 2,000 |
| 80 | 6,000 | 5,000 |
| 95 | 12,000 | 8,000 |

The operation of unit 2 is related to the efficiency of unit 1. One type of unit is being considered which may operate at one of three different temperatures, 370°C, 400°C, or 430°C, with the following data:

| % efficiency of unit 1 | Cost, $/year | | |
|---|---|---|---|
| | 370°C | 400°C | 430°C |
| 60 | 2,000 | 10,000 | 16,000 |
| 80 | 1,800 | 8,000 | 12,000 |
| 95 | 1,000 | 6,000 | 10,000 |
| Fixed cost, $ | 5,000 | 20,000 | 20,000 |

Unit 3 is related to the temperature of unit 2:

| Temp. °C, unit 2 | Income unit 3, $ for 5 years |
|---|---|
| 370 | $530,000 |
| 400 | 630,000 |
| 430 | 680,000 |

Assume a 5-year basis for costs. Find the optimal operation using DP.

**12.19.** Three types of purification units are available, as indicated in the following scheme:

Each of the three units may be operated in any one of three modes of operation, $A$, $B$, and $C$, with the following data:

| Unit 1 | | |
|---|---|---|
| Mode of operation | Purity stream 1, % | Cost, $/year |
| $A_1$ | 99 | 150,000 |
| $B_1$ | 60 | 70,000 |
| $C_1$ | 50 | 50,000 |

| Unit 2 | | | |
|---|---|---|---|
| Purity stream 1, % | Operation unit 2 | Purity stream 2, % | Cost, $/year |
| 50 | $A_2$ | 99 | 66,000 |
|  | $B_2$ | 60 | 10,000 |
|  | $C_2$ | 50 | 0 |
| 60 | $A_2$ | 99 | 40,000 |
|  | $B_2$ | 60 | 0 |
| 99 | $A_2$ | 99 | 0 |

| Unit 3 | | |
|---|---|---|
| Purity stream 2, % | Operation unit 3 | Cost, $/year |
| 50 | $A_3$ | 60,000 |
| 60 | $A_3$ | 40,000 |
| 99 | $A_3$ | 0 |

It is not required that all three units be used. Find the optimal operation policy.

**12.20.** The total cost $C_t$ in dollars per year for a piping installation is given by

$$C_t = \frac{A}{D^{4.8}} + BD^{1.5}$$

where $D$ is the diameter of the pipe in centimeters and $A$ and $B$ are constants. The first term represents the annual operating cost, and the second term represents the fixed cost. Find the percentage of optimal total costs that arise from the operating costs and the fixed costs, respectively, using GP:

(a) For the equation as written
(b) For the equation with $A$ doubled
(c) For the equation as written, with the exponent on $D$ changed from 4.8 to 5.2
(d) For part (c) with $A$ doubled

**12.21.** A cost function with $A$, $B$, and $C$ constants is given by

$$Z = AX_1 + \frac{B}{X_1^2 X_2^3} + CX_2^2$$

(a) Find a minimum value for $Z$ using GP.
(b) Find the value for $X_1$ and $X_2$ in terms of $A$, $B$, and $C$ at minimum $Z$.

# REFERENCES

## Linear programming

1. Charnes, A., and W. W. Cooper: *Management Models and Industrial Application of Linear Programming,* John Wiley & Sons, Inc., New York, 1960.
2. Dantzig, G. B.: *The Simplex Method,* Rand Corporation Report P-891, 1956.
3. Garvin, W. W.: *Introduction to Linear Programming,* McGraw-Hill Book Company, New York, 1960.
4. Gass, S. I.: *Linear Programming: Methods and Applications,* 3d ed., McGraw-Hill Book Company, New York, 1969.
5. Hadley, G.: *Linear Programming,* Addison-Wesley Publishing Company, Inc., Reading, Mass., 1962.
6. Humphreys, K. K., ed.: *Project and Cost Engineers' Handbook,* 2d ed., Marcel Dekker, Inc., New York, 1984, pp. 163–174.
7. Vajda, S.: *Readings in Linear Programming,* John Wiley & Sons, Inc., New York, 1958.

## Dynamic programming

8. Bellman, Richard E.: *Dynamic Programming,* Princeton University Press, Princeton, N.J., 1957.
9. ———, and S. Dreyfus: *Applied Dynamic Programming,* Princeton University Press, Princeton, N.J., 1962.
10. Hadley, G.: *Nonlinear and Dynamic Programming,* Addison-Wesley Publishing Company, Inc., Reading, Mass., 1964.
11. Humphreys, K. K., ed.: Project and Cost Engineers' Handbook, 2d ed., Marcel Dekker, Inc., New York, 1984, pp. 174–184.
12. Nemhauser, G. L.: *Introduction to Dynamic Programming,* John Wiley & Sons, Inc., New York, 1966.

## Geometric programming

13. Duffin, R. J., S. F. Paterson, and C. Zener: *Geometric Programming,* John Wiley & Sons, Inc., New York, 1967.
14. Holzman, A. G., ed.: *Mathematical Programming for Operations Researchers and Computer Scientists,* Marcel Dekker, Inc., New York, 1981.
15. Kermode, R. I.: "Geometric Programming—A Simple, Efficient Optimization Technique," *Chemical Engineering,* vol. 74, 1967, p. 26.
16. LaMonte, R. R., and P. B. Lederman: "The Uses and Limitations of Geometric Programming," *British Chemical Engineering and Process Technology,* vol. 17, no. 1, 1972, p. 34.
17. Wilde, D. J., and C. S. Beightler: *Foundations of Optimization,* 2d ed., Prentice-Hall, Inc., Englewood Cliffs, N.J., 1979.

# SPECIAL MATHEMATICAL TECHNIQUES

*A.H. Boas*

## UNIVARIABLE SEARCH METHODS

## 13.1 SEARCH TECHNIQUES

Very often a problem is not defined in specific analytical form. An objective function, such as a maximum or minimum value, may be evaluated, however, by trial. This may entail direct laboratory measurement or experimentation, a computer run, an iterative procedure involving graphs and tables, etc. The problem is to select the proper values of input data to optimize a specified objective function by using search techniques. A base point is given or assumed, the objective function is evaluated, and the process is repeated until no further improvement is possible.

After a certain number of experiments are performed the interval within which the optimum value must lie is a measure of the efficiency.

## 13.2 UNIFORM SEARCH

Assume that we are concerned only with one independent variable and that the experimental error of measuring or computing is negligible. Furthermore, the important assumption of unimodality (one optimum value only) is made.

This section develops the uniform-search method in which experimental or trial points are spaced equally within the allowable range.

**Example 13.1.** A function $F$ is given by the relationship

$$F = 1.3X^2 - 4X + 3 \qquad (13.1)$$

It is known that the function is unimodal and that the range of the answer lies between 0 and 5. Four experiments, or computations, are permitted. Find the minimum value of $F$.

**Solution.** It should be emphasized that the form of the function $F$ as given by Eq. (13.1) may or may not be known to the analyst. If known, the solution proceeds by computation; if unknown, by experimentation. The method of analysis is the same, however: For each value of $X$ the corresponding value of $F$ can be determined by some means.

Spacing four experimental points equidistantly within the interval leads to a choice $X = 1, 2, 3, 4$, and experiment gives

| X | F |
|---|-----|
| 1 | 0.3 |
| 2 | 0.2 |
| 3 | 2.7 |
| 4 | 7.8 |

The results are plotted in Fig. 13.1. The minimum may lie between 1 and 2, as indicated by curve $A$, or between 2 and 3, as indicated by curve $B$. It must lie between 1 and 3. Originally the range covered five units, and after four experiments the range has been narrowed down to two units, or 40 percent of the original range.

**FIGURE 13.1**
The unimodal search of
$F = 1.3X^2 - 4X + 3$.

If $N$ is the number of experiments allowed in an entire interval $L$, there are $N + 1$ intervals and the final answer spans two intervals. If $F_r$ is the fraction of the original interval in which the optimum lies, then

$$F_r = \frac{2}{N + 1} \tag{13.2}$$

For Example 13.1 $N = 4$ and $F_r = 0.40$.

## 13.3 UNIFORM DICHOTOMOUS SEARCH

In this procedure experiments are performed in pairs to establish whether the function is increasing or decreasing at a point.

**Example 13.2.** Repeat Example 13.1 using a dichotomous search.

*Solution.* Four experiments, or two pairs, are allowed. Divide the range 0 to 5 into three equal intervals. Thus let

$$X = \tfrac{1}{3}(5) = 1.667 \qquad X = \tfrac{2}{3}(5) = 3.333$$

One pair of experiments is used in the neighborhood of $X = 1.667$ to establish the trend of $F$ at that point. Similarly, the second pair of experiments establishes the trend at $X = 3.333$. For the latter pair

| X | F |
|------|------|
| 3.30 | 3.96 |
| 3.36 | 4.24 |

At $X = 3.333$ the function is increasing; hence the minimum must be below that value of $X$. The other pair of experiments gives

| X | F |
|------|--------|
| 1.60 | −0.072 |
| 1.70 | −0.043 |

so that the function is increasing with increasing $X$. The minimum must be below $X = 1.667$. The four experiments have established, at best, that the minimum must lie in the range 0 to 1.667, which is one-third of the total range 0 to 5.

The procedure will now be generalized. For $N$ experiments there will be $N/2$ pairs, and the entire interval is divided into $(N/2) + 1$ intervals, each of width $L/(N/2 + 1)$. The optimum is located over the width of one interval, $L/(N/2 + 1)$.

Therefore,

$$F_r = \frac{L/(N/2 + 1)}{L} = \frac{2}{N + 2} \qquad (13.3)$$

For the example $N = 4$, $F_r = \frac{2}{6} = 0.33$.

This method has reduced the original interval to 33 percent and is seen to be superior to a uniform search, which reduced the original interval to 40 percent.

## 13.4   SEQUENTIAL DICHOTOMOUS SEARCH

In a sequential search the information available from previous experiments is used before the next one is performed. A sequential dichotomous search involves two experiments run very closely together (closely enough to distinguish between the two outcomes) near the middle of the region under consideration. In Example 13.1 two experiments close to $X = 2.5$ show it to be increasing at $X = 2.5$; therefore, the region above $X = 2.5$ is eliminated. Another pair of experiments near the middle of the remaining region at $X = 1.25$ shows that the function is decreasing at $X = 1.25$. The minimum lies between $X = 1.25$ and 2.50, or one-quarter of the range 0 to 5.

Generalizing this procedure, it is seen that each pair of experiments bisects the previous interval. At the end of the first pair of experiments the remaining interval is one-half of the original interval; after the next pair it is one-quarter of the original interval. After $N/2$ pairs of experiments the remaining interval is $\frac{1}{2}^{N/2}$ of the original interval:

$$F_r = \frac{1}{2^{N/2}} \qquad (13.4)$$

Referring again to the example, $N = 4$, $F_r = \frac{1}{2}^{\frac{2}{2}} = 0.25$. The original interval has now been reduced to 25 percent, which is the lowest value obtained thus far.

## 13.5   FIBONACCI SEARCH TECHNIQUES

It has been shown in the articles by Johnson [13] and Keifer [14] that the most efficient sequential technique to follow for the case of one variable and unimodality is the Fibonacci search. Fibonacci numbers $F_i$ corresponding to an index $N$ are related by the recurring relationship

$$F_{i,N+2} = F_{i,N} + F_{i,N+1}$$

where $F_{i0} = 1$ and $F_{i1} = 1$.

Table 13.1 gives some Fibonacci numbers. Each number is the sum of two previous Fibonacci numbers. The use of Fibonacci numbers is illustrated in the next example.

**TABLE 13.1**
**Some Fibonacci numbers**

| N | $F_{i,N}$ | N | $F_{i,N}$ |
|---|-----------|----|-----------|
| 0 | 1 | 8 | 34 |
| 1 | 1 | 9 | 55 |
| 2 | 2 | 10 | 89 |
| 3 | 3 | 11 | 144 |
| 4 | 5 | 12 | 233 |
| 5 | 8 | 13 | 377 |
| 6 | 13 | 14 | 610 |
| 7 | 21 | 15 | 987 |

$F_{i,N+2} = F_{i,N} + F_{i,N+1}$
$F_{i0} = 1$
$F_{i1} = 1$

**Example 13.3.** Repeat Example 13.1 using a Fibonacci search technique.

*Solution.* First, a pair of experiments is run equidistant from each end of the interval. The distance $D_1$ is determined from the expression

$$D_1 = \frac{F_{i,N-2}}{F_{i,N}} L \qquad (13.5)$$

where $F_{i,N}$ = Nth Fibonacci number
$N$ = number of experiments
$L$ = interval length

For the example, $N = 4$, $L = 5$, $F_{i,N-2} = F_{i2} = 2$, and $F_{i,N} = F_{i4} = 5$. From Eq. (13.5),

$$D_1 = \tfrac{2}{5}(5) = 2$$

Therefore, two experiments are run two units from each end of the interval; i.e.,

$$X = 0 + 2 = 2$$
$$X = 5 - 2 = 3$$

The experiments give

| X | F |
|---|------|
| 2 | 0.200 |
| 3 | 2.700 |

The region for X greater than 3 is eliminated.

Second, the procedure is repeated for the remaining interval, changing $N$ to $N - 1$ because one experiment less remains. Equation (13.5) becomes

$$D_2 = \frac{F_{i,N-2-1}}{F_{i,N-1}} L = \frac{F_{i1}}{F_{i3}} L = \frac{1}{3}(3) = 1$$

The next two experiments should be run 1 unit from each end of the remaining interval:

$$X = 0 + 1 = 1$$

$$X = 3 - 1 = 2$$

However, one of these two new experiments has already been run (at $X = 2$). Hence only one new experiment is added, at $X = 1$. It will always turn out that one of the previous experiments is a Fibonacci experiment for the next run. Performing the new experiment gives the pair

| $X$ | $F$ |
|-----|-------|
| 1 | 0.300 |
| 2 | 0.200 |

The region for $X$ less than 1 is eliminated.

Third, the procedure is repeated again with $L = 3 - 1 = 2$, changing $N$ to $N - 2$. Equation (13.5) becomes

$$D_3 = \frac{F_{i,N-2-2}}{F_{i,N-2}} L = \frac{F_{i0}}{F_{i2}} L = \frac{1}{2}(2) = 1$$

This indicates that the last experiment should be run at

$$X = 1 + 1 = 2$$

$$X = 3 - 1 = 2$$

This experiment is run around $X = 2$ to determine in which half of the remaining region the answer lies. Choose $X = 2 \pm \varepsilon$, where $\varepsilon$ is a small number but large enough to detect a difference in $F$. Taking $\varepsilon = 0.05$, we have

| $X$ | $F$ | |
|------|-------|-----------------|
| 2.05 | 0.263 | new experiment |
| 2.00 | 0.200 | old experiment |

The function is increasing in the neighborhood of $X = 2$; thus the minimum is below $X = 2$. The fraction of the original interval is $\frac{1}{5} = 0.20$, or 20 percent, which is the lowest obtained. For the general case the result is

$$F_r = \frac{1}{F_{i,N}} \tag{13.6}$$

For the example, $N = 4$, $F_{i4} = 5$, $F_r = \frac{1}{5} = 0.20$.

Table 13.1 provides a method of determining the number of experiments needed to obtain a certain value of $F_r$, the fractional ratio of interval left to the original interval. For example, to narrow the original interval to at least 1 percent,

$$\frac{1}{F_{i,N}} \le 0.01 \quad \text{or} \quad F_{i,N} \ge 100$$

At least 11 Fibonacci experiments would have to be performed since $F_{i11}$ is the first Fibonacci number above 100.

**TABLE 13.2**
**Comparison of methods for univariable search techniques**

| Search technique | Interval left after $N$ experiments | |
| --- | --- | --- |
| | Original interval | For $N = 4$ |
| Uniform | $\dfrac{2}{N + 1}$ | 0.40 |
| Uniform dichotomous | $\dfrac{2}{N + 2}$ | 0.33 |
| Sequential dichotomous | $\dfrac{1}{2^{N/2}}$ | 0.25 |
| Fibonacci | $\dfrac{1}{F_{i,N}}$ | 0.20 |

## 13.6   COMPARISON OF METHODS

Table 13.2 summarizes the results obtained by the various methods described. As $N$ gets larger, the advantage of the Fibonacci technique over all the others increases.

## MULTIVARIABLE FUNCTIONS

## 13.7   ONE-AT-A-TIME METHOD

More common and much more complex is the multivariable optimization problem. A review of some widely used methods follows.

Friedman and Savage [11] described the elementary one-at-a-time method. All variables except one are kept constant, and that one is varied to obtain an improvement in the objective function. Referring to Fig. 13.2, an initial search is started, changing $Y$ only (constant $X$) along the line $AB$ until the lowest value of the objective function is obtained at point $C$. A new search is started along line $CD$ until point $D$ is reached, and the process is continued until the optimum is approached very closely. When the contours come to a sharp point, this method might not work.

**Example 13.4.** Solve the two-variable problem given by Eq. (10.17):

$$Z = 1265S^{0.5} + 4 \times 10^8 P^{-1}S^{-0.5} + 137.2P$$

minimizing $Z$ by the one-at-a-time method. Start at $P = 250$ and $S = 1000$.

**Solution.** With $P = 250$ and not varied, find the best value of $S$:

$$\frac{\partial Z}{\partial S} = 0 = \frac{0.5(1265)}{S^{0.5}} - \frac{0.5 \times 4 \times 10^8}{PS^{1.5}}$$

$$S = \frac{3.1621 \times 10^5}{P} = \frac{3.1621 \times 10^5}{250} = 1265$$

**FIGURE 13.2**
The one-at-a-time method.

Now for $S = 1265$ find the best value of $P$:

$$\frac{\partial Z}{\partial P} = 0 = \frac{-4 \times 10^8}{P^2 S^{0.5}} + 137.2$$

$$P = \left(\frac{4 \times 10^8}{137.2 S^{0.5}}\right)^{0.5} = \left(\frac{4 \times 10^8}{137.2(1265)^{0.5}}\right)^{0.5} = 286$$

Continue until no further improvement results:

| $P$ | | $S$ |
|---|---|---|
| 250 | $\longrightarrow$ | 1265 |
| 286 | $\longleftarrow$ | 1265 |
| 286 | $\longrightarrow$ | 1106 |
| 296 | $\longleftarrow$ | 1106 |
| ... | | ... |
| 299.5 | $\longleftarrow$ | 1057 |
| 299.5 | $\longrightarrow$ | 1056 |
| 299.5 | $\longleftarrow$ | 1056 |

Thus the minimum is at $P = 299.5$ and $S = 1056$, which checks the answer in Sec. 10.14.

## 13.8 METHOD OF STEEPEST ASCENT (OR DESCENT)

The direction of the gradient is the direction that gives the greatest response of the objective function per unit length of the independent variable. The incremental change in each variable is taken proportional to its partial derivative, which determines the gradient direction. This direction is called the *direction of steepest ascent* (or *descent*).

**Example 13.5.** Given the following objective function:

$$Z = X^2 + 2Y^2 + XY \tag{13.7}$$

find the minimum value of this function and the values of $X$ and $Y$ at the minimum, starting at $X = 2$, $Y = 2$.

*Solution.* The starting point is represented by $M$ in Fig. 13.3. The partial derivatives must first be calculated and evaluated at the starting point:

$$\frac{\partial Z}{\partial X} = 2X + Y \tag{13.8}$$

$$\frac{\partial Z}{\partial Y} = 4Y + X \tag{13.9}$$

At the starting point these partial derivatives become

$$\frac{\partial Z}{\partial X} = 6 \quad \text{and} \quad \frac{\partial Z}{\partial Y} = 10 \tag{13.10}$$

Because both partial derivatives are positive, the objective function varies in the same direction as $X$ and $Y$. Therefore, as this is a minimization problem, both $X$ and $Y$ should be decreased in order to decrease the value of $Z$. The ratio of the decrease in $X$ to the decrease in $Y$ is taken to be proportional to their respective partial derivatives, i.e., $\frac{6}{10}$. An arbitrary step size $\Delta Y$ of $-0.5$ is chosen initially; the corresponding $\Delta X$ would then be $0.60(-0.50) = -0.30$.

Initially $X_0 = 2$; $Y_0 = 2$; $Z_0 = 16$. After the first step

$$X_1 = X_0 + \Delta X = 2 - 0.30 = 1.70$$

$$Y_1 = Y_0 + \Delta Y = 2 - 0.50 = 1.50$$

$$Z_1 = 1.70^2 + 2(1.50^2) + 1.70(1.50) = 9.94$$

As long as the objective function is decreasing, the procedure is continued along the same gradient line. Table 13.3 summarizes the results of the calculation.

**TABLE 13.3**
**Calculations along the first gradient**

| $i$ | $X_i$ | $Y_i$ | $Z_i$ |
|---|---|---|---|
| 0 | 2.00 | 2.00 | 16.00 |
| 1 | 1.70 | 1.50 | 9.94 |
| 2 | 1.40 | 1.00 | 5.36 |
| 3 | 1.10 | 0.50 | 2.26 |
| 4 | 0.80 | 0.00 | 0.64 |
| 5 | 0.50 | -0.50 | 0.50 |
| 6 | 0.20 | -1.00 | 1.84 |

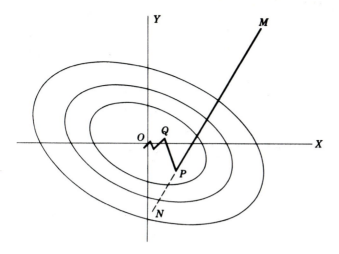

**FIGURE 13.3**
The method of steepest ascent.

Referring to Fig. 13.3, point $P$ corresponds to $(X_5, Y_5)$ and point $N$ to $(X_6, Y_6)$. Table 13.3 indicates that at point $N$ the objective function is no longer decreasing. Point $P$, in fact, represents the last successful calculation. A new gradient direction is now calculated at this point and the entire procedure repeated. At point $P$ the gradient direction is

$$\frac{\partial Z}{\partial X} = 2X + Y = 2(0.5) - 0.5 = 0.50$$

$$\frac{\partial Z}{\partial Y} = 4Y + X = 4(-0.5) + 0.5 = -1.50$$

It is noted that $\partial Z/\partial X$ is positive but $\partial Z/\partial Y$ is negative. Therefore, in order to decrease the value of $Z$, $X$ should be decreased and $Y$ should be increased. Inasmuch as the solution should now be close at hand, the step size in $Y$ will be arbitrarily reduced from 0.5 to 0.3. From the recently calculated partial derivatives, the ratio of $\Delta X/\Delta Y = -\frac{1}{3}$ so that $\Delta X = -0.1$.

Step 6 is calculated as follows:

$$X_6 = X_5 + \Delta X = 0.50 - 0.1 = 0.40$$

$$Y_6 = Y_5 + \Delta Y = -0.50 + 0.3 = -0.20$$

$$Z_6 = 0.40^2 + 2(-0.20^2) + 0.40(-0.20) = 0.16$$

Continuing along this new gradient line, the following points are obtained:

| $i$ | $X_i$ | $Y_i$ | $Z_i$ |
|---|---|---|---|
| 6 | 0.40 | −0.20 | 0.16 |
| 7 | 0.30 | 0.10 | 0.14 |
| 8 | 0.20 | 0.40 | 0.44 |

Since $Z_8$ is greater than $Z_7$, it is obvious that the extrapolation has gone too far along this line. The last successful step, point $Q$ $(X_7, Y_7)$, is indicated on Fig. 13.3. The direction is changed once again, the step size decreased, etc., until the optimum point is finally approached.

The method of steepest ascent requires two decisions — when to calculate a new gradient direction and what step size to use. Experience and familiarity with the problem increase the efficiency of the solution.

If an analytical expression is not available, it is necessary to explore by experimentation the value of $Z$, the objective function, in various directions from $M$ (Fig. 13.3) and to establish the steepest gradient experimentally.

The method can be applied to more than two variables by vector analysis techniques.

## 13.9  CONSTRAINED OPTIMIZATION

Suppose now that there are bounds or constraints imposed on problems, such as restrictions on the independent variables or on some calculated values. Equality constraints can be handled by the technique of Lagrange multipliers (Sec. 10.15), while the Kuhn–Tucker conditions can be used for handling inequality constraints.

Constraints can cause serious mathematical difficulties, and various methods have been proposed. The work of Fiacco and McCormick [6–10] is recommended, along with that of Box [5].

## SUMMATION OF SERIES

## 13.10  RECURRING POWER SERIES

Frequently it is necessary to sum a discrete power series, such as a present-value series:

$$P = \frac{A_0}{(1 + i)^0} + \frac{A_1}{(1 + i)^1} + \cdots + \frac{A_m}{(1 + i)^m} + \cdots + \frac{A_n}{(1 + i)^n} \qquad (13.11)$$

The series generally is more complicated than a geometric series, and it is desirable to develop a method for summing these more complicated types. The subject of series summation is involved, but one method, summation of recurring power series, depends on algebraic methods and is easily understood upon demonstration. Moreover, the method will accommodate most series that arise in economic analysis.

If $U = 1/(1 + i)$, then Eq. (13.11) can be written more compactly:

$$\sum = A_0 + A_1 U^1 + \cdots + A_m U^m + \cdots + A_n U^n \qquad (13.12)$$

Each $A$ is a coefficient, and the series is a *recurring power series* if and only if successive coefficients are connected by a recurring relationship.

For example, consider the series

$$\sum = 1 + 3U + 7U^2 + 15U^3 + 31U^4 + \cdots \tag{13.13}$$

Every three successive coefficients in the series are related by the equation (where the constant multiplying the highest coefficient has been made equal to unity):

$$2A_m - 3A_{m+1} + 1A_{m+2} = 0 \tag{13.14}$$

Thus

$$2(1) - 3(3) + 1(7) = 0$$

$$2(3) - 3(7) + 1(15) = 0$$

Equation (13.14) has the algebraic form

$$B_2 A_m + B_1 A_{m+1} + 1A_{m+2} = 0 \tag{13.15}$$

where each $B$ is constant and independent of $m$ and the right-hand side is zero. Equation (13.15) is known as the *scale of relation,* and the $B$ multipliers are referred to as the *constants of the scale.* Equation (13.14) has two independent constants of the scale; hence Eq. (13.13) is a recurring series of the second order.

Recurring power series were first investigated by the French mathematician de Moivre (1667–1754). A recurring power series is readily summed when its scale of relation and an expression for the general coefficient are known. The general term for Eq. (13.13), as will be shown, is

$$(2^{m+1} - 1)U^m \tag{13.16}$$

The summation is accomplished by adding the original series and others derived from it in a way that will utilize the scale of relation and compel most of the terms in the addition to become zero and vanish. For example, to find the sum $\Sigma$ for Eq. (13.13), use the numbers $+1$, $-3$, and $+2$, taken by reading from right to left in the scale of relation, to form the series

$$+1 \sum - 3U \sum + 2U^2 \sum$$

and add the series so formed. The work is shown in Table 13.4.

## 13.11  GENERAL TERM FROM A GENERATING FUNCTION

In summing Eq. (13.13) by Table 13.4, we needed an expression for the general term coefficient $A_m$. Consider the following fraction taken from Table 13.4:

$$\frac{A_0 + (A_1 - 3A_0)U}{1 - 3U + 2U^2} = \frac{1}{1 - 3U + 2U^2}$$

$$= 1 + 3U + 7U^2 + 15U^3 + 31U^4 + \cdots$$

**TABLE 13.4**
**Summing a recurring power series**

$$\Sigma = A_0 + \qquad A_1 U + A_2 U^2 + \cdots + A_n \quad U^n$$

$$-3U\Sigma = \qquad -3A_0 U - 3A_1 U^2 + \cdots - 3A_{n-1}U^n - \qquad 3A_n U^{n+1}$$

$$+2U^2\Sigma = \qquad\qquad + 2A_0 U^2 + \cdots + 2A_{n-2}U^n + \qquad 2A_{n-1}U^{n+1} + 2A_n U^{n+2}$$

$$(1 - 3U + 2U^2)\Sigma = A_0 + (A_1 - 3A_0)U + 0 \quad + \cdots + 0 \qquad + (2A_{n-1} - 3A_n)U^{n+1} + 2A_n U^{n+2}$$

with $A_0 = 1$ $\quad A_1 = 3$ $\quad A_n = 2^{n+1} - 1$

$$\Sigma = \frac{1 + [2(2^n - 1) - 3(2^{n+1} - 1)]U^{n+1} + 2(2^{n+1} - 1)U^{n+2}}{(1 - 2U)(1 - U)}$$

$$\Sigma = 2\frac{1 - (2U)^{n+1}}{1 - 2U} - \frac{1 - U^{n+1}}{1 - U}$$

$$\Sigma = 1 + 3U + 7U^2 + \cdots + (2^{m+1} - 1)U^m + \cdots + A_n U^n$$

Scale of relation $= 2A_m - 3A_{m+1} + 1A_{m+2} = 0$

The fraction is known as the *generating function*, which can be used to find the general term, inasmuch as it generates the series. By the use of partial fractions developed in standard algebra textbooks with $C$ and $D$ constants,

$$\frac{1}{1 - 3U + U^2} = \frac{1}{(1 - 2U)(1 - U)} = \frac{C}{1 - 2U} + \frac{D}{1 - U}$$

This expression is true for all values of $U$. Placing $U = 0$ gives $1 = C + D$, and placing $U = -1$ gives $1/6 = C/3 + D/2$. Solved simultaneously, it yields $C = 2$ and $D = -1$. Thus

$$\frac{1}{1 - 3U + 2U^2} = \frac{2}{1 - 2U} - \frac{1}{1 - U}$$

and

$$\frac{2}{1 - 2U} = 2 + 4U + \cdots + 2^{m+1}U^m + \cdots$$

$$-\frac{1}{1 + U} = -1 - U + \cdots \quad - 1U^m + \cdots$$

The general term is $(2^{m+1} - 1)U^m$, as used previously.

## 13.12   SOME RECURRING POWER SERIES

The following series all have the scale of relation $A_m - 2A_{m+1} + A_{m+2} = 0$:

$$\Sigma = U + 2U^2 + 3U^3 + \cdots + nU^n \tag{13.17}$$

$$\Sigma = U^2 + 2U^3 + 3U^4 + \cdots + (n - 1)U^n \tag{13.18}$$

$$\Sigma = BU + (B + A)U^2 + (B + 2A)U^3 + \cdots + [B + (n - 1)A]U^n \tag{13.19}$$

Summations, up to and including the term in $U^n$, are, respectively, where $A$ or $B$ can be negative:

$$\Sigma = U\frac{1 - (n + 1)U^n + nU^{n+1}}{(1 + U)^2} \tag{13.20}$$

$$\Sigma = U^2\frac{1 - nU^{n-1} + (n - 1)U^n}{(1 - U)^2} \tag{13.21}$$

$$\Sigma = U\frac{B - (B - A)U - (B + A_n)U^n + (B - A + A_n)U^{n+1}}{(1 - U)^2} \tag{13.22}$$

The series

$$\Sigma = GU^2 + 2GU^3 + \cdots + (n - 1)GU^n$$

is particularly important because it arises in connection with a machine for which the annual expense increases at a constant rate, $G$ per year. The summation is given by Eq. (13.21), which when multiplied by $G$ and with $1/(1 + i)$ replacing $U$ reduces to

$$\Sigma = \frac{G}{i(1 + i)^n}\left[\frac{(1 + i)^{n-1}}{i} - n\right]$$

The uniform-gradient series was discussed in Sec. 9.4.

Another series of considerable importance is the present value of $1 of depreciation taken with the sum-of-the-years-digits method of depreciation. Sum-of-the-years-digits is algebraically $0.5n(n + 1)$, and the series is

$$\Sigma = \frac{n}{0.5n(n + 1)}U + \frac{n - 1}{0.5n(n + 1)}U^2 + \cdots + \frac{1}{0.5n(n + 1)}U^n$$

The series is then summed by Eq. (13.22), which after multiplication by $1/[0.5n(n + 1)]$ and with $B = n$, $A = -1$, and $U = 1/(1 + i)$, becomes

$$\Sigma = \frac{1}{0.5n(n + 1)i}\left[n - \frac{(1 + i)^n - 1}{i(1 + i)^n}\right] \tag{13.23}$$

or

$$\Sigma = \frac{n - F_{RP,i,n}}{0.5n(n + 1)i} \tag{13.24}$$

## NOMENCLATURE

| | |
|---|---|
| $B$ | Constant in scale of relation, Sec. 13.10 |
| $D$ | Distance from the end of the interval in the Fibonacci search |
| $F$ | Objective function to be optimized |
| $F_i$ | Fibonacci number |
| $F_r$ | Fractional ratio of the interval left to the original interval |
| $F_{RP,i,n}$ | The factor $[(1 + i)^n - 1)]/i(1 + i)^n$; see Sec. 2.2 |
| $G$ | Gradient amount, or constant increase or decrease |
| $i$ | Decimal rate of return in Secs. 13.10 to 13.12 |
| $L$ | Range of a variable |
| $m$ | The $m$th or general term |
| $n$ | The $n$th or last term |
| $N$ | Number of experiments or calculations allowed |
| $P$ | Present value, $ |
| $U$ | The factor $1/(1 + i)$ |

## PROBLEMS

**13.1.** Assume that the model is really the evaluation of the square root of 3. Assume that the solution is known to lie in the range 0 to 5. Evaluate $\sqrt{3}$ to an interval that is 0.05 wide. Use the Fibonacci search technique. Hint: $|X^2 - 3|$ is a minimum if $X = \sqrt{3}$.

**13.2.** Repeat Prob. 13.1 using a uniform search method. For a condensation tabulate only in the neighborhood of the minimum.

**13.3.** Repeat Prob. 13.1 using a sequential dichotomous search.

**13.4.** Suppose

$$Z = 2X^3 - 3X^2 - 14X + 2$$

Using six trials or experiments, find a minimum value for $Z$ in the range $X = 1$ to $X = 3$ using (a) a uniform search, (b) a uniform dichotomous search, (c) a sequential dichotomous search, and (d) a Fibonacci search. Also, compare the fractional ratio of the interval left to the original interval for these methods.

**13.5.** Using the method of steepest ascent, find the minimum value of

$$Z = X^2 + 2Y^2 + XY$$

starting at the point $X = 2$ and $Y = 3$. Use an initial step on $Y$ of 0.5 until no further improvement results; then shift to a step size on $Y$ of 0.05.

**13.6.** Repeat Prob. 13.5 using the one-at-a-time method.

**13.7.** In the method of steepest ascent surfaces with spherical contours give the fastest convergence rates. Given the objective function $Z = X^2 + 25Y^2$, how can this problem be transformed into one giving spherical contours?

**13.8.** The one-at-a-time method can be applied to a three-variable problem by the technique used in Example 13.4. Given the function

$$f = X^2 + Y^2 + Z^2 - XY - YZ - 4X$$

find the values of $X$, $Y$, and $Z$ that make $f$ a minimum. Hint: Start at $X = Y = 0$ and find the best value of $Z = 0$. Then, for $Y = 0$ and $Z = 0$, find the best value of $X = 2$. Then for $X = 2$ and $Z = 0$ find the best value of $Y = 1$. Then for $X = 2$ and $Y = 1$ find the best value of $Z = 0.5$. Continue until no further improvement results.

**13.9.** Confirm the answer to Prob. 13.8 by solving analytically.

**13.10.** Repeat Prob. 13.8 using the method of steepest ascent. Start at point $X, Y, Z = (0, 0, 0)$.

**13.11.** For the series

$$\sum = -1 + X - 3X^2 + 5X^3 - 11X^4 + 21X^5 + \cdots$$

find (a) the scale of relation, (b) the generating function, (c) the coefficient for the term in $X^n$, and (d) the sum of the terms through $X^n$.

**13.12.** Find the scale of relation and the next term of the series

$$\sum = 1 + 3U + 6U^2 + 10U^3 + 15U^4 + 21U^5 + 28U^6 + \cdots$$

**13.13.** Find the generating function for Prob. 13.12.

**13.14.** Verify Eq. (13.21).

**13.15.** Verify Eq. (13.23).

**13.16.** Let $F_N$ denote the $n$th Fibonacci number and form the following series, where $U = 1$:

$$\sum = F_0 + F_1 U + F_2 U^2 + \cdots + F_N U^N$$

Knowing the scale of relation given by

$$-F_N - F_{N+1} + F_{N+2} = 0$$

find (a) a generating function for the series, and (b) an expression for the $n$th Fibonacci number.

**13.17.** Find an expression for the limit of the ratio of two successive Fibonacci numbers as $N$ becomes very large.

**13.18.** Find an expression for the sum of the first $N$ Fibonacci numbers.

# REFERENCES

**Search methods**

1. Abadie, J.: *Integer and Non-linear Programming,* North-Holland Publishing Company, New York, 1970.
2. Beveridge, G. S. G., and R. S. Schecter: *Optimization: Theory and Practice,* McGraw-Hill Book Company, New York, 1970.
3. Boas, A. H.: "How Search Methods Locate Optimum in Univariable Problems," *Chemical Engineering,* February 4, 1963, p. 105.
4. ———: "Optimizing Multivariable Functions," *Chemical Engineering,* March 4, 1963, p. 97.
5. Box, M. J.: "A New Method of Constrained Optimization and a Comparison with Other Methods," *Computer Journal,* vol. 8, April 1965, pp. 42–52.
6. Fiacco, A. V., and G. P. McCormick: Programming Under Nonlinear Constraints by Unconstrained Minimization: A Primal-Dual Method, Research Analysis Corporation Technical Paper RAC-TP-96, September 1963.
7. ———: "Computational Algorithm for the Sequential Unconstrained Minimization Technique for Non-linear Programming," *Management Science,* vol. 10, no. 4, 1964, pp. 601–617.
8. ———: "The Sequential Unconstrained Minimization Technique for Nonlinear Programming, a Primal-Dual Method," *Management Science,* vol. 10, no. 2, 1964, pp. 360–366.
9. ———: *SUMT Without Parameters,* Northwestern University Technical Institute Systems Research Memo 121, April 1965.
10. ———: "Extensions of SUMT for Non-linear Programming: Equality Constraints and Extrapolation," *Management Science,* vol. 12, no. 11, 1966, pp. 816–828.

11. Friedman, M., and L. S. Savage: *Selected Techniques of Statistical Analysis,* McGraw-Hill Book Company, New York, 1947.

12. Himmelblau, D. M.: *Applied Non-linear Programming,* McGraw-Hill Book Company, New York, 1972.

13. Johnson, S. M.: *Best Exploration for Maximum Is Fibonaccian,* U.S. Air Force Project Rand Research Memorandum, RM-1590, November 18, 1955.

14. Keifer, J.: "Sequential Minimax Search for a Maximum," *Proceedings of the American Mathematics Society,* vol. 4, no. 3, 1953.

15. Koester, J. L., and J. H. Mize: *Optimization Techniques with Fortran,* McGraw-Hill Book Company, New York, 1973.

16. Umeda, T., and A. Ichikawa: "Modified Complex Method for Optimization," Ind. & Eng. Chem., *Process Design and Development,* vol. 10, no. 229, 1971.

17. Wilde, D. J.: *Optimum Seeking Methods,* Prentice-Hall, Inc., Englewood Cliffs, N.J., 1964.

18. ——, and C. S. Beightler: *Foundations of Optimization,* 2d ed., Prentice-Hall, Inc., Englewood Cliffs, N.J., 1979.

**Series summation**

19. Chrystal, G.: *Algebra,* Dover Publications, Inc., New York, 1961, pt. 2, chap. 31.

20. Fine, H. B.: *College Algebra,* Dover Publications, Inc., New York, 1961, chap. 35.

21. Jelen, F. C.: "Series Summation for Economic Analysis," *Hydrocarbon Processing,* December 1967, pp. 153–157.

22. Nazen, S. M.: *Applied Time Series Analysis for Business and Economic Forecasting,* Marcel Dekker, Inc., New York, 1988.

# PART

# III

# COST ESTIMATION, COST/ SCHEDULE CONTROL, AND PROJECT MANAGEMENT

# CHAPTER
# 14

## CAPITAL
## INVESTMENT
## COST
## ESTIMATION

*H.C. Nelson, O.P. Kharbanda, W.A. Janda, J.H. Black, K.K. Humphreys,
K.K. Bentil, and D.F. McDonald, Jr.*

## 14.1  ESTIMATING BASICS

Estimating is the key to a successfully conceived, managed, and completed project. Estimating is not limited to construction alone, but rather is a function common to a broad spectrum of projects whereby one desires to manage cost and time. The project may be a high-rise office building, a scenic highway, the manufacturing of a product, a world tour of a symphony orchestra, or your next vacation.

As the term implies, *estimating* is an approximation procedure which provides answers that have significantly less than 100 percent probability of being correct or even close. Unlike the discussion in earlier chapters of this book wherein most calculations were reduced to precise formulas and equations, estimating is imprecise and cannot be reduced to a sequence of mathematical expressions. Thus understanding the accuracy limitations of the estimate and its potential variation is extremely important to the user.

## 14.2  DEFINITIONS

A *cost estimate* may be defined as a compilation of all the costs of the elements of a project or effort included within an agreed upon project scope. To a contractor this is the cost that will most likely be incurred to complete the project as defined in the contract documents and to turn it over to the owner. The contractor's cost is made up of the contractor's internal costs as well as those of subcontractors, suppliers, and third parties. The owner's costs are made up of the owner's own costs for administering the project, the cost that the contractor(s) charges the owner for the work performed, and costs of the owner's consultants, engineers, and suppliers. Also included in an owner's estimate must be costs of land, provisions for interim and permanent financing, plus the numerous elements of life cycle cost associated with ownership and operation.

Decisions made by management pertaining to capital investments affect the financial stability and future earnings of the firm. These earnings are usually obtained by converting capital into facilities or plants that yield a return exceeding the amount of the capital investment. This capital investment is composed of two parts:

1. The fixed capital investment to provide the physical facilities.
2. The working capital investment, a revolving fund to keep the facilities operating. An investment cost is a one-time cost, as opposed to operating costs, which are continuous or repeated.

The *fixed capital* investment is the capital required to provide *all* the facilities needed for the project regardless of how remote they might be to manufacturing operations, e.g., a rolling mill or a refrigerator in the employees' cafeteria. Fixed capital investment can be divided into various categories, such as design and engineering, land purchase and improvement, manufacturing and allied buildings, equipment, utilities and services, receiving and shipping, and start-up.

Although land does involve a capital investment, some prefer to include in fixed capital only those items for which depreciation is allowed by the Internal Revenue Service, thus excluding land. However, if land costs are included in the total investment, it makes little difference in what subcategory they appear.

*Working capital* in finance and accounting is the difference between current assets and current liabilities. Current assets are chiefly cash, accounts receivable, inventories, and negotiable securities. Current liabilities are chiefly accounts payable, short-term debts, and the portion of long-term debts due within 1 year. In this book working capital comprises the value of funds, in addition to fixed capital and start-up costs, needed to get a project started and meet subsequent obligations. Working capital is regarded as funds frozen in the project that will be recovered in full at the end of the project or even if the project is terminated at any time. A list of the components of working capital is given in Table 14.1. Inventory, from raw material to finished goods, comprises one major group. Accounts receivable usually allows a 30-day payment and thus ties up funds. On the other hand, accounts payable, which can be delayed about 30 days, works contrary to accounts receivable, and reduces the working capital requirements.

In general working capital is based on requirements for a 1-month period. In manufacturing industries working capital is about 10 to 20 percent of the fixed capital investment. In some industries with very high inventories of finished goods the figure can exceed 40 percent. As a rough estimate, working capital may be taken as 15 percent of the total investment.

**TABLE 14.1**
**Components of working capital**

Raw materials inventory
Work-in-progress inventory (semifinished goods)
Finished-products inventory
Supplies for product manufacture
Prepaid expenses
Taxes payable
Accounts receivable
Cash or equivalent on hand for salaries, wages, etc.
Accounts payable

Before a new investment is made, a capital-cost estimate is usually developed as part of calculating the economic return for the venture. This return is used as partial justification for the project. Cost estimates are required before and after a decision is made to go ahead with the investment. During the course of a capital project various types of cost estimates will be necessary for a variety of reasons and purposes.

Cost estimates for capital projects consist of two major cost elements:

1. *Direct costs* are the costs of all permanent equipment, materials, labor, and other resources involved in the fabrication, erection, and installation of the permanent facilities.

2. *Indirect costs* are all costs other than direct costs which do not become a permanent part of the facilities but are required for the orderly completion of the project. These may include, but are not limited to, construction management, start-up costs, fees, insurance, and taxes.

The process of capital-cost estimating is accomplished in steps:

Takeoff      A measurement of the quantities of work and cataloging of cost items derived from the scope document.

Costing      From the takeoff and the information presented in the scope documents, the assignment of cost values to the elements of work and related items in the project or endeavor.

Pricing      The determination of the amount to be charged the client by the contractor so as to include fully profit and contingency in addition to direct- and indirect-cost items.

## 14.3 BASIC COST ESTIMATING TERMINOLOGY

The following, while not all-inclusive, are some of the basic terms encountered in the estimating process:

Addendum      This is a written change made by the architect/engineer to contract drawings and/or specifications prior to the scheduled

|  |  |
|---|---|
|  | bid date. More than one addendum (plural) are known as addenda. |
| Allowance | A sum of money arbitrarily allotted to an item (usually defined by the architect/engineer) instead of an estimated amount. |
| Alternate | This is a request by the architect/engineer for the cost of adding or deleting an item or work element from the basic bid. The cost of adding an item is usually known as an *additive alternate,* whereas the cost of deleting an item is known as a *deductive alternate.* |
| Change order | This is a written change made by the architect/engineer to the contract drawings and/or specifications after contract award. |
| Contingency | This is a specific provision (dollar amount) to provide some protection for uninsured risk elements within the defined project's scope. Typical risks are quantity and productivity variations, unexpected site conditions, materials delivery problems, deductible portions of insurance coverage, etc. It is particularly important where previous experience relating estimates and actual costs have shown that unforeseeable events or conditions which increase costs are likely to occur. |
| Contract documents | These are the contract forms, general and special conditions, drawings, specifications, and addenda describing project scope and contract terms used in the preparation of an estimate. |
| Cost | The amount a contract item is known or estimated to cost the contractor. (See Price.) |
| Cost index | The ratio of the cost of an item at a specified point in time to a datum established in the past. Useful to take known past costs for an item and relate them to the present or future time. |
| Direct cost | Those costs which can be directly attributed to a particular item of work or activity. |
| Distributable | A cost item that is spread over other cost items, rather than managed as a separate account. |
| Escalation | The provision for an increase in the cost of equipment, material, labor, etc., over those specified in the contract, due to continuing price level changes over time. |
| Field costs | The engineering and construction costs associated with the project site rather than with the home office. |
| General conditions | A specific portion of the contract documents. These state all the responsibilities and relations of all parties to the contract as well as any conditions applicable to the contract. |
| Indirect costs | (1) In construction, all costs which do not become a final part of the installation, but which are not limited to, field administration, direct supervision, capital tools, start-up |

| | |
|---|---|
| | costs, contractor's fees, insurance, taxes, etc. (2) In manufacturing, costs not directly assignable to the end product or process, such as overhead and general purpose labor, or costs of outside operations, such as transportation and distribution. |
| Labor cost | The salary plus all fringe benefit costs associated with labor which can be definitely assigned to one item of work, product, process area or cost center. |
| Mark-up | This is the allowance for a contractor's overhead, profit, and contingency that is added to an estimate to arrive at an overall project price. It is added by management, not the estimators, during the pricing of the estimate. |
| Overhead | A cost or expense inherent in the performing of an operation, i.e., engineering, construction, operating or manufacturing, which cannot be charged to or identified with a part of the work, product, or asset, and therefore must be allocated on some arbitrary basis believed to be equitable, or handled as a business expense independent of the volume of production. |
| Price | The amount charged a customer for completion of a contract item or total contract. It incorporates direct costs, indirect costs, general overhead, profit, and contingency. |
| Productivity | This is a relative measure of labor efficiency, either good or bad, when compared to an established base or norm as determined from an area of great experience. Productivity changes may cause either an increase or decrease in cost. |
| Scope of work | Total work to be accomplished and administrative requirements associated with a contract as described within the contract documents, which may include the general conditions, special conditions, specifications, drawings, and addenda. |
| Special conditions | These are modifications to the general conditions and are applicable and unique to a specific project. |
| Time-sensitive | A term applied to those elements of cost which will be expended or incurred on a time unit basis (monthly, weekly, hourly, etc.) and are a subset of indirect costs. A cost engineer's salary on a project is a time-sensitive cost as long as that engineer is on the project. |

## 14.4 TYPES OF ESTIMATES

Capital-cost estimating is essentially an intuitive process which attempts to predict the final outcome of a future capital expenditure program even though not all parameters and conditions concerning a project are known or are not fully defined when the cost estimate is prepared.

Estimate types vary considerably depending on the available information, time demands, and purposes of the estimate. Lists of estimate types prepared by various authors, companies, professional and industrial organizations, and government agencies are available [5, 51, 60, 69]. These lists identify the various types, or classifications, of estimates most commonly prepared. The following four lists are typical examples.

Edmunds [17] gives six basic types of estimates for building construction used by architects and engineers:

1. Quickie
2. Original
3. Preliminary
4. Official or budget
5. Final or definitive
6. Revised

Another classification of estimate types [25] based on probable accuracy is:

1. Detailed unit cost: $\pm 3$ percent
2. Material takeoff: $\pm 6$ percent
3. Defined equipment ratio: $\pm 12$ percent
4. Preliminary equipment ratio: $\pm 25$ percent
5. Cost-capacity curve: $\pm 40$ percent

The AACE (American Association of Cost Engineers) [1] has proposed five classifications of estimating types:

1. Order of magnitude (ratio estimate)
2. Study (factored estimate)
3. Preliminary (budget authorization estimate)
4. Definitive (project control estimate)
5. Detailed (firm estimate)

A more recent and more condensed classification by AACE [31] shows the following types, with their probable accuracies and their method of preparation:

1. *Order of magnitude: −30 to +50 percent*. This type of estimate is made without any detailed engineering data. The estimate is often developed using cost-capacity curves, scale-up or -down factors, or ratio estimating techniques.
2. *Budget: −15 to +30 percent*. This type of estimate refers to the owner's budget, not to the budget as a project control document. Flow sheets, layouts, and equipment details are necessary to perform this type of estimate.

**3.** *Definitive: −5 to 15 percent.* This type of estimate is developed from very defined engineering data including plot plans and elevations, piping and instrument diagrams, one-line electrical diagrams, equipment data sheets and quotations, structural sketches, soil data and sketches of major foundations, building sketches, and a complete set of specifications.

This list of AACE suggested estimate categories is consistent with the original AACE list if it is recognized that the preceding "budget" category includes the previous categories of "study" and "preliminary." Similarly, the current "definitive" category includes the prior category of the same name as well as the "detailed" category.

The specified accuracy levels in the revised AACE list reflect the general fact that estimates are more likely to be low than high due to numerous factors, including inflationary trends and incomplete data, among others.

The various types of estimates are designed to identify the method of preparation and the range of accuracy. However, even with this diversity of names and classifications, cost estimates generally fall into two major categories, preliminary and detailed, as shown in Table 14.2.

Preliminary estimates are used during the formative stages of a capital expenditure program when there is a lack of firm or verifiable information for the initial evaluation of the project. In this situation various estimating techniques are used, ranging from experience and judgment as the dominant characteristics to techniques using historical cost charts, rule of thumb, and simple mathematical calculations to arrive quickly at a relatively inexpensive cost estimate. The accuracy of a preliminary estimate is usually very low. Obviously, accuracy is dependent on the amount and quality of information and the time available to prepare the estimate. Other terms commonly used in practice to identify preliminary estimates include *order of magnitude, conceptual, factored, quickie,* and *feasibility estimates* [51].

Detailed estimates are used after the scope and schedule of a project have been stabilized. Methods used for developing detailed estimates are based on more quantitative information in conjunction with formal, exacting procedures in arriving at estimated values. Techniques for detailed cost estimates vary with circumstances and the industry. However, emphasis is on factual data and mathematical modeling, the latter ranging from simple manual calculations to complex computer simulation programs.

**TABLE 14.2**
**Accuracy of cost estimates**

| Types of estimates | % accuracy |
| --- | --- |
| Preliminary | |
|    Order of magnitude | +50 to −30 |
|    Conceptual | +30 to −15 |
| Detailed | |
|    Semidefinitive | +25 to −15 |
|    Definitive | +15 to −10 |
|    Check | +10 to −5 |

These cost estimates are used for many purposes: bid proposals, establishment of budgets, fair-price estimates for bid evaluations, contract change orders, extra work orders, legal claims, appropriation of funds, and obtaining government permits and approvals. Generally these estimates consist of firm prices for materials and equipment. Quantitative data are obtained from design drawings and specifications; engineering, procurement, and construction execution plans are developed to arrive at the estimated costs for those functions.

## 14.5 AVAILABLE INFORMATION AGAINST ACCURACY

As stated earlier, the accuracy of an estimate depends on the quality and quantity of available project information. A guide for judging estimate accuracy and the type of estimate to be used, based on available information, is presented in Fig. 14.1. This guide reflects the five estimate classifications originally suggested by the AACE [1].

## 14.6 COST OF MAKING ESTIMATES

The cost of making an estimate depends on the desired level of accuracy and on the amount of information which is available. An order-of-magnitude estimate can cost very little to prepare whereas a highly accurate estimate can be quite costly. Bauman [6] reported, for example, that the cost of preparing estimates manually at a −5 to +15 percent level of accuracy can be as much as 25 to 30 times greater than the cost of manually performing an estimate at the −30 to +50 percent level of accuracy.

The introduction of computers as an estimating tool has reduced the ratio somewhat, but the cost of obtaining −5 to +15 percent estimate accuracy still exceeds the cost of obtaining −30 to +50 percent accuracy by a factor of 5 or more, as shown in Table 14.3 [61].

## 14.7 FUNCTIONS OF CAPITAL-COST ESTIMATES

Estimates of capital costs are made for various purposes, the most common of which follow.

**TABLE 14.3**
**Typical costs for preparing estimates in kilodollars**

| | | Cost of project in M$ | | |
| --- | --- | --- | --- | --- |
| Type of estimate | Accuracy range, % | Less than 1.0 | 1.0–5.0 | 5.0–50.0 |
| Order of magnitude | −30 to +50 | 7.5 to 20 | 17.5 to 45 | 30 to 60 |
| Budget | −15 to +30 | 20 to 50 | 45 to 85 | 70 to 130 |
| Definitive | −5 to +15 | 35 to 85 | 85 to 175 | 150 to 330 |

*Source:* Based on A. Pikulik and H. E. Diaz: "Cost Estimating for Major Process Equipment" [62], McGraw-Hill, Inc., New York, 1977, p. 106. Used with permission.

**Figure 14.1**

A guide for capital investment estimating. (R. H. Perry and C. H. Chilton: *Chemical Engineers' Handbook*, 5th ed., McGraw-Hill Book Company, New York, 1973.)

Required information

**Estimate types (columns):**

- Order-of-magnitude estimate >±30% range
- Study estimate ±30% range
- Budget authorization estimate ±20% range
- Project control estimate ±10% range
- Firm estimate ±5% range

**Project scope standard processes**
- Product, capacity, location & site requirements.
- Utility & service requirements. Building & auxiliary requirements. Raw materials & finished product handling & storage requirements.

**Man-hours**
- Supervision
- Labor by craft
- Engineering & drafting

**Electrical**
- Well-developed drawings
- Engineered single-line diagrams (power & light)
- Preliminary interlock, control, & instrument wiring specs.
- Preliminary lighting specifications
- Distribution specifications
- Substations, number & sizes, specifications
- Engineered list & sizes
- Preliminary motor list—approximate sizes

**Instrumentation**
- Well-developed drawings
- Engineered list & flow sheet
- Preliminary instrument list

**Insulation**
- Well-developed drawings or specifications
- Insulation specifications & schedules
- Preliminary list of equipment & piping to be insulated
- Rough specifications

**Piping**
- Piping layouts & schedules
- Engineered flow sheets
- Preliminary flow sheet & specifications

**Utility requirements**
- Well-developed drawings
- Engineered flow sheets
- Engineered heat balance
- Preliminary flow sheets
- Preliminary heat balance
- Rough quantities (steam, water, electricity, etc.)

**Building and structures**
- Detailed drawings
- General arrangements & elevations
- Preliminary structural design
- Architectural & construction
- Foundation sketches
- Approximate sizes & type of construction

**Equipment list**
- (b) Engineered
- (a) Preliminary
- General arrangement
- Vessel sheets
- Engineered specifications
- Preliminary sizing & material specifications

**Process flow sheet**
- Engineered
- Preliminary
- Rough sketches

**Site**
- Well-developed site facilities
- Well-developed site plot plan & topographical map
- Location & dimensions RR, roads, impounds, fences
- Soil bearing
- General description
- Location

Error range %

Most probable cost

40  30  20  10  0  10  20  30  40

Preliminary estimates:

1. To study project feasibility
2. To obtain designs and economic studies
3. To obtain financing
4. To obtain government permits or appraisals

Detailed estimates:

1. To seek funding
2. To prepare bid proposals
3. To prepare fair-price estimates for bid evaluations
4. To prepare changes and extra work orders for contracts
5. To bolster legal claims

Before private funds are appropriated, feasibility studies must show that the expected profit is sufficient to justify the risk of capital expenditure. Order-of-magnitude studies are used for screening purposes. Further studies can optimize the project design and the plan of execution. Cost estimates play a very important role in obtaining financing from lending institutions, if such arrangements are required.

After the decision has been made to go ahead with a project, execution of the capital expenditure program begins. Contractors to perform services in management, engineering, procurement, and construction must be selected. Usually such services are obtained through competitive bidding, which requires bid estimates. If the bid is too high, the contractor may not be awarded the job; but if the bid is too low, the contractor may be awarded the job and take a substantial loss. During the course of the project, changes in the scope of the work may be required; estimates are necessary to determine these costs. Upon completion of the project, legal claims may arise from conflicts that occurred during the project's execution. Cost estimates may be involved in presenting the claim and agreeing to a settlement.

For the allocation of public funds, e.g., to explore space or improve military preparedness, there are available measures of feasibility rather than measures of profitability. Such funds are limited, so some kind of feasibility study is important, as discussed in Chap. 5.

## 14.8 DIRECT- AND INDIRECT-COST ITEMS

It is important to keep the direct and indirect costs separate in an estimate. Many estimates are based on a historical database compiled from previous projects' cost data. The less encumbered that direct-cost data are with indirect costs, the more reliable the data are for use in a new project. Keeping indirects separate also allows one to account more accurately for the unique conditions that will be encountered in the new project. The following is a list of indirect-cost items normally seen on construction projects:

## Job indirects checklist

**Salaries, supervision**
  Project Director
  Construction Manager
  General Superintendent
  Excavation Superintendent
  Concrete Superintendent
  Carpenter Superintendent
  Rigging Superintendent
  Welding Superintendent
  Electrical Superintendent
  Chief Warehouseman

**Salaries, engineering**
  Chief Engineer
  Office Engineer
  Project Controls Manager
  Project Controls Engineers
  Materials Engineer
  Draftsmen
  Field Engineers
  Party Chiefs
  Instrumentmen

**Salaries/wages, other**
  QA Engineers
  Safety Engineers
  Mechanics
  Plant Operators
  First Aid Workers
  Secretaries and Clerks
  Computer Operators
  Warehousemen
  Guards
  Janitors
  Runners

**Automotive**
  Automobiles/cars
  Pickups
  Ambulances
  Tractor-trailers
  Special-purpose vehicles
  Fuel/lube trucks

**Temporary horizontal construction**
  Access roads
  Construction bridges
  Drainage structures
  Rail spurs
  Laydown areas
  Fencing and gates
  Parking areas
  Environmental protection

**Buildings and major equipment**
  Project office
  Warehouses
  Brass alley
  Change house
  Carpenter shop
  Pipe fabrication shop
  Welder test facility
  Electrical shop
  Rigging loft
  First aid station
  Tool sheds
  Powder house
  Cap house
  Resteel fabrication
  Machine shop
  Equipment maintenance shop
  Concrete batch plant
  Quarry
  Hoisting equipment
  Training building

**Support systems**
  Water supply
  Compressed air
  Electrical
  Site communications
  Inert gas
  Oxygen

**General expense**
  General office supplies
  Office furniture
  Engineering supplies
  Engineering equipment
  Printing/reproduction
  Computer terminal
  CPM scheduling
  Telephone/telegraph/
    radios
  Utilities
  Portable toilets
  Signs
  Safety equipment
  Permits and licenses
  Advertising
  Contributions
  Job travel expenses
  Testing and laboratory
  Legal fees
  Audit fees
  Medical supplies
  Progress photos
  Sanitary facilities
  Building rental
  Building and grounds
  Maintenance
  Exterior lighting
  Drinking water and ice
  Taxes
  Bonds
  Insurance
  Payroll burden costs
  Back charges from
    others
  Consultants
  Weather protection

*Source:* James M. Neil: *Construction Cost Estimating for Project Control,* © 1982, pp. 260–261. Reprinted by permission of Prentice-Hall, Inc., Englewood Cliffs, New Jersey.

## 14.9 ELEMENTS OF EQUIPMENT, LABOR, AND MATERIAL COSTS

Figure 14.2 is a hierarchical chart for a construction cost estimate. Other types of project cost estimates will take a similar form but may include more or less than that of the construction cost estimate. The various items listed in Fig. 14.2 are now discussed.

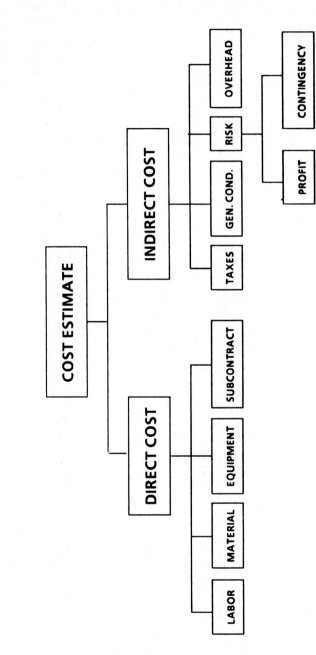

**Figure 14.2**
Components of a cost estimate.

374

# Direct Costs

**LABOR COSTS.** *Labor costs* are the total of the actual amounts paid to the field personnel who perform the actual work on the project such as carpenters, laborers, masons, painters, etc.

Labor costs can be broken into two main components: (1) the basic wage (the wage rate multiplied by the number of hours worked), and (2) labor burden, which includes items such as taxes, insurance, etc., that employers are required by law to pay as well as fringe benefits negotiated between employers and employee either individually or collectively. The labor burden portion of the total labor cost is not always treated as a direct-cost element since it is seldom directly proportional to hours worked. Normally one will treat the labor burden according to company practice unless contract conditions dictate otherwise.

**MATERIAL COSTS.** *Material costs* usually include the prices of materials or parts incorporated into a project. In some cases one has to solicit material prices from manufacturers and suppliers. In other instances the manufacturers and suppliers circulate prices to contractors who they know to be working on a specific project. Usually these prices are given to provide specific amounts of specific material within a given period at a specified location. Most contractors include delivery charges as part of the direct cost and taxes as part of indirect cost.

**EQUIPMENT COSTS.** *Equipment costs* refer to the cost of the equipment that a contractor uses to perform a contract (such as backhoes, cranes, bulldozers, etc.) and not equipment that is installed permanently as part of a contract. Equipment costs may be calculated in a variety of ways depending on whether the contractor leases or owns the equipment. If the equipment is leased, then it consists of the lease costs and the cost of fuel. If the equipment is owned by a contractor, then those costs include owning costs (investment, insurance, depreciation, etc.) and operating costs (fuel, maintenance, repairs, etc.). The costs of operators for operating equipment, even though part of the operating costs, are included by most contractors in the direct labor costs.

**SUBCONTRACT COSTS.** *Subcontract costs* of services are provided by subcontractors for performing a specific portion of a given project that the general contractor does not perform with its own forces.

# Indirect Costs

**TAXES.** These vary significantly from location to location and by tax-exempt status of the owner. Thus they are usually cataloged separately to facilitate accounting. Statutory salary/wage taxes (FICA, Workers Compensation, etc.) are often included here.

**GENERAL CONDITIONS.** This includes the cost of items that one has to furnish in order to perform a given contract as a whole. Usually these items cannot be charged to a specific element of the work and include such items as supervision, temporary facilities, office trailers, toilets, utilities, permits, photographs, small tools, clean-up, etc.

**RISK ELEMENTS.** This may be considered in two categories.

**Profit.** *Profit* may be regarded as the amount of money included by a contractor in his or her estimate as compensation for his or her risk, efforts, and endeavor to undertake a project. It is usually the "residue" of money left after a contractor has met all the costs (both direct and indirect) on a project. The amount of profit to be added is a very subjective and individual matter. It, however, usually depends on some considerations such as competition, how badly one needs the project, the job market, local conditions, and the economy.

**Contingency.** In cost estimating there is always an element of uncertainty since estimates are based on assumed values of cost and productivity, which actually are subject to considerable variability. There may be other risk items in the contract as well—weather, labor problems, transportation difficulties, etc. Thus real costs can be either higher or lower than the estimate; unfortunately, in most cases they turn out to be higher. While a contractor cannot price a contract to cover all eventualities, some protection is usually called for. A contingency markup is added by management in those cases using subjective reasoning and/or structured risk analysis.

**OVERHEAD.** This is in two categories.

**Home office overhead.** This includes the fixed costs and expenses that are incurred in the course of doing business by a company, regardless of the amount of work completed or contracts received. Home office overhead includes the costs of items such as home office rent or lease, utilities, communication (telephone, etc.), advertising, salaries of home office employees (estimators, secretaries, etc.), salaries of executives, donations, legal costs, accounting expense, etc., which are not chargeable to any specific project.

One method of calculating home office overhead is to use a percentage of the total annual home office overhead costs, with that percentage being a ratio of this specific project's total cost to the entire company/division's annual projects' costs.

**Job overhead.** This is another name for general conditions or job indirects and includes costs that are incurred at the job level in performing a specific contract but not chargeable to any specific item or element of that contract.

## 14.10   ESTIMATING TECHNIQUES

Cost estimates are prepared by a number of techniques, based on cost of similar projects or the cost of project components. Many companies have extensive files of cost data derived from previous projects. Published cost information is voluminous, but its accuracy is not always known.

The conference estimating technique is commonly used for developing quick preliminary estimates. This method draws on the collective judgment of a group of individuals with experience in cost estimating for similar projects in the past. The goal is to arrive at an estimate based on a group consensus [52].

The comparison estimating technique is a common method for estimating new capital projects and determining appraisals for existing facilities. This technique is similar to the conference method except that a more formal logic is used in arriving at a final figure. Estimating by comparing, weighing, and relating the costs of similar past projects requires adjusting for the new location and time period, which usually requires judgment or well-defined adjustment factors [52].

Graphic relationships and tabular relationships are two more estimating techniques. These methods are more quantitative and more accurate than the conference and comparison techniques, and hence are used in conceptual cost estimates. In these techniques limited historical data are used to develop cost relationships. Examples of these simple techniques are tabulations of data such as Table 14.4, exclusion charts such as Fig. 14.3, band charts such as Fig. 14.4, and upper- and lower-limit charts such as Fig. 14.5. Unfortunately, when using published cost data, it is not always clear what is included in the cost. Some equipment costs refer to purchased prices, whereas others are installed costs, which may or may not include auxiliary equipment. A similar difficulty exists with plant cost data. Some costs are for complete

**TABLE 14.4**
**Example of tabular estimating data**

Work-hours for fabrication and erection of 3/4-in. carbon steel piping

| Operation | Source A | Source B (6) | Source C (12) | Source D | Source E |
|---|---|---|---|---|---|
| Butt weld — std. wt. | 1.1 | 0.8 | 0.4 | — | 0.5 |
| Butt weld — X.H. | 1.3 | 0.9 | 1.2 | 0.7 | 0.6 |
| Lin. ft. std. pipe | 0.13 | 0.07 | 0.2 | — | 0.19 |
| Lin. ft. X.H. pipe | 0.15 | 0.08 | 0.25 | 0.16 | 0.20 |
| Std. weld ells | 2.2 | 1.6 | 0.9 | 2.1 | 1.3 |
| Std. weld tees | 3.3 | 2.4 | 1.2 | 3.0 | 1.8 |
| X.H. weld ells | 2.6 | 2.7 | 1.0 | 2.1 | 1.5 |
| X.H. weld tees | 3.9 | 2.7 | 1.5 | 3.0 | 2.1 |
| Std. weld reducer | 2.0 | 1.4 | 0.5 | 2.1 | 1.30 |
| X.H. weld reducer | 2.2 | 1.5 | 0.9 | 2.1 | 1.5 |
| 150# slip-on flange | 1.1 | — | 0.9 | — | 0.80 |
| 300# slip-on flange | 1.2 | — | 1.0 | — | 0.90 |
| 600# slip-on flange | 1.3 | — | 1.1 | — | |
| 150# weld neck flange | 1.1 | 1.1 | 0.4 | — | 0.80 |
| 300# weld neck flange | 1.2 | 1.1 | 1.3 | — | 0.95 |
| 600# weld neck flange | 1.3 | 1.1 | 1.4 | — | |
| 150# flange bolt-up | — | 0.6 | 1.5 | — | |
| 300# flange bolt-up | — | 0.7 | 1.6 | — | |
| 600# flange bolt-up | — | 0.7 | — | — | |
| Handle 150# valve | — | 0.3 | 1.5 | 0.8 | |
| Handle 300# valve | — | — | 1.6 | — | |
| Handle 600# valve | — | — | — | 1.1 | |

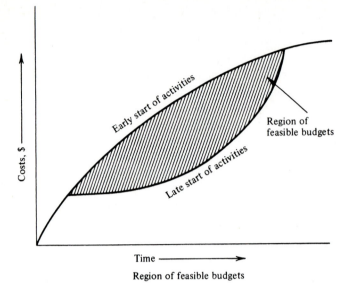

**Figure 14.3**
Example of an exclusion chart. (H. Kerzner: *Project Management: A System Approach to Planning, Scheduling and Controlling*, Van Nostrand Reinhold Company, New York, 1989. Used with permission.)

Nominal pipe size, in
Work-hours to handle and install 150-lb valves

**Figure 14.4**
Example of a band chart.

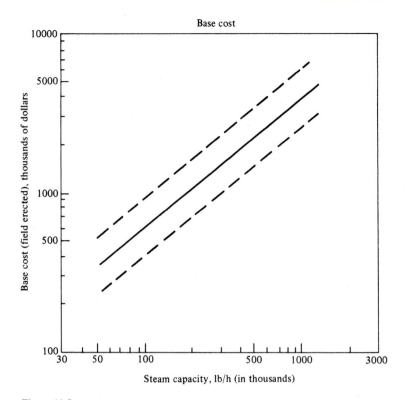

**Figure 14.5**
Example of an upper- and lower-limit chart. (K. M. Guthrie: *Process Plant Estimating, Evaluation, and Control*, Craftsman Book Company, Carlsbad, Calif., 1974. Used with permission.)

new plants, including land, utilities, and auxiliary facilities. Others are restricted to battery-limits costs. Ordinarily all manufacturing facilities are within the battery limits, but administrative offices, storage areas, utilities, and other auxiliary facilities are outside these limits. The term *battery limits* is a geographical designation for the process area of a proposed project. Published cost data should be used only for order-of-magnitude estimates since their accuracy is seldom known, and discrepancies of 50 percent are not uncommon. Some cost data are based on specific information and are not generally applicable. Other cost data are average values, and an average is questionable when applied to a particular case.

The unit rate technique is one of the most widely used approaches in preliminary and detailed estimating. This method takes the form of either the so-called bare unit cost or the complete unit cost. Examples of bare unit rates are wage cost per work-hour, purchased material cost per unit of weight or volume, and machinery cost per piece. Examples of complete unit cost are delivered price of fabricated structural steel per unit of weight, fully erected storage tank cost per unit of capacity, and installed cost of pipeline per unit of diameter and length. Table 14.4 is an example of bare unit work-hours needed to fabricate and erect piping elements.

**Example 14.1.** Using Table 14.4, source $C$, estimate the work-hours required to install 1000 ft of standard pipe, eight standard-weld ells, and one 150-lb valve.

*Solution*

$$1000 \times 0.2 = 200$$
$$8 \times 0.9 = 7.2$$
$$1 \times 1.5 = \underline{1.5}$$

209 work-hours

The unit rate technique is used in building construction. It is generally used in heavy civil construction, such as tunneling and pipeline projects.

The resource allocation method is popular in developing cost estimates for less tangible or nonrepetitive items, such as applications of new technology, research and development projects, technical services, management services, or consulting or advisory services. This technique is used in conceptual and detailed estimates when it is difficult to quantify the end product or output. Therefore, a plan is developed in terms of the times required for the various scheduled activities. The available resources — labor, equipment, and materials — are allocated against each planned activity at the appropriate unit cost to arrive at an estimated cost for the project [33]. An example of labor resource allocation is shown in Table 14.5.

**Example 14.2.** Estimate the labor cost for equipment construction during the month of April 1977 from Table 14.5.

*Solution*

Cost = (work-h) (rate) (1 + overhead + indirect costs + corporate costs + profit)
Program management = 20(11) (1 + 1.2 + 0.14 + 0.01 + 0.12)     =    543.40
Engineering = 154(10) (1 + 1.2 + 0.14 + 0.01 + 0.12)     =   3803.80
Fabrication = 154(10) (1 + 1.25 + 0.14 + 0.01 + 0.12)     =   <u>3880.80</u>

                         $8228.00

Factor estimating is a technique generally used to arrive at a preliminary capital-cost estimate inexpensively and quickly with a reasonable accuracy of 15 to 30 percent. In this method a cost is estimated by multiplying one cost by a factor to get another cost. Thus the cost of a building complete with electrical, plumbing, heating, ventilation, and air-conditioning systems might be estimated by multiplying the cost of the shell by 1.6. Such factors are called *ratio-cost factors* [83]. A common use of ratio-cost factors is the estimation of the installed cost of equipment from the purchase cost, applying an *equipment installation cost ratio*. The installed cost is the sum of the FOB cost, transportation, cost of foundations, erection, and making connections to service facilities. Although each of these costs can be estimated separately, it is much easier to multiply the FOB cost by a factor to give the installed cost. On the average, the installed cost is about 1.5 times the purchase cost, but it can vary considerably. Factor estimating has numerous synonyms, such as *ratio estimating, parameter estimating, module estimating, Lang factor estimating,* and *percentage estimating*. These methods involve substantially the same technique, with

**TABLE 14.5**
**Manpower resource allocation, cost engineering method example**

Columns 1–12 (Jan–Dec) are **1977**; columns 13–21 (Jan–Sept) are **1978**.

| Projects/tasks | Jan | Feb | Mar | Apr | May | June | July | Aug | Sept | Oct | Nov | Dec | Jan | Feb | Mar | Apr | May | June | July | Aug | Sept |
|---|---|---|---|---|---|---|---|---|---|---|---|---|---|---|---|---|---|---|---|---|---|
| **I. Program plans and building design** | | | | | | | | | | | | | | | | | | | | | |
| 1. Program management | 102 | 50 | 48 | | | | | | | | | | | | | | | | | | |
| 2. Engineering | 20 | 174 | 283 | | | | | | | | | | | | | | | | | | |
| **II. Equipment purchase** | | | | | | | | | | | | | | | | | | | | | |
| 1. Program management | | 52 | 40 | 40 | 30 | 30 | 30 | 30 | 30 | 30 | 20 | 20 | 20 | 20 | 20 | | | | | | |
| 2. Engineering | | 267 | 232 | 200 | 165 | 131 | 95 | 60 | 40 | | | | | | | | | | | | |
| 3. Testing and inspection | | | | | 154 | 154 | 154 | 154 | 154 | 266 | 266 | 266 | 154 | 112 | | | | | | | |
| 4. Shipping | | | | | | | | | | | | 154 | 154 | 308 | 308 | | | | | | |
| 5. Procurement | | | | 154 | 154 | 154 | 154 | 77 | 77 | | | | | | | | | | | | |
| **III. Equipment construction** | | | | | | | | | | | | | | | | | | | | | |
| 1. Program management | | | 20 | 20 | 30 | 30 | 30 | 30 | 30 | 30 | 20 | 20 | 20 | 20 | 20 | | | | | | |
| 2. Engineering | | | 154 | 154 | 154 | 154 | 154 | 154 | 154 | 154 | 77 | | | | | | | | | | |
| 3. Procurement | | | | 154 | 154 | 308 | 308 | 308 | 308 | 308 | 308 | 308 | 308 | 308 | 308 | | | | | | |
| 4. Shipping | | | | | | | | | | | | | 47 | 100 | 154 | | | | | | |
| 5. Fabrication | | | | | 308 | 462 | 616 | 770 | 924 | 1078 | 1386 | 1848 | 1848 | 1848 | | | | | | | |
| **IV. Training** | | | | | | | | | | | | | | | | | | | | | |
| 1. Program management | | | | | | | | | | | 20 | 20 | 20 | 20 | | | | | | | |
| 2. Engineering | | | | | | | | | | | 60 | 60 | 60 | 60 | | | | | | | |
| 3. Training | | | | | | | | | | | 154 | 154 | 154 | 154 | | | | | | | |
| **V. Plant start-up** | | | | | | | | | | | | | | | | | | | | | |
| 1. Program management | | | | | | | | | | | | | | 20 | 5 | 5 | 5 | 5 | 5 | 20 | |
| 2. Field engineering | | | | | | | | | | | | | | 154 | 154 | 154 | 308 | 154 | 154 | 154 | |

**II. Other costs** (Sept 1978)

| | |
|---|---|
| Purchased goods | $121,981 |
| Freight | 1,988 |
| Other | 3,049 |
| Overseas packing | 6,242 |
| | $133,260 |

**III. Other costs**

| | |
|---|---|
| Purchased mat'ls | 39,527 |
| Subcontracts | 28,082 |
| Overseas packing | 3,520 |
| Freight | 2,598 |
| | $73,727 |

**IV. Other costs**

| | |
|---|---|
| Supplies | $980 |

| | Rate | Oh |
|---|---|---|
| Program management | 11.00 | 120% |
| Engineering | 10.00 | 120% |
| Testing | 8.00 | 117% |
| Procurement | 8.00 | 110% ⎫ |
| Shipping | 5.70 | 100% ⎬ 125% |
| Fabrication | 10.00 | 125% |
| Training | 10.50 | 120% ⎫ 12% |
| Field engineering | 9.00 | 80% ⎭ |

| | |
|---|---|
| Indirect costs | 14% |
| Corporate costs | 1% |
| Profit | 12% |

| | |
|---|---|
| Raw materials escalation | 10% |
| Demanding ratio | 10% |
| Termination liability | 0% |
| Salary increases | 6% |

Note: One-man month = 154 Manhours

Source: H. Kerzner. *Project Management: A System Approach to Planning, Scheduling, and Controlling*. Van Nostrand Reinhold Company, New York, 1989. Used with permission.

refinements added to increase the accuracy and identification of the various cost elements comprising the estimate [22].

An exponent estimating technique is sometimes used to prepare preliminary cost estimates. The costs of similar plants or pieces of equipment of different sizes vary with the size raised to some power. This relationship can be expressed mathematically as follows:

$$\frac{C_2}{C_1} = \left(\frac{Q_2}{Q_1}\right)^X \qquad (14.1)$$

Here, $C_2$ is the desired cost of capacity $Q_2$ and $C_1$ is the known cost of capacity $Q_1$. A frequent value for $X$ is 0.6, so this relationship is often referred to the *six-tenths (0.6) factor rule* [14, 77]. The exponent $X$ can be determined by plotting actual historical costs for the equipment or plant as the ordinate on log-log paper and the equipment or plant size as the abscissa. The slope of the resulting line through the data will be the cost-capacity factor.

> **Example 14.3.** An ethylene plant with a capacity of 100,000 t/year costs 16 M$. Estimate the cost of a 200,000 t/year plant.
>
> *Solution.* Using the 0.6 factor rule for Eq. (14.1), we obtain
>
> $$C_2 = 16\left(\frac{200,000}{100,000}\right)^{0.6} = 24.2 \text{ M\$}$$

A standard unit estimating technique is frequently used in preparing detailed cost estimates. There are various forms for these standard estimating units, such as cost units, work-hour units, crew size units, construction equipment-use units, manpower in units per day, and manpower per units of operation. Standard unit data are available as a list, by industry, of standard units required to accomplish a specified task. Such data are particularly useful in systematizing procedures to give consistent estimates. The following equation explains the method:

Estimated value = (standard estimating unit) (quantity) (adjustment)     (14.2)

The adjustment in this equation takes care of variations in the units with such factors as location, time frames, and productivity.

## 14.11   COMPUTER-AIDED ESTIMATING

The application of computer technology to the preparation of cost estimates has become commonplace. Considerable investment has been made by firms in developing computer hardware and software for proprietary use.

With the advent of affordable microcomputers and sophisticated software, the world of estimating is being continually reduced from that of an art form to more of an applied science. All aspects of estimating are being computerized, with many new applications being marketed [13, 19, 63, 76, 79, 81]. The following list presents only a small example of the possibilities where the microcomputer can help in the preparation of definitive estimates:

Quantity Takeoffs using digitizers to "read" the architect/engineer's plans

CAD (computer-aided design) programs, which can automatically generate quantity takeoffs and measurements

CAD programs, which can integrate all the various trades, work to check for completeness, and easily determine feasibility costs of various options

Historical cost database generation and retrieval

Risk analysis

The spreadsheet with which one can compile and generate information, the extent of which is limited only by one's imagination.

The microcomputer is a cost-effective tool if used properly. One should not rely solely on the information generated by a computer since it is totally dependent on the accuracy of the data from which it is derived. Learn to spot check regularly and validate computer results by hand calculators.

## 14.12   COST INDEXES

A *cost index* is a dimensionless number used to adjust the cost of an item from one time period to another. The adjustment is necessary because of the changing value of money with time.

To convert costs from one time period to another, we use the following relationship:

$$C_2 = C_1 \frac{\text{index}_2}{\text{index}_1} \tag{14.3}$$

**Example 14.4.** A project was built in 1986 at a cost of 750 k$. The index for this project in 1986 was 238. What would the project cost in 1989 if the index in 1989 is 320?

*Solution.* From Eq. (14.3),

$$750 \frac{320}{238} = 1008 \text{ k\$}$$

There are numerous cost indexes covering many types of construction and many related disciplines. Some of the more common indexes are the *Engineering News-Record (ENR) Building Cost Index, Engineering News-Record Construction Cost Index, Chemical Engineering Plant Cost Index, Marshall and Swift Equipment Cost Index,* and the *Nelson-Farrar Refinery Cost Index.* The *Chemical Engineering Plant Cost Index* and the *Marshall and Swift Equipment Cost Index* are published regularly in *Chemical Engineering* magazine. The remaining indexes and many others are published in *Engineering News-Record* on a regular basis.

It is important when using indexes to be certain that the proper index is applied. Indexes are designed to measure relative cost changes for various items and generally will give erroneous results if applied to something else. The bases of the aforementioned indexes are:

1. *Engineering News-Record (ENR)* indexes for building and construction. These indexes have a base value of 100 in the year 1913. The indexes are based on costs of structural steel, portland cement, lumber, common labor (the construction index), and skilled labor (the building index) [18].
2. *Chemical Engineering* plant cost index. This index covers process-plant construction costs and has a 1957–1959 base [4, 39].
3. *Marshall and Swift Equipment Cost Index.* This index covers the installed cost of process equipment and has a 1926 base. Specialized subindexes are compiled for 47 industries [18].
4. *Nelson-Farrar Refinery Cost Index.* Formerly called the *Nelson Refinery Construction Cost Index,* this index projects the present-day cost of duplicating a 1946 refinery [18].

All indexes are calculated from a base year and vary depending on the index and when it began. To convert all indexes to a common base year, one may use the following relationship [18]:

$$\text{Index value on new base} = \frac{\text{index value (old base) to be converted}}{\text{index value (old base) for new base year}}(100)$$

$$(14.4)$$

**Example 14.5.** Convert the ENR Building Cost Index of March 1989 based on 1913 = 100 to a base year of 1967. The index for March 1989 is 2616.

*Solution.* The index value for the new base year (1967 average value) on the old base (1913 = 100) is 676. From Eq. (14.4),

$$\frac{2616}{676}(100) = 387, \text{ the ENR Building Cost Index for March 1989 (1967 = 100)}$$

The Bureau of Labor Statistics of the Department of Labor publishes much data on price changes in all industries. Some of these data, such as the Producers Price Index, formerly the Wholesale Price Index, are components of many cost indexes, and others are useful for the construction of highly specialized indexes.

The ENR indexes [18] are usually based on 1913 = 100, although the ENR and Nelson-Farrar indexes have been converted to a 1967 base because the values have become so large over the years. The components of the ENR indexes are shown in Table 14.6.

**TABLE 14.6**
**Components of the ENR indexes**

| Component | Construction Index | Building Index |
|---|---|---|
| Structural steel shapes, base mill price, cwt | 25 | 25 |
| Portland cement 20-cities average, bulk, cwt | 22.56 | 22.56 |
| 2 × 4 lumber, 20-city average, thousand board-ft (MBF) | 1.088 | 1.088 |
| Common labor, 20-city average, h | 200 | |
| Skilled labor, 20-city average, h | | 68.38 |

It is apparent from this breakdown that the only difference between the two indexes is the use of common labor, 200 h, in the Construction Index, and skilled labor, 68.38 h, in the Building Index.

The Marshall and Swift Equipment Cost Index is the average of installed costs in 47 industries and is obtained by a complex procedure involving equipment appraisals, modifying factors, and a judgment about current economic conditions [67]. The Nelson-Farrar Refinery Construction Index is based on a 40 percent material and 60 percent labor distribution, as explained in the literature [51]. The Chemical Engineering Plant Cost Index is a complex index consisting of four major components [4, 39], as shown in Table 14.7. The equipment component is made up of seven subcomponents, as shown in Table 14.8. The Chemical Engineering Index allows for a productivity factor of 1.75 percent per year compounded. Prior to 1982 a productivity factor of 2.5 was used [39]. The various components are calculated from Producers Price Indexes published by the Bureau of Labor Statistics [73].

In addition to indexes published in the literature, there are special indexes published by econometric companies. These are client-oriented indexes and are available on an annual-fee basis. Some of the more notable are:

Stanford Research Institute (SRI) Index

Data Resources Inc. (DRI) Index

Chase Econometric Inflation Index

**TABLE 14.7**
**Components in the Chemical Engineering Plant Cost Index**

| Component | Percent |
|---|---|
| Equipment machinery and supports | 61 |
| Erection and installation labor | 22 |
| Buildings, material, and labor | 7 |
| Engineering and supervision manpower | 10 |
| Total | 100 |

**TABLE 14.8**
**Equipment subcomponents in the Chemical
Engineering Plant Cost Index**

| Component | Percent |
|---|---|
| Fabricated equipment | 37 |
| Process machinery | 14 |
| Process instruments | 7 |
| Pumps and compressors | 7 |
| Electrical equipment | 5 |
| Pipes, valves, and fittings | 20 |
| Structural supports, insulation, and paint | 10 |
| | 100 |

When all cost indexes are converted to a common base year and plotted, the curves are very irregular. It is imperative that cost indexes be used with caution and an index selected that most nearly suits the project being adjusted. One of the major pitfalls of indexes is the tendency to use them over too long a period. A rule of thumb is to use indexes only to adjust costs for a 4- to 5-year time span. Using them over longer time spans greatly reduces the accuracy of the results.

Cost indexes can reflect only average costs and must be used with judgment for the specific purpose intended. The results can vary with the index selected, as shown in Example 14.6.

**Example 14.6.** An air-conditioning system was purchased in 1983 for $160,000. Estimate its cost in 1988 using both the Marshall and Swift Index and the Chemical Engineering Plant Cost Index.

*Solution.* Marshall and Swift Index:

$$160,000 \frac{857 \ (1988 \ \text{index})}{761 \ (1983 \ \text{index})} = \$180,000$$

Chemical Engineering Index:

$$160,000 \frac{345 \ (1988 \ \text{index})}{317 \ (1983 \ \text{index})} = \$174,000$$

The error for the 5-year period using two different indexes is approximately 3 percent. Using indexes over a larger time span would sharply increase the margin of error.

The American construction industry is changing from a strictly domestic business to an international one. With this change in emphasis, more indexes are being established by foreign countries. This is more specifically true for the Chemical Process Industry (CPI). One of the broadest indexes has been developed in Canada by Statistics Canada [66]. It is updated quarterly for both residential and nonresidential construction. The Chemical Plant Index is patterned after the Chemical Engineering Plant Cost Index, as shown here:

| | | |
|---|---|---|
| Annual plant cost | | 100% |
| Machinery and equipment | | 54.2 |
| Fabrication and equipment | 23.6 | |
| Process machinery | 26.7 | |
| Pumps and compressors | 7.5 | |
| Pipe, valves, and fittings | 10.2 | |
| Process instruments | 6.2 | |
| Electrical equipment | 9.6 | |
| Other utilities equipment | 7.8 | |
| Structural supports | 8.4 | |
| | 100.0 | |
| Field erection | | 16.6 |
| Buildings | | 13.5 |
| Engineering and administration | | 15.7 |
| Total annual plant cost | | 100.0 |

The base year is 1971 = 100, and no credit is taken for labor productivity.

Indexes for other countries are generally compiled and published locally. The *Cost Engineer* [10, 69], a publication of the U.K. Association of Cost Engineers, gives a chemical plant cost index for the United Kingdom. *Process Economics International* (PEI) [59] publishes plant cost indexes for 36 countries around the world. All the PEI indexes have a base value of 100 in 1980; however, the indexes cannot be applied from nation to nation. Similar values of indexes in two countries do not imply that costs are the same but merely that total cost increases since 1980 were of similar percentage magnitude as compared to 1980.

## 14.13  PLANNING ESTIMATES

Planning the preparation of cost estimates helps to establish the purpose of the estimate, the time available for its completion, the type of estimate required, the estimating technique to use, and the range of accuracy desired. When these elements are determined, a time schedule can be developed to help decide whether a preestimate survey is needed and whether manual or computer methods are to be used. It is important to have a written description of the proposed facilities, the bases for the costs, and any underlying assumptions. The final estimate must be reviewed by management to check the results against similar or past project experiences.

The planning of capital estimates provides data for gauging the cost of estimate preparation for future projects. Because the cost of preparing an estimate of a given type and accuracy can vary over a large range (see Table 14.3), standard cost factors are not usually applicable. Currently, contractors have been giving considerable attention to lowering estimate preparation costs through the use of advanced estimating techniques and computer applications. Figure 14.6 shows an example of an estimate preparation schedule or plan.

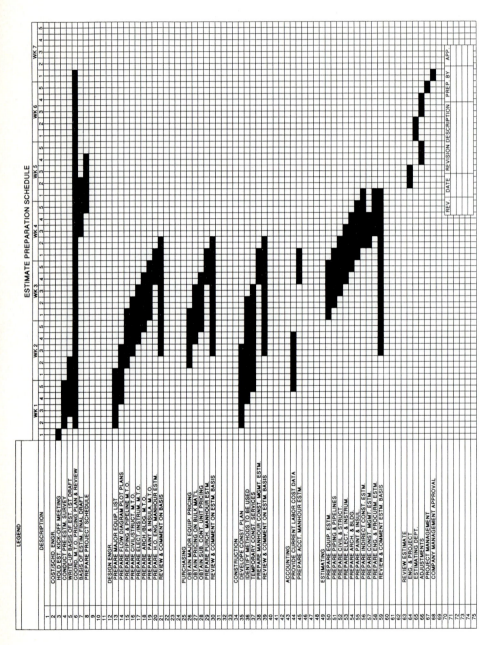

**Figure 14.6**
Example of an estimate preparation plan.

388

## 14.14   DETAILED ESTIMATES AGAINST PRELIMINARY ESTIMATES

As pointed out in Sec. 14.4, detailed estimates are expensive and time-consuming to prepare, although they are the most accurate. The checklist in Table 14.9 gives some idea of the scope of items involved in a detailed estimate. Each builder or contractor has lists of this type to help minimize omissions. Because of the expense, time, and lack of information in the early project stages, many more preliminary estimates than detailed estimates are prepared. Because not all projects pass through the screening procedure successfully, much time, effort, and expense are saved by making preliminary estimates. The next several sections will discuss and provide examples of the common preliminary estimate types. Subsequent sections discuss detailed estimates.

**TABLE 14.9**
**Checklist of items for fixed capital cost estimating**

Land
  Surveys
  Fees
  Property cost
Site development
  Site clearing
  Grading
  Roads, access and on-site
  Walkways
  Railroads
  Fence
  Parking areas
  Other paved areas
  Wharves and piers
  Recreational facilities
  Landscaping
Process buildings
  (List as required) Include in each as required
    substructure, superstructures, platforms,
    supports, stairways, ladders, access ways,
    cranes, monorails, hoists, elevators
Auxiliary buildings
  Administration and office
  Medical or dispensary
  Cafeteria
  Garage
  Product warehouse(s)
  Parts or stores warehouse
  Maintenance shops — electric, piping, sheet
    metal, machine, welding, carpenters,
    instrument
  Guard and safety
  Hose houses
  Change houses
  Smoking stations (in hazardous plants)

Process equipment
  (List carefully from checked flow sheets)
Nonprocess equipment
  Office furniture and equipment
  Cafeteria equipment
  Safety and medical equipment
  Shop equipment
  Automotive heavy maintenance and yard
    material-handling equipment
  Laboratory equipment
  Lockers and locker-room benches
  Garage equipment
  Shelves, bins, pallets, hand trucks
  Housekeeping equipment
  Fire extinguishers, hoses, fire engines
Process appurtenances
  Piping — carbon steel, alloy, cast iron,
    lead-lined, aluminum, copper, asbestos-
    cement, ceramic, plastic, rubber, reinforced
    concrete
  Pipe hangers, fittings, valves
  Insulation — piping, equipment
  Instruments
  Instrument panels
  Electrical — panels, switches, motors, conduit,
    wire, fittings, feeders, grounding,
    instrument and control wiring
Utilities
  Boiler plant
  Incinerator
  Ash disposal
  Boiler feed-water treatment
  Electric generation
  Electrical substations
  Refrigeration plant

**TABLE 14.9** *continued*

Auxiliary buildings (*continued*)
  Personnel building
  Shipping office and platforms
  Research laboratory
  Control laboratories
Building services
  Plumbing
  Heating
  Ventilation
  Dust collection
  Air conditioning
  Sprinkler systems
  Elevators, escalators
  Building lighting
  Telephones
  Fire alarm
  Paging
  Intercommunication systems
  Painting
Yard distribution and facilities (outside battery
  limits)
  Process pipe lines—steam, condensate, water,
    gas, fuel oil, air, fire, instrument, and
    and electric lines
  Raw material and finished-product handling
    equipment—elevators, hoists, conveyors,
    airveyors, cranes
  Raw material and finished-product
    storage—tanks, spheres, drums, bins, silos
  Fuel receiving, blending, and storage
  Product loading stations
  Track and truck scales
Miscellaneous
  Demolition and alteration work
  Catalysts
  Chemicals (initial charge only)
  Spare parts and noninstalled equipment spares
  Surplus equipment, supplies and equipment
    allowance
  Equipment rentals (for construction)
  Premium time (for construction)
  Inflation cost allowance
  Freight charges
  Taxes and insurance
  Duties
  Allowance for modifications and extra
    construction work during startup
Engineering costs
  Administrative

Utilities (*continued*)
  Air plant
  Wells
  River intake
  Primary water treatment—filtration,
    coagulation, aeration
  Secondary water treatment—deionization,
    demineralization, pH and hardness control
  Cooling towers
  Water storage
  Effluent outfall
  Process-waste sewers
  Process-waste pumping stations
  Sanitary-waste sewers
  Sanitary-waste pumping stations
  Impounders, collection basins
  Waste treatment, including gases
  Storm sewers
  Process, project, and general engineering
  Drafting
  Cost engineering
  Procurement, expediting, and inspection
  Travel and living expense
  Reproductions
  Communications
  Scale model
  Outside architect and engineering fees
Construction expense
  Construction, operation, and maintenance
    of temporary sheds, offices, roads,
    parking lots, railroads, electrical, piping,
    communication, and fencing
  Construction tools and equipment
  Warehouse personnel and expense
  Construction supervision
  Accounting and timekeeping
  Purchasing, expediting, and traffic
  Safety and medical
  Guards and watchmen
  Travel and transportation allowance for craft
    labor
  Fringe benefits
  Housekeeping
  Weather protection
  Permits, special licenses, field tests
  Rental of off-site space
  Contractor's home office expense and fees
  Taxes and insurance, interest

*Source:* R. H. Perry and C. H. Chilton: *Chemical Engineers' Handbook*, 5th ed., McGraw-Hill Book Company, New York, 1973, p. 25–13.

## 14.15 EQUIPMENT COSTS FOR PRELIMINARY ESTIMATES

Preliminary estimates can range from very quick, inaccurate estimates to reasonably detailed, accurate estimates, as shown in Fig. 14.1. In many preliminary estimating methods the term *purchased equipment* or *delivered equipment* is used. These terms refer to the cost of the uninstalled equipment delivered to the construction site.

The most accurate method of determining equipment costs is to obtain firm bids from fabricators or suppliers [61]. Often these same suppliers can provide quick equipment prices orally which are close to a bid price, although not to be taken as such. Next in accuracy are equipment costs from the owner's files of past purchases, adjusted by cost indexes to current costs. Last in accuracy are the equipment costs published in the literature. One is never certain, however, about what is included in published costs.

## 14.16 EQUIPMENT COSTS BY SCALING

The most common method for obtaining delivered equipment costs for equipment of a desired size is to scale the cost from data of another size, such as the exponential estimating technique discussed in Sec. 14.10. Table 14.10 gives exponents for various types of equipment [61]. An exponent of 0.6 should be used only in the absence of more specific information [45, 46]. Table 14.11 gives the magnitude of the error by using the 0.6 factor in place of various actual cost-capacity factors.

## 14.17 TURNOVER RATIOS

Probably the least accurate but most rapid method of estimating plant costs is the use of the turnover ratio, which is defined as follows:

$$\text{Turnover ratio} = \frac{\text{gross annual sales}}{\text{fixed capital investment}} \qquad (14.5)$$

Gross annual sales is the product of unit selling price and annual production rate.

Turnover ratios vary from a low of about 0.2 to a high of 8.0 [70]. A rough figure for the chemical industry would be in the range 1.00 to 1.25. Values less than 1.00 are generally found in large-volume, capital-intensive industries which use raw basic materials, such as steel making or power generation [70]. The reciprocal of the turnover ratio, or capital ratio, is sometimes used [38].

**Example 14.7.** Estimate the capital investment for a plant to produce 12 Mlb/year of chemical *Y*, which sells for $0.40/lb. M is mega-.

**Solution.** Assume a turnover ratio of 1.00 as typical of the chemical industry. From Eq. (14.5),

$$\text{Fixed capital investment} = \frac{12 \times 10^6 \times 0.40}{1.00} = \$4.8 \times 10^6$$

**TABLE 14.10**
**Exponent values for typical process equipment**

| Equipment | Size range | Exponent |
|---|---|---|
| Blender, double cone rotary, c.s. | 50–250 ft$^3$ | 0.49 |
| Blower, centrifugal | $10^3$–$10^4$ ft$^3$/min | 0.59 |
| Centrifuge, solid bowl, c.s. | 10–$10^2$ hp drive | 0.67 |
| Crystallizer, vacuum batch, c.s. | 500–7000 ft$^3$ | 0.37 |
| Compressor, reciprocating, air-cooled, two-stage, 150 psi discharge | 10–400 ft$^3$/min | 0.69 |
| Compressor, rotary, single-stage, sliding vane, 150 psi discharge | $10^2$–$10^3$ ft$^3$/min | 0.79 |
| Dryer, drum, single vacuum | 10–$10^2$ ft$^2$ | 0.76 |
| Dryer, drum, single atmospheric | 10–$10^2$ ft$^2$ | 0.40 |
| Evaporator (installed), horizontal tank | $10^2$–$10^4$ ft$^2$ | 0.54 |
| Fan, centrifugal | $10^3$–$10^4$ ft$^3$/min | 0.44 |
| Fan, centrifugal | $2 \times 10^4$–$7 \times 10^4$ ft$^3$/min | 1.17 |
| Heat exchanger, shell and tube, floating head, c.s. | 100–400 ft$^2$ | 0.60 |
| Heat exchanger, shell and tube, fixed sheet, c.s. | 100–400 ft$^2$ | 0.44 |
| Kettle, cast iron, jacketed | 250–800 gal | 0.27 |
| Kettle, glass lined, jacketed | 200–800 gal | 0.31 |
| Motor, squirrel cage, induction, 440 volts, explosion proof | 5–20 hp | 0.69 |
| Motor, squirrel cage, induction, 440 volts, explosion proof | 20–200 hp | 0.99 |
| Pump, reciprocating, horizontal, cast iron (includes motor) | 2–100 gpm | 0.34 |
| Pump, centrifugal, horizontal, cast steel (includes motor) | $10^4$–$10^5$ gpm $\times$ psi | 0.33 |
| Reactor, glass lined, jacketed (without drive) | 50–600 gal | 0.54 |
| Reactor, S.S., 300 psi | $10^2$–$10^3$ gal | 0.56 |
| Separator, centrifugal, c.s. | 50–250 ft$^3$ | 0.49 |
| Tank, flat head, c.s. | $10^2$–$10^4$ gal | 0.57 |
| Tank, c.s., glass lined | $10^2$–$10^3$ gal | 0.49 |
| Tower, c.s. | $10^3$–$2 \times 10^6$ lb | 0.62 |
| Tray, bubble cup, c.s. | 3–10 ft diameter | 1.20 |
| Tray, sieve, c.s. | 3–10 ft diameter | 0.86 |

*Source:* M. S. Peters and K. D. Timmerhaus: *Plant Design and Economics for Chemical Engineers,* 3d ed., McGraw-Hill Book Company, New York, 1980, p. 167.

**TABLE 14.11**
**Error introduced by use of the 0.6 factor**

| Ratio of $Q_2/Q_1$ | Actual cost-capacity factor | | | | | | | | |
|---|---|---|---|---|---|---|---|---|---|
| | 0.2 | 0.3 | 0.4 | 0.5 | 0.6 | 0.7 | 0.8 | 0.9 | 1.0 |
| | Percent error | | | | | | | | |
| 5:1 | +89 | +61 | +37 | +17 | 0 | −16 | −28 | −39 | −48 |
| 10:1 | +150 | +100 | +59 | +26 | 0 | −21 | −37 | −50 | −60 |

## 14.18   INVESTMENT COST PER UNIT OF CAPACITY

It is possible to make an order-of-magnitude estimate from data expressed as the fixed capital investment per unit of annual production capacity. These costs are for average capacities, and hence there will be inaccuracies since the plant capacities deviate from average. Table 14.12 shows some typical costs per unit data for chemical plants and refinery units. One should note that the cost-capacity factor, or six-tenths rule, applies to total plants as well as to items of equipment [14]. However, some authorities recommend a seven-tenths factor for entire plants.

**TABLE 14.12**
**Typical unit capital cost data for process plants, December 1988**

| Product or process | Units of production | Typical plant size | Fixed capital investment, $ per unit | Cost-capacity factor |
|---|---|---|---|---|
| Chemical plants | 1000 ton/year | | Per ton/year | |
| Acetic acid | | 10 | 600 | 0.68 |
| Acetone | | 100 | 300 | 0.45 |
| Ammonia | | 100 | 225 | 0.53 |
| Ammonium nitrate | | 100 | 45 | 0.65 |
| Butanol | | 50 | 750 | 0.40 |
| Chlorine | | 50 | 520 | 0.45 |
| Ethylene | | 50 | 245 | 0.83 |
| Ethylene oxide | | 50 | 940 | 0.78 |
| Formaldehyde (37%) | | 10 | 1500 | 0.55 |
| Glycol | | 5 | 2740 | 0.75 |
| Methanol | | 60 | 200 | 0.60 |
| Nitric acid | | 100 | 65 | 0.60 |
| Polyethylene (high density) | | 5 | 3000 | 0.65 |
| Propylene | | 10 | 300 | 0.70 |
| Sulfuric acid | | 100 | 30 | 0.65 |
| Urea | | 60 | 120 | 0.70 |
| Refinery units | 1000 bbl/day | | Per bbl/day | |
| Alkylation | | 10 | 1800 | 0.60 |
| Coking (delayed) | | 10 | 2425 | 0.38 |
| Coking (fluid) | | 10 | 1500 | 0.42 |
| Cracking (fluid) | | 10 | 1500 | 0.70 |
| Cracking | | 10 | 450 | 0.70 |
| Distillation (atm.) | | 100 | 300 | 0.90 |
| Distillation (vac.) | | 100 | 180 | 0.70 |
| Hydrotreating | | 10 | 300 | 0.65 |
| Reforming | | 10 | 2740 | 0.60 |
| Polymerization | | 10 | 450 | 0.58 |

*Source:* Based on M. S. Peters and K. D. Timmerhaus: *Plant Design and Economics for Chemical Engineers,* 3d ed., McGraw-Hill Book Company, New York, 1980, pp. 184–185.

**Example 14.8.** Estimate the capital cost in December 1988 for a plant to produce 80 Mlb/year of butanol. M is mega-.

*Solution.* Plant size is

$$\frac{80,000,000}{2000} = 40,000 \text{ t/year}$$

From Table 14.12,

$$\text{Capital cost} = 40,000 \times 750 = 30 \text{ M\$}$$

## 14.19   POWER FACTORS APPLIED TO PLANT COSTS

The use of cost-capacity factors, as in Eq. (14.1), along with cost indexes to adjust for inflation, is probably the easiest way of estimating plant costs [5]. There are numerous literature references for plant costs of various types and for various capacities. Adjustments for the desired plant size can be made with cost-capacity factors of the type shown in Table 14.12.

**Example 14.9.** Estimate the capital cost as it would have been in 1987 for the butanol plant in Example 14.8 if the capacity were 20 Mlb/year instead of 80 Mlb/year. Use 349 for the Chemical Engineering Plant Construction Index for 1988 and 324 for 1987. M is mega-.

*Solution.* From Eq. (14.1) and Eq. (14.3),

$$\text{Capital cost} = 30\left(\frac{20}{80}\right)^{0.4}\left(\frac{324}{349}\right) = 16.0 \text{ M\$}$$

The method is improved by separating out the direct and indirect costs and applying the cost-capacity relationship to just the direct-cost portion of the fixed capital investment. The fixed cost thus becomes

$$C_F = f[D(R)^X + I] \tag{14.6}$$

where $C_F$ is the fixed cost, $f$ is the ratio of cost indexes, $D$ is the direct plant cost, $R$ is the ratio of plant sizes, $X$ is the cost-capacity factor, and $I$ is the indirect-cost portion of the fixed capital cost [61].

## 14.20   FACTOR METHODS OF ESTIMATING

The factor method as a technique was mentioned in Sec. 14.10. The fixed capital investment, or in some cases the total capital investment, is estimated by multiplying the delivered or purchased equipment cost by a factor. The use of a single constant factor cannot possibly allow for all the variations among projects, geographical locations, or site-specific conditions that a detailed estimate would consider.

The earliest of these methods was the Lang factor method [36]. Improvements and refinements have been made to increase the accuracy without complicating the procedure unduly. The improvements introduce flexible factors and allow the exercise of the estimator's judgment in place of an inflexible constant factor. A discussion of the more frequently used methods follows.

## 14.21  THE LANG FACTOR METHOD

This technique, originally proposed by Lang, has been used in the past for quick order-of-magnitude cost estimates. Lang suggested multiplying the delivered cost of equipment by the following factors to obtain total process plant costs [36]:

3.10 for solid process plants
3.63 for solid-fluid plants
4.74 for fluid process plants

It is obvious that such factors do not allow the estimator to exercise any judgment.

> **Example 14.10.** Estimate the total process plant cost for a refinery addition for which the delivered cost is 8 M$. The refinery is a fluids-processing plant.
>
> *Solution.* Using the appropriate Lang factor, we obtain
>
> $$\text{Total plant cost} = 8 \times 4.74 = 38 \text{ M\$}$$

## 14.22  HAND FACTORS

Greater accuracy can be achieved by using factors based on different types of equipment. Hand [27, 28] recommends the following factors, applied to delivered equipment costs, to obtain the battery-limits cost:

4      For fractionating columns, pressure vessels, pumps, and instruments
3.5    For heat exchangers
2.5    For compressors
2      For fired heaters

These factors are very close to those in a more complete list developed later by Wroth [79] and shown in Table 14.13.

**TABLE 14.13**
**Process-plant cost ratio from individual equipment**

| Equipment | Factor* |
|---|---|
| Blender | 2.0 |
| Blowers and fans (including motor) | 2.5 |
| Centrifuges (process) | 2.0 |
| Compressors: | |
|    Centrifugals, motor-driven (less motor) | 2.0 |
|      Steam turbine (including turbine) | 2.0 |
|    Reciprocating, steam and gas | 2.3 |
|      Motor-driven (less motor) | 2.3 |
| Ejectors (vacuum units) | 2.5 |
| Furnaces (package units) | 2.0 |
| Heat exchangers | 4.8 |
| Instruments | 4.1 |
| Motors, electric | 8.5 |
| Pumps: | |
|    Centrifugal, motor-driven (less motor) | 7.0 |
|      Steam turbine (including turbine) | 6.5 |
|    Positive displacement (less motor) | 5.0 |
| Reactors — factor as approximate equivalent type of equipment | |
| Refrigeration (package unit) | 2.5 |
| Tanks: | |
|    Process | 4.1 |
|    Storage | 3.5 |
|    Fabricated and field-erected (50,000+ gal) | 2.0 |
| Towers (columns) | 4.0 |

*Source:* W. F. Wroth, "Factors in Cost Estimation," *Chemical Engineering,* McGraw-Hill, Inc., vol. 67, October 1960, p. 204.

*Multiply the purchase cost by a factor to obtain the installed cost, including the cost of site development, buildings, electrical installations, carpentry, painting, contractor's fee and rentals, foundations, structures, piping, installation, engineering, overhead, and supervision.

**Example 14.11.** Estimate the battery-limits cost of the refinery in Example 14.10 if the equipment distribution and costs are as follows:

| Equipment | Cost, k$ |
|---|---|
| Blowers and fans | 117 |
| Compressors, reciprocating | 584 |
| Furnaces | 1168 |
| Heat exchangers | 934 |
| Instruments | 584 |
| Motors, electric | 700 |
| Pumps, centrifugal, motor driven | 234 |
| Tanks: | |
|    Process | 292 |
|    Field erected | 1051 |
| Towers | 2336 |
| | 8000 |

Using the individual cost ratios from Table 14.13, we get

| Equipment | Cost, k$ | |
|---|---|---|
| Blowers and fans | $117 \times 2.5 =$ | 290 |
| Compressors, reciprocating | $584 \times 2.3 =$ | 1,340 |
| Furnaces | $1,168 \times 2.0 =$ | 2,340 |
| Heat exchangers | $934 \times 4.8 =$ | 4,480 |
| Instruments | $584 \times 4.1 =$ | 2,390 |
| Motors, electric | $700 \times 8.5 =$ | 5,950 |
| Pumps, centrifugal, motor driven | $234 \times 7.0 =$ | 1,640 |
| Tanks: | | |
|    Process | $292 \times 4.1 =$ | 1,200 |
|    Field erected | $1,051 \times 2.0 =$ | 2,100 |
| Towers | $2,336 \times 4.0 =$ | 9,340 |
| | | 31,070 |

The cost is approximately 31 M$, not too different from the 38 M$ estimated by the Lang factor method.

An improvement in equipment factors has been proposed by Cran [15]. The breakdown in types of equipment is more extensive, a distinction is made in factors for materials of construction, and the instrument and indirect costs are factored separately. In addition, Cran has introduced accuracy determinations into his estimates. His cost relationship is

$$\text{Total plant cost} = \left( \sum EF_D + IF_I \right)(1 + F_0) \qquad (14.7)$$

where $E$ is the delivered equipment cost, $F_D$ is a direct-cost factor that varies with the type of equipment and its material of construction, $I$ is the sum of the costs of all instruments, $F_I$ is the direct-cost factor for instruments, and $F_0$ is the indirect-cost factor. By this variation on Lang factors, Cran is able to increase his accuracy considerably, resulting in an error of only 6 percent [15].

## 14.23 CHILTON METHOD

Using this method, the estimator must exercise some judgment because the factors are given as ranges. In contrast to the Lang and Hand methods, this method begins with equipment installation costs. As pointed out in Sec. 14.10, these costs are about 50 percent of the delivered equipment costs. Aries and Newton [3] give 43 percent, Woods gives a range of 40 to 120 percent [78], and Peters and Timmerhaus give an average of 43.7 percent [61]. This factor changes, of course, as the relative costs of installation labor and materials vary with changes in the economy. A figure of 43 percent is common, however. Table 14.14 is a summary of the factors used in the Chilton method. Again, these factors are designed to cover the items in a detailed es-

**TABLE 14.14**
## Chilton factors for estimating total plant costs*

| Item no. | Item | Multiplying factor | Operating on item no. |
|---|---|---|---|
| 1. | Delivered equipment cost | 1.00 | 1 |
| 2. | Installed equipment cost | 1.40–2.20 | 1 |
| 3. | Process piping | | |
| | Type of plant — solid | 0.07–0.10 | 2 |
| | solid/fluid | 0.10–0.30 | 2 |
| | fluid | 0.30–0.60 | 2 |
| 4. | Instrumentation | | |
| | Amount — little or none | 0.02–0.05 | 2 |
| | some | 0.05–0.10 | 2 |
| | extensive | 0.10–0.15 | 2 |
| 5. | Buildings and site development | | |
| | Type of plant — existing | 0.00 | 2 |
| | outdoor | 0.05–0.20 | 2 |
| | outdoor/indoor | 0.20–0.60 | 2 |
| | indoor | 0.60–1.00 | 2 |
| 6. | Auxiliaries (power, steam, water) | | |
| | Extent — none | 0.00 | 2 |
| | minor addition | 0.00–0.05 | 2 |
| | major addition | 0.05–0.25 | 2 |
| | new facilities | 0.25–1.00 | 2 |
| 7. | Outside lines | | |
| | Unit — closely integrated | 0.00–0.05 | 2 |
| | separated | 0.05–0.15 | 2 |
| | scattered | 0.15–0.25 | 2 |
| 8. | Total physical plant costs ($\Sigma$ cost of items 2–7) | | |
| 9. | Engineering and construction | | |
| | Complexity — simple | 0.20–0.35 | 8 |
| | difficult | 0.35–0.50 | 8 |
| 10. | Contingency and contractor's fee | | |
| | Process — firm | 0.10–0.20 | 8 |
| | subject to change | 0.20–0.30 | 8 |
| | speculative | 0.30–0.50 | 8 |
| 11. | Size factor | | |
| | Unit — large | 0.00–0.05 | 8 |
| | small | 0.05–0.15 | 8 |
| | experimental | 0.15–0.35 | 8 |
| 12. | Total plant cost ($\Sigma$ cost of items 8–11) | | |

*Based on C. H. Chilton: "Cost Data Correlated," *Chemical Engineering*, McGraw-Hill, Inc., vol. 56, June 1949, pp. 97–106.

timate, as shown in Table 14.9, to give a total plant cost. The use of ranges for the factors should give more accurate estimates.

> **Example 14.12.** Estimate the cost of the refinery addition in Examples 14.10 and 14.11 by the Chilton method.

> *Solution.* This fluids plant is to be closely integrated with existing facilities and is to be highly instrumented with automatic controls. The refinery is located in the Gulf Coast area, where outdoor construction exists, and the addition involves well-known processes. There need be only a minor addition—a large commercial unit—to the auxiliary facilities.
> The Chilton method is tabulated in Table 14.15. As one might expect, the estimate is lower than that obtained with the Lang and Hand methods because this is a plant addition rather than a new, or "grassroots" plant.

Another method, proposed by Peters and Timmerhaus [61], is a combination of the Lang and Chilton methods. The method is based on *delivered* equipment costs but retains Lang's classification of solid, solid-fluid, and fluid processing plants. Table 14.16 shows the factors to be used in estimating major process plant additions to an existing site.

The text accompanying this table gives ranges for each item in the table so that the estimator can modify the factors to fit special project conditions. The table gives average factor values, and the end results with average values are reasonably close to those obtained using Lang factors.

Another variation on factor methods is one proposed by Rudd and Watson [65]. In this method the fixed capital cost is again calculated from the delivered equipment

**TABLE 14.15**

**Example of the use of Chilton factors for capital cost estimation of a refinery addition**

| Item no. | Item | Multiplying factor | Operating on item no. | Cost of item |
|---|---|---|---|---|
| 1. | Equipment costs, delivered | 1.0 | 1 | $\Sigma E$ |
| 2. | Installed equipment cost | 1.43 | 1 | $1.430 \Sigma E$ |
| 3. | Piping (includes insulation) | 0.40 | 2 | $0.572 \Sigma E$ |
| 4. | Instrumentation | 0.15 | 2 | $0.215 \Sigma E$ |
| 5. | Buildings and site development | 0.10 | 2 | $0.143 \Sigma E$ |
| 6. | Auxiliaries | 0.05 | 2 | $0.072 \Sigma E$ |
| 7. | Outside lines and site development | 0.05 | 2 | $0.071 \Sigma E$ |
| 8. | Total physical cost (items 2–7) | — | — | $2.503 \Sigma E$ |
| 9. | Engineering and construction | 0.25 | 8 | $0.626 \Sigma E$ |
| 10. | Contingencies and contractor's fee | 0.15 | 8 | $0.376 \Sigma E$ |
| 11. | Size factor | 0.03 | 8 | $0.075 \Sigma E$ |
| 12. | Total plant cost (items 8–11) | — | — | $3.580 \Sigma E$ |

**TABLE 14.16**
## Peters-Timmerhaus ratio factors for capital-cost estimation

| Item | Percent of delivered-equipment cost for | | |
|---|---|---|---|
| | Solid-processing plant | Solid-fluid-processing plant | Fluid-processing plant |
| *Direct costs* | | | |
| Purchased equipment delivered (including fabricated equipment and process machinery) | 100 | 100 | 100 |
| Purchased equipment installation | 45 | 39 | 47 |
| Instrumentation and controls (installed) | 9 | 13 | 18 |
| Piping (installed) | 16 | 31 | 66 |
| Electrical (installed) | 10 | 10 | 11 |
| Buildings (including services) | 25 | 29 | 18 |
| Yard improvements | 13 | 10 | 10 |
| Service facilities (installed) | 40 | 55 | 70 |
| Land (if purchase is required) | 6 | 6 | 6 |
| Total direct plant cost | 264 | 293 | 346 |
| *Indirect costs* | | | |
| Engineering and supervision | 33 | 32 | 33 |
| Construction expenses | 39 | 34 | 41 |
| Total direct and indirect plant costs | 336 | 359 | 420 |
| Contractor's fee (about 5% of direct and indirect plant costs) | 17 | 18 | 21 |
| Contingency (about 10% of direct and indirect plant costs) | 34 | 36 | 42 |
| Fixed capital investment | 387 | 413 | 483 |
| Working capital (about 15% of total capital investment) | 68 | 74 | 86 |
| Total capital investment | 455 | 487 | 569 |

*Source:* M. S. Peters and K. D. Timmerhaus: *Plant Design and Economics for Chemical Engineers,* 3d ed., McGraw-Hill Publishing Company, New York, 1980, p. 180.

cost, but the multiplying factors are all greater than 1. The method can be expressed mathematically as

$$\text{Fixed cost} = \phi_1\phi_2\phi_3\cdots\phi_n \qquad (14.8)$$

in which the $\phi$'s are the factors for the various items, such as in Table 14.14 or 14.16. Note that in the Rudd-Watson method the factors are multiplied together, whereas in the Chilton and Peters-Timmerhaus methods the factors are added together.

# 14.24  PLANT COST BY ANALYTICAL PROCEDURE

Today, with the emphasis on computers, there is interest in the use of equations, instead of tables and charts, for cost estimation. Hirsch and Glazier [30] developed the following equation, suitable for computer calculation:

$$I = E[A(1 + F_L + F_P + F_M) + B + C] \tag{14.9}$$

where $I$ = total battery-limits investment, \$

  $A$ = total purchased equipment cost on an FOB basis less the incremental cost for corrosion-resistant alloys, \$

  $B$ = installed equipment cost, \$*

  $C$ = incremental cost of alloy materials used only for their corrosion-resisting properties, \$

  $E$ = indirect-cost factor representing contractors' overhead and profit, engineering, supervision, and contingencies; $E$ is normally assumed to be 1.4

  $F_L$ = cost factor for field labor; $F_L A$ is the total cost for field labor, less supervision, and excluding the labor charges in item $B$

  $F_M$ = cost factor for miscellaneous items; $F_M A$ includes the materials cost for insulation, instruments, foundations, structural steel, buildings, wiring, painting, and the cost of freight and field supervision

  $F_P$ = cost factor for piping materials; $F_P A$ is the total cost of piping materials, including pipe, fittings, valves, hangers, and supports but excluding insulation and installation charges

The three factors $F_L$, $F_P$, and $F_M$ are not simple ratios, but are defined by the equations

$$\log F_L = 0.635 - 0.154 \log A_0 - 0.992\frac{e}{A} + 0.506\frac{f}{A} \tag{14.10}$$

$$\log F_P = -0.266 - 0.014 \log A_0 - 0.156\left(\frac{e}{A}\right) + 0.556\frac{p}{A} \tag{14.11}$$

$$F_M = 0.344 + 0.033 \log A_0 + 1.194\frac{t}{A} \tag{14.12}$$

---

*All vessels larger than 12 ft in diameter should be considered as field erected unless specifically designated as shop fabricated.

where $A_0 = A/1000$, k$

 $e$ = total heat exchanger cost, less the incremental cost of the alloy, $

 $f$ = total cost of field-fabricated vessels, less the incremental cost of the alloy, $; ordinarily all vessels larger than 12 ft in diameter are field erected

 $p$ = total pump plus the cost for the driver less the incremental cost of the alloy, $

 $t$ = total cost of the tower shells less the incremental cost of the alloy, $

These equations are easily solved, but the effort can be reduced by a nomograph prepared by Walas [75].

**Example 14.13.** Estimate the total battery-limits investment for a process plant using the estimated equipment costs shown in Table 14.17.

*Solution.* From the data given, the following factors are derived:

$$A = \$4,310,000$$

$$A_0 = \$4310$$

$$B = \$371,000$$

$$C = \$280,000 + \$100,000 + \$20,000 = \$400,000$$

$$\frac{e}{A} = \frac{\$1,200,000}{\$4,310,000} = 0.278$$

$$\frac{f}{A} = \frac{\$60,000 + \$105,000 + \$350,000}{\$4,310,000} = 0.119$$

$$\frac{p}{A} = \frac{\$140,000 + \$120,000}{\$4,310,000} = 0.060$$

$$\frac{t}{A} = \frac{\$200,000 + \$60,000}{\$4,310,000} = 0.060$$

from which Eqs. (14.10) to (14.12) give $F_L$, $F_P$, and $F_M$:

$$\log F_L = 0.635 - 0.154 \log A_0 - 0.992\frac{e}{A} + 0.506\frac{f}{A}$$

$$= 0.635 - 0.154 \log 4310 - 0.992(0.278) + 0.506(0.119) = -0.140$$

$$F_L = 0.724$$

$$\log F_P = -0.266 - 0.014 \log A_0 - 0.156\frac{e}{A} + 0.556\frac{p}{A}$$

$$= -0.266 - 0.014 \log 4310 - 0.156(0.278) + 0.556(0.060) = -0.327$$

**TABLE 14.17**
**Estimated major equipment costs for Example 14.13**

| Item | Cost calculated on carbon steel basis, k$ | Total calculated cost, including incremental alloy, k$ | Incremental cost of alloy, k$ |
|---|---|---|---|
| Towers: | | | |
| Shells, 12 ft diam. or less | $ 200 | | |
| Over 12 ft diam. | 60 | 340 | 280 |
| Trays and internals | 400 | 500 | 100 |
| Drums, 12 ft diam. or less | 80 | | |
| Over 12 ft diam. | 105 | | |
| Miscellaneous vessels: | | | |
| Shells, 12 ft diam. or less | 40 | | |
| Over 12 ft diam. | 350 | | |
| Total internals | 55 | 75 | 20 |
| Heat exchangers | 1200 | | |
| Pumps | 140 | | |
| Pump drives | 120 | | |
| Compressors | 1500 | | |
| Miscellaneous FOB equipment | 60 | | |
| Total FOB equipment cost | $4310 | | |
| Erected equipment costs: | | | |
| Fired heaters | 360 | | |
| Tanks | 11 | | |
| Total erected equipment costs | $371 | | |

$$F_P = 0.471$$

$$F_M = 0.344 + 0.033 \log A_0 + 1.194 \frac{t}{A}$$

$$= 0.344 + 0.033 \log 4310 + 1.194(0.060)$$

$$= 0.536$$

Finally, from Eq. (14.9),

$$I = 1.4[4,310,000(1 + 0.724 + 0.471 + 0.536) + 371,000 + 400,000]$$
$$= \$17,600,000$$

or approximately $18 million.

It is interesting to note that the ratio of total investment to major equipment costs is

$$\frac{17,600,000}{4,310,000 + 371,000 + 280,000 + 100,000 + 20,000} = 3.5$$

and the ratio of total investment to FOB costs of equipment is

$$\frac{17,600,000}{4,310,000 + 280,000 + 100,000 + 20,000} = 3.7$$

These factors are very close to overall plant investment factors given previously.

In the absence of other information, multiplying the cost of major equipment items by 4 seems to be the easiest way to get a reasonable figure for the total process plant investment cost.

## 14.25 COMPARTMENTALIZATION METHODS OF ESTIMATING

It has been possible to improve cost-estimating techniques by building up the estimate by parts to which separate factors are applied. In these methods the equipment is grouped, or compartmentalized, by being broken down into its components. A factor is applied to each component, and then the factored components are summed. The method proposed by Cran and described in Sec. 14.22 is a combination of the Lang method and a compartmentalized method.

Compartmentalization methods can be grouped into four types of estimates. These are (1) the average unit-cost estimate, (2) the functional unit-cost estimate, (3) the unit-operations estimate, and (4) the modular estimate [61].

Each of these approaches involves the same breakdown into components. In the average unit-cost method, in choosing the factors, the size of the equipment, the materials of construction, the operating pressures, and the process are considered. The factors are then applied to the basic equipment cost to get the battery-limits cost, from which the total plant cost is derived by using additional factors [42–44].

In the functional unit-cost method, the equipment is grouped according to function, such as distillation, evaporation, or filtration. The auxiliary equipment, such as pumps and exchangers, is grouped with the functional equipment, and a factor is applied to the group to calculate the final capital investment. A method of this type has been proposed by Bridgewater [11].

In the unit-operations estimate it is assumed that similar units have similar accessory equipment and therefore should have similar costs [5]. As an example, nearly all distillation columns have a condenser, a reflux drum, a reboiler, and reflux and bottom pumps. Instead of treating this distillation operation as consisting of six pieces of equipment, it is treated as a single-unit operation. In this method the cost for a specific operation is a function of size, materials of construction, pressure, and temperature. The method was originally proposed by Zevnik and Buchanan [82], and more recently, improvements in the method have been proposed by Hensley

[29] and Viola [74]. Each of these authors gives examples of the method, but so far, this procedure suffers from insufficient suitable data because of the wide range of possible process operations modules [70].

In the modular estimating method the focus is on individual modules in the total system, with each module consisting of a group of similar items [61]. For example, all furnaces would be included in one module, all process vessels in another, all pumps in a third, etc. The total module cost is made up of the sum of the direct material cost $M$ (itself a sum of equipment items $E$ and field installation materials $m$), and field labor costs $L$ for each piece of equipment, then multiplied by indirect-cost factors to give the total module cost. To find the fixed capital investment, the extra costs for adjuncts and auxiliaries are added to the sum of the total module costs. The module estimating method was first proposed by Dodge et al. [16] but has been expanded and improved on by Guthrie [20, 23]. The method as proposed by Guthrie is the most complicated of the factor methods and so should yield more reliable results because of the sounder basis for the installation charges. It is a popular method, but the types of equipment for which the necessary cost data are correlated are limited [70].

## 14.26   THE MILLER METHOD

Because it seems to be the most complete of the compartmentalized methods, an example will be worked out by the Miller average unit-cost method.

To make a factored estimate more accurate, the Lang factor can be broken down into parts, permitting a different factor for each project [5]. In Miller's method this idea is expanded by suggesting that it is difficult to choose a single value for each subfactor; hence the estimator should determine a low, a probable, and a high value for each factor. From these factors would come a minimum, a most likely, and a maximum plant investment cost.

Miller also proposes that because large equipment, high-pressure equipment, and alloy equipment are all very expensive per piece, the installation, piping, foundation, and other such factors should be lower than the corresponding factors for less expensive equipment. For this reason Miller developed a sliding scale for factors depending on the average cost of the "main plant items" (MPI) [44]. The average unit cost is calculated by adding all the costs of all the MPIs, through pumps, and dividing by the total number of such items. All equipment costs are based on delivered prices in 1958 in order to choose the proper factors, but these factors are applied to current costs.

In this method the plant is divided into four areas: battery limits (B/L), storage and handling (S&H), utilities (U), and services (S). The last three areas can be estimated separately because they are a function of the size of the facility, rather than of the type of product being made [5].

The battery-limits cost is calculated from the factors given in Table 14.18. The costs for storage and handling, utilities, and services are calculated from the factors given in Table 14.19. The following example illustrates the use of these tables.

**TABLE 14.18**
**Ratio factors for estimating battery-limits costs**

| | | Range of factors as percent of basic equipment | | | | | | |
| | | Average unit cost of MPI in 1958 dollars | | | | | | |
| | | Under $3,000 | 3,000–5,000 | 5,000–7,000 | 7,000–10,000 | 10,000–13,000 | 13,000–17,000 | Over 17,000 |
|---|---|---|---|---|---|---|---|---|
| Basic equipment (delivered to site, excluding sales taxes and catalyst) | MPI (main plant items) | X | X | X | X | X | X | X |
| | MUE (miscellaneous unlisted items) | | | | | | | |
| | Early flowsheet stage | 20–10% of MPIs in all categories | | | | | | |
| | Scope of work well defined | 10–1% of MPIs in all categories | | | | | | |
| | Note: Top of ranges: | | | | | | | |
| | Complicated processes | | | | | | | |
| | Many process steps | | | | | | | |
| | Bottom of ranges: | | | | | | | |
| | Simple processes | | | | | | | |
| | Few process steps | | | | | | | |
| | Basic equipment = MPI + MUE | 100 | 100 | 100 | 100 | 100 | 100 | 100 |
| Field erection of basic equipment | High percentage of equipment involving high field labor | 23/18 | 21/17 | 19.5/16 | 18.5/15 | 17.5/14.2 | 16.5/13.5 | 15.5/13 |
| | Average (mild steel equipment) | 18/12.5 | 17/11.5 | 16/10.8 | 15/10 | 14.2/9.2 | 13.5/8.5 | 13/8 |
| | High percentage of corrosion materials and other high unit-cost equipment involving little field erection | 12.5/7.5 | 11.5/6.7 | 10.8/6 | 10/5.5 | 9.2/5.2 | 8.5/5 | 8/4.8 |

**TABLE 14.18** (*Continued*)

| | Average unit cost of MPI in 1958 dollars | | | | | | |
|---|---|---|---|---|---|---|---|
| | Under $3,000 | 3,000–5,000 | 5,000–7,000 | 7,000–10,000 | 10,000–13,000 | 13,000–17,000 | Over 17,000 |
| **Equipment foundations and structural supports** | | | | | | | |
| High—predominance of compressors or mild steel equipment requiring heavy foundations | | | 17/12 | 15/10 | 14/9 | 12/8 | 10.5/6 |
| Average—for mild steel fabricated equipment | | | 12.5/7 | 11/6 | 9.5/5 | 8/4 | 7/3 |
| Average—for predominance of alloy and other high unit-price fabricated equipment | 7/3 | 8/3 | 8.5/3 | 7.5/3 | 6.5/2.5 | 5.5/2 | 4.5/1.5 |
| Low—equipment more or less sitting on floor | 5/0 | 4/0 | 3/0 | 2.5/0 | 2/0 | 1.5/0 | 1/0 |
| Piling or rock excavation | Increase above values by 25 — 100% | | | | | | |
| **Piping (includes ductwork excludes insulation)** | | | | | | | |
| High—gases and liquids, petrochemicals, plants with substantial ductwork | 105/65 | 90/58 | 80/48 | 70/40 | 58/34 | 50/30 | 42/25 |
| Average for chemical plants—liquids, electrolytic plants | 65/33 | 58/27 | 48/22 | 40/16 | 34/12 | 30/10 | 25/9 |
| Liquids and solids | 33/13 | 27/10 | 22/8 | 16/6 | 12/5 | 10/4 | 9/3 |
| Low—solids | 13/5 | 10/4 | 8/3 | 6/2 | 5/1 | 4/0 | 3/0 |
| **Insulation of equipment only** | | | | | | | |
| Very high—substantial mild steel equipment requiring lagging and very low temperatures | 13/10 | 11.5/8.5 | 10/7.4 | 9/6.2 | 7.8/5.3 | 6.8/4.5 | 5.8/3.5 |
| High—substantial equipment requiring lagging and high temperatures (petrochemicals) | 10.3/7.5 | 9/6.3 | 7.8/5.2 | 6.7/4.2 | 5.7/3.4 | 4.7/2.8 | 4.8/2.5 |
| Average for chemical plants | 7.8/3.4 | 6.5/2.6 | 5.5/2.1 | 4.5/1.7 | 3.6/1.4 | 2.9/1.1 | 2.2/0.8 |
| Low | 3.5/0 | 2.7/0 | 2.2/0 | 1.8/0 | 1.5/0 | 1.2/0 | 1/0 |

TABLE 14.18 (*Continued*)

| | Average unit cost of MPI in 1958 dollars | | | | | | |
|---|---|---|---|---|---|---|---|
| | Under $3,000 | 3,000–5,000 | 5,000–7,000 | 7,000–10,000 | 10,000–13,000 | 13,000–17,000 | Over 17,000 |
| **Insulation of piping only** | | | | | | | |
| Very high — substantial mild steel piping requiring lagging and very low temperatures | 22/16 | 19/13 | 16/11 | 14/9 | 12/7 | 9/5 | 6/3.5 |
| High — substantial piping requiring lagging and high temperatures (petrochemicals) | 18/14 | 15/12 | 13/10 | 11/8 | 9/6 | 7/4 | 4.5/2.5 |
| Average for chemical plants | 16/12 | 14/10 | 12/8 | 10/6 | 8/4 | 6/2 | 4/2 |
| Low | 14/8 | 12/6 | 10/5 | 8/4 | 6/3 | 4/2 | 2/1 |
| **All electrical (except building lighting and instrumentation)** | | | | | | | |
| Electrolytic plants (includes rectification equipment) | 26/17 | 55/42 | 50/38 | 45/33 | 40/30 | 35/26 | |
| Plants with mild steel equipment, heavy drives, solids | | 22.5/15 | 19.5/12.5 | 17/10 | 14/8.5 | 12/7 | 10/6 |
| Plants with alloy or high unit-cost equipment, chemical and petrochemical plants | 18/9.5 | 15.5/8.5 | 13/6.5 | 11/5.5 | 9/4.5 | 7.3/3.5 | 6/2.5 |

Note: Above figures include 1 to 3% for B/L outside lighting, which is not covered in building services

| | Under $3,000 | 3,000–5,000 | 5,000–7,000 | 7,000–10,000 | 10,000–13,000 | 13,000–17,000 | Over 17,000 |
|---|---|---|---|---|---|---|---|
| **Instrumentation** | | | | | | | |
| Substantial instrumentation, central control panels, petrochemicals | | 58/31 | 46/24 | 37/18 | 29/13 | 23/10 | 18/7 |
| Miscellaneous chemical plants | | 32/13 | 26/10 | 20/7 | 15/5 | 11/3 | 8/2 |
| Little instrumentation, solids | | 21/9 | 17/7 | 13/5 | 10/3 | 7/2 | 5/1 |

Note: Total instrumentation cost does not vary a great deal with size and hence is not readily calculated as a percent of basic equipment. This is particularly true for distillation systems. If in doubt, detailed estimates should be made.

# TABLE 14.18 (*Continued*)

| | Average unit cost of MPI in 1958 dollars | | | | | | |
|---|---|---|---|---|---|---|---|
| | Under $3,000 | 3,000–5,000 | 5,000–7,000 | 7,000–10,000 | 10,000–13,000 | 13,000–17,000 | Over 17,000 |
| Miscellaneous (includes site preparation, painting, and other items not accounted for above) | Range for all values of basic equipment 6–1% | | | | | | |
| Top of range: large complicated processes<br>Bottom of range: smaller, simple processes | | | | | | | |

Buildings—architectural and structural (excludes building services)

Note: When building specifications and dimensions are known, a high-speed building cost estimate is recommended especially if buildings are a significant item of cost. If a separate estimate is not possible, evaluate the buildings as follows before selecting the factors.

### Building evaluation
when most of process units are located inside buildings

| | High:<br>Brick and steel | Medium | Low:<br>Economical | Evaluation |
|---|---|---|---|---|
| Quality of construction | +4 | +2 | 0 | |
| Type of equipment | Very high unit cost equipment<br>−3 | Mostly alloy steel<br>−2 | Mixed materials<br>−1 | Mostly carbon steel<br>0 |
| Operating pressures | Very high<br>−2 | Intermediate<br>−1 | Atmosphere<br>0 | |
| | | | | Building class = algebraic sum = |

**TABLE 14.18 (Continued)**

| | Building class | Average unit cost of MPI in 1958 $ | | | | | | |
|---|---|---|---|---|---|---|---|---|
| | | Under $3,000 | 3,000–5,000 | 5,000–7,000 | 7,000–10,000 | 10,000–13,000 | 13,000–17,000 | Over 17,000 |
| Most of process units inside buildings | +2 | 92/68 | 82/61 | 74/56 | 67/49 | 59/44 | 52/39 | 46/33 |
| | +1 to −1 | 72/49 | 62/43 | 56/38 | 51/33 | 45/29 | 41/26 | 36/21 |
| | −2 | 50/37 | 44/33 | 40/29 | 35/25 | 30/21 | 27/18 | 23/15 |
| Open-air plants with minor buildings | | 37/16 | 32/13 | 28/11 | 24/8 | 20/6 | 17/4 | 14/2 |

Note: The following factors are for Battery Limit (process) buildings only and are expressed in percent of the Building-Architectural & Structural cost. They are not related to the basic equipment cost.

| Building services | High | Normal | Low |
|---|---|---|---|
| Compressed air for general service only | 4 | 1½ | 5 |
| Electrical lighting | 18 | 9 | 5 |
| Sprinklers | 10 | 6 | 3 |
| Plumbing | 20 | 12 | 3 |
| Heating | 25 | 16 | 8 |
| Ventilation: without air conditioning | 18 | 8 | 0 |
| with air conditioning | 45 | 35 | 25 |
| Total overall average* | 85 | 55 | 20 |

The above factors apply to those items normally classified as building services. They do not include:
1. Services located outside the building such as substations, outside sewers, outside water lines, etc., all of which are considered to be outside the Battery Limit, as well as outside the building.
2. Process services.

*The totals provide the ranges for the type of buildings involved and are useful when the individual service requirements are not known. Note that the over-all averages are not the sum of the individual columns.

*Source:* C. A. Miller, "Factor Estimating Refined for Appropriation of Funds," *AACE Bulletin*, vol. 7, September 1965, pp. 92–118.

**TABLE 14.19**
**Ratio factors for auxiliary costs**

| Item | Grassroots plant | Battery-limit addition to existing site |
|---|---|---|
| Storage and handling (S&H) (% of B/L cost) | | |
| Low—raw material by pipeline, little warehouse space | 2 | 0 |
| Average—average raw material storage and finished product warehousing | 15–20 | 2–6 |
| High—tank farm for raw material, substantial warehousing | 70 | 20 |
| Utilities (U) (% of B/L cost) | | |
| Low | 10 | 3 |
| Average | 20–30 | 6–14 |
| High | 50 | 30 |
| Services (S) (% of [B/L + (S&H) + U]) | | |
| Low | 5 | 0 |
| Average | 10–16 | 2–6 |
| High | 20 | 15 |

*Source:* C. A. Miller: "Factor Estimating Refined for Appropriation of Funds," *AACE Bulletin*, vol. 7, September 1965, pp. 92–118.

**Example 14.14.** Repeat Examples 14.10, 14.11, and 14.12. Assume a date of December 1980. The refinery addition for these examples is to be made up predominantly from mild steel equipment, and the processes involved in the addition are simple, involving few petrochemical-type process steps. The delivered equipment cost for 250 pieces of process equipment is 8 M$. Estimate the cost of the refinery addition by the Miller method.

*Solution.* The first step is to calculate the average unit cost in 1958 dollars. The Marshall and Swift index is 688 for December 1980 and 229 for 1958. The average unit cost in 1958 is

$$\frac{8 \times 10^6}{250} \frac{229}{688} = \$10,700$$

and therefore the factors in the fifth column of Table 14.18 are used.

Table 14.20 shows the development of the estimate. In the Miller method miscellaneous unlisted items (MUE) are added to the MPI, and low, average, and high figures are obtained by taking $\pm 10$ percent on each side of the average. In addition, an adjustment of $\pm 10$ percent is made in the factored items just before the B/L costs are calculated. This is done because it is unlikely that all costs would be at their low or high estimates. One should note that storage, utilities, and services are based on a factor times the average B/L cost and the average (B/L + S + U) cost because these auxiliaries should not vary widely.

From the results of this example, the refinery addition would probably cost 31.2 M$. It should not cost more than 37.7 M$ and is not likely to cost less than 24.4 M$.

**TABLE 14.20**
## Total plant cost for Example 14.14

| Average unit cost in 1958 = $10,700 | | | | Date: December 1980 | | |
|---|---|---|---|---|---|---|
| **No. of** | **M & S indexes** | | **Factors** | **Thousands of dollars** | | |
| **MPIs** | **1958** | **Current** | **or** | | | |
| **250** | **229** | **688** | **adjustments** | **Low** | **Average** | **High** |
| MPI (main plant items) | | | — | — | $8,000 | — |
| MUE (misc. unlisted items) | | | ±2% | | 160 | |
| Basic equipment | | | ±10% | 7,344 | 8,160 | 8,976 |
| Field erection — average | | | 9.2/11.7/14.2 | | | |
| Foundations and supports — average steel | | | 5.0/7.3/9.5 | | | |
| Piping — high, petrochemicals | | | 34.0/46.0/58.0 | | | |
| Insulation of equipment — high | | | 3.4/4.6/5.7 | | | |
| Insulation of pipe — high | | | 6.0/7.5/9.0 | | | |
| Electrical — low, petrochemical | | | 4.5/6.8/9.0 | | | |
| Instrumentation — substantial | | | 13.0/21.0/29.0 | | | |
| Miscellaneous — simple | | | 1.0/2.0/3.0 | | | |
| Buildings — minor, open air | | | 6.0/13.0/20.0 | | | |
| Building services — low — 20% buildings | | | 1.2/2.6/4.0 | | | |
| Subtotal, factored items | | | 83.3/122.5/161.4 | | | |
| Adjustments, ±10% | | | +8.3/0.0/−16.1 | | | |
| Total factored items | | | 91.6/122.5/145.3 | 6,727 | 9,996 | 13,042 |
| B/L cost | | | | 14,071 | 18,156 | 22,018 |
| Storage — high, addition | | | 10.0/15.0/20.0 | | | |
| Utilities — average, addition | | | 6.0/10.0/14.0 | | | |
| Subtotal — S + U − %(average B/L) | | | 16.0/25.0/34.0 | 2,905 | 4,539 | 6,173 |
| Subtotal — B/L + S + U | | | | 16,976 | 22,695 | 28,191 |
| Services — average, addition | | | 2.0/4.0/6.0 | 454 | 908 | 1,362 |
| Total direct cost | | | | 17,430 | 23,603 | 29,553 |
| Indirect costs | | | | | | |
|    Engineering and construction | | | 20% | 4,721 | 4,721 | 4,721 |
| Total direct and indirect costs | | | | 22,151 | 28,324 | 34,274 |
| Contingency (including 3% construction fee) | | | 10% | 2,215 | 2,832 | 3,427 |
|    Total plant cost | | | | 24,366 | 31,156 | 37,701 |

Table 14.21 is a summary of the four example estimates for the refinery addition. The estimates group about a cost of 31 M$. If 31 M$ is assumed correct, the estimates are within about ±20 percent.

## 14.27 BUILDING COST

Preliminary cost estimates for building costs can be made by multiplying the square feet of floor area or the cubic feet of volume by a unit cost. Median costs for several building types are shown in Table 14.22. The data in the table are based on average unit costs for the major cities in the United States. Variations in labor costs, local efficiency, and materials prices will result in local fluctuations from these values [41].

A more accurate method of estimating is to use unit costs for each component of a building, such as walls, floors, and foundations; however, such figures can still be in error for a specific case. Table 14.23 shows a detailed breakdown of costs for a particular type of building and gives the unit costs for various components.

**TABLE 14.21**
**Results of selected estimates for a refinery addition**

| Method | Cost estimate, M$ |
|---|---|
| Lang | 38 |
| Hand | 31 |
| Chilton | 28.6 |
| Miller—average | 31.2 |
|    High | 37.7 |
|    Low | 24.4 |

**TABLE 14.22**
**Median square foot and cubic foot costs for several classes of buildings**

| Type of building | $/ft$^2$ of floor area | $/ft$^3$ |
|---|---|---|
| Apartments, midrise | 51.15 | 4.70 |
| Auditoriums | 68.25 | 4.41 |
| Banks | 94.20 | 7.20 |
| Churches | 62.60 | 3.98 |
| Colleges, classrooms | 82.70 | 6.20 |
| Colleges, laboratories | 96.55 | 7.40 |
| Courthouses | 86.25 | 6.40 |
| Department stores | 38.10 | 1.88 |
| Dormitories, midrise | 79.20 | 7.25 |
| Factories | 33.35 | 2.30 |
| Fire stations | 66.30 | 4.24 |
| Garages (commercial) | 44.65 | 2.62 |
| Gymnasiums | 56.35 | 2.78 |
| Hospitals | 114 | 8.40 |
| Housing (public) | 48.20 | 3.79 |
| Jails | 119 | 9.70 |
| Libraries | 72.60 | 5 |
| Motels | 51.55 | 5.10 |
| Offices, midrise | 59.05 | 4.25 |
| Police stations | 92.30 | 6.85 |
| Restaurants | 82.30 | 7.10 |
| Retail stores | 40 | 2.88 |
| Schools, jr. high and middle | 60.50 | 3.98 |
| Supermarkets | 39.40 | 2.24 |
| Telephone exchanges | 104 | 6.90 |
| Terminals (bus) | 52.50 | 2.96 |
| Theatres | 55.55 | 3.25 |
| Town halls | 66.50 | 5.30 |
| Warehouses | 26.40 | 1.58 |

*Source:* Selected from information copyrighted by the R. S. Means Company, Inc., Kingston, Mass. It is reproduced from *Building Construction Cost Data,* 47th ed., 1989, and used with permission.

**TABLE 14.23**
**Detailed breakdown of midrise apartment building costs**

**171** — SF, CF, and % of Total Costs

| | | | Unit | Unit costs | | | % of total | | |
|---|---|---|---|---|---|---|---|---|---|
| | | | | ¼ | Median | ¾ | ¼ | Median | ¾ |
| 020 | 0010 | Apartments midrise (4 to 7 story) | SF | 42.35 | 51.15 | 63.55 | | | |
| | 0020 | Total project costs | CF | 3.35 | 4.70 | 5.50 | | | |
| | 0100 | Sitework | SF | 2.55 | 3.33 | 6.75 | 5.70% | 6.70% | 10.10% |
| | 0500 | Masonry | | 2.61 | 3.65 | 5.40 | 5.80% | 7.50% | 10.60% |
| | 1500 | Finishes | | 5.30 | 6.65 | 8.45 | 10.30% | 11.60% | 16.10% |
| | 1800 | Equipment | | 1.41 | 1.93 | 2.48 | 2.80% | 3.50% | 4.40% |
| | 2500 | Conveying equipment | | .97 | 1.19 | 1.43 | 2% | 2.20% | 2.60% |
| | 2720 | Plumbing | | 2.53 | 3.18 | 4.30 | 6.20% | 7.40% | 8.90% |
| 020 | 2900 | Electrical | SF | 2.87 | 3.86 | 4.73 | 6.60% | 7% | 8.90% |
| | 3100 | Total: Mechanical & electrical | " | 7.95 | 9.80 | 11.90 | 17.20% | 20.10% | 21.60% |
| | 9000 | Per apartment unit, total cost | Apt. | 39,700 | 51,500 | 70,700 | | | |
| | 9500 | Total: Mechanical & electrical | " | 9,500 | 11,100 | 16,300 | | | |

*Source:* This information is copyrighted by the R. S. Means Company, Inc., Kingston, Mass. It is reproduced from *Building Construction Cost Data*, 47th ed., 1989, and used with permission.

## 14.28 DEFINITIVE ESTIMATING

Construction firms constitute over 8 percent of the business firms in the United States, but the rate of failure within the industry is double that—over 16 percent of business failures. While some of the reasons for failures may be attributed to causes such as overextension, poor management, lack of proper financial controls, etc., most failures result from unrealistic bids due to bad estimates. Therefore, accurate and realistic estimates are needed by those contractors who bid competitively as well as those who obtain jobs through negotiation.

Accurate estimates are critical to the survival of any contractor because, if estimates are too high, he or she will be noncompetitive and there is little or no chance of getting jobs. On the other hand, if estimates are too low, there is the potential to get jobs with little or no chance of making any profit. In other words, the estimate has to be low enough to get the job but high enough to make a profit.

The following sections discuss the orderly development of a complete estimate for a construction project based on fairly complete scope documents. It should be noted that the underlying themes and techniques are not limited to contractors bidding on construction projects, but can be used by all parties in the detailed estimate of any project or endeavor.

## 14.29 ORGANIZATION OF THE DEFINITIVE ESTIMATE

Proper estimating requires a comprehensive, systematic approach because a detailed estimate is not only needed for bidding, but will also be useful during construction and for future reference. The approach involves these steps:

**A THOROUGH REVIEW OF ALL THE CONTRACT DOCUMENTS INCLUDING ADDENDA IN ORDER TO GET A CLEAR UNDERSTANDING OF THE PROJECT.** It should be noted that these documents dictate how the estimate is prepared and how the job will be executed. In reviewing the documents, emphasis should be placed on ensuring that the documents are complete (no missing pages, etc.) and that there are no unusual items that are totally unfamiliar. One should have access to and/or develop checklists to ensure that "gotcha" items are not overlooked.

**THE PREPARATION OF THE PROJECT SUMMARY OF ESTIMATE.** After the review of the contract documents is completed, it is necessary for the estimator to break the project into its cost elements. This may be done by preparing a summary sheet which may serve as a checklist and also help in determining which items will be performed with the firm's own forces and those which will be subcontracted. The summary sheet normally has three sections—direct costs, indirect costs, and a summary of alternates (if applicable). It is good practice at this point to contact all potential subcontractors for bids. Sometimes it may be necessary to mail out contract documents to subcontractors, suppliers, etc. who may be interested in bidding but have no access to the contract documents.

**PREPARATION OF A BID FILE.** The next step in organizing an estimate is to set up a portfolio to store the information related to the project in an organized fashion and in one place. In these days of computerized databases it is desirable to implement and maintain an electronic database with the bid data incorporated for ease of use at a later date. Typically, a bid portfolio should include a folder for each of the following:

The contract document (plans, specification, addenda, etc.).

All the quantity takeoff and pricing sheets

All the correspondence related to that project

All the telephone or written quotes for the job, preferably arranged in order in accordance with the Construction Specifications Institute (CSI) format

The project summary and cut/add sheets

The owner's proposal forms

Bonds

Bid form

**PREPARATION FOR "TAKEOFF."** Once the summary sheets and the project bid file are organized, certain preparations should be made prior to beginning the "takeoff." These will include:

A further review of the project documents with particular attention to items like wage rates, insurance and bonding requirements, general and special conditions, and any other items in the plans or specifications that may seem problematic and have cost implications.

The assembling of all the required forms such as those needed for "takeoff," telephone quotations, etc. Each of these forms should be properly identified. As an example, they should state the name of the project, architect/engineer's name, bid date, section of the job (usually by specification section), date of the estimate, the name of the estimator, page number, etc.

Forms and formats used in estimating vary from one company to another. However, each firm should have a standardized set of forms that is used by all its estimators and as a minimum should include:

Summary sheets

Pads for notetaking

Takeoff and costing forms

Worksheets

General conditions sheets

A schedule for preparing the estimate should be done to ensure that the estimate will be completed on time. Responsibilities for various sections of the estimate should be shown by name.

After the organization of the proposed cost estimate and after a thorough review of the plans and specifications but prior to starting the takeoff, it is recommended that a visit be made to the proposed site.

## 14.30  ESTIMATING WORK ITEMS

**THE QUANTITY TAKEOFF.** Having organized the estimate, reviewed the contract documents, and visited the site, the measurement of the work can begin. The process of measuring quantities from a set of plans using the specification as a guide is called the takeoff. There are four main parts to a takeoff:

The classification of the work
Description of items of the work
Dimensions of the items of the work
Extensions of dimensions

The takeoff should be organized in an order similar to the general sequence in which the project will be built because it:

Helps the estimator to learn the structure
Helps to eliminate gross error
Permits more than one estimator to work on the takeoff
Simplifies the job planning and scheduling process
Makes the process of cost feedback and cost comparison easier at a later date

A suggested sequence of taking-off may be as follows:

Foundations
Structural
Piping/electrical
Major equipment
Site work
Finish work
Start-up and turnover
General conditions

Subcontracts. It is important to develop an estimate, not necessarily a detailed estimate, for work that may be subcontracted. First, this serves as a check of the adequacy and responsiveness of any potential subcontract bids received. Second, it may become prudent for the contractor to do the work instead of subcontracting as a cost-saving approach. Third, even though the contractor does not have first-hand responsibility for the subcontracted work, the contractor is still ultimately responsible for it, and so must plan and manage that work within its total responsibility.

**COSTING FOR LABOR AND MATERIAL.** With the takeoff completed and the dimensions extended and checked, the next step is the costing for labor and material. Material costs are usually obtained from suppliers and/or pricing catalogs. Computing the cost of labor is probably the most difficult portion of the estimate. Depending on the contractor and the contract involved, a variety of crafts/trades may be directly hired by the contractor and their work directly supervised by staff members of the contractor. The contractor subcontracts for those work items it cannot perform efficiently or which may be too risky (for the contractor to perform).

Wage rates may be established either in the specification as part of the contract documents (government area standards, e.g.) or, on a union project, as a part of current local wage agreements. It is imperative for the estimator to know rates before starting the estimate since they may vary considerably from what the firm has been paying. The labor estimate normally has two parts:

The gross amount of paychecks issued to the field employees (include both regular and premium pay)

The labor burden (fringe benefits paid by the employer on behalf of the employee)

The cost of labor is usually a substantial part of the total project costs. Therefore, care and accuracy in costing labor is a necessity. The most common practice by most estimators is the use of unit costs (cost per unit of work). The three main methods of determining unit costs are from experience, historical productivity data, and crew size calculations. Determination of unit costs by experience should be used only if the estimator has had prior extensive experience in the field either performing or observing the task to be costed.

Information for deriving unit costs from historical productivity data may be obtained from two main sources—the firm's historical cost records or published reference books. Example 14.15 illustrates the calculation of a unit price from productivity data.

**Example 14.15.** To form 100 ft$^2$ (SF) of wall requires 6 h of carpenter time and 5 h of common laborer time: The wage rate for carpenters is \$15.00/h; the wage rate for common laborers is \$10.00/h. Estimate the unit labor cost per square foot.

*Solution.* The unit cost may be calculated as follows:

$$
\begin{array}{llll}
\text{Carpenter} & \cdot & \text{6 h at \$15.00/h} = & \$90.00 \\
\text{Laborer} & \cdot & \text{5 h at \$10.00/h} = & \$50.00 \\
\text{Total labor cost for 100 ft}^2 & & = & \$140.00
\end{array}
$$

$$
\text{Labor cost per ft}^2 = \frac{\$140.00}{100} = \underline{\underline{\$1.40}}
$$

The work-hour unit rates (work-h/yd$^3$, work-h/ton, work-h/ft$^2$, etc.) are derived from a company's historical productivity data by dividing the total work-hours required to perform a specific task by the total units of work accomplished. The following cautionary notes are in order when using the resultant data:

Under what job conditions were the data generated — ideal, average, or what?

What elements of work are included in each rate? For example, are the 100 tons of steel erected on one project and 250 tons erected on another comparable in scope — does one project include miscellaneous steel and scaffolding and the other not?

The estimator should be aware that unit cost and work-hour unit rates each have their place; but if it is possible, the work-hour unit rate calculation should be used as it eliminates problems caused by labor cost variability. After the work-hours are calculated, the labor rates for the particular project can be applied. An additional benefit created is that estimate will be in a form that can be easily transferred to the project control estimate for project management.

Determining unit costs using crew size calculations requires experience as well as the ability to visualize construction operations, site conditions, the material quantities involved, the size of crew needed to accomplish the task, and the amount of time needed to perform the task. Example 14.16 illustrates this method.

**Example 14.16.** Seventy-five cubic yards of concrete need to be placed. Due to the site conditions it is determined that the safest and best way is to use a crane and a 2 yd$^3$ bucket. It is determined that to perform the task efficiently four laborers are needed — one at the concrete truck, two at the point of placement, and one on the vibrator. It is assumed that supervision is by the superintendent. No finishing is required. Wage rate for laborers is \$10.00/h. What is the unit cost per cubic yard of concrete placed?

*Solution.* Time needed:

| | | | |
|---|---|---|---|
| Setup | | = | 30 min |
| Cycle | | | |
| Load | = 3 min | | |
| Swing, dump & return | = 5 min | | |
| | 8 min | | |

No. of cycles = 75/2 = 37.5

| | | |
|---|---|---|
| Total cycle time = 37.5 × 8 min | = | 300 min |
| Disassembly | = | 15 min |
| | | 345 min |
| Inefficiency (labor, delays, etc.) 20% | = | 69 min |
| Total operation time | = | 414 min |
| Amount of time needed | = | $\dfrac{414 \text{ min}}{60 \text{ min}}$ = 6.9 h (say, 7 h) |
| Laborers — four for 7 h at \$10.00/h | = | \$280.00 |
| Cost per 75 yd$^3$ | = | \$280.00 |
| Cost per cubic yard | = | $\dfrac{\$280.00}{75 \text{ yd}^3}$ = \$3.73 |

There are many sources of cost information for the estimator. Many of the professional societies, contractor, subcontractor, trade, and governmental organizations, as well as publishing companies, print estimating handbooks to aid the estimator. Whichever source is used, it is recommended that some judgment be exercised to match the specifics of a particular project since all the sources are based on averages and/or other situations that may not be applicable to all projects.

## 14.31 ESTIMATING OTHER COST ELEMENTS

Besides labor and material costs, the other elements of an estimate include:

**CONSTRUCTION EQUIPMENT.** The cost of equipment to be used in performing the job is usually obtained from equipment rental firms or a company's equipment division. In addition, there are many reference books published by equipment vendors, equipment associations, and others to enable one to estimate equipment costs.

**COSTS OF SUBCONTRACT ITEMS.** These are usually furnished by subcontractors and vendors either by telephone (in which case they are recorded by the estimator on a telephone bid form) or in writing (written quotations). Related costs, such as freight charges, should be included as part of the direct material costs. As stated earlier, the estimator needs to be able to develop, if not a detailed subcontract estimate, an estimate against which one can compare the potential subcontractor bids.

**JOB INDIRECTS.** As defined previously, job indirects are those items of cost that cannot be directly attributed to a particular item of work or endeavor. Refer to Section 14.8 for a job indirects checklist. Job indirects can logically be broken into the following two categories:

**General Conditions.** It is a good practice to develop a standard list of general condition items based on what is indicated in the specifications as well as other items that may not be indicated anywhere but which are needed by the contractor to perform the work. Once such a master list is developed, then it is a question of an estimator assigning costs to only the items that pertain to a specific project. The master list should include, but not be limited to, items under the broad headings of:

Bonds
Permits
Professional services
Safety equipment
Small tools
Supervision
Temporary facilities
Travel and lodging

Miscellaneous costs (e.g., cleanup, punch list, etc.)

Demobilization

All items under the preceding headings which pertain to a specific project should be costed, extended, and summed up to obtain the total cost of general conditions.

**Overhead.** Every contractor has certain fixed costs incurred regardless of the volume of contracts performed. The cost of these items are normally estimated for a year and then reduced to a percentage based on estimated total volume per year. Overhead costs may include but are not limited to the following:

Salaries (home office)

Employee benefits

Professional fees

Insurance

Office lease or rent

Office stationery and supplies

Depreciation

Maintenance

Job procurement and marketing

Home office travel and entertainment

Advertising

For instance, if a contractor has an overhead of $480,000.00 per year and annual volume is approximately $6,000,000.00, then overhead is about 8 percent of annual volume.

**ALTERNATES AND ALLOWANCES.** Some projects may have alternates (either additive or deductive) in addition to the base bid. The owner normally has the option of accepting or rejecting any or all of the alternates. Care should, therefore, be taken in pricing alternates, particularly where subcontractor costs are involved. Any markup should be added to or deducted from the subcontractor's price for an alternate price as applicable.

Sometimes an architect/engineer may include certain lump sum amounts in the contract documents for items such as landscaping, light fixtures hardware, carpet, etc. These predetermined amounts are called allowances and are intended to be included by the contractor in the estimate in the amount stated. It should, however, be noted that, unless specifically indicated, these amounts may include only the cost of the materials. The estimator should, therefore, add the cost of labor for installation, overhead, and profit to the given allowance to obtain the total price.

**RISK-MARKUP.** It is the role of the estimators to identify all potential cost elements in a project and to estimate a cost associated with each. They should also identify any risk elements in the contract so that these can be evaluated during bid review. However, it is not the estimator's job to cost the risk elements or to add the profit — these two items are discretionary and are determined by management. Hence the term *markup* is usually applied to the pricing operation of adding an amount for contingency and for profit.

**Contingency.** A markup for contingency should be based on a structured evaluation of the risks identified in the contract. There are many ways of doing this, a popular one being a "range-estimating" approach. This system utilizes a Monte Carlo approach to evaluating the composite effect of those elements in the contract subject to significant variability. Some contractors use a "gut feel" approach for handling contingency, but this is dangerous. There are no reasonable rules of thumb for contingency markup.

**Profit.** A markup for profit may be taken as a percent of labor costs or of all direct and indirect costs — each contractor has his or her own system. It can vary anywhere from 0 to 15 percent depending on the situation, size of project, competition, the economy, the client, etc. Occasionally a contractor will "buy" a project. That is, the contractor will anticipate not making a profit or may even anticipate a loss. This is not uncommon on a project where the project contains multiple, sequential contracts of the type the contractor can perform well. If the contractor gets on site early, the mobilization costs will have been absorbed, he or she will have a better handle on real costs at the location and will be in a position to be very competitive while still making a profit on subsequent contracts. In times of economic downturn a contractor may knowingly accept a loss on the project, that loss being less than the loss of maintaining a company without work. Obviously, a contractor must make a profit in the long run. The dollar value of the company's investment should yield a return that reflects current investment yields, and the contractor deserves additional profit in reward for the risk assumed.

## 14.32   BID WRAP-UP

The type of project and the form of the contract will determine the final form the bid will take. Many private projects are based on negotiated proposals from a select few, prequalified, bidders. The majority of public projects, due to state and federal regulations, are competitively bid, open to all, and have a set time and place for "bid opening." The industry has traditionally put may hours and dollars into the estimate prior to bid wrap-up only to see all that effort jeopardized on "bid day" by hectic last minute cuts and adds and misplaced decimal points. To minimize the confusion, scramble, and panic on bid day, the following procedure is suggested:

Even though it is not always possible, the final bid should be tabulated no later than the day before the bid date. To do this, the takeoffs must be extended, costed, and transferred to the summary sheet. Subcontractor bids should be evaluated to select the most appropriate and responsive bids. The selected subcontractor bids should be entered on the summary sheet.

Telephone bids from subcontractors should be recorded and filed in some logical order, usually in accordance with the logical sequence developed from the project plans and specifications and/or company format.

The total direct cost is obtained by adding up the columns on the summary sheet for labor, material, equipment, and subcontract costs.

Taxes, overhead, general conditions, labor burden, contingency, and profit are added to determine the base bid. This should be checked for errors by another person.

If there are alternates on the project, these should also be summarized.

Once the bid is finalized, all changes are made on a cut/add sheet. The cut/add sheet is used on bid day to tabulate all last minute changes in costs as well as new bids received from subcontractors, suppliers, etc.

At an appropriate time the estimator should stop accepting phone calls in order to determine the net effect of the adds and deducts on his base bid.

If the bid form has already been completed, the change is usually made on the bid envelope and submitted. If the bid form has not been completed, then the new bid price is entered thereon.

Before the bid is submitted it should receive one final check to ensure that it complies with all the bid instructions.

Finally, plans should be made to submit the bid on time.

The bid closeout process is greatly simplified if a contractor utilizes a computerized bid preparation system that is so structured that last minute changes can be made and their effect automatically reflected throughout the bid. It has been said that the "successful low bidder" is the one who has made the most and biggest mistakes — do not leave anything to chance and make sure that simple items such as formulas in spreadsheets are correct.

## 14.33 INTERNATIONAL COSTS

The operations of multinational corporations cut across national boundaries, and many major contractors work worldwide. As a result of fast and inexpensive air travel, the world is shrinking. For these and other political and social reasons comparative international costs assume great significance.

In a completely free world economy competition should tend to equalize prices throughout the world. Such, however, is not the case; in fact, price may not be related to cost. The domestic and export prices of the same item may differ widely, e.g., as a result of fiscal policies intended to protect new industries. This is particularly true of developing countries, such as India, that want to encourage local industries and also maintain an increasing share of world exports. These policies have not always been successful.

A good portion of future capital expenditure programs will be related to natural resources or forms of energy supply. However, most of the supplies of natural resources are located in remote areas, such as the Arctic, mountainous areas, deserts, offshore areas, or underdeveloped countries. Underdeveloped countries are eager to reap the benefits of their mineral wealth and become part of the modern world.

It is not feasible for estimators to prepare capital-cost estimates for projects located in areas of the world with which they are unfamiliar. Preparing cost estimates for such projects is extremely difficult, especially if the organization has no experience in the prospective geographical area. The cost estimator, together with other engineers, is responsible for developing capital estimates as accurately as possible. To do so requires the evaluation of all cost and scheduling factors that will affect the capital estimate.

Obviously, the estimator can rely on published location adjustment factors supplied by estimating services and trade journals, which consider the effects of labor, materials, construction equipment, and indirect costs applied to a particular location. This method is commonly used for preliminary estimates. Table 14.24 presents such a list of international adjustment factors [12]. This list is the result of literature searches and discussions with informed people from many countries. Miller [47] has published location factors for the United States, the United Kingdom, and Canada. A wider range of countries is represented in an earlier list by Kharbanda [34].

> **Example 14.17.** The cost of a 1000 ton/day ammonia plant, based on natural gas, was 30 M$ in the United States in 1976. Estimate the cost of the same plant in Britain in the same year.
>
> **Solution.** From Table 14.24, the location adjustment factors are 1.0 for the United States and 0.9 for Britain. The cost for the plant in Britain would be
>
> $$30\frac{0.9}{1.0} = 27 \text{ M\$}$$

The costs of plants in different countries vary with the exchange rate, productivity, and inflation. If inflation must be taken into account, Kharbanda [34] lists exchange rates and inflation indexes for various currencies in different years. This situation arises when the base plant and the new plant are built in different years. Fluctuating currency rates add a new dimension to overseas plant costs. An objective discussion appears in an article by Kumar [35].

The major economic factors affecting overseas costs are:

1. Political system
2. Financial system
3. Legal system
4. Social system
5. Geography
6. Type of industry

**TABLE 14.24**
**International adjustment factors**

Location factors for chemical plants of similar function

| | | U.K. = 1.0 | U.S. = 1.0 |
|---|---|---|---|
| Australia | | 1.4 | 1.3 |
| Austria | | 1.1 | 1.0 |
| Belgium | | 1.1 | 1.0 |
| Canada | | 1.25 | 1.15 |
| Central Africa | | ~2.0 | ~2.0 |
| Central America | | 1.1 | 1.0 |
| China | (imported element) | 1.2 | 1.1 |
| | (indigenous element) | 0.6 | 0.55 |
| Denmark | | 1.1 | 1.0 |
| Eire | | 0.9 | 0.8 |
| Finland | | 1.3 | 1.2 |
| France | | 1.05 | 0.95 |
| Germany (West) | | 1.1 | 1.0 |
| Greece | | 1.0 | 0.9 |
| Holland | | 1.1 | 1.0 |
| India | (imported element) | 2.0 | 1.8 |
| | (indigenous element) | 0.7 | 0.65 |
| Italy | | 1.0 | 0.9 |
| Japan | | 1.0 | 0.9 |
| Malaysia | | 0.9 | 0.8 |
| Middle East | | 1.2 | 1.1 |
| Newfoundland | | 1.3 | 1.2 |
| New Zealand | | 1.4 | 1.3 |
| North Africa | (imported element) | 1.2 | 1.1 |
| | (indigenous element) | 0.8 | 0.75 |
| Norway | | 1.2 | 1.1 |
| Portugal | | 0.8 | 0.75 |
| South Africa | | 1.25 | 1.15 |
| South America (North) | | 1.5 | 1.35 |
| South America (South) | | 2.5 | 2.25 |
| Spain | (imported element) | 1.3 | 1.2 |
| | (indigenous element) | 0.8 | 0.75 |
| Sweden | | 1.2 | 1.1 |
| Switzerland | | 1.2 | 1.1 |
| Turkey | | 1.1 | 1.0 |
| United Kingdom | | 1.0 | 0.9 |
| United States | | 1.1 | 1.0 |
| Yugoslavia | | 1.0 | 0.9 |

Notes:
1. Increase a factor by 10% for each 1000 miles, or part of 1000 miles, that the new plant location is distant from a major manufacturing or import center, or both.
2. When materials or labor, or both, are obtained from more than a single source, prorate the appropriate factors.
3. Investment incentives have been ignored.

*Source:* A. V. Bridgewater: "International Construction Cost Location Factors," *Chemical Engineering,* McGraw-Hill, Inc., vol. 86, November 5, 1979, p. 5.

These factors must be considered from one or more of the following points of view:

1. International
2. National
3. Local
4. Company
5. Project

No generalizations are possible, so each case must be considered individually and in depth.

If capital estimates are required for projects in a location for which there are no published adjustment factors or in-house historical information, the estimator must obtain estimating data by making a preestimate survey. The survey requires careful study of local market conditions and the many components of a capital estimate. Estimating data may be obtained by talking with local or foreign contractors who have worked that area. A prepared checklist containing questions on major materials, labor, and contracting cost elements is an excellent guide. By relating these questions to the normal estimating base, the estimator will understand local costs and any special conditions which may exist. Table 14.25 is a list of factors important in framing questions for a preestimate survey [12, 57].

There is another important reason for a preestimate survey. Conditions are constantly changing, and international cost comparisons are far from simple. Climatic conditions, e.g., can have a pronounced effect both directly (concrete must be heated in Canada but cooled in Saudi Arabia) and indirectly through productivity. Until recently the United States was noted for its high productivity in various trades, but today certain trades abroad may be equally productive, or perhaps more so, in their own environments.

*Productivity,* defined as the ratio of output to input, enters into estimating overseas costs. Productivity is an abstract and controversial subject, and there is no mathematically correct way of measuring it. Productivity applies to the use of land, labor, and capital, but cost estimating generally is more concerned with labor productivity.

Statistical data on labor productivity are usually based on how much actual work is performed for a given amount of time, effort, or resources expended compared with a standard amount of work accomplished. In process plant estimating the standard base generally used is the U.S. Gulf Coast area. It is usually difficult to get statistical data on productivity because bidding is a competitive situation and contractors are reluctant to share this type of information. If in-house productivity data are not available, the estimator must consult the limited published data, such as those shown in Fig. 14.7, or obtain assistance from local contractors or labor unions. For accuracy in evaluating productivity data, one should consider individual crafts rather than the entire work force. Coupling actual productivity data with sound judgment is the best approach to use in determining productivity factors.

Many forces affect labor productivity, such as the lack of experienced craft labor, restrictive labor regulations, other projects in the area, bad weather, poor con-

**TABLE 14.25**
## Preestimate survey checklist

Social, economical, and political
  Local culture and customs
  Language problems
  Religion and its holidays
  Attitude of community toward project
  Political climate
  Type of government
  Government agencies
  General economical climate
  Taxes and duties
  Currency fluctuations
  Inflation rate
  Cost of living

Site and environmental
  Soil conditions
  Terrain
  Site access
  Site elevation
  Earthquake factors
  Offshore wind, wave, and water depth
  data
  Climate
  Weather storms
  Environmental requirements

Material and equipment
  Local available materials
  Materials required to import
  Special selected material suppliers
  Freight and logistic cost
  Export preparation cost
  Local fabrication capability
  Acceptable foreign fabricators and
  suppliers

Contracting
  Prime contractors
  Joint ventures
  Subcontractors availability
  Contracting through agents or brokers
  Types of contracts in common use

Legal and financial
  Legal assistance
  Cost of doing business
  Multiple taxation
  Insurance
  Letters of credit
  Bid bonds
  Financing cost
  Overseas premium

Labor
  Labor cost
  Working hours practiced
  Workweek practiced
  Labor availability
  Amount of trained labor
  Labor productivity
  Unionized labor
  Labor contracts
  Training requirements

Procurement
  Local practices
  Worldwide procurement
  Public bid opening practices
  Trading company participation
  Expediting and inspection practices

Support
  Construction facilities
  Temporary facilities
  Utilities
  Housing
  Communications
  Security
  Warehousing
  Medical facilities
  Sanitary facilities
  Food and catering
  Recreational facilities
  Education facilities
  Permanent community

*Source:* A. Patrascu, *Construction Cost Engineering,* Craftsman Book Company of America, Carlsbad, Calif., 1978. Used with permission.

struction planning, inefficient use of manpower, lack of supervision, and low morale. Some factors can be controlled and others cannot, but all must be evaluated during the project and after its completion. This information will then serve as a historical basis for future estimates [57].

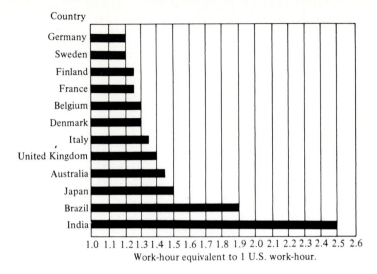

**Figure 14.7**
Typical worldwide construction-labor productivities. (J. T. Gallagher: "Analyzing Field Construction Costs," in H. Popper (ed.): *Modern Cost Engineering Techniques*, McGraw-Hill Book Company, New York, 1970, p. 227.)

## 14.34   STAYING WITHIN LIMITATIONS

Cost estimation requires aptitude, but experience is necessary for professionalism. No estimator should work in unfamiliar fields. What may be an easy estimation for one analyst is difficult, if not impossible, for another. Even competent estimators can overextend themselves. Inflation has greatly increased the estimator's task, and information must be continuously updated. Costs may be changing continuously, but estimators must limit their work to the area they know.

## NOMENCLATURE

Nomenclature for Chap. 14 is restricted for use within the chapter and conflicts, in part, with nomenclature used in other chapters.

$A$     Total cost of all battery-limits investments, Eqs. (14.9) to (14.12), $
$A_0$    $A/1000$, Eqs. (14.10) to (14.12), k$
$B$     Cost of all erected equipment, Eq. (14.9), $
$C$     Cost, Eqs. (14.1) and (14.6), $
$C$     Incremental cost of alloys used for corrosion resistance, Eq. (14.9), $
$D$     Direct plant cost, Eq. (14.6), $

| | |
|---|---|
| $e$ | Total heat exchanger cost less the incremental cost of the alloy, Eqs. (14.10) and (14.11), $ |
| $E$ | Delivered equipment cost, Eqs. (14.7) and (14.8), $ |
| $E$ | Indirect-cost factor, Eq. (14.9), dimensionless, normally assumed to be 1.4 |
| $f$ | Ratio of cost indexes, Eq. (14.6), dimensionless |
| $f$ | Total cost of field-fabricated vessels, Eq. (14.10), $ |
| $F_I$ | Direct-cost factor for instruments, Eq. (14.7), dimensionless |
| $F_L$ | Cost factor for field labor, Eqs. (14.9) and (14.10), dimensionless |
| $F_M$ | Cost factor for miscellaneous items, Eqs. (14.9) and (14.12), dimensionless |
| $F_0$ | Indirect-cost factor, Eq. (14.7), dimensionless |
| $F_P$ | Cost factor for piping materials, Eqs. (14.9) and (14.11), dimensionless |
| $I$ | Total battery-limits investment, Eq. (14.9), $ |
| $I$ | Indirect costs, Eq. (14.9), $ |
| $I$ | Instrument costs, Eq. (14.7), $ |
| $p$ | Total pump plus driver cost less the incremental cost of the alloy, Eq. (14.11), $ |
| $Q$ | Capacity, Eq. (14.1), any convenient dimensions |
| $R$ | Ratio of plant sizes, Eq. (14.6), dimensionless |
| $t$ | Total cost of tower shells less incremental cost of alloys, Eq. (14.12), $ |
| $X$ | Total cost-capacity factor, Eqs. (14.1) and (14.6), dimensionless |
| $\phi$ | Cost factors, Eq. (14.8), dimensionless |

## PROBLEMS

**14.1.** A construction project cost 950 k$ in 1985. Estimate the cost in 1988. The *Engineering News-Record Construction Cost Index* was 4195 in 1985 and 4519 in 1988.

**14.2.** Equipment for a process plant cost 250 k$ in 1983. Estimate the cost in 1988. The *Marshall and Swift Installed Equipment Cost Index* was 761 in 1983 and 857 in 1988.

**14.3.** The *Chemical Engineering Plant Cost Index* is based on 1957–1959 = 100. The index had a value of 261 in 1980 and 343 in 1988. Convert the 1988 index to a base year 1980 = 100.

**14.4.** A construction project cost 1 M$ in 1985. Find the cost in 1988.

**14.5.** Equipment installed in a project in 1985 cost 800 k$. Find the cost in 1988.

**14.6.** Describe the difference between costing and pricing.

**14.7.** You are an owner and desire to put into production a new line of pain killers (pills). A young engineer on your staff is charged with the responsibility of developing an estimate for the indirect costs associated with this endeavor. What elements of indirect cost would you expect that engineer to take into account?

**14.8.** Review the job indirects checklist found in Sec. 14.8. Are there times that an estimate might treat some of the items listed as direct costs? If so, in what situations may this occur?

**14.9.** The delivered cost of the major equipment for a solid-fluid process plant is 2 M$. Estimate the total cost of the plant.

**14.10.** Estimate the total battery-limits investment for a process plant with an estimated equipment cost based on carbon steel as follows:

| | k$ | Incremental cost of corrosion-resisting alloy, k$ |
| --- | --- | --- |
| Tower shells over 12 ft diameter | 200 | 500 |
| Trays and internals | 350 | 600 |
| Drums over 12 ft diameter | 150 | 250 |
| Miscellaneous vessels over 12 ft diameter | 400 | |
| Heat exchangers | 600 | |
| Pumps and drivers | 165 | 300 |
| Miscellaneous FOB equipment | 135 | |
| Erected equipment cost | 225 | |

**14.11.** The delivered equipment for a plant to beneficiate iron ore cost 6 M$ in 1980 and consists of 140 individual pieces of equipment. The plant is a "grassroots" plant to be built in northern Minnesota. The equipment involves little field erection and is subject to erosive action. The process is relatively simple and does not involve very high or very low temperatures. Pressures are atmospheric, and the quality of building construction is medium. Estimate the total plant cost by the Lang factor method.

**14.12.** Repeat Prob. 14.11 using the Chilton method.

**14.13.** Repeat Prob. 14.11 using the Peters and Timmerhaus factors.

**14.14.** Repeat Prob. 14.11 using the Miller method. The Marshall and Swift Installed Equipment Cost Index had a value of 688 in 1980 and 229 in 1958.

**14.15.** What is the labor unit cost for forming a 10 ft 0 in. high wall, 10 ft long, given the following data:
The crew consists of 3 carpenters and 2 laborers. It will take the carpenters 4 h and the laborers 2 h to complete the task. Each carpenter makes $18.00/h, and each laborer makes $12/h.

**14.16.** You are an estimator working on your general conditions estimate. List five items you will consider under the following major categories:
(a) Mobilization
(b) Temporary facilities
(c) Professional services
Which of the preceding items could be considered to be time-sensitive?

**14.17.** As the chief estimator, the Accounting Department has given you the following information:
(a) Annual volume = $58,000,000.00
(b) Total overhead expense:
Salaries = $1,900,000.00
Legal and accounting fees = $350,000.00
Insurance = $2,200,000.00
Rents = $900,000.00
Utilities = $200,000.00
Other = $500,000.00
One of the estimators requested to know what figure to use for overhead on an estimate. What figure would you recommend?

# REFERENCES

1. *AACE Bulletin*, vol. 1, November 1958, p. 12.
2. *AACE Standard Cost Engineering Terminology*, American Association of Cost Engineers, Morgantown, W.Va., 1986.
3. Aries, R. S., and R. D. Newton: *Chemical Engineering Cost Estimation*, McGraw-Hill Book Company, New York, 1955.
4. Arnold, T. H., and C. H. Chilton: "New Index Shows Plant Cost Trends," *Chemical Engineering*, vol. 70, Feburary 18, 1963, pp. 143–152.
5. Baasel, Q. D.: *Preliminary Chemical Engineering Plant Design*, American Elsevier Publishing Company, Inc., New York, 1976.
6. Bauman, H. C.: *Fundamentals of Cost Engineering in the Chemical Industry*, Reinhold Publishing Corporation, New York, 1964.
7. Bentil, K. K., and D. F. McDonald, Jr.: "Conceptual Estimating," *Skills and Knowledge of Cost Engineering*, 2d ed., edited by J. McNeil, American Association of Cost Engineers, Morgantown, W.Va., 1988, pp. 27–35.
8. _____: "Costing Estimating Basics," *Skills and Knowledge of Cost Engineering*, 2d ed., edited by J. M. Neil, American Association of Cost Engineers, Morgantown, W.Va., 1988, pp. 3–11.
9. _____: "Definitive Estimating," *Skills and Knowledge of Cost Engineering*, 2d ed., edited by J. M. Neil, American Association of Cost Engineers, Morgantown, W.Va., 1988, pp. 13–25.
10. Boyd, N.: "Cost and Price Indices," *Engineering Costs and Production Economics*, Elsevier Scientific Publishing Company, Amsterdam, vol. 6, April 1982, pp. 267–273; vol. 14, December 1988, pp. 247–258.
11. Bridgewater, A. V.: "The Functional-Unit Approach to Rapid Cost Estimation," *AACE Bulletin*, vol. 18, September–October 1976, p. 153.
12. _____: "International Construction Cost Location Factors," *Chemical Engineering*, vol. 86, November 5, 1979, p. 5.
13. *CCAD: Construction Computer Applications Directory*, Construction Industry Press, San Rafael, Calif., published annually.
14. Chilton, C. H.: "Six-Tenths Factor Applies to Complete Plant Costs," *Chemical Engineering*, vol. 57, April 1950, pp. 112–114.
15. Cran, J.: "Improved Factored Method Gives Better Preliminary Cost Estimates," *Chemical Engineering*, vol. 88, April 6, 1981, pp. 65–79.
16. Dodge, W. J. et. al.: "The Module Estimating Technique as an Aid in Developing Plant Capital Costs," *AACE Bulletin*, vol. 6, no. 4, September 1964, pp. 116–121.
17. Edmunds, K. B.: "Cost Estimating Building Service System, 1," *ACHV*, vol. 62, February 1965, pp. 93–94.
18. "ENR Indexes Track Costs over the Years," *Engineering News-Record*, vol. 222, March 23, 1989, pp. 52–63.
19. Gido, J.: *Project Management Software Directory*, Industrial Press Inc., New York, 1985.
20. Guthrie, K. M., "Capital Cost Estimating," *Chemical Engineering*, vol. 76, March 24, 1969, pp. 114–142.
21. _____: "Capital and Operating Costs for 54 Chemical Processes," *Chemical Engineering*, vol. 77, June 15, 1970, p. 140.
22. _____: *Process Plant Estimating, Evaluation, and Control*, Craftsman Book Company, Solana Beach, Calif., 1974, pp. 115, 334.
23. _____: *Process Plant Estimating, Evaluation, and Control*, Craftsman Book Company, Solana Beach, Calif., 1974, p. 23.
24. _____, et al.: *Managing Capital Expenditures for Construction Projects*, Craftsman Book Company, Solana Beach, Calif., 1977, p. 80.
25. Hackney, J. W.: *Control and Management of Capital Projects*, 2d ed., McGraw Book Company, New York. (In preparation)
26. Hall, R. S., W. M. Vatavuk, and J. Matley: "Estimating Process Equipment Costs," *Chemical Engineering*, November 21, 1988, pp. 66–75.

27. Hand, W. E., "From Flow Sheet to Cost Estimate," *Petroleum Refiner*, vol. 37, September 1958, pp. 331–334.

28. ———: "Estimating Capital Costs from Process Flow Sheets," *AACE Cost Engineers' Notebook*, January 1964.

29. Hensley, E. F.: "The Unit Operations Approach," *AACE Transactions*, June 1967.

30. Hirsch, J. H., and E. M. Glazier: "Estimating Plant Investment Costs," *Chemical Engineering Progress*, vol. 60, December 1964, pp. 23–25.

31. Humphreys, K. K., ed.: *Project and Cost Engineers' Handbook*, 2d ed., Marcel Dekker, Inc., New York, 1984.

32. ———, and P. Wellman: *Basic Cost Engineering*, 2d ed., Marcel Dekker, Inc., New York, 1986.

33. Kerzner, H.: *Project Management: A System Approach to Planning, Scheduling and Controlling*, 3d ed., Van Nostrand Reinhold Company, New York, 1989.

34. Kharbanda, O. P.: *Process Plant and Equipment Cost Estimation*, Craftsman Book Company, Solana Beach, Calif., 1979.

35. Kumar, P.: "Fluctuating Currency Rate Effects on an International Project," *Engineering and Process Economics*, vol. 3, 1978, pp. 115–121.

36. Lang, H. J.: "Simplified Approach to Preliminary Cost Estimates," *Chemical Engineering*, vol. 55, June 1948, pp. 112–113.

37. LeBlanc, L.: "Platform Price Tag Climbs," *Offshore*, vol. 24, September 1978, pp. 86–88.

38. Lynn, L., and R. F. Howland: "Use Capital Ratio," in H. Popper (ed.), *Modern Cost Engineering Techniques*, McGraw-Hill Book Company, New York, 1970.

39. Matley, J.: "CE Plant Cost Index — Revised," *Chemical Engineering*, April 19, 1982, pp. 153–156.

40. McGlamery, G. G., R. L. Torstrick, W. J. Broadfoot, J. P. Simpson, S. V. Tomlinson, and J. F. Young: *Detailed Cost Estimates for Advanced Effluent Desulfurization Processes*, EPA-600/2-75-006, IERL, U.S. Environmental Protection Agency, Research Triangle Park, N. C., January 1975.

41. *Means Building Construction Cost Data*, R. S. Means Company, Inc., Kingston, Mass., published annually.

42. Miller, C. A.: "Factor Estimating Refined for Appropriation of Funds," *AACE Bulletin*, vol. 7, September 1965, pp. 92–118.

43. ———: "New Cost Factors Give Quick, Accurate Estimates," *Chemical Engineering*, vol. 72, September 13, 1965, p. 226.

44. ———: "New Cost Factors Give Quick, Accurate Estimates," in H. Popper (ed.), *Modern Cost Engineering Techniques*, McGraw-Hill Book Company, New York, 1970, p. 58.

45. ———: "Current Concepts in Capital Cost Forecasting," *Chemical Engineering Progress*, vol. 69, May 1973, p. 77.

46. ———: "Process Plant Capital Cost Estimating — A Science Rather Than an Art," *AACE Cost Engineers' Notebook*, June 1978.

47. ———: "Converting Construction Costs from One Country to Another," *Chemical Engineering*, vol. 80, July 2, 1979, pp. 89–93.

48. *Modern Cost Engineering: Methods and Data*, McGraw-Hill Book Company, New York, 1979.

49. *Modern Cost Engineering Methods and Data*, vol. II, McGraw-Hill Publishing Company, New York, 1984.

50. Neil, J. M.: *Construction Estimating for Project Control*, Prentice Hall, Inc., Englewood Cliffs, NJ, 1982.

51. Nelson, W. L.: "Refinery Construction Cost Indexes," *Oil and Gas Journal*, vol. 54, October 1, 1956.

52. Ostwald, P. F.: *Cost Estimating for Engineering and Management*, Prentice-Hall, Inc., Englewood Cliffs, N.J., 1974, chaps. 6, 7.

53. Page, J. S., *Conceptual Cost Estimating Manual*, Gulf Publishing Company, Houston, Tex., 1984.

54. ———: *Cost Estimating for Pipelines and Marine Structures*, Gulf Publishing Company, Houston, Tex., 1977, p. 15.

55. Page, J. S.: *Estimator's Piping Man-Hour Manual*, 4th ed., Gulf Publishing Company, Houston, Tex., 1987.

56. Park, W. R.: *Construction Bidding for Profit*, John Wiley & Sons, New York, 1979, p. 39.

57. Patrascu, A.: *Construction Cost Engineering*, Craftsman Book Company, Carlsbad, Calif., 1978, p. 55.
58. _____: *Construction Cost Engineering Handbook*, Marcel Dekker, Inc., New York, 1988.
59. "PEI Plant Cost Indices," *Process Economics International*, published quarterly.
60. Perry, R. H., and C. H. Chilton, eds., *Perry's Chemical Engineers' Handbook*, 5th ed., McGraw-Hill Book Company, New York, 1973.
61. Peters, M. S., and K. D. Timmerhaus: *Plant Design and Economics for Chemical Engineers*, 3d ed., McGraw-Hill Book Company, New York, 1980
62. Pikulik, A., and H. E. Diaz: "Cost Estimating for Major Process Equipment," *Chemical Engineering*, vol. 84, October 10, 1977, p. 106.
63. Popescu, C., and A. Hamiani: *Directory of Microcomputer Software for Cost Engineering*, Marcel Dekker, Inc., New York, 1985.
64. Popper, H.: *Modern Cost Engineering Techniques*, McGraw-Hill Book Company, New York, 1970.
65. Rudd, D. F., and C. C. Watson, *Strategy of Process Engineering*, John Wiley & Sons, Inc., New York, 1968.
66. *Statistics Canada Bulletin*, vol. 2, no. 7, June 1973, p. 59.
67. Stevens, R. W.: "Equipment Cost Indexes for Process Industries," *Chemical Engineering*, vol. 54, November 1957, pp. 124–126.
68. Stewart, R. D., and R. M. Wyskida: *Cost Estimator's Reference Manual*, John Wiley & Sons, Inc., New York, 1987.
69. *The Cost Engineer*, Journal of the Association of Cost Engineers, published monthly.
70. Uhl, V. W.: *Standard Procedure for Cost Analysis of Pollution Control Operations—Vol. I*, EPA-600/8-79-018a, IERL, U.S. Environmental Protection Agency, Research Triangle Park, N.C., June 1979.
71. U.S. Department of Labor, Bureau of Labor Statistics: *Consumer Price Index*, Washington, D.C., published monthly.
72. U.S. Department of Labor, Bureau of Labor Statistics, *Monthly Labor Review*, Washington, D.C., published monthly.
73. U.S. Department of Labor, Bureau of Labor Statistics, *Producer Prices and Price Indexes*, Washington, D.C., published monthly.
74. Viola, J. L., Jr.: "Estimate Capital Costs via a New, Shortcut Method," *Chemical Engineering*, vol. 88, April 6, 1981, pp. 80–86.
75. Walas, S. M.: "Plant Investment Costs by the Factor Methods," *Chemical Engineering Progress* vol. 57, June 1961, pp. 68–69.
76. Webster, F. M., *Survey of Project Management Software Packages*, Project Management Institute, Drexel Hill, Pa., 1985.
77. Williams, R., Jr.: "Six-Tenth Factor Aids in Approximating Costs," *Chemical Engineering*, vol. 54, December 1947, pp. 124–125.
78. Woods, D. R.: Financial Decision Making in the Process Industry, Prentice-Hall, Inc., Englewood Cliffs, N.J., 1975.
79. Woodward, W. A., A. C. Elliott, H. L. Gray, and D. C. Matlock: *Directory of Statistical Microcomputer Software*, Marcel Dekker, Inc., New York, 1988.
80. Wroth, W. F.: "Factors in Cost Estimation," *Chemical Engineering*, vol. 67, October 1960, p. 204.
81. Yahdav, D.: *Microcomputer Project Management Software Comparison Report*, 1 Soft Decision, Mill Valley, Calif., 1988.
82. Zevnik, F. C., and R. F. Buchanan: "Generalized Correlation of Process Investment," *Chemical Engineering Progress*, vol. 59, February 1963, p. 70.
83. Zimmerman, O. T.: "Use of Ratio Cost Factors in Estimating," *Cost Engineering*, vol. 10, October 1965, pp. 13–17.

# CHAPTER

# 15

## OPERATING-
## COST
## ESTIMATION

*J.H. Black*

## 15.1 DEFINITIONS

*Operating costs* differ from capital costs mainly in that operating costs must be paid repeatedly, in contrast to *capital costs,* which involve a one-time payment. Because they recur repeatedly, operating costs are often called *repeated costs,* whereas capital costs are often referred to as *first costs*. It is important to understand these two types of costs. The operating cost, or manufacturing cost, is the expense involved in keeping a project, operation, or piece of equipment running and producing. The terminology and the connection between the component cost items are shown in Table 15.1.

In this chapter the terms *operating cost* and *manufacturing cost* are used synonymously. In a broad sense, this cost is incurred at the factory level. In addition, a company incurs costs above the factory level associated more closely with management; here it is termed *general expense,* but it is also known as *administrative, sales,* and *general expense*. The sum of the operating cost, or manufacturing cost, and the general expense is designated here as the *total product cost*.

The arrangement shown in Table 15.1 is subject to variation depending on accounting procedures and circumstances. Thus royalty payments may be directly proportional to throughput and treated as a direct production cost. In some cases, however, the purchase of a patent in connection with a project may be regarded as an expense in lieu of research and development and may appear as part of general expense. Alternatively, the royalty can be a paid-up royalty and hence will be included in the capital investment. Nevertheless, Table 15.1 illustrates the classification of operating costs and will serve well for defining the terms and showing their relationship.

**434**

**TABLE 15.1.**
# Components of total product cost

I. Operating cost or manufacturing cost
  A. Direct production costs
    1. Materials
      a. Raw materials
      b. Processing materials
      c. By-product and scrap credit
      d. Utilities
      e. Maintenance materials
      f. Operating supplies
      g. Royalties and rentals
    2. Labor
      a. Direct operating labor
      b. Operating supervision
      c. Direct maintenance labor
      d. Maintenance supervision
      e. Payroll burden on all labor charges
        i. FICA tax
        ii. Workers' compensation coverage
        iii. Contributions to pensions, life insurance, hospitalization, and dental plans
        iv. Vacations, holidays, sick leave, overtime premium
        v. Company contribution to profit sharing
  B. Indirect production costs
    1. Plant overhead or burden
      a. Administration
      b. Indirect labor
        i. Laboratory
        ii. Technical service and engineering
        iii. Shops and repair facilities
        iv. Shipping department
      c. Purchasing, receiving, and warehousing
      d. Personnel and industrial relations
      e. Inspection, safety, and fire protection
      f. Automotive and rail switching
      g. Accounting, clerical, and stenographic
      h. Communications — telephone, mail, and teletype
      i. Plant custodial and protective
      j. Plant hospital and dispensary
      k. Cafeteria and clubrooms
      l. Recreational activities
      m. Local contributions and memberships
      n. Taxes on property and operating licenses
      o. Insurance — property, liability
      p. Nuisance elimination — waste disposal and pollution control
    2. Depreciation
  C. Contingencies
  D. Distribution costs
    1. Containers and packages
    2. Freight
    3. Operation of terminals and warehouses
      a. Wages and salaries — plus payroll burden
      b. Operating materials and utilities
      c. Rental or depreciation

**TABLE 15.1 (Continued)**

II. General expense
  A. Marketing or sales costs
    1. Direct
      a. Salespersons' salaries and commissions
      b. Advertising and promotional literature
      c. Technical sales service
      d. Samples and displays
    2. Indirect
      a. Sales supervision
      b. Travel and entertainment
      c. Market research and sales analysis
      d. District office expenses
  B. Administrative expense
    1. Salaries and expenses of officers and staff
    2. General accounting, clerical, and auditing
    3. Central engineering and technical
    4. Legal and patent
      a. Inside company
      b. Outside company
      c. Payment and collection of royalties
    5. Research and development
      a. Own operations
      b. Sponsored, consultant, and contract work
    6. Contributions and dues to associations
    7. Public relations
    8. Financial
      a. Debt management
      b. Maintenance of working capital
      c. Credit functions
    9. Communications and traffic management
    10. Central purchasing activities
    11. Taxes and insurance

*Source:* Adapted from R. H. Perry et al., *Chemical Engineers' Handbook*, 5th ed., McGraw-Hill Book Company, New York, 1973, pp. 25–27.

Operating costs fall into two major classifications: direct and indirect. *Direct costs* tend to be proportional to throughput, such as raw materials, and are called *variable costs.* *Indirect costs* tend to be independent of throughput, such as local property taxes, and are called *fixed costs.* Some costs are neither fixed nor directly proportional to output and are known as *semivariable costs* or *regulated costs.* These concepts were developed in Chap. 7 in connection with break-even and minimum-cost analysis.

## 15.2 OPERATING-COST ESTIMATION — GENERAL

Operating-cost estimates are made for a variety of reasons. Preliminary estimates act as a screening procedure to minimize the expenditure of time, effort, and money on uneconomical projects and to choose the more economical route among alternatives to the same end result. The use of preliminary estimates to guide research efforts is common practice. Detailed cost estimates are made after screening estimates when the choice has been narrowed to a relatively few alternatives.

Some judgment is required in determining how far to investigate individual operating-cost items. The estimate will indicate which costs dominate and deserve more study and which costs, even if drastically misjudged, will not produce significant changes in the estimate.

It is important to calculate costs at full and reduced levels inasmuch as in practice operating costs are not a linear function of the production rate. Calculations should be made and reported in such a way that they can be reviewed in the future without ambiguity.

Frequently the operating cost to be calculated is connected with changes in and additions to existing equipment and processing, and an incremental analysis can be made. Costs based on an incremental analysis should be used only in connection with the specific project for which they were intended. Even so, it might be well to make an integral evaluation for the entire project. The increment might pay off, yet it may be wise to abandon the entire project.

Operating costs are commonly calculated on one of three bases: daily, unit-of-product, or annual. Unit-of-product cost is usually expressed as dollars per unit of end product. The annual basis has some advantages over the daily basis because (1) the effect of seasonal variations is evened out, (2) the on-stream time or equipment operating factor is considered, (3) it is more adaptable to less than full capacity operation, (4) it provides an expeditious way of considering infrequently occurring large expenses, such as the regularly scheduled replacement of parts or equipment or the periodic turnaround costs in a refinery, and (5) the calculations are more directly usable in a profitability analysis.

It is best to use a prepared form for estimating operating costs, which should conform as closely as possible to that used for recording and controlling costs. Ideally the form should provide such supplementary information as the date, the capital investment, a cost index, identification of the product or service, plant location, plant capacity, and operating rate or operating days. The supplementary information is important because these factors affect the component costs and hence the finished estimate.

Table 15.10 illustrates a prepared form and Table 15.11 presents an outline or checklist to match it. Other suggested forms appear elsewhere in the literature [16, 20, 43].

The best source of information for use in operating-cost estimates is the data from similar or identical projects within the company. Most companies have extensive records of their operations, so that reliable estimates of operating costs for new projects can be made from the existing records if the new project is similar to installations now operated by the company. Adjustments for increased costs as a result of inflation must be made, and differences in plant site and geographical location must also be considered. Company experience, however, probably yields the most accurate information, particularly for preliminary estimates.

Literature references also provide much information on costs [9, 34]. Many engineering journals and magazines also publish cost information [1, 2, 3, 8, 15, 22, 23].

Methods for estimating cost elements in the absence of specific information are discussed in the following sections. The elements of cost are presented essentially in the order shown in Table 15.11.

Perhaps the most serious sources of error in estimating operating costs result from overlooking elements of cost. A tabular form and checklist cannot always guide the analyst to every cost component. Detailed cost estimation requires that the cost engineer be much more the engineer than a cost accountant. Technical familiarity with the project is essential, and no analyst should undertake detailed cost estimation in fields not directly within his or her specialty.

Many computer programs have been developed for performing operating-cost calculations. While the initial program may be expensive, its use is justified when a number of variations are to be explored by repetitive calculations. Manual calculation is still used, however, because appraising the data takes more effort than making the actual calculations. Care must be exercised in using computer programs so that the resulting estimates are not mechanically derived to the extent that judgment, so necessary in evaluating alternatives, is lost.

## 15.3  RAW MATERIALS

Raw materials may constitute one of the largest items in the manufacturing cost, varying from 10 to 50 percent of the total product cost [33]. The operating-cost estimate should list raw materials, the units in which the materials are purchased, unit cost, and cost per unit of time per unit of production. Credit should be given for by-products and scrap.

Information on the quantities of the raw materials required can come from published articles, from experiments, or from calculations to which reasonable yields or waste allowances are applied.

Prices of raw materials purchased from outside sources are available on request from the sources of supply. Quality is an important consideration when buying raw materials, such as concentration of an ore, surface finish of a metal, etc. Generally unit price decreases with quantity. However, if the material is available at a low price as a by-product, it may be available in large quantities, if at all, only at a substantially higher unit price.

Prices suitable for estimating purposes can be obtained from catalogs, price lists, or sources such as spot prices published in *Iron Age, Chemical Marketing Reporter, Chemical and Engineering News*, and *European Chemical News* for European prices. Actual prices are negotiated and will vary from the quotations depending on quantity, quality, and such factors as the state of competition, the duration of a contract, and penalties for contract violations. If the raw material is in a competitive market, the published price could be discounted by 20 percent [16].

Raw materials produced captively present a special situation. If the product is now sold, the transfer price is normally the going market price, sales price, less any direct sales costs not incurred, plus freight and other transportation costs for the transfer. If the transferred material is at an intermediate stage that has no going market price, the transfer price can be taken as the going price of the nearest downstream

**TABLE 15.2**
**Approximate shipping costs (1990)**

| Transportation mode | Approximate shipping costs, $/ton · mi |
|---|---|
| Pipeline | 0.004–0.006 |
| Barge | 0.009–0.02 |
| Tanker | 0.003–0.01 |
| Rail | 0.025–0.08 |
| Truck | 0.09 –0.13 |

converted product having a going price, less any direct sales cost not incurred, less standard manufacturing costs for any operations which are avoided by the transfer, plus the distribution cost necessary for the transfer.

Periodic makeup of losses, such as for a catalyst in a chemical process, is included as a raw material. If the initial fill of these materials has a life of more than 1 year, its cost is treated as nondepreciable capital investment or working capital. If the life is less than 1 year, the cost is included as a start-up expense. In any case the periodic makeup is an operating cost.

Fuels used by a project, such as natural gas for a heat-treating furnace, may be considered as a raw material or a utility, depending on the accounting procedure and circumstances. Costs of fuels are discussed in the following section under utilities. In some chemical processes fuels may be used as a bona fide raw material, e.g., cracking refinery gases for the production of ethylene, and also as a fuel. The two uses may be combined as one raw material or segregated into two costs as desired.

The quantities of by-products are estimated in much the same fashion as the raw material quantities. The prices of by-products can be estimated from market or selling prices of salable materials less purification, packaging, selling, and transportation costs. If the by-products are intermediate products for which no market exists, they may be credited at their net value to downstream or subsequent operations at a cost value equal to their replacement value.

Many price quotations for raw materials will be FOB the vendor's plant or some basing point, and freight to the plant and local handling costs must be added. Shipping rates are a complicated topic. The company's traffic department can help, or the carriers can be consulted.

Large-volume, long-haul (over 250 mi) rates in dollars per ton-mile are shown in Table 15.2. The lower values in each range apply to longer distances and larger shipments, while the higher values apply to shorter distances and smaller shipments [5]. Rates for distances less than 250 mi may be two to five times higher. Small-volume (less than carload lots) rates may be two to three times higher.

## 15.4 UTILITIES

Utility prices can be obtained from company cost records for nominal utility increments. Otherwise they will have to be determined by inquiry from outside sources of a study of the company-owned system if the utility is self-generated. Unit costs often decrease substantially as demand goes up. Rates for most utilities vary considerably with location. Bauman [5] gives a comprehensive study of rates. Table 15.3 gives typical values for common utilities.

Steam can be used at several pressure levels. High-pressure steam can be reduced to lower pressures by the use of steam turbines that are used for the production of motive power. This is an economical practice if use at various pressure levels can be kept in balance. The allocation of costs between the different pressure levels can become complicated, and the balance can be upset.

Steam is generally charged on the basis of thousand pounds used but can be charged on the basis of million Btus. Credit should be given for the heating value of condensate returned to the boiler. If the condensate is frequently contaminated by processing leaks, a penalty may be assessed.

The cost of steam depends principally on the cost of the fuel and the capacity and pressure level of the generating units. The full cost of steam includes fuel, boiler-water treatment, operating labor, depreciation, and maintenance. A rough estimate is two or three times the cost of fuel.

**TABLE 15.3**
**Rates for industrial utilities**

| Utility | Cost, $ (1990) | Unit |
|---|---|---|
| Steam | | |
| High pressure (400–600 psig) | 2.20–5.50 | 1000 lb |
| Medium pressure (150–300 psig) | 2.10–5.50 | 1000 lb |
| Low pressure (30–100 psig) | 1.80–5.50 | 1000 lb |
| Low low pressure (5–25 psig) | 0.70–1.10 | 1000 lb |
| Electricity | | |
| Purchased | 0.03–0.07 | kWh* |
| Self-generated | 0.017–0.05 | kWh |
| Water | | |
| Cooling water | 0.02–0.35 | 1000 gal |
| City water | 0.80–1.80 | 1000 gal |
| Filtered water | 0.10–0.35 | 1000 gal |
| Softened water | 0.15–0.60 | 1000 gal |
| Demineralized water | 0.95–2.00 | 1000 gal |
| Boilerfeed water | 1.95–5.60 | 1000 gal |
| Condensate | 1.45–4.60 | 1000 gal |
| Compressed air | | |
| Process air | 0.10–0.25 | 1000 ft$^3$ |
| Instrument air | 0.18–0.40 | 1000 ft$^3$ |
| Natural gas | 1.75–6.75 | 10$^6$ Btu |
| Fuel oil | 0.60–1.20 | gal |
| Coal | 22.00–60.00 | ton |

*Highly dependent on the load factor and the location.

Water costs vary widely depending on the quality and quantity required. If the water can become contaminated, some provision for purification cost before disposal should be provided. Table 15.3 gives some representative water costs. Water rates may be obtained from the local water supplier or through the American Water Works Association. River water on a once-through basis without treatment costs only a few cents per 1000 gal, while most industrial water falls in the range of $0.80 to $1.80/1000 gal.

All major contracts for electrical energy contain a demand component, an energy component, and a variable-fuel-cost component. The demand charge is based on the maximum power requirement drawn by a plant during a measurable period, usually 15 or 30 min in any 1 month. The energy charge is based on kilowatt-hours used. The ratio of average usage to demand is termed the *load factor*. Most electric-power rate schedules are designed to yield greater savings preferably by improving the load factor more than by increasing the load. The fuel-adjustment charge ties the cost, say, to the price of coal, but it is an escalation clause and is used even for electricity generated from water power.

Electric rates may be obtained from the utilities. The Federal Power Commission (FPC) publishes and distributes at a nominal charge the published rates of all public utilities in the United States. Bauman [5] illustrates how rates vary with the location, load, and load factor. Table 15.3 shows some typical power costs.

Cost of fuels varies with the type of fuel and the number of Btus per unit of fuel. Special preheating and firing equipment is required to burn the cheaper but heavier fuel oils. An analysis of fuel economy is sometimes necessary to select the type and quality of fuel for a particular location and situation. Choice of fuel must allow for the required storage, which can make a difference in coal versus oil versus gas.

The estimator must check carefully to make sure that proper provisions are made for delivery of utilities to the site. A common estimating error is the assumption that utilities are available at the site when, in fact, they may not be.

The required utilities can sometimes be estimated in the early stages of project development from available information about similar operations [13, 15, 37] or estimated from a preliminary design. Utility consumption does not vary directly with the production rate, and variation to the 0.9 power of the capacity ratio might be a better relationship. A factor should be applied to the estimated quantities of utilities to allow for equipment efficiencies, other losses, and contingencies. Appropriate factors for the various utilities are suggested in the literature [11], as shown in Table 15.4. The exact value to use in each range depends on the load and the site conditions.

**TABLE 15.4**
**Utility factors for losses and contingencies**

| Utility | Steam | Water | Fuel | Electricity |
|---------|-------|-------|------|-------------|
| Factor | 1.25–2.00 | 1.10–1.50 | 1.10–1.50 | 1.10–1.25 |

## 15.5 OPERATING LABOR

The most accurate means of establishing operating labor requirements is with a complete manning table, but shortcut methods are satisfactory for most cost estimates.

If the project is a novel one, work-hour requirements can be estimated using elemental time values as developed by Haines [17] or using labor requirements as given in the literature [13, 15, 33]. Table 15.5 shows typical labor requirements for some types of process equipment [33]. Ostwald [29] gives examples and references for estimating labor for such nonprocess industries as metalworking, sheet metal operations, craft work, construction, and machine shop operations. Wessell [46] relates labor requirements to process plant capacity by the equation

$$\frac{\text{Operating work-hour}}{\text{Tons of product}} = t \frac{\text{no. of process steps}}{(\text{capacity in tons/day})^{0.76}} \qquad (15.1)$$

where $t$ is determined by the type of process as follows:

Batch operations with a maximum of labor, 23

Operations with average labor requirements, 17

Well-instrumented continuous-process operations, 10

The equation recognizes the improvement in labor productivity as plants increase in throughput, and can be used to extrapolate known work-hour requirements from one plant to another of different capacity.

It has been suggested that labor requirements vary to about the 0.2 to 0.25 power of the capacity ratio when processing-plant capacities are scaled up or down [21, 28].

**TABLE 15.5**
**Typical labor requirements for process equipment**

| Type of equipment | No. of workers per unit shift |
|---|---|
| Dryer, rotary | 0.5 |
| Dryer, spray | 1 |
| Dryer, tray | 0.5 |
| Centrifugal separator | 0.25–0.5 |
| Crystallizer, mechanical | 1/6 |
| Filter, vacuum | 1/8–1/4 |
| Evaporator | 0.25 |
| Reactor, batch | 1 |
| Reactor, continuous | 0.5 |
| Steam plant (100,000 lb/h) | 3 |

*Source:* M. S. Peters and K. D. Timmerhaus, *Plant Design and Economics for Chemical Engineers,* 3d ed., McGraw-Hill Book Company, New York, 1980, p. 195.

**Example 15.1.** A plant to produce 50 tons/day of product *A* has been designed. The plant is a modern, highly automated fluid-processing plant. The raw material for product *A* goes through the four processing steps of heat transfer, chemical reaction, solvent extraction, and distillation. Estimate the annual labor requirement for continuous operation of the plant for 330 days/year.

*Solution.* From Eq. (15.1),

$$\frac{\text{Operating work-hour}}{\text{Tons of product}} = t\frac{\text{no. of process steps}}{(\text{capacity, tons/day})^{0.76}}$$

$$= 10\frac{4}{(50)^{0.76}} = 2.0457$$

$$\text{Operating work-hour/year} = 50 \times 330 \times 2.0457 = 33,750$$

The average hourly rate for labor can be obtained from company records or data in such sources as *Engineering News-Record,* and the monthly *Employment and Earnings* from the U.S. Department of Labor. Table 15.6 gives labor rates in selected industries as of April 1989. Labor rates, such as those shown in Table 15.6, can be adjusted for geographical differences in rates and productivity and for rate escalation using the methods and information given in Peters and Timmerhaus [33].

Labor rates have been increasing, seemingly on an accelerated basis; hence care must be taen to get the current figures. Average rates should include the effect of shift and scheduled overtime premiums, but care is required to avoid the overlap with indirect payroll items which are added separately.

Supervision of labor is generally estimated as a percentage of operating labor, a typical value being 20 percent. The range is usually 10 to 25 percent.

**TABLE 15.6**
**Average labor rates in selected industries, March 1989**

| Industry | Earnings, $/h |
| --- | --- |
| Mining | 13.10 |
| Construction | 13.26 |
| Primary metals | 12.28 |
| Manufacturing | 10.40 |
| Fabricated metal products | 10.44 |
| Machinery, except electrical | 11.21 |
| Electrical and electronic equipment | 10.30 |
| Food | 9.31 |
| Paper | 11.83 |
| Chemicals | 12.93 |
| Petroleum | 15.62 |
| Transportation and public utilities | 12.48 |

*Source: Employment and Earnings,* U.S. Department of Labor, Bureau of Labor Statistics, Washington, D.C., vol. 36, no. 4, April 1989.

## 15.6   PAYROLL CHARGES

Payroll charges involve the cost of workers' compensation, pensions, group insurance, paid vacations and holidays, social security and unemployment taxes, contributions to profit sharing, and a growing list of other fringe benefits. Payroll charges are computed as a percentage of the labor and supervision cost, and generally the same rate is used for both labor and supervision. The best figures are from the company's records, but an accounting definition is important to avoid counting some items that may be included under general expense. A rough estimate is 30 to 45 percent of the labor-plus-supervision cost. This figure is rising and may eventually reach 60 percent or more, as it does in parts of Europe today.

## 15.7   MAINTENANCE

Records for the company's existing plants are the only reliable source of maintenance cost. Little has been published on the subject, but with experience, maintenance can be estimated as a function of investment. Maintenance cost as a percentage of the investment per year ranges from 11 percent for a complex plant with severe corrosive conditions to 2 percent for a relatively simple plant with mild, noncorrosive conditions. For example, average maintenance in a petroleum refinery is about 5 percent of the investment per year, but for individual units it ranges from about 3 percent for dewaxing units to about 7 percent for sulfuric acid alkylation units.

Generally maintenance can be taken as 60 percent labor cost and 40 percent material cost. When a unit is operating at 75 percent of capacity, the maintenance cost will be about 85 percent of the maintenance cost at 100 percent of capacity. When operating at 50 percent of capacity, the maintenance cost will be about 75 percent of the maintenance cost at 100 percent of capacity.

Maintenance is known to increase with the age of equipment, but the estimator must use an average figure. The procedure is equalized, in part, by the use of average value for depreciation, whereas accelerated depreciation may, in fact, be used for tax purposes. Thus the overall cost estimate is balanced because average values for maintenance and depreciation are used.

A project may create the need for additional maintenance supervisors and foremen. For major projects maintenance manning and cost tables may be required to establish the additional maintenance supervision cost; but for small additions no additional maintenance supervision may be required.

## 15.8   OPERATING SUPPLIES

Operating supplies include such things as lubricating oil, instrument charts, custodial supplies, and other items normally used by the project exclusive of those included as raw, repair, or packaging materials. Company records should be used for this cost when available. If they are not, operating supplies may be assumed to be about 6 percent of operating costs. Some authorities base such costs on a percentage of investment, such as 0.5 to 1.0 percent of the plant investment [33].

## 15.9 LABORATORY AND OTHER SERVICE COSTS

Depending on company practice and the type of project, operating costs may include service charges by other units of the company. Charges made by a central control laboratory are one example.

Laboratory costs can be estimated (1) on the basis of past experience; (2) on the basis of work-hours required; (3) as a percentage of the operating labor cost, in the range of 10 to 20 percent; and (4) from the literature [9, 17, 33].

## 15.10 WASTE DISPOSAL AND AVOIDANCE OF NUISANCES

An increasingly important item is the cost for the reduction of various nuisances. Common nuisances are the pollution of air and water, but any such problem, such as odor or noise, must be considered. Even if the nuisance is confined to company property, it must not be overlooked. For example, a moderately high noise level may appear to be harmless, but it may become a costly liability as a result of lawsuits brought by employees for alleged impairment of health. There has been much publicity about chemical-waste disposal problems at sites such as Love Canal and the Chemical Control Corporation's dump in Elizabeth, New Jersey [40]. These too are serious nuisances which have become costly liabilities to the owners and the public.

There are excellent discussions of the laws, regulations, pollutants, and methods of pollution control in the literature [10]. Peters and Timmerhaus [33] give an excellent presentation of waste disposal of all types, including the sources of regulations, key aspects of environmental regulations, and pollution-abatement methods for various types of pollutants, including, air, water, land, noise, and thermal pollution sources. Baasel [4] presents an excellent discussion of the various pollution-abatement problems and gives some costs.

The costs associated with pollution control have been published widely in the literature by the Environmental Protection Agency, Tennessee Valley Authority, research institutes, and private corporations [26, 27, 35, 39, 42]. Table 15.7 shows some typical incremental costs for selected control technologies in power stations, and Example 15.2 illustrates the use of these incremental data.

**Example 15.2.** Suppose we wish to get a rough approximation for the increase in capital costs from adding $SO_x$ (limestone), $NO_x$ (combustion modifications), and particulate matter (cold ESP) controls to a power plant. The plant has a 1000-MW boiler, and the costs are to be based on 1980 costs.

**TABLE 15.7**
**Typical incremental costs for selected control technologies, mid-1980 dollars**

| | 500 MW | | 1000 MW | |
|---|---|---|---|---|
| | Capital, $/kW | Annualized, mills/kWh | Capital, $/kW | Annualized, mills/kWh |
| Base generation system (typical heat rate, Btu/kWh) | | | | |
| Conventional fossil fuel-fired boilers | | | | |
| High-sulfur eastern coal (9800) | 640–775 | 20–24 | 575–700 | 17–21 |
| Low-sulfur western coal (9200) | 670–820 | 20–25 | 605–740 | 18–22 |
| Liquid fuel (9200) | 395–490 | 12–15 | 360–440 | 11–13 |
| Conventional nuclear light-water reactor (10,400) | 1000–1220 | 30–36 | 900–1100 | 26–33 |
| Combined cycle | | | | |
| Liquid fuel (7500) | 360–440 | 12–15 | 320–395 | 11–13 |
| Low-Btu gasification (8400) | 870–1175 | 28–36 | 780–1060 | 24–32 |
| Medium-Btu gasification (8200) | 880–1190 | 28–36 | 790–1070 | 24–33 |
| Fluidized bed combustion (FBC) | | | | |
| Atmospheric FBC (9500) | 690–930 | 21–28 | 625–840 | 20–25 |
| Pressurized FBC (8800) | 835–1260 | 26–38 | 750–1130 | 24–34 |
| | | | | |
| Incremental costs | | | | |
| Fuel* | | | | |
| High-sulfur eastern coal | | 10–16 | | 11–16 |
| Low-sulfur western coal | | 15–20 | | 14–20 |
| Physically cleaned coal | | 13–18 | | 13–18 |
| Chemically cleaned coal | | 18–29 | | 18–29 |
| Solvent-refined coal | | 33–40 | | 33–40 |
| Liquefied coal | | 40–46 | | 40–46 |
| Liquid fuel | | 20–33 | | 20–33 |
| Uranium | | 7–8 | | 7–8 |

Environmental control technology

| | | | | |
|---|---|---|---|---|
| Sulfur control | | | | |
| Flue gas desulfurization | | | | |
|   Limestone | 100–135 | 4.6–5.5 | 90–120 | 4.1–5.0 |
|   Wellman Lord | 115–155 | 4.6–5.8 | 105–140 | 4.2–5.3 |
|   Magnesia | 120–180 | 4.7–6.5 | 110–165 | 4.3–5.8 |
|   Dual alkali | 125–185 | 5.9–7.2 | 110–165 | 5.4–7.0 |
| Fuel gas cleanup | | | | |
|   Low-Btu gas | 185–250 | 6.7–8.6 | 170–225 | 6.1–7.6 |
|   Medium-Btu gas | 165–230 | 5.5–7.2 | 150–200 | 5.0–6.5 |
| Fluidized bed combustion | | | | |
|   Limestone or dolomite | 30–45 | 1.6–2.1 | 25–40 | 1.4–2.0 |
| Particulate control | | | | |
|   ESP-cold | 35–45 | 1.2–1.4 | 30–40 | 1.2–1.3 |
|   ESP-hot | 40–55 | 1.6–2.0 | 40–50 | 1.5–1.8 |
|   Fabric filter | 60–80 | 2.2–2.8 | 55–70 | 2.0–2.5 |
|   Wet scrubber | 70–85 | 2.5–3.0 | 60–80 | 2.2–2.6 |
| $NO_x$ control | | | | |
|   Combustion modifications | 0.5–10.0 | 0.05–0.8 | 0.5–9.0 | 0.05–0.7 |
|   Selective catalytic reduction | 35–60 | 1.7–2.5 | 35–55 | 1.6–2.2 |
|   Water injection for turbines | 4.5–9.0 | 0.8–0.9 | 4.5–7.5 | 0.8–0.9 |
| Thermal discharge control | | | | |
| Evaporative cooling tower | | | | |
|   Fossil | 25–30 | 1.1–1.2 | 20–25 | 0.9–1.1 |
|   Nuclear | 30–35 | 1.6–1.7 | 30–35 | 1.4–1.6 |

*Source:* Based on M. G. Klett, *Typical Costs for Electric Energy Generation and Environmental Controls,* EPA-600/7-79-026, IERL, U.S. Environmental Protection Agency, Research Triangle Park, N.C., January 1979.

*Based on a heat rate of 10,000 Btu/kWh.

*Solution.* To solve this simple problem, one might use average costs. The boiler is to burn high-sulfur eastern coal. Thermal discharge control is also to be installed:

| | | |
|---|---|---|
| Average cost of conventional boiler, eastern coal | | $635/kW |
| Average cost of limestone $SO_x$ control facilities | $105/kW | |
| Average cost of cold ESP control facilities | 35/kW | |
| Average cost of boiler combustion modifications | 5/kW | |
| Average cost of fossil thermal discharge control facilities | 22/kW | |
| Average cost of control facilities | | $167/kW |

From these figures, we get

$$\text{Average percent increase in capital cost} = \frac{\text{environmental}}{\text{base}}(100)$$

$$= \frac{167}{635}(100)$$

$$= 26.3 \text{ percent, or a 26 percent increase}$$

Solid residues from some operations must be disposed of by impounding if they are initially slurries or by hauling to industrial dumps if they are in solid form. The expense of disposal is considered part of the main operation and is included in the operating expense or as a service charge for hauling and further processing by another organization if this is required [43].

## 15.11 ROYALTIES AND RENTALS

Royalties and licensing costs are generally an operating expense. If the charge is proportional to production, or a running royalty, it can be treated as a direct cost. A single-sum payment, or a paid-up royalty, will become part of the project's capital investment, although it is not part of the depreciable investment. A tax specialist may be required to handle specific situations.

Royalty payments may range from 1 to 5 percent of the product's sales price, but the agreement may involve so many combinations that such an approximation should be used only when no data are available.

Certain royalty payments and patent purchase costs may be treated by the accountants under general expense, since they can be regarded as a replacement for research and development expense.

The cost of rentals can be obtained through the purchasing department. Payments for rentals, along with taxes and insurance on the items, are a current expense and should not be overlooked in estimating operating expense.

## 15.12 CONTINGENCIES

An allowance for contingencies should be made for both the direct and indirect costs. Contingencies for the direct costs can be applied as a percentage of the direct cost in the range of 1 to 5 percent.

## 15.13   INDIRECT COSTS

As pointed out earlier, costs independent of the production rate are called *indirect costs*. Such costs include depreciation, real estate taxes, insurance costs, and general plant overhead. These costs must be paid whether or not there is any plant output.

Depreciation was discussed in detail in Chap. 3, along with its various methods of calculation. Suffice it to say, depreciation is generally taken as straight-line depreciation over the useful life of the project, although accelerated depreciation is ordinarily used in determining taxes. Balancing of the estimate by using straight-line depreciation and level maintenance cost over the life of the project was discussed under maintenance. The useful life for cost estimation is not necessarily the life required for tax purposes. The depreciation charge does not include any return on investment.

Real estate taxes depend on the region of the country and the local laws. A figure of 2 to 4 percent of the fixed capital investment could be used for heavily populated areas, while a figure of 1 to 2 percent of the fixed capital investment could be used for less populated areas [33].

Insurance rates depend on the type of product being manufactured and the availability of protection facilities. A figure of 1 percent of the fixed capital investment could be used to cover the costs of fire, explosion, tornado, and product liability insurance [7].

Plant overhead costs cover most of the items shown under plant overhead in Table 15.1. If data are not available from company records, a figure often used is 50 to 70 percent of the total expense for operating labor, supervision, and maintenance [33].

A separate contingency charge for indirect costs may not be necessary. A contingency cost is included in the investment cost. If the factory overhead is based chiefly on the investment, a contingency allowance is automatically included.

Interest on borrowed capital is also an indirect cost if borrowing for the plant is necessary. In such instances the rate is established at the time the money is borrowed. Company records should be a good source of rates for preliminary estimates. A space is provided on the form shown in Table 15.10 in case interest should be included in the estimate.

## 15.14   DISTRIBUTION COSTS

Loading, packing, and shipping costs are best estimated from other similar company operations, modified for the situations under study. Most companies have cost-control systems and standard costs which provide such information. Such costs are sometimes estimated as 1 to 3 percent of sales.

The cost of containers can be obtained from suppliers or published sources. Table 15.8 gives the sizes and costs of some representative containers as of May 1989. There is a growing market for plastic barrels, especially for chemical transport because of their ability to be cut up, incinerated, or recycled into other products such as sewer pipe. Unlike steel barrels, however, there are no industry standard dimensions for plastic barrels and dimensions vary with the manufacturer.

TABLE 15.8
**Sizes and approximate costs of representative containers, May 1989**

| Container size, description | Approx. Unit Cost | Usable Volume, ft$^3$ |
|---|---|---|
| *55-gal plastic drum | $26.00 | 7.35 |
| *30-gal plastic drum | 19.00 | 4.00 |
| *15-gal plastic drum | 11.00 | 2.00 |
| 55-gal fiber drum, 400 lb capacity | 12.00 | 7.35 |
| 30-gal fiber drum, 300 lb capacity | 9.00 | 4.00 |
| 15-gal fiber drum, 225 lb capacity | 6.50 | 2.00 |
| †55-gal steel drum, 20/18 tight head | 22.00 | 7.35 |
| †55-gal steel drum, 20/18 open head | 23.00 | 7.35 |
| ‡85-gal steel salvage drum, 12 gauge, open head | 72.00 | 11.35 |
| 200-lb corrugated shipping boxes, 12 × 12 × 12 in. | 1.00 | 1.00 |
| 200-lb corrugated shipping boxes, 16 × 16 × 9 in. | 1.60 | 1.33 |
| Heavy-duty shipping bags, dbl. strength, padded, $8\frac{1}{2}$ × 12 in. | 0.25 | — |
| Shipping bags, plastic bubble-lined, $8\frac{1}{2}$ × 12 in. | 0.21 | — |
| Utility shipping bags, $8\frac{1}{2}$ × 12 in. | 0.18 | — |

*Meets DOT 34
†Meets DOT 17E.
‡Meets applicable DOT specifications for containing/transporting leaky 55-gal drums.

Fiber barrels are also useful for dry to semiliquid material transport. Because of the ability to add laminated linings, vapor barriers or to change adhesives to make them weatherproof, they are the economic choice for handling material that must be disposed of by incineration. Smaller sizes, such as two 20-gal fiber drums, can be placed in a steel drum for transport and the steel drum can then be reused.

Transportation costs were discussed briefly in Sec. 15.3. Railroad-car rental rates can be obtained from railroads or leasing companies. When railroad cars are owned or leased by the company, the railroads credit the company with about $0.25 to $0.35 per loaded car-mile traveled by the unit, depending on the age and the depreciated value of the car.

Distribution charges are segregated from other manufacturing costs to show the effect of various containers and the method of shipment. The cost without distribution costs is the bulk cost, or manufacturing cost before (packaging and) distribution.

## 15.15 SUMMARY OF OPERATING-COST ESTIMATION

Table 15.9 summarizes the procedures for estimating operating costs when company records are not available. Such methods are useful in the early stages of project evaluation for preparing preliminary estimates. The items in the table follow the order shown in Table 15.10.

**TABLE 15.9**
**Preliminary operating cost estimates**

A. Direct production cost
  1. Materials
    a. Raw materials — estimate from price lists
    b. By-product and scrap credit — estimate from price lists
  2. Utilities — from literature or similar operations
  3. Labor — from manning tables, literature, or similar operations
  4. Supervision — 10 to 25 percent of labor
  5. Payroll charges — 30 to 45 percent of labor plus supervision
  6. Maintenance — 2 to 10 percent of investment per year
  7. Operating supplies — 0.5 to 1.0 percent of investment per year
  8. Laboratory — 10 to 20 percent of labor per year
  9. Waste disposal — from literature, similar operations, or separate estimate
  10. Royalties — 1 to 5 percent of sales
  11. Contingencies — 1 to 5 percent of direct costs
B. Indirect costs
  1. Depreciation — 5 to 10 percent of investment per year
  2. Real estate taxes — 1 to 2 percent of investment per year
  3. Insurance — 0.5 to 1.0 percent of investment per year
  4. Interest — 10 to 12 percent of investment per year
  5. General plant overhead — 50 to 70 percent of labor, supervision, and maintenance
C. Distribution costs
  1. Packaging — estimate from container costs
  2. Shipping — from carriers or 1 to 3 percent of sales

**Example 15.3.** Estimate the total manufacturing cost to produce $2 \times 10^6$ lb/year of product $A$ by the oxidation process. The plant operates 25 h/day for 330 day/year and costs $580,000 to build. The process is batch-operated and consists of a mixing step, a reaction step, a filtration step, and a distillation step. In producing 1 lb of product $A$, 0.83 lb of product $B$, costing $0.60/lb, and 0.62 lb of product $C$, costing $0.45/lb, are consumed. In addition, 0.4 lb of by-product $D$, worth $0.20/lb, is produced. The balance of the raw material is lost up the stack as carbon dioxide. The plant has average maintenance problems, and the process requires average supervision. There are no interest, royalty, or waste disposal costs. The only utilities requirements in the plant are 150 gal of process water, 2 kWh of electricity, and 13 lb of 100-psig steam per pound of $A$ produced.

*Solution.* The solution to this example is shown in Table 15.10, a prepared form, and Table 15.11, an outline or checklist to match it. Labor is found, from Eq. (15.1), as 39.6 work-h/ton.

From Table 15.10, the total manufacturing cost is estimated to be $2,785,000/year, or $1.39/lb.

## 15.16  SHORTCUT METHODS

There is an obvious advantage to shortcut methods of estimating operating costs. One method involves writing an equation which contains all the cost elements discussed in

**452**

**TABLE 15.10**
**Solution to Example 15.3**

Date __1/23/81__

By __JHB__

**Preliminary Manufacturing Cost Estimate**

Location: __Not specified__

Product: __Product A__

Process: __Oxidation__

Manufacturing capital:

|  |  |
|---|---|
| | — |
| | — |
| Total | $580,000 |

Cost index: __270__    Operating days: __330__

Annual production: __2,000,000 lb__

| Raw materials | Annual quantity | Unit cost | $/year |
|---|---|---|---|
| B | 1,660,000 lb | $0.60/lb | 996,000 |
| C | 1,240,000 lb | $0.45/lb | 558,000 |
| | | Gross raw material cost: | 1,554,000 |

$/lb

| By-product (Credit) | | | |
|---|---|---|---|
| D | 800,000 lb | $0.20/lb | 160,000 |
| | | Total credit: | 160,000 |
| | | Net raw material cost: | 1,394,000 |

| Direct cost | Unit | Quantity | $/Unit | |
|---|---|---|---|---|
| Steam, hp | thou. lb | none | | |
| Steam, lp | thou. lb | 26,000 | $2.00 | 52,000 |
| Water, plant | thou. gal | none | | |
| Water, process | thou. gal | 300,000 | $0.60 | 180,000 |

A preliminary or shortcut estimate can be made from a few basic data and estimating other items as a percentage of these. A suggested form and the range of values for a process industry are given in Tables 15.10 and 15.9, respectively.

## QUESTIONS

**15.1.** What does the manufacturing cost include?

**15.2.** What expenses in addition to the manufacturing cost must be considered in computing the total product cost?

**15.3.** Define incremental cost.

**15.4.** Why are incremental costs significant in project appraisal?

**15.5.** Why is it sometimes necessary to compute the return of the business as a whole after the proposed project is completed if the proposed project itself is profitable on an incremental basis?

**15.6.** Why are average figures justified for the maintenance cost and the depreciation expense although neither may be constant with time?

**15.7.** The chapter included some comment on situations in which items are considered to be capital investments rather than operating costs. Why is this distinction important in figuring profitability?

**15.8.** Your company is already making a product at a raw material cost of precisely $0.512/lb out of a total operating cost of $1.623. You are responsible for investigating a proposed new project to quadruple the present capacity. Should the price of raw materials be investigated? Why?

**15.9.** Why is the cost of raw materials or fuels obtained from another company division included as part of the operating cost of the project using them?

**15.10.** What is the procedure for computing transfer prices between one company division and another?

**15.11.** Name at least three project characteristics which affect operating labor.

**15.12.** Why do maintenance cost estimates depend on the size of the project?

**15.13.** The text mentions that when a unit is being operated at 50 percent capacity, the repair costs will be about 75 percent of the repair cost at 100 percent capacity. Why not 50 percent?

**15.14.** What project characteristics make high ratios of supervision to labor necessary?

**15.15.** Why is it desirable to examine the original agreement documents specifying royalty and other patent right payments?

**15.16.** Define general works expense.

**15.17.** What is the difference between general works expense and administrative, sales, and general expense?

**15.18.** What is the difference between bulk cost and total manufacturing cost? Why are both computed?

## PROBLEMS

**15.1.** Reconcile Eq. (15.1) with the statement in the text that labor requirements vary to about the 0.2 to 0.25 power of the capacity ratio.

**15.2.** Calculate the operating cost for the plant in Example 15.3 if the plant is running at one-half capacity.

**15.3.** Calculate the operating cost for the plant in Example 15.3 if the plant is twice as large.

**15.4.** Calculate the operating cost for the plant in Example 15.3 if the plant is only one-half as large.

**15.5.** Repeat Prob. 15.3, assuming that the plant is operating at only one-half of design capacity.

## REFERENCES

1. American Association of Cost Engineers: *AACE Bulletin,* Morgantown, W.Va., 1958–1978.
2. ———: *Cost Engineering,* Morgantown, W.Va., 1978 to date.
3. ———: *Cost Engineers' Notebook,* Morgantown, W.Va., 1964 to date.
4. Baasel, W. D.: *Preliminary Chemical Engineering Plant Design,* American Elsevier Publishing Company, Inc., New York, 1976.
5. Bauman, H. C.: *Fundamentals of Cost Engineering in the Chemical Industry,* Reinhold Publishing Corporation, New York, 1964.
6. Berry, E. M.: "Estimate Manufacturing Cost," *Chemical Engineering,* vol. 67, June 27, 1960, p. 123.
7. Black, J. H.: "Estimating Operating Costs for New Ventures," *AACE Bulletin,* vol. 4, December 1962, p. 42.
8. Chemical Engineering: *Modern Cost Engineering Methods and Data,* McGraw-Hill Book Company, New York, 1979.
9. Chilton, C. H.: *Cost Engineering in the Process Industries,* McGraw-Hill Book Company, New York, 1960.
10. "Deskbook: Environmental Engineering," *Chemical Engineering,* vol. 84, October 17, 1977, 212 pp.
11. Dybdal, E. C.: "Engineering and Economic Evalution of Projects," *Chemical Engineering Progress,* vol. 46, February 1950, pp. 57–66.
12. *Employment and Earnings,* U.S. Department of Labor, Bureau of Labor Statistics, U.S. Government Printing Office, Washington, D.C., published monthly.
13. Faith, W. L., D. B. Keyes, and R. L. Clark: *Industrial Chemicals,* 4th ed., John Wiley & Sons, Inc., New York, 1975.
14. Garcia-Borras, T.: "Research-Project Evaluations," *Hydrocarbon Processing,* vol. 55, December 1976, p. 137; vol. 56, January 1977, p. 171.
15. Guthrie, K. M.: "Capital and Operating Costs for 54 Chemical Processes," *Chemical Engineering,* vol. 77, June 15, 1970, p. 140.
16. Happel, J., and D. G. Jordan: *Chemical Process Economics,* 2d ed., Marcel Dekker, Inc., New York, 1975.
17. Haines, T. B.: "Direct Operating Requirement for Chemical Processes," *Chemical Engineering Progress,* November 1957, p. 556.
18. Holland, F. A., F. A. Watson, and J. K. Wilkinson: "Manufacturing Costs and How to Estimate Them," *Chemical Engineering,* vol. 81, April 15, 1974, p. 91.
19. Horton, R. L.: "Manufacturing Costs for Batch Produced Chemicals," *Chemical Engineering,* vol. 72, October 11, 1965, p. 236.
20. Humphreys, K. K., and P. Wellman: *Basic Cost Engineering,* 2d ed., Marcel Dekker, Inc., New York, 1987.
21. Isard, W., and E. W. Schooler: *Location Factors in the Petrochemical Industry,* U.S. Department of Commerce, Washington, D.C., 1955.
22. Jenckes, L. C.: "Estimation and Optimization of Operating Costs," *AACE Transactions,* 1970, p. 48.
23. ———: "Developing and Evaluating a Manufacturing Cost Estimate," *Chemical Engineering,* vol. 78, January 11, 1971, p. 168.
24. Klett, M. G.: *Typical Costs for Electric Energy Generation and Environmental Controls,* EPA-600/7-79-026, IERL, U.S. Environmental Protection Agency, Research Triangle Park, N.C., January 1979.

25. Malstrom, E., ed.: *Manufacturing Cost Engineering Handbook,* Marcel Dekker, Inc., New York, 1984.

26. Maxwell, J. D., T. A. Burnett, and H. L. Faucett: *Preliminary Economic Analysis of $NO_x$ Flue Gas Treatment Process,* EPA-600/7-80-021, IERL, U.S. Environmental Protection Agency, Research Triangle Park, N.C., February 1980.

27. McGlamery, C. G., R. L. Torstrick, W. J. Broadfoot, J. P. Simpson, S. V. Tomlinson, and J. F. Young: *Detailed Cost Estimates for Advanced Effluent Desulfurization Processes,* EPA-600/2-75-006, IERL, U.S. Environmental Protection Agency, Research Triangle Park, N.C., January 1975.

28. O'Connell, F. P.: "Chart Gives Operating Labor for Various Plant Capacities," *Chemical Engineering,* vol. 69, February 19, 1962, p. 150.

29. Ostwald, P. F.: *Cost Estimation for Engineering and Management,* Prentice-Hall, Inc., Englewood Cliffs, N.J., 1974.

30. ———: *Cost Estimating,* Prentice-Hall, Inc., Englewood Cliffs, N.J., 1984.

31. ———: *American Machinist's Cost Estimator,* McGraw-Hill Book Company, New York, 1985.

32. Perry, R. H., et al, eds.: *Perry's Chemical Engineers' Handbook,* 5th ed., McGraw-Hill Book Company, New York, 1973.

33. Peters, M. S., and K. D. Timmerhaus: *Plant Design and Economics for Chemical Engineers,* 3d ed., McGraw-Hill Book Company, New York, 1980.

34. Popper, H.: *Modern Cost Engineering Techniques,* McGraw-Hill Book Company, New York, 1970.

35. Radian Corporation: *Sulfur Oxides Control Technology Series: Flue Gas Desulfurization-Dual Alkali Process,* IERL, U.S. Environmental Protection Agency, Research Triangle Park, N.C., October 1980.

36. *Railroad Facts,* American Association of Railroads, Washington, D.C., 1988.

37. Rudd, D. F., and C. C. Watson: *Strategy of Process Engineering,* John Wiley & Sons, Inc., New York, 1968.

38. Salvendy, G., ed.: *Handbook of Industrial Engineering,* John Wiley & Sons, Inc., New York, 1982.

39. Scher, C., and D. Netzer: "FGD Good with Coal Conversion," *Hydrocarbon Processing,* vol. 59, December 1980, p. 117.

40. Snyder, T. A., Jr.: "Waste Disposal: What Is Industry's Responsibility?," *Chemical Engineering Progress,* vol. 76, June 1980, p. 24.

41. Sommerville, R. F.: "A New Method Gives Quick, Accurate Estimate of Distillation Costs," *Chemical Engineering,* vol. 79, May 1, 1972, p. 71.

42. Tomlinson, S. V., F. M. Kennedy, F. A. Sudhoff, and R. L. Torstrick: *Definitive $SO_x$ Control Process Evaluations: Limestone, Double Alkali, and Citrate FGD Processes,* EPA-600/7-79-177, IERL, U.S. Environmental Protection Agency, Research Triangle Park, N.C., August 1979.

43. Uhl, V. W.: *Standard Procedure for Cost Analysis for Pollution Control Operations,* vols. I and II, EPA-600/8-79-018 a, b, U.S. Environmental Protection Agency, Research Triangle Park, N.C., June 1979.

44. U.S. Department of Labor, Bureau of Labor Statistics, *Employment and Earnings,* published monthly.

45. Vernon, I. R., ed.: *Realistic Cost Estimating for Manufacturing,* Society of Manufacturing Engineers, Dearborn, Mich., 1968.

46. Wessell, H. E.: "New Graph Correlates Operating Labor Data for Chemical Processes," *Chemical Engineering,* July 1952, p. 209.

# CHAPTER
# 16

# PROJECT MANAGEMENT AND CONTROL

*D. L. Anderson, J. M. Neil, and T. L. Hyvonen*

## CAPITAL-COST CONTROL

## 16.1 THE NATURE OF CONTROL

Cost control of capital and operating cost is a major and critical activity of cost engineering. In building a plant, *cost control* means regulating expenditures within the authorized budget while providing a plant of the desired quality within the time limit specified. The plant should be capable of producing a quality product at the specified design capacity.

Cost control cannot end with analysis and prediction of capital costs. Plants and factories must be built, and it is essential that such costs be controlled during their design, procurement, and construction. Plants and factories must be operated, and it is equally important that operating costs be controlled.

Most heavy industries and petrochemical construction projects are now built by what is called the *fast track* method. Design, procurement, and construction are done with significant overlap, often appearing to be almost simultaneous. The fast track method is necessary because of today's rapidly developing technology. As the design of certain portions of the plant is completed, procurement can begin on pieces of equipment with long lead time. Similarly, as elements of civil design are completed, such as site work, roads, and drainage, and with a reasonable assurance of continuing production of "approved-for-construction" drawings, the construction team can move

**458**

to the site and begin. This is typical of how current industrial and petrochemical projects are being managed. If the design were allowed to be completed before the procurement and construction commenced, the final cost of the plant would be so much greater due to escalation alone that it would not be competitive. Other factors influencing costs are interest cost and obsolescence.

To coordinate the efforts of engineering, procurement, and construction groups requires a group responsible for project general management. This group is responsible for providing the catalyst which blends the engineering, procurement, and construction efforts into a homogeneous team working together, each with knowledge of the other's interface requirements, and including the vendors and subcontractors who will also be contributing to the overall effort.

## 16.2 CAPITAL EXPENDITURE CONTROL

With the type of project briefly described in Sec. 16.1, capital expenditure control must be brought into effect as soon as the contract is awarded to the design engineer. Thus the engineer may be made aware that project management will be coordinating his or her efforts with those of the construction group toward the common goal of project completion on time within the budget. Immediately after the award of the contract, the engineer should provide a preliminary budget to project management. The project management cost engineer staff should set up the control system and establish the procedures for an overall project control estimate to be made usually after 6 to 8 months of engineering design and project scope development. Timing depends on the type of project under consideration, but in no case should the timing in establishing control exceed 6 to 8 months of conceptual development. The engineer's responsibility to provide this estimate should be very clear and made a part of his or her budget. Meanwhile, staff scheduling engineers should establish some major milestones and provide a conceptual schedule so that everyone on the project team is aware of the milestone dates and the magnitude of the project in terms of construction, equipment, labor, and the time frame for completion of major components. This tends immediately to set the pace and give direction to the team members. Engineering design and procurement efforts can then be directed toward the long lead time of major equipment. Thus the elements of capital expenditure control are:

1. Cost control
2. Estimating
3. Planning and scheduling
4. Material management

## 16.3 COST CONTROL

The primary requirement of cost control is a trackable budget, which should be developed from a detailed estimate. The estimate must include a detailed definition of the project's scope. If the design of the plant has not progressed to the point where certain specific elements are defined, the estimators must complete the definition sufficiently to produce a detailed estimate. This is called *conceptual design* and *conceptual estimating*. The estimate is translated into the budget by the code of accounts. The code of accounts should be standard for project after project, and therefore well fixed in the minds of project team members. The only development that should occur in the code of accounts is the tailoring needed to permit subsequent capitalization for the owner. These capitalization accounts should be kept separate from the trackable budget used to measure progress. Table 16.1 shows a code of accounts for major activities. Activities are controlled divisions of construction efforts relating to the different types of work involved on the construction project. All direct work performed in the field will be related to a specific activity, such as building concrete forms and welding pipe by size and type.

Figure 16.1 shows a record of the cash flow of labor and the total cost for a project, but it is not in a form to indicate corrective action at any time.

Figure 16.2 is a typical example of a standard S curve showing cumulative work-hours needed to do a particular part of a project against time. The term *earned value* refers to physical accomplishment and is used in cost engineering to avoid confusion with money or labor committed. An example of progress made could be the design drawings completed through the reporting period, weighted in terms of estimated work-hours per drawing. At the November vertical line shown in Fig. 16.2 the difference between the actual work-hours and the earned work-hours, points *A* and *B*, is the work-hours over the budget. Similarly, the difference between *C*, the schedule baseline, and *D*, the actual accomplishment (earned value) at the November line discloses how far the project is behind schedule. Obviously corrective action is necessary.

**TABLE 16.1**
**Major construction activities**

| Account number | Title |
| --- | --- |
| 01XX | Site preparation, demolition, salvage, removal for relocation |
| 02XX | Site improvements |
| 03XX | Underground electrical |
| 04XX | Underground piping |
| 07XX | Piling |
| 08XX | Concrete and excavation |
| 12XX | Structural steel |
| 14XX | Architectural (manufacturing and forest products projects) |
| 15XX | Building construction (petroleum and chemical projects) |
| 16XX | Ducts |
| 17XX | Aboveground piping |
| 18XX | Aboveground electrical |
| 19XX | Instrumentation |
| 20XX | Insulation |
| 21XX | Painting |
| 23XX | Paving |
| 4XXX | Major equipment |
| 70XX | Proratables |

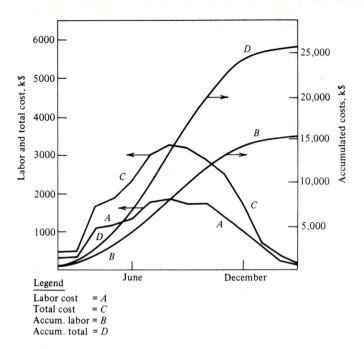

**Legend**

Labor cost   = A
Total cost    = C
Accum. labor = B
Accum. total  = D

**FIGURE 16.1**
Cash flow of labor and total cost.

 Baseline
Actual work-hours
Earned work-hours

**FIGURE 16.2**
Earned value status graph.

Variations from the schedule are important and carefully monitored by charts and tabulations. When a definite trend is noted, a variance is generated. A *variance* is any deviation from the budget. The following are examples of the types of variance which might be experienced on a typical capital project:

Change order (scope change)
Damages (fire, wind, flood)
Productivity (loss or gain)
Quantity overruns

Figure 16.3 is a sample form of a project variance notice.

Computers are now used to print out detailed information periodically, such as weekly or monthly. Each account, such as piping, structural steel, or electrical, is shown, giving information such as the following:

Previous committed cost
Added for period
Total to date
Current estimated cost at completion
Percent expended
Original budget
Over (or under) budget
Percent deviation

Finally, to complete this cost control effort, a record of final costs must be made. This is the history which is to be used in evaluating the estimate for the next similar project. When the code of accounts is standard, this estimate evaluation is then possible. This history becomes a record of the company's experience and is a valuable tool in enhancing its ability to successfully manage complex projects.

At this time, when the costs are known, a capital distribution for the owner should be made. This may consist of several types of reports, as follows:

1. Government (FPC, Federal Power Commission)
2. Insurance underwriters
3. Tax credits
   a. Investment credit
   b. Energy conservation credit
   c. Pollution-abatement devices
4. IRS (Internal Revenue Service) asset write-off (depreciation)

| | DATE | | P.V.N. NO. | REV. NO. |
|---|---|---|---|---|

**TO:**    **B&R JOB NOS.**    **VARIANCE ORIGINATION**
☐ HOME OFFICE    ☐ FIELD

**PROJECT**    **COST/SCHEDULE AREA**

**DRAWING, SPECIFICATION. EQUIPMENT NO., OR OTHER REFERENCE**

---

**SECTION 1.0 — REASON FOR AND NATURE OF VARIANCE**

**DESCRIPTION AND/OR SKETCH**

---

**SECTION 2.0**

**ORIGINATION**

**ORIGINATOR**    **SUPERVISOR**

---

**SECTION 3.0**

SECTION 3.0 TO BE COMPLETED WHEN INFORMATION IS AVAILABLE

| EFFECT ON COST | EFFECT ON SCHEDULE | EFFECT ON CONTRACT |
|---|---|---|
| | | |

**VARIANCE EVALUATION (By Control Team)**

| HOME OFFICE | | MATERIALS $ | SUB $ | FIELD | | | TOTAL $ |
|---|---|---|---|---|---|---|---|
| MANHOURS | AMOUNT $ | | | MANHOURS | LABOR $ | OTHER $ | |
| | | | | | | | |

---

**SECTION 4.0**

**VARIANCE CLASSIFICATION**

**CLIENT**
☐ PROCESS MODIFICATION
☐ DESIGN MODIFICATION
☐ SPECIFICATION MODIFICATION
☐ IMPROVE OPERABILITY
☐ SAFETY CONSIDERATIONS
☐ OTHER _____
  (SPECIFY)

**INTERNAL**
☐ ESTIMATE
☐ DESIGN
☐ PROCESS

☐ OTHER _____
  (SPECIFY)

**ECONOMIC**
☐ CHANGE IN MATERIAL COST
☐ CHANGE IN LABOR COST
☐ RESULTS OF PROCUREMENT
  ACTION
☐ FIELD CONDITIONS
☐ OTHER _____
  (SPECIFY)

---

**SECTION 5.0**

☐ APPROVED
☐ APPROVED-REQUEST CHANGE ORDER - TO CLIENT
☐ DISAPPROVED
☐ ADDITIONAL INFORMATION REQUIRED
☐ REQUEST CONFERENCE
☐ OTHER (SPECIFY) _____

**SCOPE CHANGE ONLY**
☐ DISAPPROVED
☐ PREPARE CHANGE ORDER
  REQUEST ESTIMATE

| RESPONSIBLE MANAGER | CLIENT RESPONSIBLE MANAGER |
|---|---|

**FIGURE 16.3**
Project variance notice.

The following is an outline of the essential elements of cost control:

1. Project setup
   a. Budget estimate (trackable)
   b. Code of accounts
2. Costing
   a. Labor cost
   b. Material cost
   c. Construction equipment and tools cost
   d. Subcontracts cost
   e. Overhead and indirect cost
3. Reporting
4. Analysis
5. Variancing
6. Finalization
   a. Final cost report — history
   b. Capital distribution — client use

## 16.4   ESTIMATING

Most capital projects are divided into two phases. The initial phase provides financing of conceptual design up to a point at which a detailed conceptual estimate can be made — usually 6 to 8 months, depending on the type and size of the project. During this time purchase orders may be issued for equipment that requires a long lead time. These purchase orders usually have a cancellation clause in the event that project funds are not subsequently released to execute the detail/design/procurement second phase.

The information provided in the initial phase is enough to use in developing a trackable estimate. Engineering is required to provide scope definition in order to define size, to rate equipment, and to provide foundation sizes, structural steel requirements, instruments by loop, piping, and valves by line. The owners may require this scope development for funding of the second phase of project execution. Design deviation from this conceptual detail forms the basis for variances and affects control of design, engineering, procurement, and construction during the second phase.

The basic elements of a detailed conceptual estimate used to produce a trackable budget are as follows:

1. Estimate scope development
2. Factored allowances
3. Estimate backup or support documentation
4. Estimate review — engineering — construction — owner
5. Qualifications to the estimate

The estimate scope development has been covered somewhat in the previous description of information to be provided by the engineering team in the initial phase. However, part of the scope is included in the code of accounts. The standard code of accounts is used, and the detail definition of those accounts determines the needed scope of each account. A listing of requirements from the engineer, with a description, outlines the responsibilities of each design function. Each estimator should have preprinted estimate summary sheets by the code of accounts. If the estimate is computerized, this of course would not be required since the computer would automatically format the estimate in the standard code of accounts. One computer printout is called an *ESLUM spread sheet,* with *ESLUM* an acronym for *EStimated Labor Units and Materials.* For various items of equipment it gives the account number and quantity, work-hours, labor work-hours, permanent materials, and construction materials, among others.

Factored allowances can be used in a conceptual estimate but only in determining such costs as temporary facilities, small tools, guard service, fuels, lubricants, and repair parts. Although these factors are part of the company's historical records and are generally accepted, they should never be used to substitute for detail of the facility that is to be built. Other elements must also be applied to a conceptual estimate with the best judgment that a project and its company can muster. Some of these factors, which are more difficult to define and sometimes not generally accepted, are essential to complete a detailed conceptual estimate. These are:

Design allowance
Quantity allowance
Escalation
Contingency
Risk

The *design allowance* is an adjustment to the estimated purchased equipment cost to provide for historically normal, routine changes in the design from estimate to delivery. The allowance will vary with the type of equipment and the amount of design information available at the time of the estimate.

The *quantity allowance* is an adjustment applied to the quantities, other than those of process equipment, to provide for the differences which invariably occur between takeoffs at estimate time and installed quantities due to incomplete design and for fabrication scrap or waste. It provides an adjustment to labor, materials, and supplies and is usually based partially on the constructor's work records.

*Escalation* is an allowance for inflation and is applied to all elements of cost which are affected.

*Contingency* is a percentage of the escalated base cost to cover errors in human calculation or judgment in estimating the scope of the project. It is a measure of the accuracy of the estimator. Contingencies are anticipated to be spent and form an integral part of the cost of the project.

*Risk* is an amount of money or a percent of the base cost plus lower-level (above listed) additives required to cover the probability that the base cost plus additives does not adequately define all the expenditures to be covered in seeking funding for the project. It is a judgment of the quality of the estimate made by the person applying the risk.

The estimate backup and support consist of five main elements, all of which contribute to the total capital expenditure:

Construction direct cost
Construction overhead and indirect cost
Engineering cost
Management cost
Owner cost

Each of these elements must be qualified and properly documented. The largest and most time-consuming to develop and qualify is the construction direct cost. Each activity, such as site development, area preparation, underground electrical, concrete, and piping, must be supported in detail. Each foundation sketch, each isometric pipeline, each piece of equipment by number must be supported by price, quantity, and reference number. All these data should be bound and referenced for the estimate backup and support of changes in scope and other variances.

In the estimate review an estimator may have to support elements of the estimate with these data. This type of support also assures that the estimate cannot be arbitrarily reduced. If the quantities contained in the estimate are correct and supported, and the unit price to extend the quantities is developed from historical data, then logically the only reduction that could be made in the estimate would be in the factors defined earlier or in the facilities or scope. This is one reason the estimate support is so important. It becomes the definition of the original scope of the project and the basis for all changes in scope and other variances; therefore, it supports control.

The reviews of the estimate are another important step in communicating the scope definition to all members of the project team. The reviews clarify the estimate for the team, and where misunderstandings may occur, they are cleared up in these reviews. Mistakes, omissions, errors, etc. are found and corrected. Weak areas in the estimate are highlighted, and additional work may be required to strengthen a portion or element of detail.

After the estimating team managers are satisfied, through meetings and internal reviews, that the estimate properly reflects the scope of the project, four reviews are usually required:

Engineering review
Construction review
Management review
Owner review

## 16.5   WHAT IS PLANNING?

A stitch in time saves nine.

Prior planning prevents poor performance.

When all else fails, follow the instructions.

Each of these expressions has been heard many times and they repeat in their own way a well-known fact — an operation that has the benefit of up-front planning and is executed according to plan stands the best chance of success. Yet, how often do we feel rushed, ignore preplanning, and stumble ahead planning as we go? When we do, we usually end up mumbling another often heard expression:

We never have time to do it right the first time, but always find time to do it over.

And when redoing it, we invariably suffer time and monetary losses, to say nothing of the embarrassment and loss of credibility.

Planning can be defined as making decisions now with the objective of influencing the future. It should be part of both our personal and business lives. This future-oriented decision process involves (1) setting objectives, (2) gathering information, (3) determining feasible alternatives, (4) choosing the best alternative, and (5) communicating the plan. To these steps three more must be added to provide a complete planning cycle — (6) implementing the plan, (7) making in-process adjustment of the plan to meet new conditions, and (8) after-action reviewing of results.

## 16.6   THE IMPORTANCE OF PLANNING

It would be convenient if cost-benefit studies were available to prove the value of planning. In theory, planning payoff could be measured by relating the value of planning results to the cost of planning. Unfortunately, there are seldom two identical opportunities which can be tested — one with planning, one without — so we must look to other indicators. Some years ago in a formal study, Stanford Research Institute found that companies supporting planning programs have shown superior growth rates as compared to companies without such programs.

The reason companies that plan are the ones most likely to grow seems fairly obvious when we think about the cyclic process that leads to growth, namely:

Planning
Monitoring and controlling
Feedback

Given any project or opportunity, a company develops a plan for maximizing the opportunity. In developing this plan, they use the best information available. When the plan is implemented, all activity is carefully monitored and controlled using the plan as a reference baseline. Complete records are maintained throughout this phase. Then in the third phase the experience gained is fed back to the company databases to increase the base of knowledge for the next planning action. This cycle is the *learning curve* in action — each repetition makes planning for and achievement of the next opportunity that much easier. Lacking commitment to such a cycle, a company is continually "reinventing the wheel," wasting time and money, and probably sealing its doom in the competitive marketplace.

## 16.7   ESTABLISHING A PLANNING CULTURE

A proactive, structured approach to planning will yield very significant benefits as compared to a reactive or random approach. The first benefit is the minimization of the potential for overlooking something. Second, the process forces definition of the scope of the problem, project, or activity being studied. Third, if handled as a team effort by its participants, it permits all viewpoints and ideas to be heard. Fourth, the resultant plan, when well documented, serves as a communication device for all participants. Fifth, the plan produces a baseline for control during the execution phase. Sixth, serious after-action reviews will greatly reduce the potential for planning errors on subsequent activities.

In business, planning is a management function. As with any operation, if those who must manage the plan are not active in its preparation, their commitment to it may be less than total. Certainly, staff personnel may contribute to and facilitate the planning process, but that is the extent of their role. An organization's planning philosophy should not reflect the statement, "I thought this up, now you take the rap for it."

Effective planning becomes a matter of routine in an organization when that organization has a planning culture. The key to establishing this culture is top management's commitment to planning and this commitment must be projected to everyone in the organization. The next key ingredient is a team approach to planning which involves all levels of management. This builds employee confidence in the organization because it stimulates communication among all affected parties and promotes their commitment to (ownership of) plans they help develop. It also shows that management has a direction, that decision making is under control, and that the total organization is working to achieve the same objectives. An additional, but no less important, benefit is the training that lower-tier managers receive in organization procedures and operations while engaged in planning. They will be in a much better position to assume higher levels of responsibility as opportunities develop and will bring with them a planning philosophy that is fully ingrained.

## 16.8   CONSTRUCTION PROJECT PLANNING

How important is construction project planning? In a study a few years ago a major U.S. contractor decided to review the factors contributing to the success or failure of its projects. From the records of hundreds of projects that had been completed, they noted that many projects successfully achieved their basic project objectives while others failed. The histories of the projects were carefully reviewed to identify factors common to successful projects, vis-à-vis those associated with less successful projects. Their study revealed that the major factor directly influencing success was the quality and depth of early planning by the project management group.

What is the planning record of the construction industry? Certainly, many of the cost overruns of the past two decades can be attributed to poor planning at some level. In a 1981 report of a study of hundreds of construction projects an index of 100 was used to represent the expected cost of a project with reasonable planning. Then the actual cost was converted to an index relative to that base. It was found that actual cost indices ranged from 60 to 500, with an average of 150. In other words, exceptional planning, in the best case, yielded 40 percent cost savings, whereas poor planning created overruns as high as 400 percent of base and, on average, projects were costing 50 percent more than they should.

Other research has shown that most owners and construction organizations are very limited in their scope of construction planning. They tend to place primary emphasis on time planning and, to a lesser extent, on resource allocation and cash flow. Work methods, materials management, and similar subjects receive minimal attention. This observation is reinforced by noting that many commercial textbooks on project management treat "planning" and "scheduling" as synonymous. Also, many organizations even use the job classification, "Planning and Scheduling Engineer," but the term "planning" is never part of the job titles of those involved in engineering, field erection, quality, materials, or budget planning. In fact, the industry has become almost preoccupied with sophisticated scheduling software packages for time planning, while very few are available or used to facilitate other types of planning. The strangest part of all this is that these critical-path-based scheduling programs, which are predicated on strict, logical interrelationships of activities with fixed durations, are being applied to an industry characterized by extreme variabilities and uncertainties.

Further emphasizing past planning weaknesses of the industry is the attention now being given by the Business Roundtable, Construction Industry Institute, and others to scope definition, value engineering, constructability, materials management, and quality management. The reason is that evidence is mounting that reasonable planning in those areas can yield savings of 5 to 40 percent of total project cost. And, with profit margins being what they are for a contractor, even a few percent savings can double or triple profit.

Several reasons for slighting the planning process for construction projects can be cited. First, planning time is often very limited. Second, there are often a number of projects being planned simultaneously so the staff is spread very thin. Third, there is always the tendency to feel that past experience can be applied directly to the new project and "our experienced field personnel can pull it off."

As for the first reason — shortage of planning time — more planning time is seldom available. However, the planning process can be made faster and easier if the organization develops standard procedures and models that can be adapted easily to each new project. Also helpful will be checklists to help ensure coverage of all considerations. To overcome the second problem — staff spread too thin — the best solution is for management to be more selective in choosing projects to chase so the staff can concentrate on those with the most potential. For the third problem—overreliance on home office databases—the answer is to be more exhaustive in the investigation of a project. No two projects are exactly alike so the information in the organization database must be adapted to each new project. Construction planners must be like investigative reporters—get to the scene, look under every stone to determine what makes that project unique, and adjust database information accordingly.

Not mentioned so far is contingency planning. It was noted earlier that construction is characterized by variability and uncertainty. Thus "what if" planning should be included for those items subject to enough variability to impact significantly project cost and time. With construction, the chances of everything going according to plan are very close to zero so any contingency planning will facilitate more timely response to variations if and when they occur.

A plan is no good if it is not made a matter of record and communicated to those with a need to know. The total project management plan for a project consists of many subplans including the estimate, schedule, materials management plan, quality management plan, etc. In addition to serving as a communication vehicle, the plan provides a baseline for control of the project and becomes a reference document in case of changes to the project and a document of record for future claims or litigation.

After-action review of projects is another frequently neglected area. Yet, such review is an important step in improving the organization's planning database.

## 16.9    CREW-LEVEL PLANNING

Thus far, we have talked about planning at the project level. But, isn't planning important for every crew foreman or independent craft worker? In a recent study by the University of California, productivity of workers was being evaluated while performing a task in a clear area on the ground as opposed to performing the same task in an elevated area. Although it was expected that productivity would be less in the elevated position, the opposite proved true. In fact, the ratio of elevated to on-ground productivity was more than 2:1. Analysis showed that workers always carefully preplanned elevated work to minimize their exposure. When on the ground they planned as they went and it proved very wasteful. This should not be unexpected since other research has shown that when planning takes place concurrently with task execution, people tend to concentrate on operating routines and to neglect planning. What is surprising is the 2:1 improvement in productivity with planning. The lesson to be learned is obvious: Plan! Plan! Plan! At all levels!

## 16.10    SPECIFIC PLANNING

Having discussed planning in general, it is appropriate to discuss specific planning. For any project manager there are numerous categories of planning, but the most important are time planning, cost planning, resource planning, quality planning, and contingency planning.

The basic approach to planning is to break the total endeavor into manageable parts, plan each of these in detail, combine the parts and test the total against project objectives, and then refine planning as necessary to eliminate variances between initial planning results and project objectives.

## 16.11   TIME PLANNING

Time planning is the development of plans, usually in the form of schedules, to accomplish all included elements of a project within an established time period. This planning starts with the need date for a project or other milestone date at which all actions must be complete and works backward from that point.

The second step in time planning is the division of the total project or effort into its components parts. In construction, this is usually referred to as development of the work breakdown structure (WBS). Such a structure is a hierarchical breakdown of the total project into successively more detailed components, each component being a well-defined part of the total. Some users refer to these components as *work packages* and a common definition of work package is

> *A well-defined scope of work that terminates in a deliverable product(s) or completion of a service. Each package may vary in size, but each must be a measurable and controllable unit of work to be performed.*

In other endeavors there may not be a WBS in the same sense as used in construction, but the total effort is still broken down into components which make up the total. For example, when time planning a vacation, the available vacation period becomes the window in which all desired activities (both travel and fun) must be compressed.

Once a project or effort is divided into its components, the next step is to array these components in the order of their accomplishment. Some activities must be handled in strict sequence; others may be executed simultaneously; for still others there may be a number of options. One of the most useful formats for handling this step is the use of the critical path logic diagram. Using arrows or nodes to represent each component (or activity), they are arrayed in what appears to be a logical sequence with dependencies shown among all activities where these dependencies constitute an actual constraint (e.g., one activity cannot start before another finishes). There may be several ways of handling the overall project, so it may be appropriate to develop two or more logic diagrams and then test each option. Figure 16.4 is a simple logic diagram for preparing a breakfast of eggs, bacon, toast, and coffee.

The next step is to associate a time (duration) with each activity. Either databases or prior experience provides this information. Then, using critical path techniques as will be discussed later in this chapter, the total time requirements for the endeavor can be determined. Should the resultant total time requirement exceed available time, planners must reevaluate the entire program and take whatever action is required to meet time objectives. Perhaps some activities are actually optional and

**FIGURE 16.4**
Simple logic diagram.

can be dropped; possibly others can be shortened by applying more resources or other schedule compression techniques.

The results of time planning take one of several common forms — a critical path network, a bar chart, or simply a time table (e.g., an itinerary).

## 16.12 COST PLANNING

Just as there is a WBS to describe the included efforts within a total endeavor, there should also be an accompanying cost breakdown structure (CBS) that will prove very useful for cost planning. A CBS is nothing more than a total cataloging of all cost elements expected within a project, the sum total of which equals the project budget. Ideally, the CBS will parallel the WBS; in fact, the objective is to have the WBS within the CBS. To explain further, using construction as an example, the WBS will include all components of the project to be constructed and each of those components can be used as a center for tracking both cost (CBS) and work progress (WBS). But there are many cost accounts in a project that are not directly related to a work account — most overhead items are in this category. So these accounts will appear in the CBS but not in the WBS.

Figure 16.5 provides a visualization of the relationship at a summary level of the WBS and CBS for an engineering-procurement-construction project.

## 16.13 RESOURCE PLANNING

Resources involved in a project typically include personnel, support equipment and tools, permanent materials and installed equipment, and expendable supplies. Some combination of these are involved in each control account of the CBS and WBS. Requirements are determined as described in Chap. 14 for construction. For nonconstruction endeavors historical experience and judgment usually provide the basis. It is essential that every resource requirement be accounted for somewhere within the CBS so that individual control account estimates of cost, as well as total estimated costs, can be generated.

For the WBS, the important resources to be planned and tracked are the workhours and quantities of major materials. For example, it is generally too much trouble

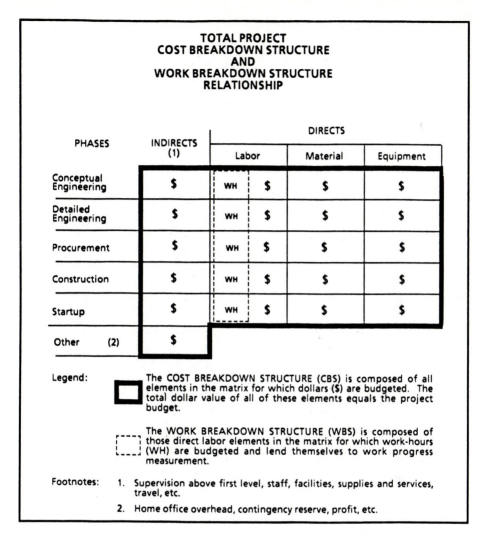

| PHASES | INDIRECTS (1) | DIRECTS | | | |
| --- | --- | --- | --- | --- | --- |
| | | Labor | | Material | Equipment |
| Conceptual Engineering | $ | WH | $ | $ | $ |
| Detailed Engineering | $ | WH | $ | $ | $ |
| Procurement | $ | WH | $ | $ | $ |
| Construction | $ | WH | $ | $ | $ |
| Startup | $ | WH | $ | $ | $ |
| Other        (2) | $ | | | | |

**TOTAL PROJECT COST BREAKDOWN STRUCTURE AND WORK BREAKDOWN STRUCTURE RELATIONSHIP**

Legend:    The COST BREAKDOWN STRUCTURE (CBS) is composed of all elements in the matrix for which dollars ($) are budgeted.  The total dollar value of all of these elements equals the project budget.

The WORK BREAKDOWN STRUCTURE (WBS) is composed of those direct labor elements in the matrix for which work-hours (WH) are budgeted and lend themselves to work progress measurement.

Footnotes:    1.  Supervision above first level, staff, facilities, supplies and services, travel, etc.

2.  Home office overhead, contingency reserve, profit, etc.

**FIGURE 16.5**
CBS and WBS.

to track expendables and tools, work account by work account in the WBS. Instead these are handled in a separate CBS account.

Identification of resource requirements is only a first step in their planning. They must be available in the quantities needed at the proper time. Thus, there will be plans such as the Materials Management Plan, Personnel Plan, Equipment and Rigging Plan, etc. to show how these items are to be made available. Certain materials, equipment items, and personnel will be identified as critical and their management will be given particular attention.

## 16.14   INTEGRATED PLANNING

Much is said these days of integrated project control. Such control is possible only if planning has been integrated. What this means is that time, cost, and resource planning has been accomplished against the same basic structure. The CBS and WBS, described earlier, provide the common structure for this planning. Thus, one level of WBS elements becomes activities for scheduling as well as centers for tracking resources and costs. This enables resource loading of the schedule, resource budgeting against time, and the development of a variety of cost budgets plotted against time.

## 16.15   QUALITY PLANNING

The objective of any project is to meet time, cost, and quality objectives. Quality objectives are met if the project delivers products conforming to established requirements without hassle. This requires development of a quality plan. This plan is composed of the requirements themselves, the plan for communicating these requirements to those responsible for compliance, any training required to assure that workers have the capability to meet requirements, and techniques for determining compliance.

## 16.16   CONTINGENCY PLANNING

A frequently neglected planning action is contingency planning. Since plans almost invariably are based on assumptions subject to some variability, the possibility of everything going according to plan is quite remote. This is recognized by the experienced planners and contingency plans developed.

Contingency planning can take at least two forms, and both are in response to "what if" questions asked by the planners. What if the weather turns bad? What if key materials are delivered late? What if our quantity estimates are low? The first form of contingency plan is an alternate plan to be implemented in the event an adverse situation comes true (What if a concrete pump breaks down in the middle of a large pour?). Obviously, planners can't plan for every eventuality, but such plans are considered for critical events.

The second form of contingency planning is the provision of an additional amount of money in a project budget or time in a project schedule to handle potential unfavorable variances in budget or schedule. To establish a contingency account, planners must first attempt to identify the project's risk elements—accidents, vandalism and theft, work quantity variances, productivity variances, unfavorable weather, adverse labor activity, etc. Certain risk elements are insurable, but seldom at the 100 percent level. Structured techniques are available for evaluating the combined exposure of a project to remaining uninsured risk elements, this exposure usually expressed in terms of a probability of loss not exceeding certain limits. Using this information, amounts are established for both cost and schedule contingency, and these accounts are managed with the same care as any other control account in the project.

## 16.17 PLANNING TOOLS

In the course of previous discussion, a number of planning tools were mentioned. These and others are summarized as follows:

**STANDARD OPERATING PROCEDURES.** Standard policies and procedures of the company should be reduced to writing. Planners can apply these policies and procedures without having to continually seek management guidance and approval.

**CHECKLISTS.** Extensive checklists should be developed to support planning. These checklists help prevent overlooking key items that may have cost and schedule implications.

**MODEL PLANS.** An organization which tends to repeat types of work is well advised to develop a model project of each type and to develop corresponding model plans. These plans are then adapted as necessary for each new project.

**HISTORICAL DATABASES.** Company experience on past projects should be cataloged in a standard format for use in planning new endeavors.

**CODES OF ACCOUNTS.** Codes of accounts structured to catalog work, cost accounts, resources and other information are essential if planning is to take advantage of modern computer systems. These should be standardized to the extent practicable to best assure consistency of data cataloging and use. For WBS and CBS, these codes should be hierarchical in format to permit capture of information at various levels of detail.

**COMMERCIAL HANDBOOKS AND SOFTWARE PROGRAMS.** A variety of commercial handbooks and commercial computer software programs are available to facilitate planning. Users of these items must understand their basis and limitations and not blindly apply them to company operations.

## 16.18 PLANNING—SUMMARY

In summary, the following are recommended:

Develop a planning culture. Through training and practice make planning a habit.

Prepare and maintain procedures, models, checklists, and other guidelines to facilitate the planning process.

Be selective in choosing projects to bid or propose on so that planning can be more complete on those chosen.

Be relentless in the site investigation of each new project. Do not rely on home office databases.

Give proper planning attention to materials management, quality management, work methods, subcontracting, etc., in addition to schedule.

Include contingency planning for major activities whose variability has the potential for significantly impacting cost and schedule.

Document all plans and make them available to those who need to know.

Train foremen and workers to plan their work.

## 16.19   THE SCHEDULE

Just as the budget is the basic tool in controlling cost, the project plan and schedule budget are the basic tools in controlling time. The *project plan* logically states the sequence for accomplishing the required work. It must consider all work phases of the project, including engineering design, procurement, construction, mechanical checkout, start-up, and special requirements such as tie-ins and early start-up of certain parts of the plant.

Since most capital projects are now being carried out in two phases, conception and execution, there are usually two phases in the planning and scheduling effort. For the first phase a conceptual schedule or major milestone schedule is developed. This dual-purpose schedule provides control of engineering development, construction planning, and procurement of equipment with long lead time and provides the estimating team with the essential timing of expenditures needed to support those elements of the detail conceptual estimate that are sensitive to timing due to interest, escalation, etc. The second phase is a definitive network schedule and again has a dual purpose. First, it provides control of the following:

1. Engineering, design, sequencing, and progress
2. Procurement
3. Construction timing
   a. 30–60–90 look-ahead schedules
   b. System sequencing for commissioning

Second, it provides for special conditions. Among these are such items as shutdowns, tie-in schedules, and utilities or service scheduling.

A *key item schedule* is a chart against time showing the sequencing of activities, the starting and completion times, and the milestones along the way. Figure 16.6 shows a key item schedule.

## 16.20   MATERIAL CONTROL

The materials management team's responsibilities are divided into three groups:

1. Procurement
2. Material control and tracking
3. Material coordination

**FIGURE 16.6**
Key item schedule.

477

The procurement group is responsible for the coordination and consolidation of purchasing and subcontracting. The bulk materials are consolidated in summarized quantity packages for more efficient procurement. These packages are coordinated with engineering and construction requirements, forming inquiries to vendors for bids. The bids are evaluated, and a successful vendor is selected after approval by the owner.

The material control and tracking group divides its responsibilities into three primary functions:

1. Major equipment control
2. Minor equipment and component control
3. Bulk materials or commodities

Major equipment consists of those items of process equipment which perform the essential functions of the manufacturing process. These items are always numbered as shown on the plot plan and equipment arrangement drawings. Examples of major equipment include heaters, exchangers, boilers, motors, pumps, coolers, blowers, fans, tanks, drums, vessels, towers, special machines, and reactors. The minor equipment and components are such items as instruments, traps, components of major equipment items such as filters, and small accessory groups. The bulk materials or commodities include such items as concrete, reinforcing steel, structural steel, pipe, electrical elements such as electrical cable, wire, and conduits, instrument tubing, insulation, and paint.

The primary function of material control and tracking is to assure that materials and equipment are available to the construction forces when required and are used as intended. Good material control substantially aids an efficient construction schedule and assures a minimum of surplus material at the conclusion of the project. Figures 16.7, 16.8, and 16.9 are examples of materials reports.

Material coordination involves inspection to provide quality control and assurance, traffic, and expediting functions. As an example, some projects require pieces of equipment which are very large, bulky, and difficult to handle. When required, the material coordination team comes in to assist. They will provide the necessary inspections, routing, and special handling required to get the equipment on site in the shortest possible time and at the least expense.

# OPERATING-COST CONTROL

## 16.21   THE REPETITIVE NATURE OF OPERATING COSTS

Operating costs continue repetitively over the life of the plant and will exceed the capital cost manyfold; thus they are very important. *Operating-cost control* means the

# MATERIALS MANAGEMENT

## MAJOR EQUIPMENT SUMMARY

C/S AREA: 8788

| ENGR. ITEM NO. | DESCRIPTION | ORDER NO. | VENDOR | ORIGINAL PROMISE | LATEST PROMISE | FIELD NEED | RCVD. DATE | SET DATE | COMPLETE DATE |
|---|---|---|---|---|---|---|---|---|---|
| CATEGORY: HEAT EXCHANGERS, CONDENSERS, ETC. | | | | | | | | | |
| 14 14–5103 | WET CRUDE EXCHANGER | P0060 | OLD DOMINION FAB. | 18APR80 | 25FEB80V | 15APR80 | | | |
| 14 14–5219 | HEAVY ENDS COOLER | P0026 | BAS–TEX CORP. | 08FEB80 | 26MAY80E | 02APR80 | | | |
| CATEGORY: PUMPS | | | | | | | | | |
| 23 23–5104 | GLYCOL PUMP | P0100 | GOULDS PUMP | 01MAR80 | 28MAY80V | 15JUN80 | | | |

**FIGURE 16.7**
Major equipment summary.

479

# MATERIALS MANAGEMENT

## MINOR EQUIPMENT SUMMARY

C/S AREA: 8788

| ENGR. ITEM NO. | DESCRIPTION | ORDER NO. | VENDOR | ORIGINAL PROMISE | LATEST PROMISE | FIELD NEED | RCVD. DATE | SET DATE | COMPLETE DATE |
|---|---|---|---|---|---|---|---|---|---|
| CATEGORY: INSTRUMENTATION | | | | | | | | | |
| IN FE-090-005 | ORIFICE PLATE & FLANGE | P61024 | DANIEL IND. | 01DEC79 | 15DEC79V | | | | |
| IN FI-090-005 | FLOW INDICATOR | P71000 | DUPONT INST. | 15DEC79 | 15DEC79S | 15DEC79 | 16DEC79 | | |
| CATEGORY: ELECTRICAL SPECIALTY | | | | | | | | | |
| ES 090-13-011-5 | 600 V SAFETY SWITCH | P62653 | NUNN ELECTRIC | 14MAR80 | 14MAR80S | 14MARCH80 | 20MAR80 | | |
| ES 090-13-011-6 | JUNCTION BOX | P62750 | NUNN ELECTRIC | 30JUNE80 | 30JUN80E | 01JUL80 | | | |

**FIGURE 16.8**
Minor equipment summary.

# MATERIALS MANAGEMENT

## PURCHASE ORDER INVENTORY

| STOCK NO. | DESCRIPTION | REQUISITIONED | PURCHASED | RECEIVED | PRICE |
|---|---|---|---|---|---|
| 50000015 | 12" Blind Flange, 150 # | | | | |
| 34–12042 | P.O.–00020 | 4 | 4 | 4 | 269.00 |
| 34–12160 | P.O.–00025 | 5 | 5 | 5 | 337.00 |
| 34–13100 | P.O.–00030 | 4 | 4 | 0 | 269.00 |

| JOB BAL. | REQUIRED | ISSUED | INV. BAL. | | | | |
|---|---|---|---|---|---|---|---|
| 3 | 10 | 8 | 1 | 13 | 13 | 9 | 875.00 |

**FIGURE 16.9**
Purchase order inventory.

regulation of expenditures on materials, labor, utilities, and associated costs to produce a quality product at a competitive price.

Operating-cost control requires a program that is updated periodically to meet the changing conditions for such factors as varying material price and quality, changing product quality, demand and price, increasing maintenance costs, technological obsolescence, and myriad other causes. It must define the objectives, devise a plan, provide a means of comparing performance with the plan, and indicate steps necessary to reduce costs.

## 16.22   CONJUNCTION WITH COST ACCOUNTING

Accounting personnel are set up to obtain information on costs. A knowledge of cost accounting, at least the principles developed in Chap. 17, will help the engineer cooperate with the accounting department.

The sections on operating-cost control are limited chiefly to direct material costs, direct labor, and direct production expense. Most of the indirect cost is not subject to control, at least from an engineering point of view. However, the control of indirect expenses is a legitimate subject for the cost engineer regarding items that are controllable, such as the material and labor which go into the indirect cost. The *allocation* of costs can strongly influence apparent costs, as explained in Chap. 17.

Cost control of operating expense, therefore, requires cooperation with the accounting department. Because engineering and technology are major factors, the cost engineer should have the responsibility for operating-cost control.

## 16.23   DEFINING OBJECTIVES

Table 16.2 gives an outline of operating-cost control. Many of the items in the table are established during the design and construction of the project; hence good operating-

**TABLE 16.2**
**Outline of operating-cost control**

1. Define objectives
    a. Production schedules
    b. Labor-utilization schedules
    c. Input-material quality specifications
    d. Input-material utilization schedules
    e. Inventory requirements
    f. Repair and maintenance requirements
    g. Quality-control requirements
    h. Utility requirements
    i. Product-quality specifications
    j. Waste-disposal requirements
    k. Off-specification material rework
2. Define standards
    a. Standard costs
3. Cost reporting
    a. Periodic reports
    b. Special investigations
4. Corrective action

cost control begins at the design stage. Good plant layout will minimize operating labor and repair and maintenance costs.

Production schedules require careful review of inventory requirements, sales forecast, input material inventory, quality control, and efficient utilization of labor equipment and material. Similarly, labor schedules should include scheduling of direct operating labor, maintenance labor, and indirect labor for housekeeping, packing, and shipping. The labor schedules should be reviewed frequently for efficient use of both straight time and overtime.

During the early stages of plant operation large expenditures for quality control may be required, but they should diminish with time. However, they should not be reduced below a reasonable frequency that will provide a continuing check on the operating personnel, material, and technical factors.

## 16.24 STANDARD COSTS

Standards are necessary to judge whether costs are satisfactory. These standards are used for comparison and like all standards they can be somewhat arbitrary — whether boys are measured by feet or yards does not change a tall boy into a short boy. However, there is some advantage in establishing standard costs in relationship to technological and optimum considerations rather than, say, to some normal historical cost. Standard costs of this nature are sometimes referred to as *industrial engineered standards*. Setting standard costs is a task that requires engineering knowledge of plant operations and is one of the most important functions of an engineer in operating-cost control since it is comparison of actual costs with standard costs which instigates action.

## 16.25   COST REPORTING

Although the engineers must set standard costs, it is the responsibility of the accounting department to report the comparison of actual costs against standard costs. Standard forms are used with periodic reporting by the various departments. The object, of course, is to reduce the collection of data to a routine basis.

There are occasions, however, when additional or more detailed information is required. Cost centers can be set up on a temporary basis. Frequently, the cost analysis requires considerable engineering effort, and experimental equipment will have to be installed to establish material and cost data.

## 16.26   CORRECTIVE ACTION

After a plant has been designed and built it must be operated so as to produce the desired quality of product at minimum cost with the plant as it exists. Modifications to correct operating inefficiencies, e.g., high maintenance cost, can be effected throughout the life of the plant, but they must be justified on the basis of savings in operating costs.

The plant supervisor or manager is responsible for operating-cost control and for overseeing proper scheduling and use of resources. It is the supervisor's responsibility to review the reports and to see that corrective action is taken where necessary.

The need for corrective action is indicated by comparison of the reported values with the standard costs. The cost engineer can be a valuable advisor in interpreting the comparison. Implementation of corrective action is the responsibility of plant supervisors, who also must follow through to see that corrective action is carried out satisfactorily.

## 16.27   CLASSIFICATION OF OPERATING-COST CONTROL PROBLEMS

Control problems can vary considerably depending on the type of project and kind of equipment involved and also on material variations, sales-forecast variations, inventory requirements, etc. Although the number, magnitude, and character of the problems encountered are seemingly without end, many can be classified. Table 16.3 gives a checklist for some common types of control problems and the control action required.

## SCHEDULING BASICS

## 16.28   BAR CHARTS

The use of schedules to make most efficient use of available project time is common management practice. The more simple projects are readily scheduled using bar

**TABLE 16.3**
**Control-problem classification and corrective action required — operating costs**

| Control problem | Types of corrective action |
| --- | --- |
| 1. Low yields — material usage too high | a. Check usage directions; see that they are being followed<br>b. Check material-weighing and metering devices<br>c. Check material quality<br>d. Check for leaks, spills, even theft |
| 2. High off-specification production and high waste, reworking, and downgrading | a. Check processing procedures<br>b. Check material quality<br>c. Check equipment cleaning and changeover procedures<br>d. Check control instruments |
| 3. Low production rates | a. Check for excessive rework<br>b. Check maintenance downtime and causes<br>c. Check for excessive changeover and downtime<br>d. Check processing procedures |
| 4. High maintenance costs | a. Review preventive and routine maintenance procedures<br>b. Check that maintenance is performed by qualified personnel<br>c. Emphasize maintenance scheduling against waiting for emergencies<br>d. Displace equipment requiring excessive maintenance |
| 5. High utility costs | a. Check for proper control and eliminate leaks<br>b. Check that utilities are shut off when not required<br>c. Check utility metering for accuracy |
| 6. High labor costs | a. Check for avoidable overtime<br>b. Check production scheduling for economic labor scheduling<br>c. Check absenteeism and lateness records for chronic offenders<br>d. Check for excessive personnel turnover causing high overtime and training expenditures |

charts; the more complicated are best scheduled through use of the critical path method (CPM). CPM is discussed in detail later in this chapter.

Bar charts, also called Gantt charts, have been used for the scheduling of construction as long as anyone can remember. They are simple to construct, very easy to interpret, and lend themselves well to schedule control. To construct a bar chart schedule, the planner first divides the work effort into its components. These are then scheduled against time using bars.

Further discussion of bar charts is reserved for Secs. 16.56 to 16.70 where the use of bar charts for planning and control is illustrated.

## 16.29 NETWORK DIAGRAMS

In the late 1950s the U.S. Navy, with the help of a consulting firm, developed a program evaluation and review technique (PERT) for scheduling the Navy's Polaris missile program. PERT was originally a method of scheduling to complete a project on time. It did not include costs but did include probabilities for the elapsed time of activities. It has since been modified to include additional factors such as cost. About the same time, Kelley and Walker [13] developed the CPM to expedite projects and included costs.

Both PERT and CPM, based on network or arrow diagrams, were an immediate success. They have been followed by many modifications but remain the most popular. Further discussion in this chapter is limited to CPM, the most common of the two methods in construction.

## 16.30 ARROW DIAGRAM METHOD (ADM)

Arrow diagram method (ADM) networks are characterized by events, activities, dummies, and hammocks.

**EVENTS.** An *event* marks a particular point in the course of a project and is represented by a node in the network. Each activity begins and ends at an event. The event from which the activity emerges is called the preceding event (PE), and the event into which it goes is called the succeeding event (SE). Each event in the network is identified by a numeric or alphanumeric coding, and an activity may be identified by the values of its preceding and succeeding events. The PE and SE of an activity are sometimes called the *i*- and *j*-nodes.

**ACTIVITIES.** An *activity* is a work item with a duration in time units and which has a clearly defined beginning and end. Activities normally have a requirement for one or more resources.

Each activity is represented by an arrow between two nodes, or events, in the network.

**PARALLEL ACTIVITIES.** If two or more activities share the same PE and SE values (in other words, join the same two nodes in the network), they are called *parallel activities*. Parallel activities may cause difficulties if the PE and SE values are used to identify activities, because two or more activities will share the same PE and SE. The solution is to use a dummy (see later example).

**DUMMIES.** A *dummy,* or dummy activity, is similar to an activity in that it joins two events, but there are important differences. Dummies do not represent work; they show logical links between events which are not joined by an activity but are nevertheless logically dependent. A dummy may have a duration, but it cannot have resources assigned to it.

A second common use of dummies is to avoid parallel activities. Dummies are generally shown by broken lines on the network.

**HAMMOCKS.** A *hammock* is a special feature in the network. It spans between events which may be separated by any number of activities and is used to measure the elapsed time between these events. It follows that the longest possible hammock is one which has the project start node as its preceding event and the project finish node as its succeeding event.

**NETWORK LOGIC.** Here is a simple illustration of the features described:

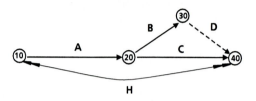

1. Event 20 is achieved when activity *A* is completed; activities *B* and *C* cannot start until event 20 is achieved. In other words, the starts of activities *B* and *C* are dependent on the completion of activity *A*.
2. *D* is a dummy, stating that event 40 is dependent on the completion of activities *B* and *C*. In this example the dummy is used to avoid having *B* and *C* as parallel activities.
3. *H* is a hammock, spanning from event 10 to event 40. Its calculated duration would give the elapsed time between these events. It is not part of the actual network.

## 16.31   A SIMPLE CPM NETWORK

Figure 16.10 shows a network with a normal completion time of 130 days.

Each small circle with a number is an *event* or *node*. The terminal numbers are 1, the start, and 7, the finish. Intermediate numbers correspond to completion of an intermediate task such as finishing the foundations or the third-story plumbing. An arrow represents an *activity*. Thus arrow 2–3 is the activity to go from event 2 to event 3, and number 45 on the arrow is the time in days for activity 2–3 to be completed. No activity can be started until all the arrows, or activities, leading to its tail event (from which it starts) have been completed. Thus activity 4–5 cannot be started before activities 1–4 and 2–4 have both been completed. Frequently it is necessary to use dummy activities or restraints, shown by dotted or solid lines, with zero time requirement. They are necessary to keep the sequence in order. The lengths of the arrows are unrelated to time or cost. The figure shown has seven events and 10 activities, including three dummy activities.

The first step in the analysis is the *forward pass* calculation to find the earliest start time (EST) and the earliest finish time (EFT) for each activity. The result is shown in Fig. 16.11. The calculation is made by starting at event 1 and working for-

**FIGURE 16.10**
CPM network.

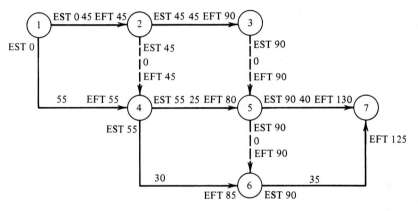

**FIGURE 16.11**
Forward-pass calculation for the network.

ward to the finish event, 7. Thus activity 1–2 has an EST of 0, a 45-day duration, and an EFT of 45. Activity 4–5 has an EST at 55 days because the *highest* EFT into event 4 is 55 days.

The next step is the *backward pass* calculation to find the latest finish time (LFT) and the latest start time (LST). The calculation is made by working backward from the finish to the start. In the backward pass calculation, when there are two or more numbers leading out of an event (LST), the *lowest* number prevails and provides the LST for all activities ending at the event. Figure 16.12 is the result of the backward pass calculation. From the forward pass calculation the network time to finish is 130 days, shown by the end of activity 5–7 event 7. Hence in the backward pass calculation the LFT for both activities 5–7 and 6–7 is at 130 days. Activity 6–7 takes 35 days, so that its LST is at 130 − 35 = 95 days. At event 5 the *lowest* LST *out* of the

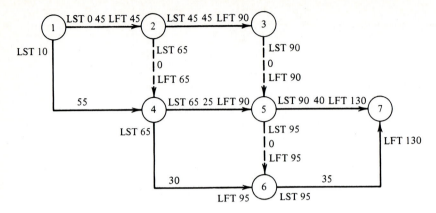

**FIGURE 16.12**
Backward-pass calculation for the network.

event is at 90 days for activity 5–7, and thus 90 days becomes the LFT for both activities, 3–5 and 4–5, ending at event 5.

Figure 16.13 shows the forward and backward pass calculations combined, and Table 16.4 is a tabulation of the results.

Of particular interest is the *total float* for an activity, which represents the time by which an activity may be delayed or extended without affecting the duration of the project. Activity 4–5 has an EFT at 80 but can be allowed to go to an LFT at 90 days without extending the duration of the project beyond 130 days. Thus activity 4–5 has a total float of 10 days, given by the difference between the LFT and the EFT. The total float for all activities is given in column 7 of Table 16.4. Float is sometimes called *slack*.

The *critical path* is the longest duration chain of activities from start to finish. There is no total float for any activity on the critical path. Any increase in duration

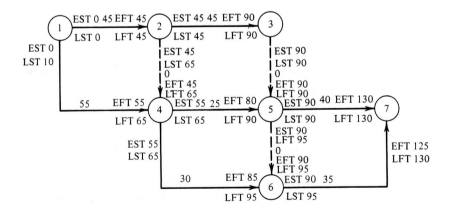

**FIGURE 16.13**
Forward and backward-pass calculations for the network.

**TABLE 16.4**
**Results of a CPM network**

| (1) | (2) | (3) | (4) | (5) | (6) | (7) | (8) |
|---|---|---|---|---|---|---|---|
| Activity | Duration, days | Earliest start time, days | Earliest finish time, days | Latest start time, days | Latest finish time, days | Total float, days | Free float, days |
| 1–2 | 45 | 0 | 45 | 0 | 45 | 0 | 0 |
| 1–4 | 55 | 0 | 55 | 10 | 65 | 10 | 10 |
| 2–3 | 45 | 45 | 90 | 45 | 90 | 0 | 0 |
| 3–5 | 0 | 90 | 90 | 90 | 90 | 0 | 0 |
| 2–4 | 0 | 45 | 45 | 65 | 65 | 20 | 10 |
| 4–5 | 25 | 55 | 80 | 65 | 90 | 10 | 10 |
| 5–6 | 0 | 90 | 90 | 95 | 95 | 5 | 0 |
| 4–6 | 30 | 55 | 85 | 65 | 95 | 10 | 5 |
| 5–7 | 40 | 90 | 130 | 90 | 130 | 0 | 0 |
| 6–7 | 35 | 90 | 125 | 90 | 130 | 5 | 5 |

along the way would delay the finish of the project. Column 7 of Table 16.4 shows that there is no total float along activities 1–2, 2–3, 3–5, and 5–7, and together they constitute the critical path with a duration of 130 days. There can be more than one critical path, but all critical paths have to be equal.

Another term used is *free float,* representing the time an activity can be delayed without affecting the EST of *any* other activity. Free float cannot exceed total float and may be less. To find the free float for an activity *A* ending at event *E*, first find the earliest start time for activities out of event *E*. Second, note the earliest finish time for activity *A*. The difference between the two is the free float. There can be no free float along the critical path. The free float for the activities is given in column 8 of Table 16.4, as determined from Fig. 16.11. For example, the total float for activity 2–4 is 20 days. However, the activity can be lengthened only 10 days without delaying the EST for activities 4–5 and 4–6. The free float is 10 days.

## 16.32  ADVANTAGES OF CPM

CPM makes critical activities known at all times, so that those requiring special attention will receive it. On the other hand, activities not on a critical path have some flexibility. CPM is not something that is used merely at the start of a project. The total float and free float along activities change with time on a large project, and CPM is kept up to date continuously. For large projects government and industry frequently require its use as part of the contract.

CPM is very adaptable to computer programming. In practice, actual dates instead of elapsed dates are printed out, thus allowing for weekends, holidays, and the like. CPM can be combined with costs and is a valuable tool in calculating the most economical way to crash a program at a cost to reduce the time for a project. CPM will keep a project going smoothly with respect to time and expenditure. It cannot correct deficiencies in estimates, but it can disclose them early enough for reassessment and reduce their frequency.

A project can be accomplished in more than one way, and therefore by more than one network diagram. Thus in setting up a network, the analyst should not use a particular network just because it is feasible. A better network may exist.

## 16.33    PRECEDENCE DIAGRAM METHOD (PDM)

Precedence networks are characterized by activities, events, and constraints.

**ACTIVITIES (ACTIVITY-ON-NODE).** An *activity* is a work item with a duration in time units and has a clearly defined beginning and end. Each precedence activity forms a node in the network.

**EVENTS.** An *event* indicates a particular point in the course of a project and is defined as having no duration. Events are often included to clarify the logic of the network. Events can be used to describe milestones in the progress of a project, such as starts, finishes, the time at which payments must be made, etc. Every PDM network must have a *start* event and a *finish* event to initiate and close out the network.

**CONSTRAINTS.** *Constraint* is the term used to define the relationship between activities and events in terms of relative start and finish times. Constraints may have a delay or a duration in time units. They do not represent any work and cannot hold resources.

There are four basic types of constraint (also called dependencies, links, and relationships), and these are:

1. Finish-to-start (FS)
2. Start-to-start (SS)
3. Finish-to-finish (FF)
4. Start-to-finish (SF)

The four constraint types are illustrated diagrammatically in the following sections.

## 16.34    FINISH-TO-START CONSTRAINTS

A finish-to-start constraint links the finish of the preceding activity (A) with the start of the succeeding activity (B). It indicates that the succeeding activity cannot start until the preceding activity has been completed. If required, a duration can be placed on the constraint to indicate that the succeeding activity cannot start until a given time after the preceding activity has finished.

In the preceding example activity $B$ cannot start until 10 time units after the finish of activity $A$.

## 16.35  START-TO-START CONSTRAINTS

A start-to-start constraint links the start of the preceding activity ($A$) with the start of the succeeding activity ($B$). It indicates that the succeeding activity cannot start until the start of the preceding activity plus any duration given for the constraint. The duration on the constraint denotes the lag between the start of activities $A$ and $B$.

In the preceding example activity $B$ cannot start until three time units after the start of activity $A$.

## 16.36  FINISH-TO-FINISH CONSTRAINTS

A finish-to-finish constraint links the finish of the preceding activity ($A$) with the finish of the succeeding activity ($B$). It indicates that the succeeding activity cannot finish until the preceding activity has finished plus any duration given for the constraint. The earliest start of the succeeding activity ($B$) is not necessarily affected by this constraint.

In the preceding example activity $B$ cannot finish until six time units after the finish of activity $A$.

## 16.37  START-TO-FINISH CONSTRAINTS

A start-to-finish constraint links the start of the preceding activity ($A$) with the finish of the succeeding activity plus any duration given for the constraint. The earliest start of the succeeding activity ($B$) may not be affected by this constraint.

In the preceding example activity $B$ cannot finish until five time units after the start of activity $A$.

**HAMMOCKS.**  A *hammock* is a special type of network record which is used to measure the elapsed time between two points in a network.

## 16.38   CALCULATING TIMES

For precedence networks the calculation of early and late times plus total and free float is considerably more complicated than with arrow networks. The following rules are used for calculating these times:

Early start (ES):   For activities the ES date of an activity is equal to the latest EF of preceding SS or FS constraints.

For SF or SS constraint the ES date is equal to the ES date of its preceding activity.

For an FS or FF constraint the ES date is equal to the EF of its preceding activity.

Early finish (EF):   For activities the EF is equal to the latest of either the ES + duration or the latest EF of any SF or FF constraint ending on that activity.

For all constraints the EF = ES + duration.

Late start (LS):   For an activity the LS date is the earliest of either its LF − duration or the earliest LS of an SS or SF constraint leading from that activity.

For all constraints the LS = LF − duration.

Late finish (LF):   For an activity the LF is the earliest LS of any succeeding FS or FF constraint leading from that activity.

For an SF or FF constraint the LF is equal to the LF date of its succeeding activity.

For an FS or SS constraint the LF is equal to the LS of its succeeding activity.

Total float (TF):   For both activities and constraints TF = LS − ES.

Free float (FF):   For an activity the FF is the least FF of any constraint leading from the start or finish of the activity. All end activities have FF = 0.

For FS or SS constraints FF = (ES of the succeeding activity) − (EF of the constraint).

For SF or FF constraints FF = (EF of the succeeding activity) − (EF of the constraint).

The calculations of float assume that an activity, once started, will continue uninterrupted. Actually there may be total float available in addition to that calculated which permits stretching of an activity's schedule without affecting the overall project completion date.

## APPLIED SCHEDULING

## 16.39   SCHEDULE DEVELOPMENT

The mechanics for developing a critical path network have been discussed earlier. In summary, the process is as follows:

**DEVELOP THE WORK BREAKDOWN STRUCTURE (WBS).** This involves the dividing up of the project into controllable parts suitable for schedule control; these parts are converted into schedule activities, where one or more activities comprise one element of the WBS.

**DEVELOP THE LOGIC DIAGRAM.** The *logic diagram* for the project is developed by taking the schedule activities and arraying them in ADM or PDM format to show their sequencing and interrelationships. The diagram contains no times.

**ASSIGNING DURATIONS TO ACTIVITIES.** Each of the activities (and constraints on a PDM diagram) is assigned a duration. A duration may be a product of the estimating process or it may be developed through consensus among the project team planners.

**CALCULATING TIMES.** Once durations have been assigned, the early, late, and float times may be calculated to create the schedule. Once this has been done, there is a schedule. In other words, the logic diagram + times will yield the schedule.

On very large projects the schedule is usually developed as described earlier, but then it is converted to a computer printout. The computer printout is then used to monitor progress.

On medium to small projects the logic diagram can be a useful tool for control in conjunction with the computer printout, particularly if drawn in a time-scaled format. On all but the smallest jobs today, computers are used to calculate early start, early finish, late start, and late finish dates plus float times. They also identify the critical path.

Summary schedules are frequently prepared for the purpose of reporting progress to management in simplified graphic and tabular computer printout format. Often these summary diagrams are in the form of bar charts.

## 16.40   SIGNIFICANCE OF FLOAT

The procedures for calculation of float times were presented earlier. Float definitions and the significance of each of the types of float—total, free, independent, and interfering—are given here. As you read them, refer to Fig. 16.14.

**TOTAL FLOAT.** Total float represents the total *potential* slack time available to an activity. It is calculated assuming all prior activities are finished as early as possible and all following activities are started as late as possible. Total float is usually shared with other activities and its availability is subject to its use by those other activities.

**FREE FLOAT.** Free float is defined as slack time available to an activity if both preceding and following activities are started as late as possible. All free float of an activity may be used by that activity without compromising final project completion. However, its use can reduce potential float for following activities.

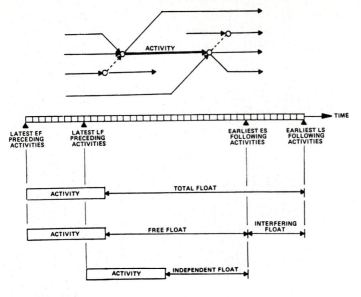

**FIGURE 16.14**
Floats.

**INDEPENDENT FLOAT.** Independent float is any slack time still available to an activity if all preceding activities start as late as possible and all following activities start as early as possible. Independent float is totally available to an activity and its use will not affect other activities in any way. In the typical network very few activities will have independent float.

**INTERFERING FLOAT.** Interfering float is the numerical difference between total float and free float and represents the potential loss of total float if following activities begin before their late start.

Project controllers must understand the significance of the different floats. Many novices assume that total float is equivalent to independent float, an assumption that could prove disastrous. On schedule summaries it is desirable to show total, free, and independent float times so there is no question as to the flexibility available. It is not necessary to show interfering float because it is simply the numerical difference between two other floats.

## 16.41 HARD AND SOFT LOGIC

In developing a critical path schedule, a planner can easily constrain a project through introduction of arbitrary logic. For example, assume that four identical foundations, $A$–$D$, are to be built. The planner may show them as being built in sequence, $A$ through $D$, on the CPM. While that is perfectly logical, perhaps the field personnel would find it easier to build $C$ first, $B$ second, $D$ third, and $A$ fourth. What the planner has done is apply hard logic to a soft logic situation. Such actions compromise the value of CPM.

Planners must learn to handle soft logic items at a summary level so that field planners have the flexibility to do the work in the most efficient manner at that time. In the preceding example, instead of four activities with labels *A–D* on them, a single activity of "four foundations, area _____" should be used. The field planners could then determine the exact sequence for doing them when developing their working schedule.

Hard logic is defined as that which covers work involving well-defined activities that have fixed interrelationships. Soft logic covers work whose component activities are less defined and whose interrelationships are conditional. A typical hard logic item is the forming, reinforcing, placing, and finishing of concrete. Another is equipment assembly. Typical soft logic items are electrical tray and conduit installation, piping system erection, and wire pulling.

## 16.42 INTEGRATION OF PROCUREMENT AND SUBCONTRACTING

A project schedule should include all activities which may influence the project completion date. Since procurement activity definitely is a determinant of when work involving those materials or equipment can be accomplished, key procurement lead times should be shown as schedule activities that begin when a requisition or purchase order is first initiated and terminate at delivery to the site. This delivery date is an event in the schedule which becomes a constraint for any activities requiring the materials or equipment on the purchase order.

Subcontracts are work activities coequal with activities performed by the direct hire work forces of the general contractor and will appear as separate activities on the project's control schedule. On the more complicated subcontracts, particularly those involving several interface points with other project activities, the general contractor will require the subcontractor to develop and present a detailed schedule for incorporation in the master control schedule. The general contractor must allow time for the solicitation-bidding-award-mobilization times for the subcontracts when developing his or her overall schedule.

## 16.43 SCHEDULE LEVELS

The various participants in a construction contract have different levels of interest in the scheduling of the project. The owner and the contractor's home office are interested in summary level schedules. Project-level personnel are interested in more detail. Thus there are various schedule levels. Although there is no universal agreement as to the number of schedule levels and their format, the following are representative:

Level 0     This is the total project and is, in effect, a single bar spanning the time from start to finish.

Level 1        This schedules the project by its major components. For example, a
               level 1 schedule for a process plant may be broken in process area,
               storage and handling area, site and services, and utilities. It is
               shown in bar chart format.

Level 2        Each of the level 1 components is further subdivided. For example,
               utility systems are broken into water, electrical, gas, sanitary, etc.
               In most cases this schedule level can only be shown in bar chart for-
               mat although a bar chart with key constraints may be possible.

Level 3        The subdivision continues. This is probably the first level that a
               meaningful critical path network can be drawn. It is also a good
               level for the project's overall *control schedule* since it is neither too
               summarized nor too detailed.

Levels 4–?     The subdivision continues to whatever level of detail is needed by the
               user. When operating at these more detailed levels, the planners
               generally work with less than the total schedule. In most cases these
               "look-ahead" schedules span periods of from 30–180 days. The user
               may utilize either bar chart or CPM format for these schedules.

## 16.44   SCHEDULE SUBMISSIONS

Contract documents will usually contain specific requirements concerning submission
of schedules by the contractor and their review by the client (or client's agent). Typically, a detailed CPM format schedule covering the first 120 days of the project may
be required for submission within 14 days of the Notice to Proceed (NTP). This
schedule is to cover procurement- and mobilization-type activities, as well as construction. All coding systems, abbreviations, etc. to be used by the contractor must
also be detailed. The contractor may be required to annotate the drawings to cross-
reference fully the schedules to the drawings.

Normally within 30–60 days after the NTP the contractor will have a requirement to submit a schedule (usually in CPM format) for the entire project. This schedule must include subcontracted work, include key procurement activity, incorporate
key milestones, and show the general plan for start-up (if applicable). This schedule
will normally be about level 3.

Throughout the project updated schedules will be required by the client. Also,
the client may require detailed, short-term schedules for the forthcoming 30–90 days
(usually in bar chart form) to show work planned in the immediate future.

## 16.45   CONTROL VS. WORKING SCHEDULE

A number of schedules will be utilized over the life of the project. When the project
is first planned, the *original control schedule* is developed. When an owner-approved
change is received which changes the schedule, the initial control schedule is modified to create the *current control schedule* (a copy of the original control schedule
will be maintained for historical purposes). The current control schedule will be sub-

sequently modified with each owner-approved change. None of the control schedules are changed to accommodate variances in field performance by the constructors — a control schedule remains a reference baseline.

Recognizing that field performance will probably differ from that planned, the project controllers will also maintain a *working schedule*. This is based initially on the original control schedule, but it is modified at least monthly to reflect actual past performance and a more current look at the future. Thus it can differ significantly from the current control schedule although that schedule remains the objective. Working schedules may be at levels more detailed than that of the control level.

## 16.46  SCHEDULE COMPRESSION

During the course of a project it is entirely possible that some critical path activities will take longer than their allotted duration, and completion of the project by the target date is jeopardized. Planners then look for a means to compress the schedule — i.e., shorten durations from those of the original plan. Several common methods for schedule compression are:

*Revise schedule logic.* Through complete review of the project's logic, it may be possible to find that some artificial constraints or faulty logic had been included which, if corrected, could permit rescheduling of remaining work within remaining time.

*Change work methods.* The durations assumed in scheduling were based on various assumptions regarding the way work would be handled. It may be possible to use a different method and save time. For example, an item scheduled for field erection by your personnel might be handled using off-site prefabrication by other personnel with only final installation by your personnel.

*Multiple-shifting.* If adequate skills are available, work may be scheduled over multiple shifts. This may cause some reduction in overall productivity because of overlap/coordination problems between shifts.

*Overtime.* Existing crews can be asked to work overtime. This has the disadvantages of higher wage rates and lowered productivity.

## PROGRESS AND COST CONTROL BASICS

## 16.47  MEASURING WORK PROGRESS

The work tasks required to complete a construction project vary from those required to clear and grade a site to start-up and turnover of the completed facility. During the course of the project it will be necessary for those executing the construction to report periodically the progress on each task. There are six methods for measuring this progress.

**Method 1   Units completed.** This method is applicable to those tasks which involve repeated production of easily measured pieces of work, each piece requiring approximately the same level of effort. Usually there is no included mix of subtasks, or if there is a mix, these are accomplished simultaneously and one of the subtasks can be used as the reference task.

Wire pulling is a task whose accomplishment is easily measured in terms of linear meters of wire pulled. If wire pulling of a certain type of wire is a control account, the *units completed* method is applicable. For example, if 10,000 LM of wire is to be pulled and 4000 LM has been pulled, the percent complete is found by dividing 4000 LM by 10,000 LM to show 40 percent complete.

Placing and finishing of a reinforced concrete slab is a type of work with multiple tasks handled simultaneously (placing and finishing), but progress would normally be reported on the basis of cubic meters (or yards) of concrete placed and finished or on the number of square meters (or feet) of finished surface.

**Method 2   Incremental milestone.** This method is applicable to any control account which includes subtasks that must be handled in sequence. For example, the installation of a major vessel in an industrial facility includes the sequential tasks or operations in the following list. Completion of any subtask or operation is considered to be a milestone, which, when completed, represents a certain percentage of the total installation. The percentage chosen to represent each milestone is normally based on the number of work-hours estimated to be required to that point in relation to the total. In the listing, the attainment of each milestone has been judged to be equivalent to having completed the percentage given of the total installation:

| | |
|---|---|
| Received and inspected | 15% |
| Setting complete | 35% |
| Alignment complete | 50% |
| Internals installed | 75% |
| Testing complete | 90% |
| Accepted by owner | 100% |

This method is called "level of effort" under the DOD/DOE cost and schedule control system criteria (C/SCSC).

**Method 3   Start/finish.** This method is applicable to those tasks which lack readily definable intermediate milestones and/or the effort/time required is very difficult to estimate. Millwright alignment work is usually in this category. For example, alignment of a major fan and motor may take from a few hours to a few days depending on the situation. Workers know when this work starts and when it is finished, but they never know the percentage completion in between.

Other examples include planning activities, flushing and cleaning, testing, and major rigging operations.

With the *start/finish* approach, one arbitrarily assigns a percent complete to the start of a task with 100 percent being recorded when it is finished. A starting percentage of 50 percent is equivalent to a task completed at a constant rate over time and is reasonable for short duration, lower value tasks. For longer or higher value tasks a contractor would probably have to use something less (20–30 percent) since an owner will hesitate to recognize too much completion in advance because it directly affects progress payments. For very short tasks the start/finish percentages are usually 0 percent/100 percent.

**Method 4   Supervisor opinion.** This is a subjective approach and should be used only for relatively minor tasks where development of a more discrete method cannot be used. Painting, dewatering, construction support facility construction, architectural trim, and landscaping are candidates for this approach.

With this method the supervisor simply makes a judgment of percent complete.

**Method 5   Cost ratio.** This method is applicable to tasks, such as project management, quality assurance, contract administration, and project controls which involve a long period of time or are continuous during the life of a project and are estimated and budgeted on bulk allocations of dollars and work-hours rather than on the basis of production.

With this method percent complete is found as follows:

$$\text{Percent complete} = \frac{\text{actual cost or work-hours to date}}{\text{forecast at completion}}$$

**Method 6   Weighted or equivalent units.** This method is applicable where the task being controlled is a major effort involving a long period of time and is composed of two or more overlapping subtasks, each with a different unit of work measurement.

Structural steel erection provides a good example for application of this method. Structural steel is normally estimated and controlled using tons of steel components as the unit of measure. But as illustrated in the following listing of subtasks included in steel erection, each subtask has a different unit of measure. To handle this, each subtask is weighted according to the estimated level of effort (usually work-hours) that will be dedicated to that subtask. As quantities of work are completed for each subtask, these quantities are converted into equivalent tons as illustrated:

| Weight | Subtask | Unit of measure | Quantity total | Equivalent steel, ton | Quantity to-date | Earned tons* |
|--------|---------|-----------------|----------------|-----------------------|------------------|--------------|
| 0.02 | Run foundation bolts | each | 200 | 10.4 | 200 | 10.4 |
| 0.02 | Shim | % | 100 | 10.4 | 100 | 10.4 |
| 0.05 | Shakeout | % | 100 | 26.0 | 100 | 26.0 |
| 0.06 | Columns | each | 84 | 31.2 | 74 | 27.5 |
| 0.11 | Beams | each | 859 | 52.0 | 0 | 0.0 |
| 0.10 | Cross-braces | each | 837 | 57.2 | 0 | 0.0 |
| 0.20 | Girts & sag rods | bay | 38 | 104.0 | 0 | 0.0 |
| 0.09 | Plumb & align | % | 100 | 46.8 | 5 | 2.3 |
| 0.30 | Connections | each | 2977 | 156.0 | 74 | 3.9 |
| 0.05 | Punchlist | % | 100 | 26.0 | 0 | 0.0 |
| 1.00 | Steel | Ton | | 520.0 | | 80.5 |

$$*\text{Earned tons to date} = \frac{(\text{quantity to date})(\text{relative weight})(520 \text{ tons})}{(\text{total quantity})} \qquad \text{Percent complete} = \frac{80.5 \text{ tons}}{520 \text{ tons}} = 15.5\%$$

A variation of this approach utilizes equivalent units for each subtask. In the preceding example each subtask item would be given a unit of measure that is an equivalent ton. For example, each beam would have an equivalent ton equal to

$$\text{Beam equivalent ton} = \frac{(0.11 \text{ weighting factor}) (520 \text{ tons})}{(859 \text{ beams})} = 0.0666$$

## 16.48   EARNED VALUE

As noted in the previous discussion, there can be many types of work and numerous ways of measuring progress on work. Having outlined methods for determining progress on single type of work, the next challenge is to develop a method for determining overall percent complete of a combination of unlike work tasks or a complete project. The system for accomplishing this is called *earned value,* although the terms *achieved value* and *accomplished value* are occasionally used.

Earned value is keyed to the budget for a project, a budget which is expressed in both work-hours and dollars. Work-hours and dollars are the only common denominators for the many accounts within a project.

Many projects are constrained by fixed budgets; others have floating or variable budgets. Earned value techniques are applicable to both situations although there are differences in detail in applying those techniques between the two situations. In this section the basics of earned value will be explained assuming the fixed budget situation. Later we advance to the variable budget situation and conclude with comparisons of the two methods.

In developing the control system for any project, we need to break down the project into its controllable parts. For purposes of control of work we develop the work breakdown structure (WBS), which comprises all work tasks that must be controlled for purposes of determining project progress. Each of these tasks will have its own dollar and work-hour budget. Add to the WBS all other accounts in the project which have either a cost or a cost and work-hour budget, but which are not used for purposes of measuring progress (e.g., management, quality control, administration) to create the cost breakdown structure (CBS) for the project. In other words, the WBS is incorporated within the CBS.

Under earned value a direct relationship is established between percent complete of an account and the budget for that account. This relationship is expressed as

Earned value = (percent complete) (budget for that account)

As can be seen from this equation, a budgeted amount is "earned" as a task is completed up to the total amount in that account. One cannot earn more than has been budgeted. For example, assume that $10,000 and 60 work-hours have been budgeted for a given account and that account is now 25 percent complete as measured by one of the methods previously described. That means that $2500 and 15 work-hours have been earned to date.

Since progress in all accounts can be reduced to earned work-hours and dollars as illustrated, this provides a method for summarizing multiple accounts and calculating overall progress. The formula for this is

$$\text{Percent complete} = \frac{\text{(earned work-hours/dollars all accounts)}}{\text{(budgeted work-hours/dollars all accounts)}}$$

## 16.49 COST AND SCHEDULE PERFORMANCE

The concepts discussed thus far provide a system for determining percent complete of single work tasks or combinations of tasks. The next challenge is to analyze the results for the purpose of determining how well things are proceeding as compared to what was planned. Fortunately, the earned value system lends itself very well to such a system.

We have spoken of budgeted and earned work-hours and dollars to this point. To these we must add *actual* work-hours or dollars. These three provide the combinations needed for analysis. This can be most easily explained if one defines each of these terms in this way:

Budgeted work-hours or $ to date represent *what I plan to do*. Under C/SCSC this is called budgeted cost for work scheduled (BCWS).

Earned work-hours or $ to date represent *what I did*. Under C/SCSC this is called budgeted cost for work performed (BCWP).

Actual work-hours or $ to date represent *what I paid for*. Under C/SCSC this is called actual cost of work performed (ACWP).

Performance against schedule is then simply a comparison of *what I planned to do* against *what I did*. In other words, I compare budgeted and earned work-hours. If budgeted work-hours are less than earned, it means I did more than planned and I am ahead of schedule. The reverse would place me behind schedule.

Performance against budget is measured by comparing *what I did* to *what I paid for*. To do this, I compare earned to actual work-hours. If I paid for more than I did, I have overrun the budget.

The preceding can be expressed in these ways:

$$\text{Schedule variance (SV)} = \text{(earned work-hours or \$)} - \text{(budgeted work-hours or \$)}$$

$$= \text{BCWP} - \text{BCWS}$$

$$\text{Schedule performance index (SPI)} = \frac{\text{(earned work-hours or \$ to date)}}{\text{(budgeted work-hours or \$ to date)}}$$

$$= \frac{\text{BCWP}}{\text{BCWS}}$$

$$\text{Cost variance (CV)} = (\text{earned work-hours or \$}) - (\text{actual work-hours or \$})$$

$$= BCWP - ACWP$$

$$\text{Cost performance index (CPI)} = \frac{(\text{earned work-hours or \$ to date})}{(\text{actual work-hours or \$ to date})}$$

$$= \frac{BCWP}{ACWP}$$

It should be noted that a positive variance and an index of 1.0 or greater is favorable performance.

## 16.50   PRODUCTIVITY

Project managers are always interested in knowing how well actual productivity compares with that used in planning and budgeting the work. Although a comparison of earned to actual work-hours may appear to provide an evaluation of productivity, it does so only if actual quantities of work exactly equal those budgeted for. Another mechanism is needed to evaluate productivity.

*Credit work-hours* (CWH), like *earned work-hours* (EWH), is a derived quantity. It provides a vehicle for handling work quantity variations between budgeted and actual without distorting crew productivity figures. CWH are equal to the budgeted productivity work-hour unit rate (work-h/unit) for a given task multiplied by the number of units completed. Since the actual units of work in a work package may vary from the budgeted (estimated) number of units, CWH may vary either way from EWH in a given work package. CWH are equal to EWH only if budgeted and actual quantities of work are equal. If any work task is omitted during the planning/estimating phase (i.e., no work-hour budget), CWH are not calculated for that package; calculations are confined to work packages for which work-hours were allocated. A productivity index (PI) may be calculated for a single work package or a combination of work packages (or the total project) using this formula:

$$\text{Productivity index (PI)} = \frac{(\text{sum of credit work-hours})}{(\text{sum of actual work-hours})}$$

## 16.51   QUANTITY—ADJUSTED BUDGETS

The concepts discussed in the preceding sections are based on a fixed budget scenario, as is most often the case in fixed-price work. However, in the case of cost-reimbursable contracts and other situations where the budget is subject to considerable variation, the fixed budget system will not be appropriate for making judgments on cost and schedule performance. In that case earned value determinations should be based on a variable budget system.

This system is particularly suited for a project which is initiated on the basis of incomplete definition and can take advantage of floating budgets. Each identified work package in the project is assigned a budget (work-hours and/or dollars) based on the best available work quantity information at that point in time. Then as each work package is fully defined, its budget is adjusted to reflect final work quantities.

A quantity-adjusted budget (QAB) is a budget that varies directly with quantity of work. It is calculated by multiplying the budgeted work-hour rates (and/or dollar rates) by the actual work quantities. For example, if the initial budget for construction of a foundation assumed 1000 yd$^3$ of concrete at 10 work-h/yd$^3$ for a total of 10,000 work-hours, but the actual design quantity turns out to be only 950 yd$^3$, then the QAB would be equal to 950 yd$^3$ times the 10 work-hours, or 9500 work-hours. In a sense this is not actually a new budget, but rather a forecast reflecting the latest designed material quantities to be installed at budgeted productivity, and becomes the yardstick for measuring field achievement. The real budgets under this system are the unit rates. Table 16.5 illustrates the calculation of a QAB.

The project's final QAB cannot be established until the completion of design engineering, and this is usually well after the start of construction; therefore, the initial QAB must be based on forecasted quantities from sampling and early takeoffs. The QAB is adjusted as better quantity data are supplied from the engineering office and these adjustments have an impact on construction progress measurement. Since frequent adjustments to the QAB can cause fluctuations in progress measurement, it is advisable to use the initial budget as the QAB until such time as reasonably firm quantity information becomes available. Quantity data are developed in a code-by-code sequence as engineering progresses and final commodity reviews are completed; earthwork is normally first, followed by concrete structural steel, etc. If the QAB is developed code by code in the same sequence as the final commodity reviews, the transition from budget to QAB becomes a smooth gradual process.

**TABLE 16.5**
**Calculating QAB (using work-hours)\***

| Work item | Unit of measure | Original Budget | | Budgeted unit rate | Design quantity | QAB work-hours (000) |
|---|---|---|---|---|---|---|
| | | Quantity (000) | Work-hours (000) | | | |
| Earthwork | CY | 234 | 193 | 0.825 | 257 | 212 |
| Concrete | CY | 94 | 2201 | 23.41 | 102 | 2388 |
| Steel erection | TN | 2.5 | 119 | 47.6 | 2.2 | 105 |
| Mechanical equipment | EA | 1.1 | 152 | 138.2 | 1.3 | 180 |
| Piping | LF | 180 | 470 | 2.61 | 210 | 548 |
| Electrical systems | LF | 84 | 220 | 2.62 | 79 | 207 |
| Total | | | 3355 | | | 3640 |

\*Note: Same approach would be used for calculating QAB dollars.

**TABLE 16.6**
**Calculating % complete using QAB**

| Work item | Design quantity (000) | QAB work-hours (000) | J-T-D* quantity (000) | % complete | Earned work-hours |
|---|---|---|---|---|---|
| Earthwork | 257 CY | 212 | 100 | 38.9 | 82 |
| Concrete | 102 CY | 2388 | 42 | 41.2 | 984 |
| Steel erection | 2.2 TN | 104 | 0.85 | 38.6 | 40 |
| Mechanical equipment | 1.3 EA | 180 | 0.4 | 30.8 | 55 |
| Piping | 210 LF | 548 | 35 | 16.7 | 92 |
| Electrical systems | 79 LF | 207 | 22 | 27.8 | 58 |
| Total | | 3639 | | | 1311 |

$$\text{Total percent complete} = \frac{\text{total earned work-hours}}{\text{total QAB}} = \frac{1311}{3639} = 36.0\%$$

*Job to date.

## 16.52   COST AND SCHEDULE PERFORMANCE

The methods previously discussed for calculation of percent complete, SV, SPI, CV, and CPI, as described under the fixed budget system, are fully applicable in the variable budget system. Earned work-hours may be calculated by multiplying percent complete by the QAB, or, for those activities tracked under the units complete method, by multiplying the units completed by the budgeted unit rate. Table 16.6 illustrates the calculation of percent complete using QAB.

## 16.53   PRODUCTIVITY ANALYSIS

Under the variable budget system the CPI is equal to the productivity index (PI) because the QAB automatically accounts for quantity variations. There is no need for a separate calculation of a PI using CWH.

## 16.54   CAUTIONARY NOTES

Reimbursable projects, where construction commences before complete design drawings are available, will tend to experience significant rework as a consequence of design change. Such rework increases the budgets of involved work packages. When determining percent complete it would be wrong to include the reworked portion of these budgets or those hours earned when doing the replaced work in the calculations even though these may be paid for by the client. To get around this, it is necessary to purge such hours from the accounts as rework occurs. QABs and actual hours wasted as a result of rework should be transferred to separate accounts outside the basic control structure so that they may later show the extent and cost of rework.

One must also keep in mind that when using the QAB method, percent complete changes with every change in the forecasted quantities and work-hours for individual accounts. A change of this type, completely independent of work accomplished in the field, confuses construction superintendents and tends to undermine the credibility of a performance tracking system.

Another pitfall to beware of is that the use of variable forecasts tends to reward incompetent supervisors—there is no longer that incentive of a budget constraint. As supervisors require additional work-hours to accomplish their work their weighting increases and their portion of the total project appears to increase. Conversely, the forecasted work-hours, and therefore the weighting, for more competent supervisors are progressively decreased, and their apparent contribution to the overall project decreases.

As with the fixed budget system, it is important when setting up databases and algorithms for manipulating data that EWH not be summed by adding those earned during the current period to those accumulated during prior periods. Instead, calculations should be made for each account and the totals generated by adding the EWH of all accounts to date in which hours or dollars have been earned.

Original budget unit rates must represent realistic, achievable objectives; otherwise the QAB is invalid for all purposes. Another factor that can invalidate the QAB is failure to keep change orders up to date. The whole premise of cost control requires that the budget be realistic and kept current. If this is done, the QAB provides a firm, fair basis for calculating percent complete.

## 16.55 WHICH SYSTEM, FIXED OR VARIABLE?

The choice of a system to be used on a project is dictated in some cases by the project itself; in other cases there is an option. On a project started on the basis of incomplete design the variable system should be used since it is the only one responsive to the inevitable quantity variations that come as the project becomes fully defined. On well-defined projects there is a choice, the choice being made on the basis of characteristics desired in the control system.

The fixed budget system has these characteristics:

Provides a direct evaluation of cost and schedule performance; requires supplementary system for productivity evaluation.

More simplified bookkeeping; less potential for operator-caused errors.

The fixed budgets provide a constant target for management to see. This is ideal for fixed-price work or other work with target budgets. Fixed budgets provide an incentive for working smarter.

Under this system the CPI and PI are not necessarily the same. Having the two separate indices provides more tools for analysis.

Performance data are susceptible to distortions if the project budget is not realistically distributed.

The variable budget system has these characteristics:

Provides direct evaluation of productivity and schedule performances (CPI and PI are the same when using work-hours; this is also true when using cost if there are no wage rate variances); requires supplementary system for evaluation of cost performance if operating against a fixed or target budget.

Provides a moving budget that varies directly with actual quantities of work and budgeted productivity rates for included tasks. This is ideal for projects with open budgets. If applied to projects with a fixed or target budget, a contingency account will be required to balance additions and deletions in the work accounts.

Requires more operator attention to database management because of continually changing baseline information.

Example 16.1 illustrates the application of the following to a simple project:

Fixed budget approach for fixed-price contract
Variable budget approach for fixed-price contract
Variable budget approach for variable budget

Incorporated within this example are calculations for SPI, CPI, and PI to illustrate further differences among the approaches.

### Example 16.1. Fixed versus variable-budget approaches

#### 1. Fixed budget approach for fixed-price contract
In the planning of a project composed of three work packages, the following data were generated:

| Package | Quantity | Unit rate | Total work-hours |
|---------|----------|-----------|------------------|
| A | 10 | 15 | 150 |
| B | 15 | 10 | 150 |
| C | 20 | 5 | 100 |
| | | | 400 |

Work was scheduled as follows during initial planning:

| Week | A | B | C |
|------|---|---|---|
| 1 | 2 | | |
| 2 | 2 | | |
| 3 | 2 | | |
| 4 | 2 | 5 | |
| 5 | 2 | 5 | |
| 6 | | 5 | |
| 7 | | | 5 |
| 8 | | | 5 |
| 9 | | | 5 |
| 10 | | | 5 |

After mobilization it was determined that two work quantities estimated were in error. Actual quantities are $A = 12$, $B = 15$, $C = 22$. As a result, work was rescheduled as follows since schedule target completion date remains:

| Week | A | B | C |
|------|---|---|---|
| 1 | 2 | | |
| 2 | 3 | | |
| 3 | 3 | | |
| 4 | 2 | 5 | |
| 5 | 2 | 5 | |
| 6 | | 5 | |
| 7 | | | 5 |
| 8 | | | 6 |
| 9 | | | 6 |
| 10 | | | 5 |

Since the project was fixed price and the budget fixed, the budgeted work-hours were redistributed as follows within available budget. Redistribution based on needed work-hours for actual quantities but spread within 400 work-hour budget follows:

| Package | Quantity | Unit rate | Needed work-hours | Allocated 400/440 |
|---------|----------|-----------|-------------------|-------------------|
| A | 12 | 15 | 180 | 164 |
| B | 15 | 10 | 150 | 136 |
| C | 22 | 5 | 110 | 100 |
| | | | 440 | 400 |

In completing the work, the following data were generated (plan/actual):

| Week | A Quantity | A Work-hours | B Quantity | B Work-hours | C Quantity | C Work-hours | Total work-hours |
|------|----------|------------|----------|------------|----------|------------|------------------|
| 1 | 2/1 | 28/16 | | | | | 28/16 |
| 2 | 3/2 | 41/31 | | | | | 41/31 |
| 3 | 3/3 | 41/40 | 0/2 | 00/22 | | | 41/62 |
| 4 | 2/3 | 27/38 | 5/4 | 46/40 | | | 73/78 |
| 5 | 2/2 | 27/24 | 5/4 | 45/42 | | | 72/66 |
| 6 | 0/1 | 00/15 | 5/4 | 45/36 | | | 45/51 |
| 7 | | | 0/1 | 00/12 | 5/3 | 23/18 | 23/30 |
| 8 | | | | | 6/5 | 27/30 | 27/30 |
| 9 | | | | | 6/6 | 27/33 | 27/33 |
| 10 | | | | | 5/5 | 23/28 | 23/28 |
| 11 | | | | | 0/3 | 00/14 | 00/14 |

From the preceding data, the following were calculated per period:

| Week | Scheduled work-hours | Actual work-hours | Earned work-hours[1] | Credit work-hours[2] | SPI[3] | CPI[4] | PI[5] |
|------|------|------|------|------|------|------|------|
| 1 | 28 | 16 | 14 | 15 | 0.50 | 0.88 | 0.94 |
| 2 | 41 | 31 | 27 | 30 | 0.66 | 0.87 | 0.97 |
| 3 | 41 | 62 | 59 | 65 | 1.44 | 0.95 | 1.05 |
| 4 | 73 | 78 | 77 | 85 | 1.05 | 0.99 | 1.09 |
| 5 | 72 | 66 | 64 | 70 | 0.89 | 0.97 | 1.06 |
| 6 | 45 | 51 | 50 | 55 | 1.11 | 0.98 | 1.08 |
| 7 | 23 | 30 | 23 | 25 | 1.00 | 0.77 | 0.83 |
| 8 | 27 | 30 | 23 | 25 | 0.85 | 0.77 | 0.83 |
| 9 | 27 | 33 | 26 | 30 | 0.96 | 0.79 | 0.91 |
| 10 | 23 | 28 | 23 | 25 | 1.00 | 0.82 | 0.89 |
| 11 | 0 | 14 | 14 | 15 | n/a | 1.00 | 1.07 |

Notes:
[1] Earned work-hours = (% complete)(budget in work-hours)
[2] Credit work-hours = (Original budgeted unit rate)(units completed)
[3] SPI = (earned work-hours) ÷ (scheduled work-hours)
[4] CPI = (earned work-hours) ÷ (actual work-hours)
[5] PI = (credit work-hours) ÷ (actual work-hours)

And the following cumulative totals were calculated:

| Week | Scheduled work-hours | Actual work-hours | Earned work-hours | Credit work-hours | SPI | CPI | PI |
|------|------|------|------|------|------|------|------|
| 1 | 28 | 16 | 14 | 15 | 0.50 | 0.88 | 0.94 |
| 2 | 69 | 47 | 41 | 45 | 0.59 | 0.87 | 0.96 |
| 3 | 110 | 109 | 100 | 110 | 0.91 | 0.92 | 1.01 |
| 4 | 183 | 187 | 177 | 195 | 0.96 | 0.95 | 1.04 |
| 5 | 255 | 253 | 241 | 265 | 0.94 | 0.95 | 1.05 |
| 6 | 300 | 304 | 291 | 320 | 0.97 | 0.96 | 1.05 |
| 7 | 323 | 334 | 314 | 345 | 0.97 | 0.94 | 1.03 |
| 8 | 350 | 364 | 337 | 370 | 0.96 | 0.93 | 1.02 |
| 9 | 377 | 397 | 363 | 400 | 0.96 | 0.91 | 1.01 |
| 10 | 400 | 425 | 386 | 425 | 0.97 | 0.91 | 1.00 |
| 11 | 400 | 439 | 400 | 440 | n/a | 0.91 | 1.00 |

**2. Variable budget approach for fixed-price contract**
Situation: In the planning of the same project composed of three work packages, the following data were generated (same as the first case):

| Package | Quantity | Unit rate | Total work-hours |
|---------|----------|-----------|------------------|
| A | 10 | 15 | 150 |
| B | 15 | 10 | 150 |
| C | 20 | 5 | 100 |
|   |    |   | 400 |

Work was scheduled as follows during initial planning (same as first case):

| Week | A | B | C |
|------|---|---|---|
| 1 | 2 | | |
| 2 | 2 | | |
| 3 | 2 | | |
| 4 | 2 | 5 | |
| 5 | 2 | 5 | |
| 6 | | 5 | |
| 7 | | | 5 |
| 8 | | | 5 |
| 9 | | | 5 |
| 10 | | | 5 |

After mobilization it was determined that two work quantities estimated were in error. Actual quantities are $A = 12$, $B = 15$, $C = 22$. As a result, work was rescheduled as follows since schedule target completion date remains (same as first case):

| Week | A | B | C |
|------|---|---|---|
| 1 | 2 | | |
| 2 | 3 | | |
| 3 | 3 | | |
| 4 | 2 | 5 | |
| 5 | 2 | 5 | |
| 6 | | 5 | |
| 7 | | | 5 |
| 8 | | | 6 |
| 9 | | | 6 |
| 10 | | | 5 |

The work package budgets were adjusted to reflect the additional work using the original unit rates. A new overrun (O/R) account was established to balance the budget to account for the fact that only 400 work-hours were really in the project's control budget:

| Package | Quantity | Unit rate | Work-hours |
|---------|----------|-----------|------------|
| A | 12 | 15 | 180 |
| B | 15 | 10 | 150 |
| C | 22 | 5 | 110 |
| Control budget | | | 440 |
| Budget O/R | | | −40 |
| Real budget | | | 400 |

In completing the work, the following data were generated (plan/actual):

| Week | A Quantity | A Work-hours | B Quantity | B Work-hours | C Quantity | C Work-hours | Total Work-hours |
|------|-----------|--------------|------------|--------------|------------|--------------|------------------|
| 1  | 2/1 | 30/16 |     |       |     |       | 30/16 |
| 2  | 3/2 | 45/31 |     |       |     |       | 45/31 |
| 3  | 3/3 | 45/40 | 0/2 | 00/22 |     |       | 45/62 |
| 4  | 2/3 | 30/38 | 5/4 | 50/40 |     |       | 80/78 |
| 5  | 2/2 | 30/24 | 5/4 | 50/42 |     |       | 80/66 |
| 6  | 0/1 | 00/15 | 5/4 | 50/36 |     |       | 50/51 |
| 7  |     |       | 0/1 | 00/12 | 5/3 | 25/18 | 25/30 |
| 8  |     |       |     |       | 6/5 | 30/30 | 30/30 |
| 9  |     |       |     |       | 6/6 | 30/33 | 30/33 |
| 10 |     |       |     |       | 5/5 | 25/28 | 25/28 |
| 11 |     |       |     |       | 0/3 | 00/14 | 00/14 |

From the preceding data, the following were calculated per period:

| Week | Scheduled work-hours | Actual work-hours | Earned work-hours[1] | SPI[2] | CPI[3] | Work-hours deficit[4] |
|------|----------------------|-------------------|----------------------|--------|--------|------------------------|
| 1  | 30 | 16 | 15 | 0.50 | 0.94 |             |
| 2  | 45 | 31 | 30 | 0.67 | 0.97 |             |
| 3  | 45 | 62 | 65 | 1.44 | 1.05 |             |
| 4  | 80 | 78 | 85 | 1.06 | 1.09 |             |
| 5  | 80 | 66 | 70 | 0.88 | 1.06 |             |
| 6  | 50 | 51 | 55 | 1.10 | 1.08 | −14 (pkg $A$) |
| 7  | 25 | 30 | 25 | 1.00 | 0.83 | −2 (pkg $B$)  |
| 8  | 30 | 30 | 25 | 0.83 | 0.83 |             |
| 9  | 30 | 33 | 30 | 1.00 | 0.91 |             |
| 10 | 25 | 28 | 25 | 1.00 | 0.89 |             |
| 11 | 0  | 14 | 15 | n/a  | 1.07 | −23 (pkg $C$) |

Notes:
[1] Earned work-hours = (budgeted unit rate)(units completed)
[2] SPI = (earned work-hours) ÷ (scheduled work-hours)
[3] CPI = (earned work-hours) ÷ (actual work-hours)
[4] Deficit calculated when each package completed = (original budget) − (actual work-hours)

And the following cumulative totals were calculated:

| Week | Scheduled work-hours | Actual work-hours | Earned work-hours | SPI | CPI | Work-hours deficit to date |
|------|------|------|------|------|------|------|
| 1 | 30 | 16 | 15 | 0.50 | 0.94 | |
| 2 | 75 | 47 | 45 | 0.60 | 0.96 | |
| 3 | 120 | 109 | 110 | 0.92 | 1.01 | |
| 4 | 200 | 187 | 195 | 0.98 | 1.04 | |
| 5 | 280 | 253 | 265 | 0.95 | 1.05 | |
| 6 | 330 | 304 | 320 | 0.97 | 1.05 | −14 |
| 7 | 355 | 334 | 345 | 0.97 | 1.03 | −16 |
| 8 | 385 | 364 | 370 | 0.96 | 1.02 | |
| 9 | 415 | 397 | 400 | 0.96 | 1.01 | |
| 10 | 440 | 425 | 425 | 0.97 | 1.00 | |
| 11 | 440 | 439 | 440 | n/a | 1.00 | −39 |

### 3. Variable approach with variable budget

This would be handled the same as for the fixed budget except that both the real and the control project budget would be 440 work-hours to reflect the increased quantities. Budget variance would then be:

Budget variance = (budget work-hours) − (actual work-hours):

Package $A$ = (180 work-hours) − (164 work-hours) = +16 work-hours

Package $B$ = (150 work-hours) − (152 work-hours) = − 2 work-hours

Package $C$ = (110 work-hours) − (123 work-hours) = −13 work-hours

Total project =                                    + 1 work-hours

## APPLIED PROGRESS AND COST CONTROL

## 16.56  BASELINES

In order to have control of any operation, there must first be a plan for that operation — this plan is the basis for control. Actually the *plan* for a construction project is a compilation of a number of interrelated planning documents — schedules, budgets, materials management plan, subcontracting plan, etc., and these comprise the baselines.

A number of formal control structures are included in the total management of the project and these are collectively grouped under the term *project control*. These are cost control, schedule control, materials control, and quality control. This section is concerned primarily with two of these — cost and schedule control.

## 16.57   COST CONTROL VS. FINANCIAL CONTROL

Cost control is obviously important on any project. In this connection it is appropriate to distinguish between *financial control* and *cost control*. Financial control is concerned with the receipts and expenditures of importance to good bookkeeping and accepted accounting practice. The structure for such accounting must be in accordance with accepted rules of accounting and serve whatever requirements that may relate to contract payment provisions, taxation, regulation, or project capitalization. Financial accounting will also reflect the pricing of a contract which may differ significantly from its costing (because of unbalancing and the tracking of indirect accounts such as profit and distributables). Field construction managers are concerned with cost—what specific operations should cost and what they do cost.

Cost control should be approached as an application of Pareto's law, which essentially states that 80 percent of the outcome of a project is determined by only 20 percent of the included elements. Thus in establishing a cost-control system, the idea is to isolate and control in detail those elements with the greatest potential impact on final cost, with only summary-level control on the remaining elements. The greatest variable in the final cost of a construction project is usually the labor cost. Most other cost elements in the project (materials, equipment, and overhead) can be predicted or established with reasonable accuracy if the project is properly planned and estimated. Labor cost is a function of worker hourly cost and worker productivity. The hourly rates are relatively easy to predict; productivity is the real variable. Thus a contractor will want to monitor work-hours expended and productivity as major elements in the cost-control program. Of course, quantity control will be included as a basis for progress reporting as well as estimate verification.

## 16.58   BUDGET BASELINES

The budget baselines for a project are generated through the estimating process. Whether or not the design documents are complete, planners must develop a cost estimate for the project using the most appropriate methods discussed in previous sections. If the project has yet to be fully defined, this estimate is approximate and subject to some variation. As the project becomes better defined, the estimates are updated to reflect the new definition.

For a fixed-price project good estimating is critical because the estimate establishes the bid price, which must incorporate all elements of cost while also providing a reasonable profit to the contractor. The estimate also provides all quantity, cost, and productivity targets to be used for detailed control. Ideally, the estimate will have been prepared using the same work breakdown structure as that used for the control schedule. This directly enables the production of quantity, cost, and productivity targets for each control work package.

## 16.59    SCHEDULE BASELINES

A major effort during the planning process is the development of the work break-down structure. This structure provides the basis for the schedule. As mentioned during the scheduling discussions, there are multiple levels of schedules and various forms of schedules (CPM, bar chart, etc.). The control schedule is, as its name implies, the schedule used for master control of the project. The control schedule can be in bar chart format, but on the larger projects it is best presented in CPM format, particularly time-scaled CPM. The control schedule is at a level of detail which can be intelligently reviewed by the planners; too much detail can go beyond human comprehension and can be assumed to contain illogical and arbitrary constraints. Detailed schedule control is best handled using bar charts.

## 16.60    THE CONTROL ACCOUNT BASELINE

To illustrate how a planner moves from the control to the detailed level, reference is made to Fig. 16.15. This figure is called a control account baseline. This document takes a control schedule work package (in this case service water piping) and plans it out in detail. Note how the piping system is broken down into the work tasks required for its completion (large pipe, valves, etc.). These are then scheduled in bar chart format. As is so often the case, these tasks are overlapping and there is some flexibility in their sequencing (soft logic). Use of the bar chart format with float shown for each bar gives the field the flexibility it needs for accomplishing the work. Also, note from other information included on the baseline, that the document provides the basis for earned value control and progress payments.

## 16.61    STATUSING

Having established the basis for control, project controllers are in a position to exercise that control. They will do this by receiving reports of actual progress and costs and then comparing these to the plan via various reports.

## 16.62    WORK STATUS

The various methods for measuring work progress were previously explained. On the control account baseline the method to be used is established under the unit of measure column for each task. These can then be rolled up using earned value to show the overall percent complete of the control account. Figure 16.16 is a representation of a reporting format using the service water piping of Fig. 16.15 as an example. The many control accounts, in turn, can be summarized at various levels or for the whole project using earned value.

# Control Account for Service Water Piping

**FIGURE 16.15**

514

# Monthly Quantity Report

Project _____   Date _____   Rev _____

## Service Water Piping

Account Code: S W P 0 0 0 0

Description _____

| Weight | Activity | U/M | Latest Estimate | Week Ending 1/6 | | 1/13 | | 1/20 | | 1/27 | | This Period / To Date | |
|---|---|---|---|---|---|---|---|---|---|---|---|---|---|
| 0.25 | Large Hangars | EA | 100 | 5 | 5 | 15 | 20 | 15 | 35 | 15 | 50 | | |
| 0.30 | Large Pipe | LF | 2000 | | | | | 50 | 50 | 50 | 50 | | |
| 0.10 | Large Valves | EA | 10 | | | | | | | | | | |
| 0.15 | Large Pipe Weld | EA | 150 | | | | | | | | | | |
| 0.20 | Small Pipe | LF | 1500 | | | | | | | | | | |
| Total Control 1.00 | Control Item Large Pipe | U/M LF | Control Quantity 2000 | 25 | 25 | 75 | 100 | 75 | 175 | 90 | 265 | | |

Field Engineer

**FIGURE 16.16**

515

## 16.63   COST STATUS

As mentioned previously, the contractor will surely be interested in the work-hour statistics for the project as a major cost-tracking tool and will utilize the CPI (cost performance index) and CV (cost variance) for work-hours as indicators.

Cost in terms of dollars also will be statused. Certain costs will be tracked control account by control account. Such costs will include at least labor and contractor-furnished installed materials. Equipment costs and costs of construction materials and supplies (materials consumed but not incorporated in the final product) may be tracked as part of the work control account, but they are more likely tracked in separate accounts for such items. The CPI and CV for dollars can be calculated for whatever cost items are tracked.

Tabular reports are appropriate for summarizing cost status in various ways. Typical summaries would be:

A cost summary that shows for each account the original control work-hours per dollars, current control work-hours per dollars, this period work-hours per dollars, job-to-date work-hours per dollars, remaining to complete work-hours per dollars, estimate at completion work-hours per dollars, and variance.

A labor rate report shows for each craft and each control account the original control figures for dollars, work-hours, and dollars per work-hour; and then provides for each category the current control, experience this period, job-to-date experience, estimate at completion, and variances.

A quantity and work-hour report which shows the original control work quantities, work-hours and the work-hours per unit of work for each control account; and then provides comparable information under the headings of current control, current period, job-to-date, and estimate at completion. This same report can show the earned work-hours this period, earned work-hours to date, and the labor CPI.

## 16.64   SCHEDULE STATUS

Schedule status is best displayed on a bar chart representation of the schedule. A sample of an excellent format for summary-level reporting to management is shown in Fig. 16.17.

## 16.65   ANALYSIS, TRENDING, AND FORECASTING

Although it is important to know the exact status of a project at a point in time, it is equally important to analyze the situation so that appropriate corrective action can be taken if needed. The following discussion provides via example ways for analyzing, trending, and forecasting.

# Monitoring Schedule

| | WORKHOURS | | | 2-1 | 3-1 | 4-1 | 5-1 | 6-1 | 7-1 | 8-1 | 9-1 | 10-1 | 11-1 | 12-1 | 1-1 | 2-1 | 3-1 | 4-1 |
|---|---|---|---|---|---|---|---|---|---|---|---|---|---|---|---|---|---|---|
| Potable Water | 502 | 0.8% | P | | | | | | | | | | | | 49 | 100 | | |
| | | | A | 0 | 16 | 49 | | | | | | | | | | 74 | | |
| Steam & Condensate | 19085 | 29.7% | P | | | | | | 0 | 4 | 8 | 18 | 30 | 52 | 68 | 86 | 98 | 100 | |
| | | | A | | | | | | | | | | 32 | | | | | | |
| Air & Vacuum | 2407 | 3.8% | P | | | | | | | | 0 | 8 | 14 | 22 | 40 | 67 | 100 | | |
| | | | A | | | | | | | | | | | 52 | | | | | |
| Process Liquors | 6587 | 10.2% | P | | | | | | | | | | | | 0 | 18 | 50 | 85 | 100 |
| | | | A | | | | | | | | | | | | | | 65 | | |
| Deaerated B.F.W. | 2995 | 4.7% | P | | | | | | | 0 | 17 | 60 | 84 | 100 | | | | | |
| | | | A | | | | | | | | 34 | | | | | | | | |
| Oil, Chemicals, Vents, Millwater & Misc. | 25330 | 39.4% | P | | | | | | 0 | 8 | 11 | 31 | 45 | 66 | 83 | 92 | 98 | 100 | |
| | | | A | | | | | | | | | | 43 | | | | | | |
| Contingency | 7362 | 11.4% | P | | | | | | | | | | | | | | | | |
| | | | A | | | | | | | | | | | | | | | | |
| | | | P | | | | | | | | | | | | | | | | |
| | | | A | | | | | | | | | | | | | | | | |
| | | | P | | | | | | | | | | | | | | | | |
| | | | A | | | | | | | | | | | | | | | | |
| | | | P | | | | | | | | | | | | | | | | |
| | | | A | | | | | | | | | | | | | | | | |
| | | | P | | | | | | | | | | | | | | | | |
| | | | A | | | | | | | | | | | | | | | | |
| Totals | 64268 | 100% | P | | | | | | 5 | 1 | 8 | 21 | 33 | 51 | 66 | 82 | 95 | 98 | 100 |
| | | | A | 0 | 2 | 5 | | | | | | | 37 | | | | | | |

LEGEND

— Planned

▪▪▪▪▪ Actual

FIGURE 16.17

517

## 16.66   THE COST AND SCHEDULE PERFORMANCE CURVES

One of the handiest presentations for quickly showing the cost and schedule status of the project is shown in Fig. 16.18. On this graph are plotted the planned budget cumulative expenditure curves (in terms of either dollars or work-hours) plus the cumulative actual and the cumulative earned to the date of the report. One can quickly see the cost and schedule variances and approximately how much the project is behind or ahead of schedule.

## 16.67   INDEX TRACKING

Figure 16.19 is an example of a graph for tracking indices of the types described in Sec. 16.50. It tracks the PI. Note on this graph the use of a "projected" curve which does not coincide with the 1.0 datum line. This curve recognizes that productivity can usually be expected to be lower during the early stages of a project, reach a peak about midway in the project, and then decrease toward closeout. Having done this, the actual productivity plot can be more meaningfully evaluated. As shown on the example chart, a productivity index of 1.06, which one normally assumes is favorable, is actually low compared to what it should be for that point in time.

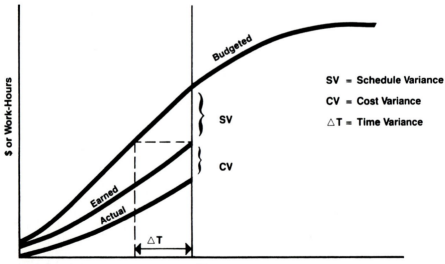

## Cost and Schedule Performance Graph

FIGURE 16.18

# Productivity Profile

FIGURE 16.19

## 16.68   OTHER TRACKING

Figure 16.20 is an interesting variation of Fig. 16.19. Note how the vertical axis is work-hours per percent complete. On this graph the cumulative plan curve is an upside-down image of the projected curve in Fig. 16.19 because of the different choice of units on the vertical axis. This graph also includes the plan for period and actual period plots to give it more usability. Note how the point identified as "(1)" shows that actual period performance equals that of planned performance. But when one looks at the actual cumulative performance, it shows that the project still has a problem because of the poor performance of prior periods, and performance must become better than planned if the project is to recover.

Figures 16.21 and 16.22 track building steel erection work-hour rates and unit wage rates, respectively, on a project and are self-explanatory.

Figure 16.23 is representative of a format for tracking bulk quantity items—in this example wire pulling and terminations. The two curves represent the plan. By superimposing actual performance, the situation and trends are readily shown in the figure.

## 16.69   ANALYSIS TECHNIQUES

Each report item has significance in itself, but it usually takes a combination of items for the total situation to be shown. For example, poor labor cost performance (CPI < 1.0) is certainly a problem, but the CPI does not point to the cause of that problem—

# Workhour Productivity Trend Chart

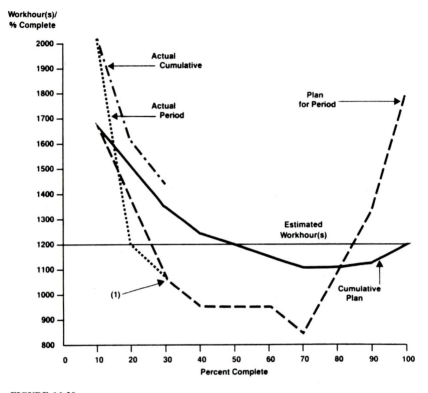

**FIGURE 16.20**

it could be any combination of low productivity, bad quantity estimate, excessive manning, or higher crew wage rates. Thus report data must be available in each of those areas to enable the manager to isolate the problem and take remedial action.

Figure 16.24 is an analysis tree involving just two report items — SPI and total float. Note the many possible combinations here. Other analysis trees using other report items are readily developed.

## 16.70   FORECASTING

Some very sophisticated forecasting techniques are explained in any management science text, but few constructors are comfortable with them and their reliability seems no better than a few simple approaches. One thing is certain, no two methods, so-

**Building Structural Steel Erection**

FIGURE 16.21

521

**FIGURE 16.22**

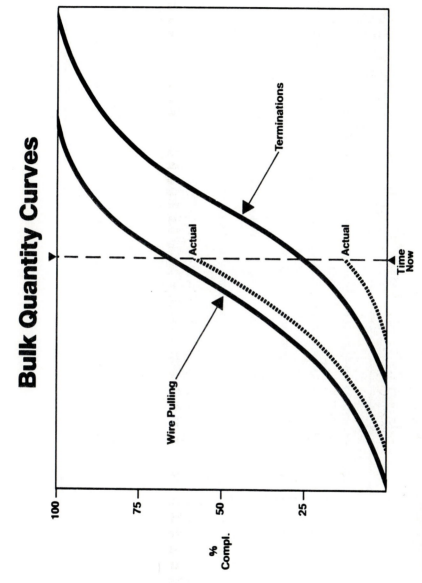

**FIGURE 16.23**

# Analysis Tree -- Total Float and SPI

TF > 0
- SPI > 1.0 — AHEAD OF SCHEDULE ON CRITICAL PATH; MORE WORK BEING DONE THAN PLANNED
- SPI = 1.0 — AHEAD OF SCHEDULE ON CRITICAL PATH; SOME SHORTFALL IN WORK ON NON-CRITICAL ACTIVITIES
- SPI < 1.0 — AHEAD OF SCHEDULE ON CRITICAL PATH; SIGNIFICANT SHORTFALL IN WORK ON NON-CRITICAL ACTIVITIES

TF = 0
- SPI > 1.0 — CRITICAL PATH ON SCHEDULE; MORE WORK BEING DONE ON NON-CRITICAL ACTIVITIES THAN PLANNED
- SPI = 1.0 — CRITICAL PATH ON SCHEDULE; TOTAL WORK VOLUME IS AS PLANNED
- SPI < 1.0 — CRITICAL PATH ON SCHEDULE; SHORTFALL IN WORK ON NON-CRITICAL ACTIVITIES

TF < 0
- SPI > 1.0 — CRITICAL PATH ACTIVITIES BEHIND SCHEDULE; TOTAL WORK MORE THAN PLANNED INDICATING EXCESS ATTENTION TO NON-CRITICAL ACTIVITIES
- SPI = 1.0 — CRITICAL PATH ACTIVITIES BEHIND SCHEDULE; TOTAL WORK VOLUME AS PLANNED MEANING TOO MUCH ATTENTION TO NON-CRITICAL ACTIVITIES
- SPI < 1.0 — CRITICAL PATH ACTIVITIES BEHIND SCHEDULE; TOTAL WORK LESS THAN PLANNED; NEED MORE OVERALL EFFORT

FIGURE 16.24

524

phisticated or otherwise, produce the same answer. Three basic approaches are provided here.

**Method 1.** This method is useful for forecasting costs and work-hours. It assumes that work from this point forward will progress at planned rates whether or not these rates have prevailed to this point. In formula form it is

$$EAC = (ACWP + (BAC - BCWP)$$

where  EAC = estimate at completion
ACWP = actual cost of work performed to date
BAC = original budget at completion
BCWS = budgeted cost of work performed to date

**Method 2.** This method assumes that the rate of progress prevailing to date will continue to prevail. This is found through this formula:

$$EAC = (BAC) \div (CPI)$$

where CPI = cost performance index

**Method 3.** This method utilizes curves such as several of those included herein and is useful for forecasting any piece of data represented by those curves. The forecaster simply makes the best extrapolation possible using the typical shapes of such curves and whatever other information may be available to the forecaster to make the projection.

It is recommended that no single forecasting method be used. Rather, one should include a forecast by each of the preceding methods because these provide a range of possibilities.

## PROBLEMS

**16.1.** Define the critical path for the following network for the production of an instrument.

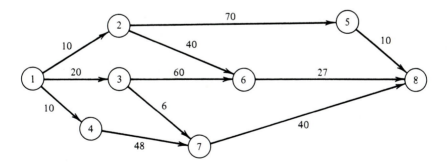

**16.2.** Define the critical path for the following process to produce an electronic device. Find the free float and total for each operation.

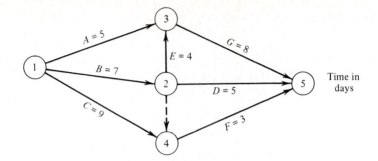

**16.3.** The following additional information applies to Prob. 16.2:

|           | **Normal**   |                        | **Crash**    |                        |
| --------- | ------------ | ---------------------- | ------------ | ---------------------- |
| Operation | Duration     | Operation cost, $      | Duration     | Operation cost, $      |
| A         | 5            | 100                    | 4            | 150                    |
| B         | 7            | 200                    | 6            | 300                    |
| C         | 9            | 400                    | 8            | 450                    |
| D         | 5            | 150                    | 4            | 200                    |
| E         | 4            | 100                    | 3            | 150                    |
| F         | 3            | 300                    | 2            | 350                    |
| G         | 8            | 700                    | 7            | 850                    |

(*a*) What is the total cost of the project?

(*b*) What would be the minimum additional cost required to finish the project 2 days earlier?

**16.4.** A network with normal time in days is as follows:

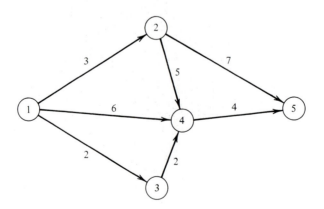

Find the critical path and free float of all activities.

**16.5.** The following additional information applies to Prob. 16.4. Assume a straight line or linear relationship between cost and reduction in time, as in the last column:

| Activity | Normal Days | Normal $ | Crash Days | Crash $ | Crash $ per day |
|----------|-------------|----------|------------|---------|-----------------|
| 1–2 | 3 | 50 | 2 | 100 | 50 |
| 1–3 | 2 | 25 | 1 | 50 | 25 |
| 1–4 | 6 | 140 | 4 | 260 | 60 |
| 2–4 | 5 | 100 | 3 | 180 | 40 |
| 3–4 | 2 | 80 | 2 | 80 | — |
| 2–5 | 7 | 115 | 5 | 175 | 30 |
| 4–5 | 4 | 100 | 2 | 240 | 70 |
| Totals | | 610 | | 1085 | |

Find the minimum cost for completing the project in 12, 11, 10, 9, 8, and 7 days.

**16.6.** For the following network with time in days, find the critical path and the free float of all activities:

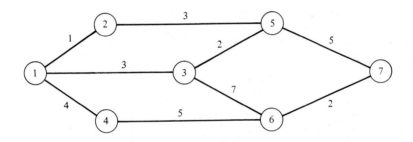

**16.7.** Return to Prob. 16.6 and assume that labor can be shifted from any activity to any other with an equal change in time, e.g., activity 3–6 can be reduced by 0.1 day by increasing activity 1–2 by 0.1 day. Find the minimum time for the network after replanning.

**16.8.** Redo Prob. 16.6 with a dummy activity arrow from event 5 to event 6.

**16.9.** Refer to the diagram for Prob. 16.6. Activity 3–6 can be started when activity 1–3 is one-third completed, and activity 3–5 can be started when activity 1–3 is two-thirds completed. Redraw the network, making sure that event 3 is completed, and find the critical path.

**16.10.** Three operations are performed in series, with time in hours as shown. *B* can be started when *A* is one-third completed, but *B* cannot be completed until 1 h after *A* is completed. *C* can be started when *B* is one-half completed, but *C* cannot be more than 50 percent completed until 1 h after *B* is completed. Show this as a network and find the time for the critical path:

**16.11.** On the following logic diagram are shown the durations of each activity. Calculate early start and finish, late start and finish, total float and free float times. Also, show the critical path. As an option, calculate any independent float times:

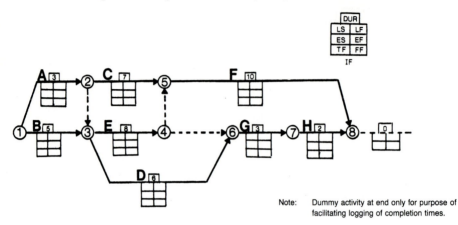

Note:   Dummy activity at end only for purpose of facilitating logging of completion times.

**16.12.** The following data are from a project's status reports at the end of a reporting period. Complete the worksheet, determine the percent complete of subaccounts 0810 and 0820, and determine the overall percent complete of summary account 0800:

| Code | accounts | Unit of measure | Total quantity | Quantity-to-date | Budget work-hours | Earned work-hours |
|------|----------|-----------------|----------------|-------------------|-------------------|-------------------|
| 0811 | Forms | SM | 500 | 500 | 5,000 | |
| 0812 | Resteel | CWT | 10 | 9 | 1,000 | |
| 0813 | Place & finish | CM | 1,000 | 750 | 10,000 | |
| 0810 | Slabs at grade | XXX | XXX | XXX | | |
| 0821 | Forms | SM | 550 | 55 | 6,000 | |
| 0822 | Resteel | CWT | 10 | 2 | 1,000 | |
| 0823 | Place and finish | CM | 2,500 | 0 | 15,000 | |
| 0820 | Elevated slabs | XXX | XXX | XXX | | |
| 0800 | Concrete | XXX | XXX | XXX | | |

**16.13.** You have summarized all control accounts in area A of a project to the end of the reporting period. You note that you had scheduled 28,000 work-hours, had earned 26,000 work-hours, and actually paid for 25,000 work-hours. Analyze the cost and schedule status in area A at the end of the reporting period.

**16.14.** In planning and budgeting a fixed-price project, a given work package was estimated to include 200 units of work. Estimators further utilized a unit rate of 4 work-hours per unit of work so budgeted for 800 work-hours in this account. In the field it was subsequently determined that there were really 240 units of work to be performed. This was strictly an estimating error and, with no contingency fund available, the budget remained at 800 work-hours. At the end of the latest reporting period work was 50 percent complete (120 units) and 432 work-hours had been paid for. Is this package overrunning or underrunning cost and is productivity better or worse than planned?

# REFERENCES

1. Ahuja, H. N.: *Construction Performance Control by Networks,* John Wiley & Sons, Inc., New York, 1976.
2. ———: *Project Management: Techniques in Planning and Controlling Construction Projects,* 2d ed., John Wiley & Sons, Inc., New York, 1984.
3. Antill, J. M., and R. W. Woodhead: *Critical Path Method in Construction Practice,* 3d ed., Wiley-Interscience, New York, 1982.
4. Clark, F. D., and A. B. Lorenzoni: *Applied Cost Engineering,* 2d ed., Marcel Dekker, Inc., New York, 1985.
5. Drigani, F.: *Computerized Project Control,* Marcel Dekker, Inc., New York, 1989.
6. Dudick, T. S.: *Cost Controls for Industry,* 2d ed., Prentice-Hall, Inc., Englewood Cliffs, N.J., 1976.
7. Fleming, Q. W.: *Cost/Schedule Control Systems Criteria,* Probus Publishing Company, Chicago, Ill., 1988.
8. Gido, J.: *An Introduction to Project Planning,* 2d ed., Industrial Press, Inc., New York, 1984.
9. Gutman, N.: *How to Keep Product Costs in Line,* Marcel Dekker, Inc., New York, 1985.
10. Hackney, J. W.: *Control and Management of Capital Projects,* 2d ed., McGraw-Hill Book Company, New York. (In preparation)
11. Horowitz, J.: *Critical Path Scheduling,* The Ronald Press Company, New York, 1967.
12. Kavanagh, T. C., F. Muller, and J. J. O'Brien: *Construction Management,* McGraw-Hill Book Company, New York, 1978.
13. Kelley, J. E. Jr., and M. R. Walker, *Critical-Path Planning and Scheduling: An Introduction,* Manchley Associates, Ambler, Pa., 1959.
14. Levin, R. I., and C. A. Kirkpatrick: *Quantitative Approaches to Management,* 2d ed., McGraw-Hill Book Company, New York, 1971.
15. McNeill, W. I.: *Effective Cost Control,* Prentice-Hall, Inc., Englewood Cliffs, N.J., 1965.
16. McNesby, E. J.: *Systematic Control of Factory and Management Costs,* Prentice-Hall, Inc., Englewood Cliffs, N.J., 1978.
17. Neil, J. M.: *Construction Cost Estimating for Project Control,* Prentice-Hall, Inc., Englewood Cliffs, N.J., 1982.
18. ———: "Applied Scheduling," *Skills and Knowledge of Cost Engineering,* 2d ed., American Association of Cost Engineers, Morgantown, W.Va., 1988, pp. 71–75.
19. ———, ed.: *Modern Management Systems for Engineered Construction: A Compilation of Construction Industry Institute Reports,* 2d ed., American Association of Cost Engineers, Morgantown, W.Va., 1989.
20. ———: "Planning," *Skills and Knowledge of Cost Engineering,* 2d ed., American Association of Cost Engineers, Morgantown, W.Va., 1988, pp. 55–62.
21. ———: "Scheduling Basics," *Skills and Knowledge of Cost Engineering,* 2d ed., American Association of Cost Engineers, Morgantown, W.Va., 1988, pp. 63–70.
22. ———, and T. L. Hyvomen: "Progress and Cost Control," *Skills and Knowledge of Cost Engineering,* 2d ed., American Association of Cost Engineers, Morgantown, W.Va., 1988, pp. 77–104.
23. O'Brien, J. J.: *CPM in Construction Management,* 3d ed., McGraw-Hill Book Company, New York, 1984.
24. Radke, M.: *Manual of Cost Reduction Techniques,* McGraw-Hill Publishing Company, Maidenhead, England, 1972.
25. Wilson, R. M. S.: *Cost Control Handbook,* John Wiley & Sons, Inc., New York, 1975.
26. Zeyher, L. R.: *Zeyher's New Guide to Cost Reduction in Plant Operations,* Prentice-Hall, Inc., Englewood Cliffs, N.J., 1977.

# PART
# IV

# ASSOCIATED
# TOPICS

# CHAPTER
# 17

## COST
## ACCOUNTING

*R. A. Rogers*

## BASIC CONCEPTS

## 17.1 COST ACCOUNTING
## AND THE COST ENGINEER

The interrelationship between the accounting function and the cost engineering function varies widely between manufacturing and processing industries, between operations and construction, and among individual companies. In some cases accounting has sole responsibility for recording and reporting cost data; in other instances, the task is performed jointly by accounting and cost engineering; and in still other environments the cost data are derived solely by cost engineering with only the requirement that the data be reconcilable with accounting records. The last situation is the most usual in the construction of new facilities and is necessitated by the requirement to project costs based on all known data rather than to rely on historical data alone. In all these situations it is essential that there be mutual understanding between the two groups. Cost data frequently must be allocated to different products, processes, or constructed facilities. This requires considerable technical and engineering familiarity with operations as well as knowledge of accounting theory and practice so that the allocations are fair and reasonably equitable.

This chapter presents a brief survey of the concepts and practices of accounting to the extent required to provide cost engineers with the basic, minimum understanding needed for their professional work.

**533**

## 17.2 ACCOUNTING DEFINITIONS

*Accounting* is a technique of measuring economic events in terms of money. A general definition is:*

> Accounting is the art of recording, classifying and summarizing in a significant manner and in terms of money, transactions and events which are, in part at least, of a financial character and interpreting the results thereof.

Accounting uses what is called a *double-entry system* for recording each transaction in monetary units. The double-entry system can be thought of as a series of equations; therefore, the two sides of the equation, the debits and the credits, must be equal for each transaction and cumulatively.

The terms *debit* and *credit* are conventional and have no particular significance as such.

Transactions are classified in accordance with a chart of accounts, which is a list of files by name and code number. Table 17.1 is a skeletal chart of accounts. Each account may be thought of as a record of all the transactions for a particular classification. Each account may have either debit or credit transactions. The difference between the sum of its debit transactions and its credit transactions is called the *balance* and is said to be a *debit balance* or a *credit balance* depending on which is greater. Certain accounts normally have debit balances, such as the cash account, and others normally have credit balances, such as the accounts payable account.

Another common equation used in accounting is

$$\text{Assets} - \text{liabilities} = \text{net worth}$$

Accounts are further categorized as balance sheet accounts or profit-and-loss accounts. *Balance sheet accounts* are essentially those identified in the preceding equation, and their balances are used to prepare a financial statement called a *balance sheet*. Table 17.2 is a pro forma balance sheet. The balance sheet reflects the financial condition of the enterprise as of a given date, usually the end of the month or end of the year. *Profit-and-loss accounts*, or *expense and revenue accounts*, are for the purpose of preparing a financial statement called an *income statement*, which reflects the operating results of the enterprise for a given period of time, usually a month or a year. Table 17.3 is a pro forma income statement.

Among the asset accounts is an account called *work-in-process*. This is where construction, manufacturing, or process costs are recorded and carried until the work is complete. This account may be further subdivided into jobs, and the jobs subdivided into elements of cost called *cost codes*. It is these cost accounts with which cost engineers are concerned.

---

*P. Grady, *Inventory of Generally Accepted Principles for Business Enterprises*, AICPA Accounting Research Study 7, American Institute of Certified Public Accountants, New York, 1965, p. 2.

**TABLE 17.1**
## Pro forma chart of accounts*

| Balance sheet accounts | |
|---|---|
| 2000 Assets | 3000 Liabilities |
|   2100 Cash |   3100 Accounts payable |
|   2200 Accounts receivable |   3200 Notes payable |
|   2300 Notes receivable |   3300 Taxes payable |
|   2400 Inventory — materials and supplies |   3400 Accrued liabilities |
|   2500 Inventory — finished products |   3500 Reserve accounts |
|   2600 Work-in-progress | |
|   2700 Equipment | 4000 Equity |
|   2800 Buildings and fixtures |   4100 Capital stock issued and outstanding |
|   2900 Land |   4200 Retained earnings |

| Profit-and-loss accounts | |
|---|---|
| 5000 Revenues | 6300 Heat, light, and power |
|   5100 Sales of finished goods | 6400 Communications expense |
|   5200 Other revenues | 6500 Reproduction expense |
| | 6600 Insurance |
| 6000 Expenses | 6700 Taxes |
|   6100 Cost of goods sold | 6800 Depreciation |
|   6200 Salaries and wages | 6900 Interest expense |

| Cost classification accounts (work-in-process) | |
|---|---|
| 7000 Construction work-in-progress | 7600 Electrical systems |
|   7100 Site preparation | 7700 Piping systems |
|   7200 Concrete work | |
|   7300 Structural steel | 8000 Manufactured goods-in-progress |
|   7400 Heavy equipment |   8100 Direct materials |
|   7500 Buildings |   8200 Direct labor |
| |   8300 Overhead |

*These are summary-level accounts, sometimes called *control accounts*. More detail can be provided by using subaccounts as required.

## 17.3  TYPES OF ACCOUNTING

One type of accounting is *financial accounting*. It is essentially a historical record of transactions which have been completed or partially completed. Its function is to report operating results and financial condition of the enterprise to stockholders, Securities and Exchange Commission, Internal Revenue Service, and other regulatory bodies.

Another type of accounting, called *managerial accounting*, is for the purpose of providing financial operating data to the management of the company to assist them in performing the management function. Cost accounting, the principal topic of this chapter, is among the most important parts of managerial accounting. Cost accounting expands the techniques of financial accounting. Its major application is in

**TABLE 17.2**

## Pro forma Balance Sheet
## The Acme Corporation
## Dec. 31, 199X
(in thousands of dollars)

### Assets

Current assets:

| | |
|---|---:|
| Cash................................................................... | $ 2,131 |
| Accounts receivable.............................................. | 36,292 |
| Notes receivable — due within 1 year....................... | 125 |
| Inventories and work-in-process ............................. | 12,614 |
| Subtotal.................................. | 51,162 |

Fixed assets:

| | |
|---|---:|
| Notes receivable — not due within 1 year..................... | 1,250 |
| Equipment — less reserve for depreciation................... | 44,768 |
| Buildings and fixtures — less reserve for depreciation...................... | 16,347 |
| Land ................................................................ | 5,000 |
| Subtotal.................................. | 67,365 |
| Total assets ....................................................... | 118,527 |

### Liabilities and net worth

Current liabilities:

| | |
|---|---:|
| Accounts payable................................................. | 13,421 |
| Notes payable — due within 1 year........................... | 1,200 |
| Taxes payable ..................................................... | 2,500 |
| Accrued liabilities ............................................... | 1,248 |
| Subtotal.................................. | 18,369 |

Long-term liabilities:

| | |
|---|---:|
| Notes payable — not due within 1 year....................... | 10,000 |
| Total liabilities ................................................. | 28,369 |

Net worth:

| | |
|---|---:|
| Capital stock issued and outstanding......................... | 72,172 |
| Retained earnings................................................ | 17,986 |
| Subtotal.................................. | 90,158 |
| Total liabilities and net worth ................................... | 118,527 |

the planning and control aspects of management, which have increased in importance during the last few decades since the introduction of scientific techniques to the managerial decision-making process.

## 17.4   THE MECHANICS OF ACCOUNTING

Accounting includes three basic steps: recording, classifying, and summarizing economic data in terms of money.

TABLE 17.3

## Pro forma Income Statement
## The Acme Corporation
## Year ended Dec. 31, 199X
(in thousands of dollars)

| | |
|---|---:|
| Revenues | $88,126 |
| Operating expenses: | |
|     Cost of goods sold | 46,218 |
|     Salaries and wages | 2,729 |
|     Heat, light, and power | 1,516 |
|     Communications expense | 127 |
|     Reproduction expense | 38 |
|     Insurance | 1,243 |
|     Taxes — other than federal income taxes | 746 |
|     Depreciation | 9,982 |
|               Subtotal | 62,599 |
| Operating income | 25,527 |
| Nonoperating expenses: | |
|     Interest expense | 1,120 |
| Income before federal income taxes | 24,407 |
| Federal income taxes | 11,715 |
| Net income | 12,692 |

The chart of accounts must be established by mutual agreement among the various groups involved, including particularly accounting, cost engineering, any group charged with responsibility for classifying any of the transactions, and any operating, construction, or management group that will be expected to interpret accounting/cost engineering reports. The chart of accounts as shown in Table 17.1 can be broken down into as much detail as desired by the use of subaccounts. In using the accounts for entry and retrieval of cost information, the cost engineer must be entirely familiar with what each subaccount includes. In practice, a compromise must be made in establishing subaccounts. Too few subaccounts create a lack of details in the cost, and too many add to the accounting cost and can confuse rather than help.

Transactions may originate from several sources, such as (1) payment of a vendor's invoice, which is recorded on a document called a *cash voucher* or an *accounts payable voucher*; (2) payment of salaries and wages based on employee time sheets; (3) transfer of materials and supplies or capital equipment from inventory accounts on warehouse tickets; (4) equipment use charges processed from equipment time sheets; and (5) calculation of depreciation, insurance, taxes, inventory adjustments, and cost allocations, which are prepared on special forms called *journal vouchers*.

Accounting transaction documents may be prepared in the departments where the transactions originate. Transactions will normally be classified by the personnel closest to the operation and in the best position to know the proper classification. All transaction classifications will be checked by an independent party before entry into the accounts. Errors can be corrected later if discovered, but the best and least expensive method is to prevent as many errors as is economically and practically feasible. Cost data are only as good as the quality of the original data and the quality of the mechanical entry. Data entry errors are kept to a minimum by various checks and balances built into the system.

The summarization step consists of closing the revenue and expense accounts to determine the profit or loss for the period (usually at the end of each month and at the end of each year) and to prepare the income statement. The result is then transferred to the equity account, or retained earnings. The asset and liability account balances are then used to prepare the other main financial statement, the balance sheet. Finally, various cost reports are prepared for both financial and control purposes. The responsibility for the latter may rest partially or wholly with the cost engineer. See Sec. 17.1, which explains the varying relationships which may exist between the cost accountant's work and that of the cost engineer.

Most accounting systems are computerized. This makes it possible to handle great volumes of data on a timely basis and to manipulate the data in various ways to produce meaningful reports and analyses.

## 17.5 DEFINITIONS AND CLASSIFICATIONS OF COST

The term "cost" has numerous meanings and definitions. The AICPA *Inventory* states:*

> Cost is the amount, measured in money, or cash expended or other property transferred, capital stock issued, services performed, or a liability incurred, in consideration of goods or services received. Costs can be classified as unexpired and expired. Unexpired costs (assets) are those which are applicable to the production of future revenues. . . . Expired costs are those which are not applicable to the production of future revenues and for that reason are treated as deductions from current revenues, or are charged against retained earnings.

Costs may be divided into as many classifications as may be useful in preparing meaningful reports for better analysis and understanding. Care should be taken not to use too many classifications; otherwise reports will be so voluminous as to be useless. It is important to select the appropriate cost type for individual objectives. No general rules are possible because of the wide variety of circumstances.

As mentioned earlier, the cost estimator needs future-directed, predetermined data rather than purely historical costs. Insufficient attention is often given to this important difference. In an era of great change, as in recent years, the assumption that the past will repeat itself in the future cannot be considered valid even in the short term. The cost estimator, therefore, must attempt to adjust recorded data for estimat-

---

* Ibid., p. 228.

ing purposes after soliciting information from knowledgeable persons within the organization and utilizing available indexes of predicted economic behavior. The more future-directed the cost estimator's figures, the more critical become the predicted activity levels and economic behavior factors.

Direct costs, also called *prime costs*, are directly traceable to the product being manufactured or processed or to the physical asset being constructed, e.g., the fabric in clothing or the concrete in a foundation.

Indirect costs are all the costs of manufacturing, processing, or construction that cannot be classified as direct costs, because it is either impossible or impractical. Each classification of indirect cost is initially accumulated in a separate account, and at the end of the accounting period the balance is allocated to the individual benefiting activities, such as cost centers, products, or construction activities. There are many bases for allocation, such as direct labor cost, square feet of floor space, or number of employees. An allocation basis for a particular situation must be carefully selected. The end result of any allocation must be reasonable and equitable. The topic is discussed more fully in Secs. 17.10 to 17.14.

## 17.6 COST ACCOUNTING FOR EXPENDITURE OF CONSTRUCTION FUNDS

Accounting for the costs of constructing new capital facilities has several purposes. The first is to account for the expenditure of the company's funds. Another is to accumulate costs in a manner which will support depreciation calculations for both financial reporting and income tax computations. Payment of progress fees to contractors also requires cost data for calculating the amount earned under the terms of the contract, such as the percent of work completed by cost classification times appropriate contractual monetary amounts.

Finally, of most interest to the cost engineer is cost data for the construction budget. This information is useful for project cost control purposes. The cost engineer is able to forecast budget overruns by adding to cost accounting's actual cost records all those anticipated costs which can be identified from information such as purchase order commitments, contract commitments, and estimated cost to complete the work.

## 17.7 COST ACCOUNTING FOR MANUFACTURING OPERATIONS

The cost of manufacturing operations is accumulated in three separate accounts:

1. Direct material: cost of materials, assemblies, and parts used to complete the project
2. Direct labor: wages of workers who are helping to complete the product
3. Overhead (also called *burden*): the cost of all other factors contributing to the completion of the product.

As the product input factors are used in the factory, they are charged to a *work-in-process* account, which may be broken down into the three preceding subaccounts. When the completed product is physically removed from the factory and transferred to the finished-product stores area, its accumulated cost is transferred from the work-in-process account to the *finished-products* account. It follows that the balance in the work-in-process account represents the accumulated cost of the products being manufactured but not yet completed.

The cost of materials is the purchase price and all incidental expenses incurred in transferring them to the purchaser's premises, such as transportation, tariff duties, or insurance. In most cases purchased materials and parts are charged to a *materials-stores* account, which is credited when the materials and parts are physically transferred to the factory floor and become part of the work-in-process account.

Labor cost is the result of several factors. These include the basic hourly rates and the hours worked, augmented by certain provisions — overtime, social security taxes, etc. — and contractual agreements such as vacation pay, health, pension payments.

The labor cost is the gross pay, although the worker receives a smaller net, or take-home, pay because of deductions for taxes, union dues, etc.

## 17.8 COST CENTERS AND UNIT COSTS

Costs are accumulated by cost centers, which may or may not coincide with operating departments. *Cost centers* are placed at points in the production process where costs can be measured and recorded as conveniently and accurately as possible. Manufacturing operations are organized by departments headed by a line supervisor who usually has an office with a small staff to care for the clerical work associated with production, such as materials requisitions and labor time control. A department may be broken down into several cost centers to obtain more detailed cost data, or several departments may be combined. The choice of cost centers is important. Cost engineers should maintain liaison with the cost accountants and those who design the cost-accumulation system. The cost engineer's requirements must be known in order to be met within the limitations of cost and practicality.

Cost data must be related to an activity factor such as volume of production. Comparison merely of total costs for different time periods is meaningless. Therefore, all cost reports should present the unit cost as well as the total cost. Selection of a suitable activity factor can be difficult in multiproduct operations when there are substantial differences in the unit costs of the various products.

## 17.9 JOB-ORDER COSTING
## AND PROCESS COSTING

There are two basic accounting techniques for accumulating unit costs: job-order costing and process costing. *Job-order costing* is used for specialized production jobs, e.g., electric generators manufactured to individual specifications. Costs can be computed accurately, but considerable paperwork and effort are required. For mass production of repetitive units a job order can cover an accumulation of units. Job orders can be issued for individual customers or for stock.

*Process costing* is used for continuous-process industries which operate 24 h a day, such as oil, chemicals, and steel. Process costing accumulates average costs for a time period, most frequently 1 month. The production cycle continues without interruption, while the cost cycle is cut for each accounting period to determine the result of operations. Process costs use averages and are less accurate than job-order costs but are simpler and less expensive to compute. One requirement is the need to estimate inventories at the point of the cost cutoff, where incomplete units are evaluated in terms of completed units.

Process costing is the more widely used technique of cost accumulation. It serves continuous-process industries, which have no alternative, but also mass-production industries, which could use either method. Efficiency is measured by statistical methods which implement the costing technique.

In the job-order costing method the cost of the basic input factors, direct material and direct labor, are charged to the specific numbered job orders by posting periodically on the job-order sheets, one for each job order. Upon completion of the job order, overhead is applied on the basis of an activity factor, as explained in the next section. Table 17.4 is an example of a job-order cost sheet.

**TABLE 17.4**
**Job-order cost sheet**

Job-Order Cost Sheet

Job order no. ———   Customer ———
Part no. ———   Work started ———
Number of units ———   Work completed ———

| Week of | Material | | Labor | | Applied overhead | Total |
|---|---|---|---|---|---|---|
| | Requisition no. | Amount | Hours | Amount | | |
| | | | | | | |
| | | | | | | |
| | | | | | | |
| | | | | | | |
| | | | | | | |
| | | | | | | |
| Total | | | | | | |

Process costs are accumulated by departments (processes) for a time period, usually 1 month. In many instances materials are introduced to the process at an early stage, and in subsequent processes only the *cost of conversion*, direct labor and overhead, has to be added. Some units may be lost, e.g., for technical reasons in chemical industries, and the cost must be absorbed by the salable units. In other instances an increase in units may occur. Costs are accumulated on process cost sheets, which show input and output for both physical units and costs, and unit costs by individual processes. Table 17.5 is an example of a process cost sheet.

**TABLE 17.5**
**Process cost sheet**

Process Cost Sheet
for the month of _____

|  | Process 1 | | Process 2 | | Process 3 | |
|---|---|---|---|---|---|---|
| **Quantity schedule** (units) | | | | | | |
| **Input:** Transferred from preceding department | | | | | | |
| Added in department | | | | | | |
| Total | | | | | | |
| **Output:** Transferred to next department | | | | | | |
| Work-in-process inventory | | | | | | |
| Lost (added) units | | | | | | |
| Total | | | | | | |
| **Cost schedule** | **Amount** | **Units** | **Amount** | **Units** | **Amount** | **Units** |
| **Input:** | | | | | | |
| Cost transferred from preceding department | | | | | | |
| Cost added in department | | | | | | |
| Material | | | | | | |
| Labor | | | | | | |
| Overhead | | | | | | |
| Adjustment for lost (added) units | | | | | | |
| Total input | | | | | | |
| **Output:** | | | | | | |
| Cost transferred to next department | | | | | | |
| Work-in-process inventory | | | | | | |
| Cost from preceding department | | | | | | |
| Cost in department | | | | | | |
| Material | | | | | | |
| Labor | | | | | | |
| Overhead | | | | | | |
| Total inventory | | | | | | |
| Total output | | | | | | |

# OVERHEAD ALLOCATION

## 17.10 TECHNIQUES OF OVERHEAD-COST ACCOUNTING

The *overhead-cost* classification includes all production costs which are not considered prime costs (direct material and direct labor). There is no limit to the number of overhead classifications; however, a few principal groups can be distinguished:

Indirect materials, also known as *supplies*, are materials such as lubricants that do not become a part of the finished product.

Indirect labor, the wages and salaries of employees who are not directly connected with the manufacture of a product, such as supervisors, maintenance workers, and internal transportation. Frequently fringe benefits are included in this classification.

Facilities costs, both short-term costs of the current year and long-term costs which are depreciated over a number of years. The former includes building maintenance, local real estate taxes, and others; the latter, buildings and equipment.

Service-department costs, for facilities which support production but are not part of production, e.g., accounting, laboratories, stores, cafeteria, and first-aid stations.

It is difficult, if not impossible, to enumerate the many different types of costs which make up the overhead-cost classification. Indeed, overhead accounting is known as one of the most complex problems of cost accounting.

The accumulated overhead costs are allocated in two stages. First, the overhead costs are allocated to the cost centers (departments). These, in turn, allocate the overhead to job-order costs and process costs.

Assuming a large number of cost classifications and a large number of cost centers, the number of computations can become quite extensive. In practice, various shortcut techniques are used to reduce the clerical work.

Cost engineers must recognize that the cost data which finally emerge from this procedure are significantly affected by the measure used for determining the indirect cost and the method used for allocation. Therefore, they should not accept these data unless they are familiar with the techniques used in computing them.

In order to select a *basis for overhead-cost allocation*, a measure must be found that reflects closely the actual benefit provided by a cost classification. Allocation must be made on this particular basis. As an example, consider building maintenance. This can be measured on the basis of square feet and allocated on the basis of floor space occupied by each cost center.

In turn, cost centers must have a measure by which the accumulated overhead is allocated to each job order or process. This may be, e.g., labor hours, or some material quantity for which appropriate data are available. Introduction of computers has greatly simplified the clerical and computational job but has not necessarily improved the quality of the resulting data.

The procedure for allocating overhead, as described thus far, has two disadvantages. It is complicated, and the overhead applicable to units made early in the month cannot be calculated until the end of the month, when the accumulated overhead becomes known. A shortcut method has evolved to overcome this disadvantage. Two estimates are required. First, an estimate is made for each overhead-cost classification covering specific periods, usually a year. Second, the estimate of some measure must be chosen which permits each job order to be charged with the appropriate share of the overhead cost. Direct labor, in the form of dollars or hours, has been the most widely used measure.

A rate can be established by dividing the total estimated overhead cost by the estimated quantity of the selected measure; this is called the *burden rate*. As an illustration, assume that the estimated total overhead cost is $50,000 and that the estimated direct labor cost is $25,000. The burden rate is $50,000/$25,000, or $2, for each direct labor dollar. If the direct labor cost of job order 750 is $2000, the applied overhead is 2000 × $2, or $4000.

It must be kept in mind that the applied burden rate is an estimate, not necessarily the actual overhead. A difference or variance must be expected and reconciled periodically. Other problems arise because overhead-cost classifications react differently to volume fluctuations, as discussed in Sec. 17.11.

There are two principal advantages in using burden rates. The first, the savings in time and cost, has been mentioned. Second, and perhaps more important, is the possibility of obtaining data which are more satisfactory for determination of operating efficiency. Use of an actual overhead cost subjects the data to fluctuations within the period, such as payment of taxes or the cost of an overhaul. An applied burden rate, on the other hand, equalizes such charges by spreading them over the year. Abrupt overhead costs are eliminated as factors influencing the determination of efficiency.

## 17.11   COST BEHAVIOR AGAINST ACTIVITY VOLUME

The importance of the volume of activity in the accumulation of cost data has already been stressed. Businesspeople have long been aware that profits increase faster than sales when activity is expanding, and vice versa. This is because different cost classifications react differently to changes in production volume and sales volume. Costs are broken down into fixed costs, variable costs, and semivariable costs, as discussed in Chap. 7. One of the principal responsibilities of cost accountants is to study the relationship of the various cost classifications to volume fluctuations.

A number of methods have been developed to accomplish this task. It must be remembered that all are approximations, often with a considerable margin of error. The widely used break-even chart was described in Chap. 7. Charts that are not elaborate incorporate a number of simplifying assumptions which limit their usefulness. Theoretically, in terms of accurate measurement, most cost classifications are semivariable. However, this circumstance can be accommodated by not limiting the charts to the simplest type.

By dividing costs into fixed and variable, one obtains a relationship which is considered applicable only within a limited range of volume fluctuations.

In computing burden rates, the estimated overhead is divided by the estimated quantity of measure, such as estimated direct labor hours for the same period. The overhead will depend on volume to some extent and the estimated labor hours much more so and in a complicated fashion. Hence the burden rate will depend on volume, i.e., a burden figure is correct only for a particular volume.

The distinction between fixed and variable costs refers to short-term considerations. In the long run all costs can be thought of as variable. Costs are fixed as a result of a previous management decision. The cost estimator should recognize that some fixed costs may be changed by new arrangements.

## 17.12  DIRECT COSTING AND ABSORPTION COSTING

The system of applying overhead described so far is known as *absorption costing* because the direct, fixed, and variable costs are absorbed by the work in process and in turn become part of the inventory. If production is greater than sales, inventory of the finished products will be increasing. Suppose volume is high enough so that the overhead applied to the work in process, which is an estimate, is greater than the actual overhead. As a consequence, inventory will be overvalued. Under such conditions the apparent profit of the organization can be increased by increasing production beyond the normal volume and adding to inventory. This contradicts the basic principle of accounting that profits should be recognized only if they result from sales to outsiders, i.e., arm's-length sales.

Another system of applying overhead known as *direct costing* attempts to overcome this disadvantage but introduces some problems of its own. In direct costing the fixed part of the overhead, and only the fixed part, is not applied to the work in process but is regarded as a period cost and is entirely absorbed as an expense against income in the period. In this way the applied overhead burden includes only variable costs, and no over- or underestimation of the inventory occurs.

There are differences between absorption and direct costing. Inventory will always be held at a lower value under direct costing. Over a period of time the two methods will give the same cumulative net income because fluctuations in inventory are smoothed out. Over a short period of time the two can differ depending on the volume and how much the inventory changes during the period.

There is a difference of opinion among professional accountants about the applicability of the direct-costing technique. It underestimates the cost of the inventory because it does not include the fixed charges. The cost estimator must be alert to the use of these alternative methods, which can have a significant effect on cost and profit data.

## 17.13  STANDARD COSTS AND BUDGETS

Up to this point most of the discussion has been concerned with the accumulation of historical costs. A very important aspect is the use of these data for future-directed costs. The true analysis of efficiency depends on the previous determination of measures or yardsticks against which the actual performance is evaluated. These future-directed and estimated costs are particularly important to cost engineers. Since they can depend on technological factors, cost engineers should cooperate with those who determine such costs.

*Standard costs* are determined by using scientific techniques and objective quantitative measurement. In contrast, estimated costs are based on individual judgment with limited significance. Standard costs do not necessarily represent expected performance, but rather desired objectives.

Materials standards developed by engineers are valued in accordance with actual or estimated purchase prices to be converted into standard costs. Labor standards, after measurement, are converted into standard costs from appropriate labor rates. Overhead-cost standards are difficult to establish and often are merely estimates based on past performance.

Actual costs are compared to standard costs for control purposes.

Budgets serve for both planning and control purposes. Budgets reflect mainly expected performance and have to be adjusted more frequently than standards. Budget data may be especially valuable to the cost engineer. Comparison of actual costs to budgeted costs indicates how closely actual operations have tracked against the plan.

## 17.14   JOINT COSTS

An intricate problem in accounting arises when data are related to more than one activity or transaction and the share of each cannot be measured accurately. When this happens, costs cannot be allocated appropriately; some activities will be overcharged and some undercharged. An interesting case arises in connection with the costing of so-called by-products and joint products.

Several products are often derived from a single source, such as crude oil. Another well-known case is the meat packing industry. The relationship between the products developed is often the result of technological rather than market conditions. Sometimes there is no market for a product that arises in connection with another product which is in great demand.

A product for which there is no demand or only a limited demand is called a *by-product*. The product produced at the same time which is in great demand is the *main product*. Several main products derived from a single source are called *joint products*.

As an illustration, consider the carcass of a slaughtered animal carried through several processes. After it has been divided into several parts a product remains that cannot be sold but for which there is a market if it is improved. This may be called the by-product. At the next stage several marketable products become available. The point of separation is called the *split-off point*. The problem arises of how to allocate the accumulated cost at that point to the several products. The most widely used techniques are:

1. In the case of a by-product the total cost at the split-off point is absorbed by the main product after any net incidental revenue from the by-product is credited.

2. The allocation for joint products is more difficult. The simplest method is related to some measurable quantity, such as the number of units, weights, volume or

vitamin content, etc., but the resulting gross-margin percentages will fluctuate unreasonably. A method based on the net realizable value is therefore the most common basis for allocation of joint costs.

The important point is that the purpose of joint cost allocation is to determine inventory valuations and to develop an aid in setting prices and estimating the resulting profit from the sales of each joint product. Control is still based on cost centers, standard costs, budgets, and the applications of cost control techniques.

The effect of the allocation method for costs is readily shown by an example.

**Example 17.1.** At a split-off point $900 must be allocated to product $A$, which amounts to 1000 liters at a sales value of $0.55/liter, and product $B$, which amounts to 500 liters at a sales value of $0.90/liter. Compare the percentage profits for $A$ and $B$ for an allocation ($a$) based on the volume of products and ($b$) based on the sales value.

*Solution*

($a$)

$$\text{Total liters} = 1000 + 500 = 1500$$

$$\text{Cost allocated to } A = 900(1000/1500) = \$600$$

$$\text{Sales value for } A = 1000 \times 0.55 = \$550$$

$$\text{Profit for } A = \$50 \text{ loss}$$

$$\% \text{ profit for } A = 100(50/550) = 9.09\% \text{ loss}$$

$$\text{Cost allocated to } B = 900(500/1500) = \$300$$

$$\text{Sales value for } B = 500 \times 0.90 = \$450$$

$$\text{Profit for } B = \$150$$

$$\% \text{ profit for } B = 100(150/450) = 33.3\%$$

($b$)

$$\text{Total sales value} = 1000 \times 0.55 + 500 \times 0.90 = \$1000$$

$$\text{Cost allocated to } A = 900(550/1000) = \$495$$

$$\text{Profit for } A = 550 - 495 = \$55$$

$$\% \text{ profit for } A = 100(55/550) = 10\%$$

$$\text{Cost allocated to } B = 900(450/1000) = \$405$$

$$\text{Profit for } B = 450 - 405 = \$45$$

$$\% \text{ profit for } B = 100(45/450) = 10\%$$

An allocation based on volume shows a loss for $A$ and a profit for $B$. An allocation based on realizable value shows a profit for each, but with a reduced profit for $B$.

Allocation costs for joint products are usually closely guarded proprietary information since companies do not wish their competitors to know the apparent profits for individual products and the manner in which they are calculated.

### 17.15   INVENTORY VALUATION

This and the following section deal with problems of financial accounting which affect cost figures.

Identical items in materials and supplies inventory may have been acquired at different times and at different prices. In such cases a rule is required to determine the value at which units will be transferred to work in process. The rule applies to pricing and has nothing to do with which unit is physically removed from the inventory. Physically, the units are all alike, and the unit removed is generally the one that is most conveniently transferred.

The most common method is called *first-in, first-out,* or *FIFO.* The cost of the oldest unit in inventory, as it appears on the record, is used. When prices are increasing, such as during inflation, the lower acquisition prices are matched with the higher selling prices, resulting in higher accounting profit figures and higher income taxes.

To avoid these unsatisfactory effects, a *last-in, first-out* or *LIFO* method was introduced. The cost of the last unit added to inventory, as it appears on the record, is used. Now the last additions which supposedly reflect current prices are matched with the selling prices of the same period; thus the *paper profits* resulting from the FIFO method are eliminated. The inventory, however, will reflect the prices of the earlier, less expensive additions, and therefore will be lower than it would be using the FIFO method.

The use of average valuation represents an effort to find a compromise between the two methods. Either a moving average can be used or new averages can be computed periodically.

In order to use material or finished-goods inventory data from the accounting records, one must be aware of the cost differences that may result from different inventory-pricing methods.

### 17.16   DEPRECIATION

Depreciation was discussed in detail in earlier chapters. It becomes part of the fixed costs of overhead. The importance of depreciation is steadily increasing because of the growing trend toward automation. Thus labor cost is replaced by machine cost, which is reflected in a depreciation charge.

It should be noted here that the manner in which depreciation is taken does affect the current period costs. Moreover, accountants generally prefer to group items together for accounting purposes, and thus a single piece of equipment loses its individuality. Again, in some processes part of the equipment may be fully depreciated, and current costs are not realistic since they make insufficient allowance for wear and tear of equipment.

Cost estimators are not bound by the depreciation expense used by the accounting method. They can set rates for their computations which they consider realistic regardless of whether they are acceptable for tax purposes and therefore accounting procedures.

In conclusion, it can be said that the introduction of more efficient equipment and scientific techniques will cause a new appraisal of present accounting practices. There is no doubt that closer cooperation of cost engineers, who are major users of accounting data, and those who are responsible for designing the systems should be encouraged.

## QUESTIONS

**17.1.** Discuss the reasons that cost engineers need to be familiar with the fundamental concepts of accounting, particularly cost accounting.

**17.2.** Distinguish between historical and managerial accounting.

**17.3.** Describe the basic steps for accumulating accounting data.

**17.4.** Discuss the reasons for the large number of cost definitions and list some of the most important ones.

**17.5.** Explain the reasons for the use of future-directed cost from the cost engineer's point of view.

**17.6.** Explain the distinction between direct (prime) costs and indirect (overhead) costs. Give examples.

**17.7.** Explain the importance of cost centers and their relationship to departments.

**17.8.** Discuss the importance of presenting unit costs in all cost statements.

**17.9.** Describe the difference between job-order costs and process costs. Give examples.

**17.10.** Explain the difficulties in dealing with overhead costs. Give examples of several kinds of overhead costs.

**17.11.** Discuss the reasons for the need of overhead allocation and the technique used.

**17.12.** Explain the technique of computing burden rates and the advantages and difficulties arising from their use.

**17.13.** Explain the importance of the concept of activity volume for cost accounting purposes. Give examples of variable, semivariable, and fixed costs.

**17.14.** Explain the benefits and shortcomings of absorption costing compared with direct-costing techniques.

**17.15.** Explain the techniques and benefits of standard costs and overhead budgets.

**17.16.** Explain the techniques used for dealing with by-product costs and joint costs. Give examples of each type of cost.

**17.17.** Discuss the relationship between cost engineers and cost accountants.

## REFERENCES

1. Adrian, J. J.: *Construction Accounting*, Reston Publishing Company, Englewood Cliffs, N.J., 1986.
2. Black, H. A., and J. D. Edwards: *Managerial and Cost Accountants' Handbook*, Dow Jones-Irwin, Inc., New York, 1975.
3. Davidson, S.: *Handbook of Modern Accounting*, McGraw-Hill Book Company, New York, 1970.
4. Eppes, B. G., and D. E. Whiteman: *Cost Accounting for the Construction Firm*, John Wiley & Sons, New York, 1984.

5. Grady, P.: *Inventory of Generally Accepted Accounting Practices for Business Enterprises,* AICPA Accounting Research Study 7, American Institute of Certified Public Accountants, New York, 1965.

6. Horngren, C. T.: *Cost Accounting, A Managerial Emphasis,* 4th ed., Prentice-Hall, Inc., Englewood Cliffs, N.J., 1977.

7. Louderback, J. B.: *Managerial Accounting,* Wadsworth Publishing Company, Inc., Belmont, Calif., 1975.

8. Lucas, P. D.: *Modern Construction Accounting Methods and Controls,* Prentice-Hall, Inc., Englewood Cliffs, N.J., 1984.

9. Moore, C. L.: *Managerial Accounting,* 3d ed., South-Western Publishing Company, Cincinnati, Ohio, 1972.

10. Nickerson, C. B.: *Managerial Cost Accounting and Analysis,* 2d ed., McGraw-Hill Book Company, New York, 1962.

11. Spiller, E. A., Jr.: *Financial Accounting, Basic Concepts,* 3d ed., Richard D. Irwin, Inc., Homewood, Ill., 1977.

12. Wixom, R., ed.: *Accountants' Handbook,* 5th ed., The Ronald Press Company, New York, 1970.

# CHAPTER 18

# COST
# ENGINEERING
# AND BEYOND

*F.C. Jelen*

## 18.1  ELEMENTS OF COMPLETE COST

This chapter discusses some topics that generally belong to management, yet must be understood at basic levels in cost engineering. Some of these topics have already been introduced, in part, but are mentioned here for varied emphasis.

The elements of the total cost of manufacturing and selling a product were outlined in Table 15.1. The cost engineer is concerned chiefly with part I of the table, the operating or manufacturing cost, but must be knowledgeable about part II, general expense. The classification in Table 15.1 is subject to variation. Hence the cost engineer must be familiar with the company's classification of costs and the accounting procedure used.

## 18.2  START-UP COSTS

When a plant or project is supplied with the complete fixed investment and working capital, it is ready for start-up. Start-up does not occur at a precise instant, but rather over a period of time; however, one point is usually selected as the start-up time. Some start-up expenses will have occurred earlier, such as training personnel and bringing individual pieces of equipment onstream. Other expenses will occur after the selected start-up time, such as making equipment changes and debugging.

The duration of the start-up can vary widely depending on the novelty of the process and various technical and personnel considerations. However, it is considered complete when the plant has demonstrated that it is capable of operating at 90 percent of capacity for a predetermined time.

Start-up costs can be divided into two groups for accounting purposes. The first group consists of costs which may be written off immediately as an expense if desired. If a pump is ruined by unexpected corrosion, the cost can be expensed since there is no increase in capitalization. The second group consists of costs which must be capitalized. If an additional processing step is required, a pump needed for that step must be capitalized. Although items in the first group *may* be expensed, they may be capitalized if desired since the government does not care if the immediate tax benefit from the expense is delayed. The trend now is to capitalize most start-up expenses in the belief that the cost of a plant should include all the costs necessary to establish essentially full production capability.

The total start-up expense seldom exceeds 10 percent of the fixed capital cost. Start-up costs will be small with an established technology for a local addition. In contrast, start-up costs can be significant for a plant using a new technology in a new location, perhaps a foreign land. Start-up time can be reduced by more extended technical investigation before the plant is built, such as more pilot plant work. Development work is less expensive in the pilot plant than it is in the actual plant, but once funds are committed to building a plant, delays in a pilot plant can become expensive. Hence the thoroughness with which a process is investigated before a plant is built can be a trade-off between the value of time spent in prior investigation and the start-up cost.

The start-up time may vary from 1 month for well-known processes to as much as 1 year for new ones. It is desirable that skilled crews of engineering supervisors who are production oriented be assigned early in the engineering phase to consult with the project and design engineers. The purpose is to build into the project features calculated to reduce the start-up time and facilitate the process. Arranging for start-up can be a complicated activity requiring extensive planning and execution. It will be necessary to train not only new operators but also perhaps new trainers.

A list of start-up components is given in Table 18.1. As preparations build to a climax, skilled engineers and operators are stationed at the site during the last stage of construction to start operator-training procedures, make adjustments for convenience in operating, and then supervise final adjustments and the actual start-up program. Pieces of equipment are tested, modified if necessary, and brought onstream individually. Some of this activity may start months ahead of the termination of construction.

Table 18.1 shows that start-up costs can be classified into two types: first, costs arising from technical difficulties requiring engineering and equipment costs; second, costs associated with personnel, involving early employment, training, and provision of temporary habitation.

**TABLE 18.1**
## Components of start-up costs

Supervisory and technical personnel hired early
Operating line supervisors hired early
Other overhead personnel, including office personnel, hired early
Sales force hired early
Labor force hired early
Overhead and miscellaneous for all above personnel
Training of personnel
Training, operating, and emergency manuals
Scale models
Outside contracts for operating assistance
Technical assistance from parent company
Consultants and outside work
Contingencies
Travel and living expenses for temporary and transferred personnel
Initial low operating efficiency
Wasted, lost, and spoiled material
Special disposal problems
Engineering cost for design modifications
Installed cost for design modifications
Cost for checking and testing equipment
Cost for supplying temporary resources, such as utilities
Extra cost for initial quality testing and control

## 18.3  PLANT LOCATION

The cost of manufacturing, selling, and distributing a product is significantly affected by the location of the plant in which it is made. In an earlier age manufacturing facilities were located in the market area in which the product was sold. Today high transportation costs may still limit the location of large bulk-product plants, such as steel mills and fertilizers, to primary market areas, but other factors have increasing weight in site determination for most other products.

The Regional Science Research Institute in Philadelphia indicates that there are about 100 factors involved in the selection of a site for an industrial plant. The most significant are listed in Table 18.2.

The location of a plant is a matter for high-level management and requires attention to economics, politics, personnel, and public relations. If building a new plant requires closing an old plant in another location, the cost of terminating the employment of personnel in the old plant can be a major problem.

For a discussion of factors affecting plant location the reader can find helpful information sources such as local chambers of commerce, utility companies, railroads, labor union organizations for labor rates and working conditions, maps from the U.S. Geodetic Survey (Department of the Interior), Department of Labor for excellent statistics on factory labor rates published in its *Statistical Abstract of the United States*, the Federal Power Commission for electric power rates, the Interstate

**TABLE 18.2**
**Significant factors in selection of a site for a plant**

1. Accessibility to market
2. Availability of raw materials
3. Qualified and productive labor force
4. Adequate utilities
   a. Electricity
   b. Water
   c. Steam
   d. Gas
5. Construction costs
6. Fuel availability
   a. Coal
   b. Oil
   c. Gas
7. Effluent disposal
   a. Local sewage plants
   b. Lagoons
   c. Deep wells
   d. Large bodies of diluting waters
   e. Legislation affecting cost of pollution
      control
8. Transportation facilities
   a. Rail
   b. Truck
   c. Water
   d. Pipeline
   e. Air
9. Legislation affecting business activities
   a. Tax concessions
   b. Restrictive ordinances
   c. Real estate values and taxation
10. Site considerations
   a. Cost of acquisition, preparation, and
      maintenance
   b. Accessibility
   c. Need for services such as parking,
      cafeteria, recreational facilities, etc.
11. Taxes
   a. Sales
   b. Business
   c. State and local
   d. Gasoline and oil
   e. Severance and depletion
12. Climate effects
   a. Absenteeism
   b. Heating
   c. Air conditioning

Commerce Commission, and the Federal Communications Commission. A complete list of the publications of the government agencies can be obtained from the Superintendent of Documents, Government Printing Office, Washington.

Private organizations such as *Fortune* magazine, *Engineering News-Record*, The Industrial Conference Board, the American Manufacturers Association, and McGraw-Hill, Inc., publish data useful in industrial site selection.

## 18.4   CONTINGENCIES

The term *contingencies* has more than one meaning. In this book it refers to costs that will probably occur based on past experience, but with some uncertainty regarding the amount. The term is not used as a catchall to cover ignorance. It is poor engineering and poor philosophy to make second-rate estimates and then try to satisfy them by using a large contingency account. The contingency allowance is designed to cover items of cost which are not known exactly at the time of the estimate but which will occur on a statistical basis. For instance, an allowance for water-batch testing of vessels and piping is necessary before process fluids are circulated and must be part of the start-up cost. However, it is known from past experience that some of the vessels will fail structurally during the test, and a contingency account exists to cover that circumstance. When a vessel fails in the test, the cost of the replacement must come out of the job contingency.

In practice, some items are risky because of ignorance or unpredictable circumstances, but the cost is not covered by a contingency. As an example, suppose a construction company is subbidding on the foundations for a project the soil characteristics of which are not known precisely. The proper engineering approach is to perform a soil test from which accurate cost estimates for the foundations can be made. However, the subbidder may not be prepared for this, since he or she will try to avoid the cost for testing the soil unless he or she is sure of the contract. In that case the subbidder may estimate the cost for what the soil is expected to be, increasing the cost to cover the risk. The bid is increased to cover the risk, not to include contingencies. Contingencies will occur independently of the nature of the soil. For example, past experience shows that about 5 percent extra must be allowed to cover emergency conditions, such as delivery of some batches of concrete after 5 P.M. That cost would be covered by a contingency.

## 18.5  FINANCING

As time proceeds a prosperous company will require additional funds for expansion or for new ventures. The funds can be raised from five broad areas:

Profits
Additional common stock
Preferred stock
Loans
Bonds

If profits are used, no financing is required, but profits must be raised on an after-tax basis. Under existing conditions of relatively low profits and high costs for investment, financing new ventures from profits is very limited.

Issuing additional common stock increases the equity of a company and dilutes the position of a stockholder. However, the existing stockholders can be given preference for the new stock to avoid dilution. Adding common stock is the safest way of obtaining new funds since the debt is not increased and dividends can be omitted when necessary. On the other hand, dividends are paid out of profits, i.e., on an after-tax basis.

Issuing preferred stock also increases the equity of a company and does not increase the debt. However, a prescribed dividend is stipulated and must be paid from profits, on an after-tax basis.

Loans are obtained principally from banks and insurance companies and may be short-term or long-term. Interest paid on loans is deductible for tax purposes. However, failure to pay interest or to repay the loan itself on time can lead to bankruptcy. Loans increase debt.

Bonds are similar to loans but are usually sold to the public. They are long-term, generally in $1000 units. The interest paid on bonds is an expense for tax purposes. Bonds increase debt, and failure to pay interest or the face value on maturity can lead to bankruptcy.

The method of financing is a consideration for high-level management, and there are many variations within the types previously listed. Cost engineering fundamentally is not concerned with the source of funds. Thus profitability of a venture should be measured independently of the financing. Acceptance or rejection of a venture can be dominated by financing considerations, but the absolute measure of profitability is the same whether or not the money is borrowed.

Consider a 3-year project requiring an investment of 1 M$ at zero time and having a uniform end-of-year income of 0.6438 M$ after expenses but before depreciation and taxes for each of the 3 years. Sum-of-the-years-digits depreciation is to be used for a 3-year tax life, and the combined tax rate is 46 percent. The discounted cash-flow rate of return (DCFRR) is the rate of return that makes the present value after taxes equal to zero, and by Table 3.6, items 1 and 2, is

$$P = 0 = -1(1 - 0.46F_{SDP, r, 3}) + 0.6438(1 - 0.46)F_{RP, r, 3}$$

and by trial and error, $r = 0.25$. The DCFRR is 25 percent.

If the 1 M$ investment is borrowed at a yearly rate of 15 percent, amounting to a before-tax expense of 0.15 M$/year, and if this is included as an extra expense, the calculation becomes

$$P = 0 = -1(1 - 0.46F_{SDP, r, 3}) + 0.6438(1 - 0.46)F_{RP, r, 3} - 0.15(1 - 0.46)F_{RP, r, 3}$$

$$= -1(1 - 0.46F_{SDP, r, 3}) + 0.26665F_{RP, r, 3}$$

and by trial and error, $r = 0.134$. The DCFRR now becomes 13.4 percent. The DCFRR of the project itself remains 25 percent but appears to be 13.4 percent only because the finance charge has been considered as an expense. If the interest on the investment is considered, then the source of the investment itself should be considered. In that case all the investment is borrowed in this venture and there is no initial investment by the company, yet a profit results. The conclusion, then, is that a profit results from no investment, and the DCFRR is infinite!

## 18.6 PRICING FOR PROFITABILITY

At times the inverse problem of the preceding section occurs, namely, to find the selling price required to produce a specified profitability. Suppose in the previous example 1 million items of product are made per year to sell at $S$ dollars per item. Find the value of $S$ that yields a DCFRR of 25 percent. Expenses are $0.5 per item. Then income less expense before taxes will be $(S - 0.5)$ megadollars per year, and

$$P = 0 = -1(1 - 0.46F_{SDP, 25\%, 3}) + (S - 0.5)(1 - 0.46)F_{RP, 25\%, 3}$$

$$S = \$1.1438$$

If the company charges $1.1438 per item, it will have $(1.1438 - 0.5) = \$0.6438$ per item, which agrees with the figure in the previous section of 0.6438 M$ for 1 million items. The project then pays off at 25 percent per year.

A company is allowed to calculate a required selling price in any way it desires. In some circumstances, as in the next section, the sales price is calculated with the inclusion of the financing cost, or anything else that is desired, such as a dividend to stockholders.

## 18.7 LEVELIZED COST

Public utilities, in appearing before regulating commissions, aim principally to establish rates. Hence they are very concerned with what they must charge for their product, kilowatt-hours. Utilities are allowed a return on investment after taxes based on their investment. Interest paid on the investment is an expense for tax purposes and is included in arriving at the price the customer should pay. For internal and economic comparisons a method based on levelized cost is sometimes used. Perhaps it should be called *levelized price* since it is the revenue needed expressed per unit of product produced, e.g., kilowatt-hour, but it is a cost to the consumer. The unit cost to the consumer will vary from year to year, but the *levelized cost* is a cost that has been leveled out for all the years. There are variations in the method, but the following simplified example will illustrate the principle.

**Example 18.1.** A public utility has an investment of 100 M$ at the start that will be written off in 3 years using straight-line depreciation. At all times 50 percent of the investment is borrowed money and 50 percent is equity. The borrowed money is financed at 10 percent per year. The gross expense before taxes and depreciation, and before the financing cost, is 150 M$/year. The tax rate is 46 percent, and each year $5000 \times 10^6$ kWh of electrical energy is sold. The company desires to make 15 percent per year on its equity and in addition 5 percent for dividends to its stockholders, a total of 20 percent on equity. Find the levelized cost.

**Solution.** Let $N$ be the kilowatt-hours in millions sold each year and $S$ the cost to the consumer in dollars per kilowatt-hour. Table 18.3 shows the calculations leading to the revenue required for each year in megadollars and the value of $S$, the unit cost to the consumer, for each year. The unit cost $S$ will not be constant; however, all the years can be combined on the basis of present value. Then, using megadollars, we obtain

Present value of $(SN)$ = present value of (revenue required each year)

If $S$ is to be constant, it becomes the levelized cost $S_L$ and can be taken outside a summation sign. Thus

$$S_L = \frac{\sum \text{revenue each year}/(1 + r)^n}{\sum \text{kWh each year}/(1 + r)^n}$$

The computation for levelized cost is

$$S_L = \frac{206.85/(1.15)^1 + 199.02/(1.15)^2 + 191.17/(1.15)^3}{5000/(1.15)^1 + 5000/(1.15)^2 + 5000/(1.15)^3} = 0.0399$$

Thus the company would seek a customer cost of $0.0399/kWh and would have a 15 percent return on its varying equity in the investment and another 5 percent for a dividend.

**TABLE 18.3**
**Costs in megadollars for Example 18.1**

|  | Year 1 | Year 2 | Year 3 |
|---|---|---|---|
| 1. Total investment | 100 | 66.67 | 33.33 |
| 2. Equity investment | 50 | 33.33 | 16.67 |
| 3. Financed investment | 50 | 33.33 | 16.67 |
| 4. Gross income | NS | NS | NS |
| 5. Gross expenses | 150 | 150 | 150 |
| 6. Interest expense | 5 | 3.333 | 1.667 |
| 7. Depreciation expense | 33.333 | 33.333 | 33.333 |
| 8. Taxable income | | | |
| (4) − (5) − (6) − (7) | NS−188.33 | NS−186.67 | NS−185.00 |
| 9. Tax at 46% | 0.46NS−86.632 | 0.46NS−85.868 | 0.46NS−85.1 |
| 10. Net profit (8) − (9) | 0.54NS−101.698 | 0.54NS−100.802 | 0.54NS−99.9 |
| 11. 15% of equity for retained earnings | 7.5 | 5.0 | 2.5 |
| 12. 5% of equity dividends | 2.5 | 1.667 | 0.834 |
| 13. Revenue required, NS | | | |
| (10) − (11) + (12) | 206.85 | 199.02 | 191.17 |
| 14. N in kilowatt-hours, millions* | 5000 | 5000 | 5000 |
| 15. Calculated S for year[†] | | | |
| (13) ÷ (14) | 0.0414 | 0.0398 | 0.0382 |

*$N$ = kilowatt-hours sold each year in millions.
[†] $S$ = cost in dollars per kilowatt-hour as required revenue.

A practical problem would be spread over many years and expenses, receipts, and energy produced each year would not be constant, but the principle remains the same. There are variations in the method used to compute levelized cost, but the example represents a common use of levelized cost in economic comparisons. In comparing two projects, the one with the lower levelized cost would be preferred.

## 18.8   FORECASTING

In industrial activities forecasting is normally done by marketing or economics personnel. Cost engineers become involved in forecasting most frequently in connection with venture profitability. Of the many factors involved, those listed in Table 18.4 are the most significant. The table gives the probable variations from forecasts over an assumed 10-year plant life.

The cost of the *fixed capital investment* may be appraised within a range of −10 percent to +25 percent. *Construction time* may vary from −5 percent to as high as +50 percent of the estimated time but generally can be predicted more accurately. *Start-up costs* have even a wider variation, ranging from −10 percent to +100 percent, but have a negligible effect on profitability if the gap left by delay in reaching the market with the product is not filled by a competitor.

Forecasts of *sales volume* and *price* for a new product are probably the most challenging factors. Such forecasts are not within the scope of this book, but they constitute the heart of the business process. Predictions of sales volume and prices

**TABLE 18.4**
**Factors affecting profitability forecasts**

| Factor | Probable variation from forecasts over 10-year plant life, % |
|---|---|
| 1. Cost of fixed capital investment | $-10$ to $+25$ |
| 2. Construction time | $-5$ to $+50$ |
| 3. Start-up costs and time | $-10$ to $+100$ |
| 4. Sales volume | $-50$ to $+150$ |
| 5. Price of product | $-50$ to $+20$ |
| 6. Depreciation method | None |
| 7. Plant replacement and maintenance costs | $-10$ to $+100$ |
| 8. Obsolescence of process or equipment | Indeterminate |
| 9. Income tax rate | $-5$ to $+15$ |
| 10. Inflation rates | $-10$ to $+100$ |
| 11. Interest rates | $-50$ to $+50$ |
| 12. Working capital | $-20$ to $+50$ |
| 13. Legislation affecting product | Indeterminate |
| 14. Raw material availability and price | $-25$ to $+50$ |
| 15. Competition | Indeterminate |
| 16. Salvage value | $-100$ to $+10$ |
| 17. Intuition or calculated risk | Indeterminate |
| 18. Profit | $-100$ to $+10$ |

are made from studies involving market and product research activities. Most companies now employ market research and commercial development specialists to assist production and sales managements in determining the extent of the market, potential sales, and probable price of product over the life of the production assets. Actual sales and price can vary widely from the predicted levels as indicated in the table.

Another factor that is difficult to assess is process or equipment *obsolescence*. Processes have been known to be obsolete within a year after start-up, and in a technological era specific pieces of equipment become obsolete years before the end of their forecasted lives.

Among the factors affecting profitability, the impact of *legislative changes* must be considered. It is difficult to forecast to what degree this factor will affect profitability. For instance, government interest in health, pollution, quotas, and tariff agreements, among others, are all factors in economic evaluation. In addition, there is the possibility of income tax changes.

Competition from abroad is an increasingly important factor in assessing profitability. Many single products in some industries have already been affected, and the outlook is that increasing competition from abroad will tend to depress prices and profits in the United States. By and large, the individual effect of this factor on profitability is indeterminate at this time.

There is still one other consideration. After all known factors affecting profitability have been carefully weighed, the *go* or *no-go* decision is sometimes reached through a complex mental process which reflects the experience of the individual making the decision. Since it is impossible to characterize this process, it has been referred to as *intuition, calculated risk,* or *educated guess* in the prediction of profitability. Some forecasting can be based on extrapolation of historical data. A good example would be the prediction of steel consumption in the United States per capita for the next 10 years based on the extrapolation of the historical data for the last 20 years. Forecasting frequently requires some subjective inferences about the future, and the value of the forecast is closely related to the correctness of the inferences.

Extrapolation of historical data can be based on a linear relationship of the form

$$y = mx + b$$

but most frequently the relationship is curved, lying above or below a linear relationship. Nonlinear relationships are frequently chosen in exponential forms such as

$$y = ax^b$$

$$y = ab^x$$

which have the advantage of being plotted as straight lines on log-log graph paper.

## 18.9   CASH FLOW FOR FINANCING

Cash flow was treated in Chap. 5 as in- and out-of-pocket flow of cash for a project and its relation to profitability. To the controller of a company cash flow has an importance unrelated to profitability, namely, its significance to funds required by a project from the beginning of construction to the termination of start-up. Expenditures for a large project can extend over years, and funds must be available at all times to meet expenses. Good planning and timing are important, for funds must be available on time. However, if they are available too soon, they must be reinvested outside the project to compensate for the interest expense.

## 18.10   PROFIT

Like all terms in common use, *profit* is not easily defined. In a broad sense, profit connotes success and is not a simple concept. It is certainly an important concept to the cost engineer, as well as to the accountant, since costs, as expenses or receipts, are the quantitative basis for defining profit.

In Chap. 5 the difference between profitability and profit was emphasized, profitability being a long-term evaluation of short-term profits. Table 5.15 listed a number of pitfalls in profitability but is applicable to short-term profits as well. Consider the case of depreciation. How depreciation is scheduled over the years influences the apparent profit for the individual years. This is true for other expenses that can be allocated over the years. For example, one power company lost an entire generating station because of a landslide at the river edge. Instead of taking the entire loss in 1 year, the company preferred to allocate the loss over a number of future years. This decision affected the apparent profits for many years ahead.

In Chap. 6 it was shown that profit and profitability are overstated when no allowance for inflation is included. True profits exist only after taxes *and* after an allowance for hidden costs arising from inflation has been included. A problem in the United States is that profits as a percentage have been declining over the years, and hidden costs arising from inflation have compounded the problem. Consequently, internal funds available for expansion and modernization for manufacturing have been disappearing.

*Profit* has become a disreputable word to the public, but in a capitalist society profit is necessary for survival. The strong desire to show short-term and immediate profits can result in a policy of patching. The long-term objective of modernization and creation of new manufacturing facilities must not be slighted. A policy aimed at improving short-term profits and neglecting the long term can only delay the time of disaster and is very costly in the end. Companies exist to prosper and grow, and the measure of this is profit; but apparent profit, particularly over a short time, can be deceiving.

## 18.11   ROLE OF THE COMPUTER

The computer has become the single most important tool in cost engineering. It is dominant in cost estimation, scheduling, and control. Computers can do the following:

1. Make complex calculations at fantastic speeds and with great accuracy
2. Rearrange with great speed
3. Compare
4. Make simple decisions
5. Store vast quantities of information
6. Retrieve with great speed
7. Print out details and summations
8. Print out graphics and illustrations
9. Have their stored information updated easily and quickly

On the other hand, despite opinions to the contrary, a computer cannot think. Every step must be programmed. A computer can operate under the guidance of statistics and probability and can even generate randomness internally, yet it only follows orders. A computer can make a very sophisticated analysis, devouring a vast amount of information, but it does so by sequencing myriad single steps. The computer appears brilliant only because of the speed with which it operates.

The computer has two basic capabilities which can be intermixed. First, the computer can be used as a tool for calculation; that was the sole purpose of the original computers. The second basic function of a computer is known as *data processing*, in which a computer serves as a library from which stored information can be retrieved and organized. Some calculation may be involved, but it is usually incidental. Data processing occurs in banking, insurance, credit card accounting, taxation, statistics, information on the spot for sports announcers, and a host of other applications. The spectacular expansion of the storage capacity of computers has enhanced their applicability beyond predictions.

Applications in cost engineering that use computers as calculators are in linear programming, inventory problems, dynamic programming, queuing problems, risk analysis, optimization studies, and profitability analysis.

Applications in cost engineering involving chiefly data processing are in the following fields:

Estimation

Scheduling

Construction and cost accounting

Material control

Cost control

In these fields computers are a necessity.

In estimation, the computer can store thousands of basic costs, adjust for size and inflation, and print out information in a well-organized manner. The information can easily be kept up to date and expanded, while confusion with deleted material is eliminated.

For scheduling and control the computer can assimilate a large array of information and reports and permit immediate retrieval in an organized arrangement. Again, the ease with which the information is kept up to date is an advantage.

Many commercial management services are available for estimating and control, and they are all computer oriented. Various programs can be purchased to be run on the user's computer. The computer applications committee of the AACE should be consulted for information about currently available software programs.

It is not necessary for cost engineers to be experts in programming, but they should know how to use computers in their work. Computers are user-oriented, and paradoxically, the more powerful they become internally, the easier they are to use.

## 18.12   MANAGEMENT ORGANIZATION

Management can be broken down into three types — functional, projectized, and matrix or hybrid. Consider a construction company in which cost engineering is part of engineering. The functional organization is shown in Fig. 18.1 and is the standard pyramid type, with the organization broken down into functional groups or departments, each reporting to an executive.

The projectized organization is shown in Fig. 18.2, and in this arrangement the organization is by projects.

The matrix organization is shown in Fig. 18.3 and is a combination of the two previous types. In this arrangement the engineering department reports directly to an executive and to various project coordinators. Engineering has teams working on various projects.

Small projects and projects in the range up to 100 M$ will generally be handled in the matrix organization, whereas projects above 100 M$ will fall in the projectized organization. The functional organization occurs chiefly where there are no projects, such as a manufacturing company.

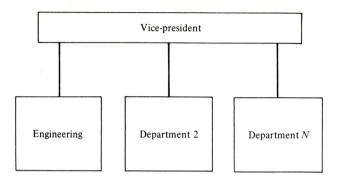

**FIGURE 18.1**
A functional organization.

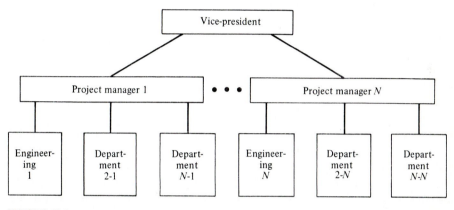

**FIGURE 18.2**
A projectized organization.

## 18.13 LEGAL ASPECTS

Although cost engineers are not lawyers, frequently their work becomes associated with legal matters, and some understanding of legal documents becomes necessary. A contract is perhaps the most important of these. One aspect of legal affairs is that whenever a dispute arises, both sides are represented by legal staffs whose only purpose is to win the case. The ultimate decision will be made by a jury composed of average people, not cost engineers. Contracts must, therefore, be carefully worded. In the end it is the wording of the contract that will be judged, not what was meant, much less hoped for.

Construction contracts can be grouped into four classifications, each with variations. The basic types of contracts are:

1. Fixed-fee
2. Cost-plus
3. Shared or target-figure contracts
4. Unit-price contracts

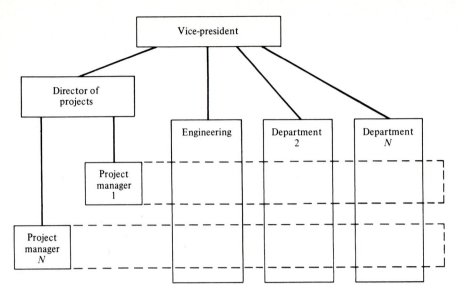

**FIGURE 18.3**
A matrix organization.

In *fixed-fee* contracts the work is designated by detailed specifications and drawings which may be supplied by the owner, or by the contractor and accepted by the owner. The cost is fixed and includes the contractor's profit. The risk is taken by the contractor, who has to consider this in bidding. If the bid contains too much for risk, the contractor may not be competitive. If the bid underrates the risk, the contractor may be a heavy loser. The owner assumes little risk. If the job is not within specifications, the contractor can be held liable. The specifications may not be adequate, and if the contractor has performed within the specifications, he or she is not liable. Owners can protect themselves by including performance specifications. Inflation is a serious factor in estimating construction bids, particularly if the construction is spread out over 3 years or more. A fixed-fee contract must include some allowance for inflation and can also include a variation in the fee. Thus there can be a penalty for a late finish and a reward for an early finish. The contract can allow for all kinds of incidental factors, including catastrophic events. Nevertheless, in principle, the fixed-fee contract gives the owner the least risk. The contractor assumes most, if not all, of the risk.

The second type of contract covers a wide variety of agreements known as *cost-plus* or *cost-plus-fee*. The work is described by detailed specifications and drawings, but the ultimate cost to the owner will not be known until the work is completed. It is known as an *open-ended* contract. Limits may be specified on some of the component costs. The contract must state accurately how the costs are to be treated and whether any limitations or exemptions apply. The fee can be a fixed amount or a percentage amount, and many variations exist. There can be a penalty for a late finish and a reward for an early finish. The owner takes most of the risk, and the contractor is well protected against loss. Cost-plus contracts become necessary when the project is unique and when significant changes in specifications are anticipated after the project is started. Many contracts for the military and the government can be handled only on a cost-plus basis.

Between a fixed-fee and a cost-plus-fee contract lies a large gray area known as a *shared* or *target-figure* contract. The fixed-fee contract places most of the risk on the contractor, and the cost-plus-fee contract places most of the risk on the owner. In the shared or target-figure contract the risk is shared between them to some extent. Here the owner obtains essentially a fixed fee for the original specifications, but the contractor generally determines the fees for variations in the specifications after the project begins. This type of contract usually contains provisions for sharing savings according to some stipulated agreement.

The unit-price contract is sometimes used in very heavy construction projects in the United States and Canada but is more frequently used abroad. The contractor's bid is based on the quantities of work to be done and a unit price for each type of work. A typical example is a contract for constructing a highway. It might be difficult to estimate in advance the exact amounts of fill and excavation required, but a contract based on unit costs, e.g., per cubic meter, would protect both the owner and the contractor. The total cost is fixed by summation of quantities multiplied by a rate.

Technology has become so complicated that it is difficult to prepare drawings and specifications which are completely free from misinterpretation. As a result, *hedge,* or *bludgeoning, clauses* have been rapidly developed. They are used by both owners and vendors. These clauses are general so as to be applicable to any unforeseen condition that may arise. The hedge clause, although general, can be enforced. A cost engineer who is bidding against hedge clauses is required to place an extra cost against every hedge clause in a set of bid documents since these clauses may affect the cost calculation. The use of hedge clauses may be necessary in part, but excessive use can be costly.

Cost is a fundamental factor in all the costs just discussed. Hence cost engineers must be familiar with specifications, drawings, and forensic vernacular, whether they are working for owners or contractors. Cost engineers have other contacts with legal matters. Appraisers frequently appear before regulatory commissions and in court, and these specialists should be masterful expert witnesses.

Every engineer has legal responsibilities to the public as well as to the employer. Thus the design of a bridge must be approved by a professional engineer. Most work done by a cost engineer does not involve public safety, at least directly. However, a cost engineer who shades estimates to favor decisions in which the public has an interest could be violating the law. Inasmuch as cost engineering involves estimating, it becomes difficult to prove guilt since being a poor estimator does not make one a criminal. However, if an estimator willfully distorts quantities and unit costs, he or she becomes subject to the law; and interpretation and enforcement of the law are becoming more stringent.

Legal clauses affecting penalties, brand names, delayed decisions, work by the owner, patents, segregated bids, etc. are so dispersed throughout the bid documents that the cost engineer must become familiar with all documents to evaluate the risks involved, usually in a comparatively short time. A knowledge of the habits of the owner—asking for bids, requiring prompt approval of shop drawings, permitting substitution of materials, attitude, etc.—can be of great value. Such information is available from past experience, from competitors, and from suppliers and subcontractors.

Practically every general condition states that the word of the engineer will be final. Many people contend that the judge's or jury's word is final. Juries, composed of laypeople, are unschooled in technical engineering and perhaps are influenced by anything except the facts of a case. There is a ray of hope in arbitration, but a cost engineer might recall that "an ounce of prevention is worth a pound of cure." It is the obligation of the cost engineer to be aware fully of all legal aspects and to advise the employer—even not to bid on the job when the bid documents have many hazards.

## 18.14  ETHICS

Most professions have a code of ethics for their members that transcends and supplements legal obligations. The AACE *Canons of Ethics of Cost Engineers* are given in Table 18.5. They can be accepted without reservation by cost engineers.

One might say that this code of behavior can be reduced to the golden rule—"Do unto others as you would have others do unto you." Inherent in the golden rule, of course, is the corollary, "Honesty is the best policy." These thoughts are the basis of the canons.

The canons are organized into rules governing the individual engineer's relationships with the public, with employer or clients, and with other engineers. Adherence to these rules is important to the individual engineer's personal reputation and to the public image and dignity of the profession.

**TABLE 18.5**
**Canons of ethics of cost engineers**

---

### INTRODUCTION

The Engineer, to uphold and advance the honor and dignity of the engineering profession and in keeping with high standards of ethical conduct:

   I. Will be honest and impartial, and will serve employer, clients, and the public with devotion;

  II. Will strive to increase the competence and prestige of the engineering profession;

 III. Will apply knowledge and skill to advance human welfare.

---

### I. Relations with the public

---

  A. Engineers will hold paramount the safety, health, and welfare of the public, including that of future generations.

  B. Engineers will endeavor to extend public knowledge and appreciation of engineering and its achievements, and will oppose any untrue, unsupported, or exaggerated statements regarding engineering.

  C. Engineers will be dignified and modest, ever upholding the honor and dignity of their profession, and will refrain from self-laudatory advertising.

  D. Engineers will express an opinion on an engineering subject only when it is founded on adequate knowledge and honest conviction.

**TABLE 18.5** *continued*

E. Engineers will issue no statements, criticisms, or arguments on engineering matters which are inspired or paid for by an interested party or parties, unless they preface their comments by identifying themselves, by disclosing the identities of the party or parties on whose behalf they are speaking, and by revealing the existence of any pecuniary interest they may have in matters under discussion.

F. Engineers will approve or seal only those documents, reviewed or prepared by them which are determined to be safe for public health and welfare in conformity with accepted engineering and economic standards.

G. Engineers whose professional judgment is overruled under circumstances where the safety, health, and welfare of the public are endangered, shall inform their clients or employers of the possible consequences.

H. Engineers will work through professional societies to encourage and support engineers who follow these concepts.

I. Engineers will work only with those who follow these concepts.

J. Engineers shall be objective and truthful in professional reports, statements, or testimony. They shall include all relevant and pertinent information in such reports, statements, or testimony.

## II. Relations with employers and clients

A. Engineers will act in professional matters as a faithful agent or trustee for each employer or client.

B. Engineers will act fairly and justly toward vendors and contractors, and will not accept any commissions or allowances from vendors or contractors, directly or indirectly.

C. Engineers will inform their employer or client of financial interest in any potential vendor or contractor, or in any invention, machine, or apparatus, which is involved in a project or work of either employer or client. Engineers will not allow such interest to affect any decisions regarding engineering services which they may be called upon to perform.

D. When, as a result of their studies, Engineers believe a project(s) will not be successful or, if their engineering or economic judgment is overruled, they shall so advise their employer or client.

E. Engineers will undertake only those engineering assignments for which they are qualified. Engineers will engage or advise their employers or clients to engage specialists whenever their employer's or client's interests are served best by such an arrangement. Engineers will cooperate fully with specialists so engaged.

F. Engineers shall treat information coming to them in the course of their assignments as confidential, and shall not use such information as a means of making personal profit if such action is adverse to the interests of their clients, their employers, or the public.

  1. They will not disclose confidential information concerning the business affairs or technical processes of any present or former employer or client or bidder under evaluation, without consent, unless required by law.

  2. They shall not reveal confidential information or finding of any commission or board of which they are members unless required by law.

  3. Designs, calculations, sketches, etc., supplied to Engineers by clients shall not be duplicated by the Engineers for others without the express permission of the client(s).

  4. Engineers shall not use confidential information coming to them in the course of their assignments as a means of making personal profit if such action is adverse to the interests of their clients, employers or the public.

**TABLE 18.5** *continued*

---

G. Engineers will not accept compensation—financial or otherwise—from more than one party for the same service, or for other services pertaining to the same work, without the consent of all interested parties.

H. Employed Engineers will engage in supplementary employment or consulting practice only with the consent of their employer.

I. Engineers shall not use equipment, supplies, laboratory or office facilities of their employers to carry on outside private practice without the consent of their employers.

J. Engineers shall not solicit an engineering contract from a governmental body on which a principal, officer, or employee of their organization serves as a member.

K. The Engineer shall act with fairness and justice to all parties when administering a construction (or other) contract.

L. Before undertaking work for others in which the Engineer may make improvements, plans, designs, inventions, or records which may justify copyrights or patents, the Engineer shall enter into a positive agreement regarding the rights of respective parties.

M. Engineers shall admit and accept their own errors when proven wrong and refrain from distorting or altering the facts to justify their decisions.

N. Engineers shall not attempt to attract an employee from another employer by false or misleading representations.

O. Engineers shall act in professional matters for each employer or client as faithful agents or trustees, and shall avoid conflicts of interest.

  1. Engineers shall avoid all known or potential conflicts of interest with their employers or clients and shall promptly inform their employers or clients of any business association, interests, or circumstances which could influence their judgment or the quality of their services.

  2. Engineers shall not solicit or accept gratuities, directly or indirectly, from contractors, their agents, or other parties dealing with their clients or employers in connection with work for which they are responsible.

  3. Engineers in public service as members, advisors, or employees of a governmental body or department shall not participate in considerations or actions with respect to services solicited or provided by them or their organization in private or public engineering practice.

---

### III. Relations with engineers

---

A. Engineers will take care that credit for engineering work is given to those to whom credit is properly due.

B. Engineers will provide a prospective engineering employee with complete information on working conditions and their proposed status of employment. After employment, they will keep the employee informed of any changes in status and working conditions.

C. Engineers will uphold the principle of appropriate and adequate compensation for those engaged in engineering work, including those in subordinate capacities.

D. Engineers will endeavor to provide opportunity for the professional development and advancement of engineers in their employ or under their supervision.

E. Engineers will not attempt to supplant other Engineers in a particular employment after becoming aware that definite steps have been taken toward the others' employment or after they have been employed.

F. Engineers shall not maliciously or falsely, directly or indirectly, injure the professional reputation, prospects, practice or employment of another Engineer, nor shall they indiscriminately criticize another's work. Proof that another Engineer has been unethical, illegal, or unfair in his practice shall be cause for advising the proper authority.

**TABLE 18.5** *continued*

G. Engineers will not compete unfairly with other Engineers.

H. Engineers will cooperate in advancing the engineering profession by interchanging information and experience with other engineers and students, by contributing to public communication media, and to engineering and scientific societies and schools.

I. Engineers will not request, propose, or accept professional commissions on a contingent basis under circumstances under which their professional judgments may be compromised.

J. Engineers will not falsify or permit misrepresentation of their, or their associates', academic or professional qualifications. They shall not misrepresent or exaggerate their degrees of responsibility in or for the subject matter of prior assignments. Brochures or other presentations incident to the solicitation of employment shall not misrepresent pertinent facts concerning employers, employees, associates, joint ventures, past accomplishments, or membership in technical societies.

K. Engineers will prepare articles for the lay or technical press which are only factual, dignified and free from ostentatious or laudatory implications. Such articles shall not imply other than their direct participation in the work described unless credit is given to others for their share of the work.

L. Engineers will not campaign, solicit support, or otherwise coerce other engineers to support their candidacy, or candidacy of a colleague, for elective office in a technical association.

### IV. Standards of professional performance

A. Engineers shall be dignified and modest in explaining their work and merit, and will avoid any act tending to promote their own interests at the expense of the integrity, honor, and dignity of the profession.

B. Engineers, when serving as expert witnesses, shall express an engineering opinion only when it is founded upon adequate knowledge of the facts, upon a background of technical competence, and upon honest conviction.

C. Engineers shall continue their professional development throughout their careers, and shall provide opportunities for the professional development of those engineers under their supervision.

  1. Engineers should keep current in their specialty fields by engaging in professional practice, participating in continuing education courses, reading in the technical literature, and attending professional meetings and seminars.

  2. Engineers should encourage their engineering employees to become registered at the earliest possible date.

  3. Engineers should encourage engineering employees to attend and present papers at professional and technical society meetings.

  4. Engineers shall uphold the principle of mutually satisfying relationships between employers and employees with respect to terms of employment including professional grade descriptions, salary ranges, and fringe benefits.

*Source:* American Association of Cost Engineers, P.O. Box 1557, Morgantown, W.Va. 26507-1557.

## 18.15 SOCIAL ASPECTS

For many years both the engineering profession and industrial concerns have been giving increasing attention to the role of the engineer and the role of the company in good citizenship. Because engineers do not live in a social vacuum, and because of the impact of the engineer's work on society, engineering curricula have been placing increasing emphasis on the study of the humanities. It is important that engineers, if they are to be more than mere technicians, give consideration to the social implications of their work.

A dramatic and well-known example of this problem is provided by the dangers of large-scale contamination by the products of atomic fission. A feeling of social responsibility prompts prominent engineers and scientists to work toward reduction and control of the contamination and toward peaceful applications of the nuclear energy developed as a wartime weapon.

Industrial leaders are also vitally interested in their role as good citizens and in their social responsibilities. The goodwill and other benefits which eventually derive from this interest probably defy profitability analysis by the methods described in previous chapters. The engineer, as an employee of these industrial concerns, is intimately involved with the activities of these concerns and with the social consequences of their activities. In general a company's social responsibilities are manifested in community relations, employee relations, and educational activities.

As civilization has progressed from the cave to the modern city, people have become increasingly interdependent. As a result, they have begun to consider the welfare of their neighbors. Thus a dog-eat-dog society eventually becomes a welfare society.

The effects of continued prosperity, with its accompanying vast increases in population, hasten the change to a welfare society. The net effect of the influences of technical developments, government, and social changes is a decline in but not elimination of the time value of money. This change, of course, influences and modifies our sense of values.

Cost engineers should, then, be aware of the fact that there are influences which affect their work other than cold mathematics and economics. They should recognize these social effects and nonmathematical aspects on the field of engineering.

## NOMENCLATURE

| | |
|---|---|
| DCFRR | Discounted cash-flow rate of return, decimal per year |
| $F_{SDP}$ | Present value of $1 for sum-of-the-years-digits depreciation |
| $F_{RP}$ | Factor to convert $R$ to $P$, years |
| $n$ | The $n$th year |
| $N$ | Units of items or service per year |
| $P$ | Present value, $ |
| $r$ | Rate of return after taxes, decimal per year |
| $R$ | Uniform end of year cost, unacost, $/year |
| $S$ | Unit cost, $ per item |
| $S_L$ | Levelized cost, $ per item |

## REFERENCES

1. ASCE: *Ethics, Professionalism and Maintaining Competence,* American Society of Civil Engineers, New York, 1977.

2. ASCE/ICE Joint Conference: *The Engineer in the Community,* The Institution of Civil Engineers, London, 1970.
3. Beauchamp, T. L., and N. E. Bowie: *Ethical Theory and Business,* Prentice-Hall, Inc., Englewood Cliffs, N.J., 1979.
4. Bent, J. A., and A. Thumann: *Project Management for Engineering and Construction,* The Fairmont Press, Lilburn, Ga., 1989.
5. Clifton, D. S., Jr., and D. E. Fyffe: *Project Feasibility Analysis,* John Wiley & Sons, Inc., New York, 1977.
6. Constance, J. D.: *How to Become a Professional Engineer,* 4th ed., McGraw-Hill Book Company, New York, 1988.
7. Cress, K. R.: "Establish a System for Contractor Proposal Evaluation," *Cost Engineering,* vol. 23, June 1981, pp. 153–156.
8. Florman, S. C.: *The Civilized Engineer,* St. Martin's Press, New York, 1987.
9. Henrickson, C., and Tung Au: *Project Management for Construction,* Prentice-Hall, Inc., Englewood Cliffs, N.J., 1989.
10. Jeynes, P. H.: *Profitability and Economic Choice,* Iowa State University Press, Ames, 1968.
11. Lock, D., ed.: *Engineer's Handbook of Management Techniques,* Grover Press, Ltd., Epping, England, 1973.
12. Ludwig, E. E.: *Applied Project Engineering and Management,* Gulf Publishing Co., Houston, Tex., 1988.
13. O'Brien, J. J.: *Management with Computers,* Van Nostrand Reinhold Company, New York, 1972.
14. Perrucci, R., and J. E. Gerstl: *The Engineer and the Social System,* John Wiley & Sons, Inc., New York, 1969.
15. *Solving Quality and Productivity Problems: Goodmeasure's Guide to Corrective Action:* American Society for Quality Control, Milwaukee, Wis., 1988.
16. Steele, L. W.: *Managing Technology,* McGraw-Hill Book Company, New York, 1989.
17. Sullivan, W. G., and W. W. Claycombe: *Fundamentals of Forecasting,* Reston Publishing Company, Inc., Reston, Va., 1977.
18. Walton, C. C., ed: *Ethics of Corporate Conduct,* Prentice-Hall, Inc., Englewood Cliffs, N.J., 1977.
19. Weston, J. F., and E. U. Brigham: *Essentials of Managerial Finance,* 5th ed., The Dryden Press, Hinsdale, Ill., 1977.

# PART
# V

APPENDIXES

# DISCRETE
# COMPOUND
# INTEREST

# Table A1.1 Discrete compound-interest table

Discrete compound interest = 2%

| | Single payment | | Uniform annual series | | | Uniform gradient series | Depreciation series | |
|---|---|---|---|---|---|---|---|---|
| | Compound-interest factor $(1+i)^n$ | Present-worth factor $\dfrac{1}{(1+i)^n}$ | Unacost present-worth factor $\dfrac{(1+i)^n-1}{i(1+i)^n}$ | Capital-recovery factor $\dfrac{i(1+i)^n}{(1+i)^n-1}$ | Capitalized-cost factor $\dfrac{(1+i)^n}{(1+i)^n-1}$ | Present-worth factor $\dfrac{F_{RP}-nF_{SP}}{i}$ | Sum-of-digits present-worth factor $\dfrac{n-F_{RP}}{0.5n(n+1)i}$ | Straight-line present-worth factor $\dfrac{1}{niF_{PK}}$ |
| | P to S | S to P | R to P | P to R | P to K | G to P | SD to P | SL to P |
| $n$ | $F_{PS}$ | $F_{SP}$ | $F_{RP}$ | $F_{PR}$ | $F_{PK}$ | $F_{GP}$ | $F_{SDP}$ | $F_{SLP}$ |
| 1 | 1.0200E 00 | 9.8039E-01 | 9.8039E-01 | 1.0200E 00 | 5.1000E 01 | | 9.8039E-01 | 9.8039E-01 |
| 2 | 1.0404E 00 | 9.6117E-01 | 1.9416E 00 | 5.1505E-01 | 2.5752E 01 | 9.6117E-01 | 9.7398E-01 | 9.7078E-01 |
| 3 | 1.0612E 00 | 9.4232E-01 | 2.8839E 00 | 3.4675E-01 | 1.7338E 01 | 2.8458E 00 | 9.6768E-01 | 9.6129E-01 |
| 4 | 1.0824E 00 | 9.2385E-01 | 3.8077E 00 | 2.6262E-01 | 1.3131E 01 | 5.6173E 00 | 9.6136E-01 | 9.5193E-01 |
| 5 | 1.1041E 00 | 9.0573E-01 | 4.7135E 00 | 2.1216E-01 | 1.0608E 01 | 9.2403E 00 | 9.5513E-01 | 9.4269E-01 |
| 6 | 1.1262E 00 | 8.8797E-01 | 5.6014E 00 | 1.7853E-01 | 8.9263E 00 | 1.3680E 01 | 9.4897E-01 | 9.3357E-01 |
| 7 | 1.1487E 00 | 8.7056E-01 | 6.4720E 00 | 1.5451E-01 | 7.7256E 00 | 1.8903E 01 | 9.4287E-01 | 9.2457E-01 |
| 8 | 1.1717E 00 | 8.5349E-01 | 7.3255E 00 | 1.3651E-01 | 6.8255E 00 | 2.4878E 01 | 9.3685E-01 | 9.1569E-01 |
| 9 | 1.1951E 00 | 8.3676E-01 | 8.1622E 00 | 1.2252E-01 | 6.1258E 00 | 3.1572E 01 | 9.3085E-01 | 9.0692E-01 |
| 10 | 1.2190E 00 | 8.2035E-01 | 8.9826E 00 | 1.1133E-01 | 5.5663E 00 | 3.8955E 01 | 9.2492E-01 | 8.9826E-01 |
| 11 | 1.2434E 00 | 8.0426E-01 | 9.7868E 00 | 1.0218E-01 | 5.1089E 00 | 4.6998E 01 | 9.1905E-01 | 8.8971E-01 |
| 12 | 1.2682E 00 | 7.8849E-01 | 1.0575E 01 | 9.4560E-02 | 4.7280E 00 | 5.5671E 01 | 9.1324E-01 | 8.8128E-01 |
| 13 | 1.2936E 00 | 7.7303E-01 | 1.1348E 01 | 8.8118E-02 | 4.4059E 00 | 6.4948E 01 | 9.0749E-01 | 8.7295E-01 |
| 14 | 1.3195E 00 | 7.5788E-01 | 1.2106E 01 | 8.2602E-02 | 4.1301E 00 | 7.4800E 01 | 9.0179E-01 | 8.6473E-01 |
| 15 | 1.3459E 00 | 7.4301E-01 | 1.2849E 01 | 7.7825E-02 | 3.8913E 00 | 8.5202E 01 | 8.9614E-01 | 8.5662E-01 |
| 16 | 1.3728E 00 | 7.2845E-01 | 1.3578E 01 | 7.3650E-02 | 3.6825E 00 | 9.6129E 01 | 8.9055E-01 | 8.4861E-01 |
| 18 | 1.4282E 00 | 7.0016E-01 | 1.4992E 01 | 6.6702E-02 | 3.3351E 00 | 1.1946E 02 | 8.7952E-01 | 8.3289E-01 |
| 20 | 1.4859E 00 | 6.7297E-01 | 1.6351E 01 | 6.1157E-02 | 3.0578E 00 | 1.4460E 02 | 8.6871E-01 | 8.1757E-01 |
| 25 | 1.6406E 00 | 6.0953E-01 | 1.9523E 01 | 5.1220E-02 | 2.5610E 00 | 2.1426E 02 | 8.4255E-01 | 7.8094E-01 |
| 30 | 1.8114E 00 | 5.5207E-01 | 2.2396E 01 | 4.4650E-02 | 2.2325E 00 | 2.9172E 02 | 8.1779E-01 | 7.4655E-01 |
| 35 | 1.9999E 00 | 5.0003E-01 | 2.4999E 01 | 4.0002E-02 | 2.0001E 00 | 3.7488E 02 | 7.9376E-01 | 7.1425E-01 |
| 40 | 2.2080E 00 | 4.5289E-01 | 2.7355E 01 | 3.6556E-02 | 1.8278E 00 | 4.6199E 02 | 7.7101E-01 | 6.8389E-01 |
| 45 | 2.4379E 00 | 4.1020E-01 | 2.9490E 01 | 3.3910E-02 | 1.6955E 00 | 5.5157E 02 | 7.4927E-01 | 6.5534E-01 |
| 50 | 2.6916E 00 | 3.7153E-01 | 3.1424E 01 | 3.1823E-02 | 1.5912E 00 | 6.4236E 02 | 7.2849E-01 | 6.2847E-01 |

Discrete compound interest = 4%

| | Single payment | | Uniform annual series | | | Uniform gradient series | Depreciation series | |
|---|---|---|---|---|---|---|---|---|
| | Compound-interest factor | Present-worth factor | Unacost present-worth factor | Capital-recovery factor | Capitalized-cost factor | Present-worth factor | Sum-of-digits present-worth factor | Straight-line present worth factor |
| | $(1+i)^n$ | $\dfrac{1}{(1+i)^n}$ | $\dfrac{(1+i)^n-1}{i(1+i)^n}$ | $\dfrac{(1+i)^n}{(1+i)^n-1}$ | $\dfrac{(1+i)^n}{(1+i)^n-1}$ | $\dfrac{F_{RP}-nF_{SP}}{i}$ | $\dfrac{n-F_{RP}}{0.5n(n+1)i}$ | $\dfrac{1}{niF_{PK}}$ |
| | $P$ to $S$ | $S$ to $P$ | $R$ to $P$ | $P$ to $R$ | $P$ to $K$ | $G$ to $P$ | $SD$ to $P$ | $SL$ to $P$ |
| $n$ | $F_{PS}$ | $F_{SP}$ | $F_{RP}$ | $F_{PR}$ | $F_{PK}$ | $F_{GP}$ | $F_{SDP}$ | $F_{SLP}$ |
| 1 | 1.0400E 00 | 9.6154E-01 | 9.6154E-01 | 1.0400E 00 | 2.6000E 01 | . . . . . . | 9.6154E-01 | 9.6154E-01 |
| 2 | 1.0816E 00 | 9.2456E-01 | 1.8861E 00 | 5.3020E-01 | 1.3255E 01 | 9.2456E-01 | 9.4921E-01 | 9.4305E-01 |
| 3 | 1.1249E 00 | 8.8900E-01 | 2.7751E 00 | 3.6035E-01 | 9.0087E 00 | 2.7025E 00 | 9.3712E-01 | 9.2503E-01 |
| 4 | 1.1699E 00 | 8.5480E-01 | 3.6299E 00 | 2.7549E-01 | 6.8873E 00 | 5.2670E 00 | 9.2526E-01 | 9.0747E-01 |
| 5 | 1.2167E 00 | 8.2193E-01 | 4.4518E 00 | 2.2463E-01 | 5.6157E 00 | 8.5547E 00 | 9.1363E-01 | 8.9036E-01 |
| 6 | 1.2653E 00 | 7.9031E-01 | 5.2421E 00 | 1.9076E-01 | 4.7690E 00 | 1.2506E 01 | 9.0222E-01 | 8.7369E-01 |
| 7 | 1.3159E 00 | 7.5992E-01 | 6.0021E 00 | 1.6661E-01 | 4.1652E 00 | 1.7066E 01 | 8.9102E-01 | 8.5744E-01 |
| 8 | 1.3686E 00 | 7.3069E-01 | 6.7327E 00 | 1.4853E-01 | 3.7132E 00 | 2.2181E 01 | 8.8004E-01 | 8.4159E-01 |
| 9 | 1.4233E 00 | 7.0259E-01 | 7.4353E 00 | 1.3449E-01 | 3.3623E 00 | 2.7801E 01 | 8.6926E-01 | 8.2615E-01 |
| 10 | 1.4802E 00 | 6.7556E-01 | 8.1109E 00 | 1.2329E-01 | 3.0823E 00 | 3.3881E 01 | 8.5868E-01 | 8.1109E-01 |
| 11 | 1.5395E 00 | 6.4958E-01 | 8.7605E 00 | 1.1415E-01 | 2.8537E 00 | 4.0377E 01 | 8.4830E-01 | 7.9641E-01 |
| 12 | 1.6010E 00 | 6.2460E-01 | 9.3851E 00 | 1.0655E-01 | 2.6638E 00 | 4.7248E 01 | 8.3812E-01 | 7.8209E-01 |
| 13 | 1.6651E 00 | 6.0057E-01 | 9.9856E 00 | 1.0014E-01 | 2.5036E 00 | 5.4455E 01 | 8.2812E-01 | 7.6813E-01 |
| 14 | 1.7317E 00 | 5.7747E-01 | 1.0563E 01 | 9.4669E-02 | 2.3667E 00 | 6.1962E 01 | 8.1830E-01 | 7.5451E-01 |
| 15 | 1.8009E 00 | 5.5526E-01 | 1.1118E 01 | 8.9941E-02 | 2.2485E 00 | 6.9735E 01 | 8.0867E-01 | 7.4123E-01 |
| 16 | 1.8730E 00 | 5.3391E-01 | 1.1652E 01 | 8.5820E-02 | 2.1455E 00 | 7.7744E 01 | 7.9921E-01 | 7.2827E-01 |
| 18 | 2.0258E 00 | 4.9363E-01 | 1.2659E 01 | 7.8993E-02 | 1.9748E 00 | 9.4350E 01 | 7.8080E-01 | 7.0329E-01 |
| 20 | 2.1911E 00 | 4.5639E-01 | 1.3590E 01 | 7.3582E-02 | 1.8395E 00 | 1.1156E 02 | 7.6306E-01 | 6.7952E-01 |
| 25 | 2.6658E 00 | 3.7512E-01 | 1.5622E 01 | 6.4012E-02 | 1.6003E 00 | 1.5610E 02 | 7.2138E-01 | 6.2488E-01 |
| 30 | 3.2434E 00 | 3.0832E-01 | 1.7292E 01 | 5.7830E-02 | 1.4458E 00 | 2.0106E 02 | 6.8322E-01 | 5.7640E-01 |
| 35 | 3.9461E 00 | 2.5342E-01 | 1.8665E 01 | 5.3577E-02 | 1.3394E 00 | 2.4488E 02 | 6.4823E-01 | 5.3327E-01 |
| 40 | 4.8010E 00 | 2.0829E-01 | 1.9793E 01 | 5.0523E-02 | 1.2631E 00 | 2.8653E 02 | 6.1607E-01 | 4.9482E-01 |
| 45 | 5.8412E 00 | 1.7120E-01 | 2.0720E 01 | 4.8262E-02 | 1.2066E 00 | 3.2540E 02 | 5.8647E-01 | 4.6045E-01 |
| 50 | 7.1067E 00 | 1.4071E-01 | 2.1482E 01 | 4.6550E-02 | 1.1638E 00 | 3.6116E 02 | 5.5917E-01 | 4.2964E-01 |

## Table A1.1 (*Continued*)

Discrete compound interest = 6%

| | Single payment | | Uniform annual series | | | Uniform gradient series | Depreciation series | |
|---|---|---|---|---|---|---|---|---|
| | Compound-interest factor | Present-worth factor | Unacost present-worth factor | Capital-recovery factor | Capitalized-cost factor | Present-worth factor | Sum-of-digits present-worth factor | Straight-line present-worth factor |
| | $(1+i)^n$ | $\dfrac{1}{(1+i)^n}$ | $\dfrac{(1+i)^n-1}{i(1+i)^n}$ | $\dfrac{(1+i)^n}{(1+i)^n-1}$ | $\dfrac{(1+i)^n}{(1+i)^n-1}$ | $\dfrac{F_{RP}-nF_{SP}}{i}$ | $\dfrac{n-F_{RP}}{0.5n(n+1)i}$ | $\dfrac{1}{niF_{PK}}$ |
| | P to S | S to P | R to P | P to R | P to K | G to P | SD to P | SL to P |
| $n$ | $F_{PS}$ | $F_{SP}$ | $F_{RP}$ | $F_{PR}$ | $F_{PK}$ | $F_{GP}$ | $F_{SDP}$ | $F_{SLP}$ |
| 1 | 1.0600E 00 | 9.4340E-01 | 9.4340E-01 | 1.0600E 00 | 1.7667E 01 | .......E-01 | 9.4340E-01 | 9.4340E-01 |
| 2 | 1.1236E 00 | 8.9000E-01 | 1.8334E 00 | 5.4544E-01 | 9.0906E 00 | 8.9000E-01 | 9.2560E-01 | 9.1670E-01 |
| 3 | 1.1910E 00 | 8.3962E-01 | 2.6730E 00 | 3.7411E-01 | 6.2352E 00 | 2.5692E 00 | 9.0830E-01 | 8.9100E-01 |
| 4 | 1.2625E 00 | 7.9209E-01 | 3.4651E 00 | 2.8859E-01 | 4.8099E 00 | 4.9455E 00 | 8.9149E-01 | 8.6628E-01 |
| 5 | 1.3382E 00 | 7.4726E-01 | 4.2124E 00 | 2.3740E-01 | 3.9566E 00 | 7.9345E 00 | 8.7515E-01 | 8.4247E-01 |
| 6 | 1.4185E 00 | 7.0496E-01 | 4.9173E 00 | 2.0336E-01 | 3.3894E 00 | 1.1459E 01 | 8.5927E-01 | 8.1955E-01 |
| 7 | 1.5036E 00 | 6.6506E-01 | 5.5824E 00 | 1.7914E-01 | 2.9856E 00 | 1.5450E 01 | 8.4382E-01 | 7.9748E-01 |
| 8 | 1.5938E 00 | 6.2741E-01 | 6.2098E 00 | 1.6104E-01 | 2.6839E 00 | 1.9842E 01 | 8.2880E-01 | 7.7622E-01 |
| 9 | 1.6895E 00 | 5.9190E-01 | 6.8017E 00 | 1.4702E-01 | 2.4504E 00 | 2.4577E 01 | 8.1419E-01 | 7.5574E-01 |
| 10 | 1.7908E 00 | 5.5839E-01 | 7.3601E 00 | 1.3587E-01 | 2.2645E 00 | 2.9602E 01 | 7.9997E-01 | 7.3601E-01 |
| 11 | 1.8983E 00 | 5.2679E-01 | 7.8869E 00 | 1.2679E-01 | 2.1132E 00 | 3.4870E 01 | 7.8614E-01 | 7.1699E-01 |
| 12 | 2.0122E 00 | 4.9697E-01 | 8.3838E 00 | 1.1928E-01 | 1.9880E 00 | 4.0337E 01 | 7.7268E-01 | 6.9865E-01 |
| 13 | 2.1329E 00 | 4.6884E-01 | 8.8527E 00 | 1.1296E-01 | 1.8827E 00 | 4.5963E 01 | 7.5958E-01 | 6.8098E-01 |
| 14 | 2.2609E 00 | 4.4230E-01 | 9.2950E 00 | 1.0758E-01 | 1.7931E 00 | 5.1713E 01 | 7.4683E-01 | 6.6393E-01 |
| 15 | 2.3966E 00 | 4.1727E-01 | 9.7122E 00 | 1.0296E-01 | 1.7160E 00 | 5.7555E 01 | 7.3441E-01 | 6.4748E-01 |
| 16 | 2.5404E 00 | 3.9365E-01 | 1.0106E 01 | 9.8952E-02 | 1.6492E 00 | 6.3459E 01 | 7.2232E-01 | 6.3162E-01 |
| 18 | 2.8543E 00 | 3.5034E-01 | 1.0828E 01 | 9.2357E-02 | 1.5393E 00 | 7.5357E 01 | 6.9906E-01 | 6.0153E-01 |
| 20 | 3.2071E 00 | 3.1180E-01 | 1.1470E 01 | 8.7185E-02 | 1.4531E 00 | 8.7230E 01 | 6.7699E-01 | 5.7350E-01 |
| 25 | 4.2919E 00 | 2.3300E-01 | 1.2783E 01 | 7.8227E-02 | 1.3038E 00 | 1.1597E 02 | 6.2649E-01 | 5.1133E-01 |
| 30 | 5.7435E 00 | 1.7411E-01 | 1.3765E 01 | 7.2649E-02 | 1.2108E 00 | 1.4236E 02 | 5.8191E-01 | 4.5883E-01 |
| 35 | 7.6861E 00 | 1.3011E-01 | 1.4498E 01 | 6.8974E-02 | 1.1496E 00 | 1.6574E 02 | 5.4237E-01 | 4.1424E-01 |
| 40 | 1.0286E 01 | 9.7222E-02 | 1.5046E 01 | 6.6462E-02 | 1.1077E 00 | 1.8596E 02 | 5.0719E-01 | 3.7616E-01 |
| 45 | 1.3765E 01 | 7.2650E-02 | 1.5456E 01 | 6.4700E-02 | 1.0783E 00 | 2.0311E 02 | 4.7575E-01 | 3.4346E-01 |
| 50 | 1.8420E 01 | 5.4288E-02 | 1.5762E 01 | 6.3444E-02 | 1.0574E 00 | 2.1746E 02 | 4.4756E-01 | 3.1524E-01 |

Discrete compound interest = 8%

| n | Single payment — Compound-interest factor $(1+i)^n$ — $P$ to $S$ — $F_{PS}$ | Single payment — Present-worth factor $\dfrac{1}{(1+i)^n}$ — $S$ to $P$ — $F_{SP}$ | Uniform annual series — Unacost present-worth factor $\dfrac{(1+i)^n-1}{i(1+i)^n}$ — $R$ to $P$ — $F_{RP}$ | Uniform annual series — Capital-recovery factor $\dfrac{i(1+i)^n}{(1+i)^n-1}$ — $P$ to $R$ — $F_{PR}$ | Uniform annual series — Capitalized-cost factor $\dfrac{(1+i)^n}{(1+i)^n-1}$ — $P$ to $K$ — $F_{PK}$ | Uniform gradient series — Present-worth factor $\dfrac{F_{RP}-nF_{SP}}{i}$ — $G$ to $P$ — $F_{GP}$ | Depreciation series — Sum-of-digits present-worth factor $\dfrac{n-F_{RP}}{0.5n(n+1)i}$ — $SD$ to $P$ — $F_{SDP}$ | Depreciation series — Straight-line present worth factor $\dfrac{1}{niF_{PK}}$ — $SL$ to $P$ — $F_{SLP}$ |
|---|---|---|---|---|---|---|---|---|
| 1  | 1.0800E 00 | 9.2593E-01 | 9.2593E-01 | 1.0800E 00 | 1.3500E 01 |            | 9.2593E-01 | 9.2593E-01 |
| 2  | 1.1664E 00 | 8.5734E-01 | 1.7833E 00 | 5.6077E-01 | 7.0096E 00 | 8.5734E-01 | 9.0306E-01 | 8.9163E-01 |
| 3  | 1.2597E 00 | 7.9383E-01 | 2.5771E 00 | 3.8803E-01 | 4.8504E 00 | 2.4450E 00 | 8.8105E-01 | 8.5903E-01 |
| 4  | 1.3605E 00 | 7.3503E-01 | 3.3121E 00 | 3.0192E-01 | 3.7740E 00 | 4.6501E 00 | 8.5984E-01 | 8.2803E-01 |
| 5  | 1.4693E 00 | 6.8058E-01 | 3.9927E 00 | 2.5046E-01 | 3.1307E 00 | 7.3724E 00 | 8.3941E-01 | 7.9854E-01 |
| 6  | 1.5869E 00 | 6.3017E-01 | 4.6229E 00 | 2.1632E-01 | 2.7039E 00 | 1.0523E 01 | 8.1971E-01 | 7.7048E-01 |
| 7  | 1.7138E 00 | 5.8349E-01 | 5.2064E 00 | 1.9207E-01 | 2.4009E 00 | 1.4024E 01 | 8.0073E-01 | 7.4377E-01 |
| 8  | 1.8509E 00 | 5.4027E-01 | 5.7466E 00 | 1.7401E-01 | 2.1752E 00 | 1.7806E 01 | 7.8242E-01 | 7.1833E-01 |
| 9  | 1.9990E 00 | 5.0025E-01 | 6.2469E 00 | 1.6008E-01 | 2.0010E 00 | 2.1808E 01 | 7.6475E-01 | 6.9410E-01 |
| 10 | 2.1589E 00 | 4.6319E-01 | 6.7101E 00 | 1.4903E-01 | 1.8629E 00 | 2.5977E 01 | 7.4771E-01 | 6.7101E-01 |
| 11 | 2.3316E 00 | 4.2888E-01 | 7.1390E 00 | 1.4008E-01 | 1.7510E 00 | 3.0266E 01 | 7.3126E-01 | 6.4900E-01 |
| 12 | 2.5182E 00 | 3.9711E-01 | 7.5361E 00 | 1.3270E-01 | 1.6587E 00 | 3.4634E 01 | 7.1537E-01 | 6.2801E-01 |
| 13 | 2.7196E 00 | 3.6770E-01 | 7.9038E 00 | 1.2652E-01 | 1.5815E 00 | 3.9046E 01 | 7.0003E-01 | 6.0798E-01 |
| 14 | 2.9372E 00 | 3.4046E-01 | 8.2442E 00 | 1.2130E-01 | 1.5162E 00 | 4.3472E 01 | 6.8521E-01 | 5.8887E-01 |
| 15 | 3.1722E 00 | 3.1524E-01 | 8.5595E 00 | 1.1683E-01 | 1.4604E 00 | 4.7886E 01 | 6.7089E-01 | 5.7063E-01 |
| 16 | 3.4259E 00 | 2.9189E-01 | 8.8514E 00 | 1.1298E-01 | 1.4122E 00 | 5.2264E 01 | 6.5704E-01 | 5.5321E-01 |
| 18 | 3.9960E 00 | 2.5025E-01 | 9.3719E 00 | 1.0670E-01 | 1.3338E 00 | 6.0843E 01 | 6.3071E-01 | 5.2066E-01 |
| 20 | 4.6610E 00 | 2.1455E-01 | 9.8181E 00 | 1.0185E-01 | 1.2732E 00 | 6.9090E 01 | 6.0606E-01 | 4.9091E-01 |
| 25 | 6.8485E 00 | 1.4602E-01 | 1.0675E 01 | 9.3679E-02 | 1.1710E 00 | 8.7804E 01 | 5.5097E-01 | 4.2699E-01 |
| 30 | 1.0063E 01 | 9.9377E-02 | 1.1258E 01 | 8.8827E-02 | 1.1103E 00 | 1.0366E 02 | 5.0382E-01 | 3.7526E-01 |
| 35 | 1.4785E 01 | 6.7635E-02 | 1.1655E 01 | 8.5803E-02 | 1.0725E 00 | 1.1609E 02 | 4.6320E-01 | 3.3299E-01 |
| 40 | 2.1725E 01 | 4.6031E-02 | 1.1925E 01 | 8.3860E-02 | 1.0483E 00 | 1.2604E 02 | 4.2798E-01 | 2.9812E-01 |
| 45 | 3.1920E 01 | 3.1328E-02 | 1.2108E 01 | 8.2587E-02 | 1.0323E 00 | 1.3373E 02 | 3.9724E-01 | 2.6908E-01 |
| 50 | 4.6902E 01 | 2.1321E-02 | 1.2233E 01 | 8.1743E-02 | 1.0218E 00 | 1.3959E 02 | 3.7026E-01 | 2.4467E-01 |

# Table A1.1 (Continued)

Discrete compound interest = 10%

| | Single payment | | Uniform annual series | | | Uniform gradient series | Depreciation series | |
|---|---|---|---|---|---|---|---|---|
| | Compound-interest factor $(1+i)^n$ | Present-worth factor $\dfrac{1}{(1+i)^n}$ | Unacost present-worth factor $\dfrac{(1+i)^n - 1}{i(1+i)^n}$ | Capital-recovery factor $\dfrac{(1+i)^n}{(1+i)^n - 1}$ | Capitalized-cost factor $\dfrac{(1+i)^n}{(1+i)^n - 1}$ | Present-worth factor $\dfrac{F_{RP} - nF_{SP}}{i}$ | Sum-of-digits present-worth factor $\dfrac{n - F_{RP}}{0.5n(n+1)i}$ | Straight-line present-worth factor $\dfrac{1}{niF_{PK}}$ |
| | $P$ to $S$ | $S$ to $P$ | $R$ to $P$ | $P$ to $R$ | $P$ to $K$ | $G$ to $P$ | $SD$ to $P$ | $SL$ to $P$ |
| $n$ | $F_{PS}$ | $F_{SP}$ | $F_{RP}$ | $F_{PR}$ | $F_{PK}$ | $F_{GP}$ | $F_{SDP}$ | $F_{SLP}$ |
| 1 | 1.1000E 00 | 9.0909E-01 | 9.0909E-01 | 1.1000E 00 | 1.1000E 01 | · · · · | 9.0909E-01 | 9.0909E-01 |
| 2 | 1.2100E 00 | 8.2645E-01 | 1.7355E 00 | 5.7619E-01 | 5.7619E 00 | 8.2645E-01 | 8.6154E-01 | 8.6777E-01 |
| 3 | 1.3310E 00 | 7.5131E-01 | 2.4869E 00 | 4.0211E-01 | 4.0211E 00 | 2.3291E 00 | 8.5525E-01 | 8.2895E-01 |
| 4 | 1.4641E 00 | 6.8301E-01 | 3.1699E 00 | 3.1547E-01 | 3.1547E 00 | 4.3781E 00 | 8.3013E-01 | 7.9247E-01 |
| 5 | 1.6105E 00 | 6.2092E-01 | 3.7908E 00 | 2.6380E-01 | 2.6380E 00 | 6.8618E 00 | 8.0613E-01 | 7.5816E-01 |
| 6 | 1.7716E 00 | 5.6447E-01 | 4.3553E 00 | 2.2961E-01 | 2.2961E 00 | 9.6842E 00 | 7.8321E-01 | 7.2588E-01 |
| 7 | 1.9487E 00 | 5.1316E-01 | 4.8684E 00 | 2.0541E-01 | 2.0541E 01 | 1.2763E 01 | 7.6128E-01 | 6.9549E-01 |
| 8 | 2.1436E 00 | 4.6651E-01 | 5.3349E 00 | 1.8744E-01 | 1.8744E 01 | 1.6029E 01 | 7.4030E-01 | 6.6687E-01 |
| 9 | 2.3579E 00 | 4.2410E-01 | 5.7590E 00 | 1.7364E-01 | 1.7364E 01 | 1.9421E 01 | 7.2022E-01 | 6.3989E-01 |
| 10 | 2.5937E 00 | 3.8554E-01 | 6.1446E 00 | 1.6275E-01 | 1.6275E 01 | 2.2891E 01 | 7.0099E-01 | 6.1446E-01 |
| 11 | 2.8531E 00 | 3.5049E-01 | 6.4951E 00 | 1.5396E-01 | 1.5396E 01 | 2.6396E 01 | 6.8257E-01 | 5.9046E-01 |
| 12 | 3.1384E 00 | 3.1863E-01 | 6.8137E 00 | 1.4676E-01 | 1.4676E 01 | 2.9901E 01 | 6.6491E-01 | 5.6781E-01 |
| 13 | 3.4523E 00 | 2.8966E-01 | 7.1034E 00 | 1.4078E-01 | 1.4078E 01 | 3.3377E 01 | 6.4798E-01 | 5.4661E-01 |
| 14 | 3.7975E 00 | 2.6333E-01 | 7.3667E 00 | 1.3575E-01 | 1.3575E 01 | 3.6800E 01 | 6.3174E-01 | 5.2619E-01 |
| 15 | 4.1772E 00 | 2.3939E-01 | 7.6061E 00 | 1.3147E-01 | 1.3147E 01 | 4.0152E 01 | 6.1616E-01 | 5.0707E-01 |
| 16 | 4.5950E 00 | 2.1763E-01 | 7.8237E 00 | 1.2782E-01 | 1.2782E 01 | 4.3416E 01 | 6.0120E-01 | 4.8898E-01 |
| 18 | 5.5599E 00 | 1.7986E-01 | 8.2014E 00 | 1.2193E-01 | 1.2193E 01 | 4.9640E 01 | 5.7302E-01 | 4.5563E-01 |
| 20 | 6.7275E 00 | 1.4864E-01 | 8.5136E 00 | 1.1746E-01 | 1.1746E 01 | 5.5407E 01 | 5.4697E-01 | 4.2568E-01 |
| 25 | 1.0835E 01 | 9.2296E-02 | 9.0770E 00 | 1.1017E-01 | 1.1017E 01 | 6.7696E 01 | 4.8994E-01 | 3.6308E-01 |
| 30 | 1.7449E 01 | 5.7309E-02 | 9.4269E 00 | 1.0608E-01 | 1.0608E 01 | 7.7077E 01 | 4.4243E-01 | 3.1423E-01 |
| 35 | 2.8102E 01 | 3.5584E-02 | 9.6442E 00 | 1.0369E-01 | 1.0369E 01 | 8.3987E 01 | 4.0247E-01 | 2.7555E-01 |
| 40 | 4.5259E 01 | 2.2094E-02 | 9.7791E 00 | 1.0226E-01 | 1.0226E 01 | 8.8953E 01 | 3.6855E-01 | 2.4448E-01 |
| 45 | 7.2890E 01 | 1.3719E-02 | 9.8628E 00 | 1.0139E-01 | 1.0139E 01 | 9.2454E 01 | 3.3949E-01 | 2.1917E-01 |
| 50 | 1.1739E 02 | 8.5184E-03 | 9.9148E 00 | 1.0086E-01 | 1.0086E 01 | 9.4889E 01 | 3.1439E-01 | 1.9830E-01 |

Discrete compound interest = 12%

| | Single payment | | Uniform annual series | | | Uniform gradient series | Depreciation series | |
|---|---|---|---|---|---|---|---|---|
| | Compound-interest factor | Present-worth factor | Unacost present-worth factor | Capital-recovery factor | Capitalized-cost factor | Present-worth factor | Sum-of-digits present-worth factor | Straight-line present worth factor |
| | $(1+i)^n$ | $\dfrac{1}{(1+i)^n}$ | $\dfrac{(1+i)^n - 1}{i(1+i)^n}$ | $\dfrac{(1+i)^n}{(1+i)^n - 1}$ | $\dfrac{(1+i)^n}{(1+i)^n - 1}$ | $\dfrac{F_{RP} - nF_{SP}}{i}$ | $\dfrac{n - F_{RP}}{0.5n(n+1)i}$ | $\dfrac{1}{ni}F_{PK}$ |
| | $P$ to $S$ | $S$ to $P$ | $R$ to $P$ | $P$ to $R$ | $P$ to $K$ | $G$ to $P$ | $SD$ to $P$ | $SL$ to $P$ |
| $n$ | $F_{PS}$ | $F_{SP}$ | $F_{RP}$ | $F_{PR}$ | $F_{PK}$ | $F_{GP}$ | $F_{SDP}$ | $F_{SLP}$ |
| 1 | 1.1200E 00 | 8.9286E-01 | 8.9286E-01 | 1.1200E 00 | 9.3333E 00 | | 8.9286E-01 | 8.9286E-01 |
| 2 | 1.2544E 00 | 7.9719E-01 | 1.6901E 00 | 5.9170E-01 | 4.9308E 00 | 7.9719E-01 | 8.6097E-01 | 8.4503E-01 |
| 3 | 1.4049E 00 | 7.1178E-01 | 2.4018E 00 | 4.1635E-01 | 3.4696E 00 | 2.2208E 00 | 8.3079E-01 | 8.0061E-01 |
| 4 | 1.5735E 00 | 6.3552E-01 | 3.0373E 00 | 3.2923E-01 | 2.7436E 00 | 4.1273E 00 | 8.0221E-01 | 7.5934E-01 |
| 5 | 1.7623E 00 | 5.6743E-01 | 3.6048E 00 | 2.7741E-01 | 2.3117E 00 | 6.3970E 00 | 7.7512E-01 | 7.2096E-01 |
| 6 | 1.9738E 00 | 5.0663E-01 | 4.1114E 00 | 2.4323E-01 | 2.0269E 00 | 8.9302E 00 | 7.4944E-01 | 6.8523E-01 |
| 7 | 2.2107E 00 | 4.5235E-01 | 4.5638E 00 | 2.1912E-01 | 1.8260E 00 | 1.1644E 01 | 7.2507E-01 | 6.5197E-01 |
| 8 | 2.4760E 00 | 4.0388E-01 | 4.9676E 00 | 2.0130E-01 | 1.6775E 00 | 1.4471E 01 | 7.0194E-01 | 6.2095E-01 |
| 9 | 2.7731E 00 | 3.6061E-01 | 5.3282E 00 | 1.8768E-01 | 1.5640E 00 | 1.7356E 01 | 6.7995E-01 | 5.9203E-01 |
| 10 | 3.1058E 00 | 3.2197E-01 | 5.6502E 00 | 1.7698E-01 | 1.4749E 00 | 2.0254E 01 | 6.5906E-01 | 5.6502E-01 |
| 11 | 3.4785E 00 | 2.8748E-01 | 5.9377E 00 | 1.6842E-01 | 1.4035E 00 | 2.3129E 01 | 6.3918E-01 | 5.3979E-01 |
| 12 | 3.8960E 00 | 2.5668E-01 | 6.1944E 00 | 1.6144E-01 | 1.3453E 00 | 2.5952E 01 | 6.2026E-01 | 5.1620E-01 |
| 13 | 4.3635E 00 | 2.2917E-01 | 6.4235E 00 | 1.5568E-01 | 1.2973E 00 | 2.8702E 01 | 6.0224E-01 | 4.9412E-01 |
| 14 | 4.8871E 00 | 2.0462E-01 | 6.6282E 00 | 1.5087E-01 | 1.2573E 00 | 3.1362E 01 | 5.8507E-01 | 4.7344E-01 |
| 15 | 5.4736E 00 | 1.8270E-01 | 6.8109E 00 | 1.4682E-01 | 1.2235E 00 | 3.3920E 01 | 5.6869E-01 | 4.5406E-01 |
| 16 | 6.1304E 00 | 1.6312E-01 | 6.9740E 00 | 1.4339E-01 | 1.1949E 00 | 3.6367E 01 | 5.5306E-01 | 4.3587E-01 |
| 18 | 7.6900E 00 | 1.3004E-01 | 7.2497E 00 | 1.3794E-01 | 1.1495E 00 | 4.0908E 01 | 5.2390E-01 | 4.0276E-01 |
| 20 | 9.6463E 00 | 1.0367E-01 | 7.4694E 00 | 1.3388E-01 | 1.1157E 00 | 4.4968E 01 | 4.9724E-01 | 3.7347E-01 |
| 25 | 1.7000E 01 | 5.8823E-02 | 7.8431E 00 | 1.2750E-01 | 1.0625E 00 | 5.3105E 01 | 4.3992E-01 | 3.1373E-01 |
| 30 | 2.9960E 01 | 3.3378E-02 | 8.0552E 00 | 1.2414E-01 | 1.0345E 00 | 5.8782E 01 | 3.9428E-01 | 2.6851E-01 |
| 35 | 5.2800E 01 | 1.8940E-02 | 8.1755E 00 | 1.2232E-01 | 1.0193E 00 | 6.2605E 01 | 3.5482E-01 | 2.3359E-01 |
| 40 | 9.3051E 01 | 1.0747E-02 | 8.2438E 00 | 1.2130E-01 | 1.0109E 00 | 6.5116E 01 | 3.2273E-01 | 2.0609E-01 |
| 45 | 1.6399E 02 | 6.0980E-03 | 8.2825E 00 | 1.2074E-01 | 1.0061E 00 | 6.6734E 01 | 2.9563E-01 | 1.8406E-01 |
| 50 | 2.8900E 02 | 3.4602E-03 | 8.3045E 00 | 1.2042E-01 | 1.0035E 00 | 6.7762E 01 | 2.7252E-01 | 1.6609E-01 |

## Table A1.1 (*Continued*)

Discrete compound interest = 14%

| | Single payment | | Uniform annual series | | Uniform gradient series | | Depreciation series | |
|---|---|---|---|---|---|---|---|---|
| | Compound-interest factor $(1+i)^n$ | Present-worth factor $\dfrac{1}{(1+i)^n}$ | Unacost present-worth factor $\dfrac{(1+i)^n-1}{i(1+i)^n}$ | Capital-recovery factor $\dfrac{(1+i)^n}{(1+i)^n-1}$ | Capitalized-cost factor $\dfrac{(1+i)^n}{(1+i)^n-1}$ | Present-worth factor $\dfrac{F_{RP}-nF_{SP}}{i}$ | Sum-of-digits present-worth factor $\dfrac{n-F_{RP}}{0.5n(n+1)i}$ | Straight-line present-worth factor $\dfrac{1}{niF_{PK}}$ |
| | P to S | S to P | R to P | P to R | P to K | G to P | SD to P | SL to P |
| $n$ | $F_{PS}$ | $F_{SP}$ | $F_{RP}$ | $F_{PR}$ | $F_{PK}$ | $F_{GP}$ | $F_{SDP}$ | $F_{SLP}$ |
| 1 | 1.1400E 00 | 8.7719E-01 | 8.7719E-01 | 1.1400E 00 | 8.1429E 00 | | 8.7719E-01 | 8.7719E-01 |
| 2 | 1.2996E 00 | 7.6947E-01 | 1.6467E 00 | 6.0729E-01 | 4.3378E 00 | 7.6947E-01 | 8.4128E-01 | 8.2333E-01 |
| 3 | 1.4815E 00 | 6.7497E-01 | 2.3216E 00 | 4.3073E-01 | 3.0767E 00 | 2.1194E 00 | 8.0758E-01 | 7.7388E-01 |
| 4 | 1.6890E 00 | 5.9208E-01 | 2.9137E 00 | 3.4320E-01 | 2.4515E 00 | 3.8957E 00 | 7.7592E-01 | 7.2843E-01 |
| 5 | 1.9254E 00 | 5.1937E-01 | 3.4331E 00 | 2.9128E-01 | 2.0806E 00 | 5.9731E 00 | 7.4615E-01 | 6.8662E-01 |
| 6 | 2.1950E 00 | 4.5559E-01 | 3.8887E 00 | 2.5716E-01 | 1.8368E 00 | 8.2511E 00 | 7.1814E-01 | 6.4811E-01 |
| 7 | 2.5023E 00 | 3.9964E-01 | 4.2883E 00 | 2.3319E-01 | 1.6657E 00 | 1.0649E 01 | 6.9176E-01 | 6.1261E-01 |
| 8 | 2.8526E 00 | 3.5056E-01 | 4.6389E 00 | 2.1557E-01 | 1.5398E 00 | 1.3103E 01 | 6.6694E-01 | 5.7986E-01 |
| 9 | 3.2519E 00 | 3.0751E-01 | 4.9464E 00 | 2.0217E-01 | 1.4441E 00 | 1.5563E 01 | 6.4343E-01 | 5.4960E-01 |
| 10 | 3.7072E 00 | 2.6974E-01 | 5.2161E 00 | 1.9171E-01 | 1.3694E 00 | 1.7991E 01 | 6.2128E-01 | 5.2161E-01 |
| 11 | 4.2262E 00 | 2.3662E-01 | 5.4527E 00 | 1.8339E-01 | 1.3103E 00 | 2.0357E 01 | 6.0035E-01 | 4.9570E-01 |
| 12 | 4.8179E 00 | 2.0756E-01 | 5.6603E 00 | 1.7667E-01 | 1.2619E 00 | 2.2640E 01 | 5.8056E-01 | 4.7169E-01 |
| 13 | 5.4924E 00 | 1.8207E-01 | 5.8424E 00 | 1.7116E-01 | 1.2226E 00 | 2.4825E 01 | 5.6182E-01 | 4.4941E-01 |
| 14 | 6.2613E 00 | 1.5971E-01 | 6.0021E 00 | 1.6661E-01 | 1.1901E 00 | 2.6901E 01 | 5.4408E-01 | 4.2872E-01 |
| 15 | 7.1379E 00 | 1.4010E-01 | 6.1422E 00 | 1.6281E-01 | 1.1629E 00 | 2.8862E 01 | 5.2725E-01 | 4.0948E-01 |
| 16 | 8.1372E 00 | 1.2289E-01 | 6.2651E 00 | 1.5962E-01 | 1.1401E 00 | 3.0706E 01 | 5.1129E-01 | 3.9157E-01 |
| 18 | 1.0575E 01 | 9.4561E-02 | 6.4674E 00 | 1.5462E-01 | 1.1044E 00 | 3.4038E 01 | 4.8173E-01 | 3.5930E-01 |
| 20 | 1.3743E 01 | 7.2762E-02 | 6.6231E 00 | 1.5099E-01 | 1.0785E 00 | 3.6914E 01 | 4.5500E-01 | 3.3116E-01 |
| 25 | 2.6462E 01 | 3.7790E-02 | 6.8729E 00 | 1.4550E-01 | 1.0393E 00 | 4.2344E 01 | 3.9840E-01 | 2.7492E-01 |
| 30 | 5.0950E 01 | 1.9627E-02 | 7.0027E 00 | 1.4280E-01 | 1.0200E 00 | 4.5813E 01 | 3.5326E-01 | 2.3342E-01 |
| 35 | 9.8100E 01 | 1.0194E-02 | 7.0700E 00 | 1.4144E-01 | 1.0103E 00 | 4.7952E 01 | 3.1667E-01 | 2.0200E-01 |
| 40 | 1.8888E 02 | 5.2943E-03 | 7.1050E 00 | 1.4075E-01 | 1.0053E 00 | 4.9238E 01 | 2.8654E-01 | 1.7763E-01 |
| 45 | 3.6368E 02 | 2.7497E-03 | 7.1232E 00 | 1.4039E-01 | 1.0028E 00 | 4.9996E 01 | 2.6140E-01 | 1.5829E-01 |
| 50 | 7.0023E 02 | 1.4281E-03 | 7.1327E 00 | 1.4020E-01 | 1.0014E 00 | 5.0438E 01 | 2.4015E-01 | 1.4265E-01 |

Discrete compound interest = 16%

| | Single payment | | Uniform annual series | | | Uniform gradient series | | Depreciation series | | |
|---|---|---|---|---|---|---|---|---|---|---|
| | Compound-interest factor | Present-worth factor | Unacost present-worth factor | Capital-recovery factor | Capitalized-cost factor | Present-worth factor | Sum-of-digits present-worth factor | Straight-line present worth factor | | |
| | $(1+i)^n$ | $\frac{1}{(1+i)^n}$ | $\frac{(1+i)^n-1}{i(1+i)^n}$ | $\frac{(1+i)^n}{(1+i)^n-1}$ | $\frac{(1+i)^n}{(1+i)^n-1}$ | $\frac{F_{RP}-nF_{SP}}{i}$ | $\frac{n-F_{RP}}{0.5n(n+1)i}$ | $\frac{1}{niF_{PK}}$ | | |
| | $P$ to $S$ | $S$ to $P$ | $R$ to $P$ | $P$ to $R$ | $P$ to $K$ | $G$ to $P$ | $SD$ to $P$ | $SL$ to $P$ | | |
| $n$ | $F_{PS}$ | $F_{SP}$ | $F_{RP}$ | $F_{PR}$ | $F_{PK}$ | $F_{GP}$ | $F_{SDP}$ | $F_{SLP}$ | | |
| 1 | 1.1600E 00 | 8.6207E-01 | 8.6207E-01 | 1.1600E 00 | 7.2500E 00 | . . . . . . | 8.6207E-01 | 8.6207E-01 | | |
| 2 | 1.3456E 00 | 7.4316E-01 | 1.6052E 00 | 6.2296E-01 | 3.8935E 00 | 7.4316E-01 | 8.2243E-01 | 8.0263E-01 | | |
| 3 | 1.5609E 00 | 6.4066E-01 | 2.2459E 00 | 4.4526E-01 | 2.7829E 00 | 2.0245E 00 | 7.8553E-01 | 7.4863E-01 | | |
| 4 | 1.8106E 00 | 5.5229E-01 | 2.7982E 00 | 3.5738E-01 | 2.2336E 00 | 3.6814E 00 | 7.5114E-01 | 6.9955E-01 | | |
| 5 | 2.1003E 00 | 4.7611E-01 | 3.2743E 00 | 3.0541E-01 | 1.9088E 00 | 5.5858E 00 | 7.1904E-01 | 6.5486E-01 | | |
| 6 | 2.4364E 00 | 4.1044E-01 | 3.6847E 00 | 2.7139E-01 | 1.6962E 00 | 7.6380E 00 | 6.8907E-01 | 6.1412E-01 | | |
| 7 | 2.8262E 00 | 3.5383E-01 | 4.0386E 00 | 2.4761E-01 | 1.5476E 00 | 9.7610E 00 | 6.6103E-01 | 5.7694E-01 | | |
| 8 | 3.2784E 00 | 3.0503E-01 | 4.3436E 00 | 2.3022E-01 | 1.4389E 00 | 1.1896E 01 | 6.3479E-01 | 5.4295E-01 | | |
| 9 | 3.8030E 00 | 2.6295E-01 | 4.6065E 00 | 2.1708E-01 | 1.3568E 00 | 1.4000E 01 | 6.1020E-01 | 5.1184E-01 | | |
| 10 | 4.4114E 00 | 2.2668E-01 | 4.8332E 00 | 2.0690E-01 | 1.2931E 00 | 1.6040E 01 | 5.8713E-01 | 4.8332E-01 | | |
| 11 | 5.1173E 00 | 1.9542E-01 | 5.0286E 00 | 1.9886E-01 | 1.2429E 00 | 1.7994E 01 | 5.6547E-01 | 4.5715E-01 | | |
| 12 | 5.9360E 00 | 1.6846E-01 | 5.1971E 00 | 1.9241E-01 | 1.2026E 00 | 1.9847E 01 | 5.4510E-01 | 4.3309E-01 | | |
| 13 | 6.8858E 00 | 1.4523E-01 | 5.3423E 00 | 1.8718E-01 | 1.1699E 00 | 2.1590E 01 | 5.2594E-01 | 4.1095E-01 | | |
| 14 | 7.9875E 00 | 1.2520E-01 | 5.4675E 00 | 1.8290E-01 | 1.1431E 00 | 2.3217E 01 | 5.0789E-01 | 3.9054E-01 | | |
| 15 | 9.2655E 00 | 1.0793E-01 | 5.5755E 00 | 1.7936E-01 | 1.1210E 00 | 2.4728E 01 | 4.9086E-01 | 3.7170E-01 | | |
| 16 | 1.0748E 01 | 9.3041E-02 | 5.6685E 00 | 1.7641E-01 | 1.1026E 00 | 2.6124E 01 | 4.7479E-01 | 3.5428E-01 | | |
| 18 | 1.4463E 01 | 6.9144E-02 | 5.8178E 00 | 1.7188E-01 | 1.0743E 00 | 2.8583E 01 | 4.4525E-01 | 3.2321E-01 | | |
| 20 | 1.9461E 01 | 5.1385E-02 | 5.9288E 00 | 1.6867E-01 | 1.0542E 00 | 3.0632E 01 | 4.1878E-01 | 2.9644E-01 | | |
| 25 | 4.0874E 01 | 2.4465E-02 | 6.0971E 00 | 1.6401E-01 | 1.0251E 00 | 3.4264E 01 | 3.6352E-01 | 2.4388E-01 | | |
| 30 | 8.5850E 01 | 1.1648E-02 | 6.1772E 00 | 1.6189E-01 | 1.0118E 00 | 3.6423E 01 | 3.2020E-01 | 2.0591E-01 | | |
| 35 | 1.8031E 02 | 5.5459E-03 | 6.2153E 00 | 1.6089E-01 | 1.0056E 00 | 3.7633E 01 | 2.8556E-01 | 1.7758E-01 | | |
| 40 | 3.7872E 02 | 2.6405E-03 | 6.2335E 00 | 1.6042E-01 | 1.0026E 00 | 3.8299E 01 | 2.5737E-01 | 1.5584E-01 | | |
| 45 | 7.9544E 02 | 1.2572E-03 | 6.2421E 00 | 1.6020E-01 | 1.0013E 00 | 3.8660E 01 | 2.3405E-01 | 1.3871E-01 | | |
| 50 | 1.6707E 03 | 5.9855E-04 | 6.2463E 00 | 1.6010E-01 | 1.0006E 00 | 3.8852E 01 | 2.1448E-01 | 1.2493E-01 | | |

# Table A1.1 (*Continued*)

Discrete compound interest = 20%

| | Single payment | | Uniform annual series | | | Uniform gradient series | Depreciation series | |
|---|---|---|---|---|---|---|---|---|
| | Compound-interest factor $(1+i)^n$ | Present-worth factor $\frac{1}{(1+i)^n}$ | Uncost present-worth factor $\frac{(1+i)^n-1}{i(1+i)^n}$ | Capital-recovery factor $\frac{i(1+i)^n}{(1+i)^n-1}$ | Capitalized-cost factor $\frac{(1+i)^n}{(1+i)^n-1}$ | Present-worth factor $\frac{F_{RP}-nF_{SP}}{i}$ | Sum-of-digits present-worth factor $\frac{n-F_{RP}}{0.5n(n+1)i}$ | Straight-line present-worth factor $\frac{1}{niF_{PK}}$ |
| | $P$ to $S$ | $S$ to $P$ | $R$ to $P$ | $P$ to $R$ | $P$ to $K$ | $G$ to $P$ | $SD$ to $P$ | $SL$ to $P$ |
| $n$ | $F_{PS}$ | $F_{SP}$ | $F_{RP}$ | $F_{PR}$ | $F_{PK}$ | $F_{GP}$ | $F_{SDP}$ | $F_{SLP}$ |
| 1 | 1.2000E 00 | 8.3333E-01 | 8.3333E-01 | 1.2000E 00 | 6.0000E 00 | ...... | 8.3333E-01 | 8.3333E-01 |
| 2 | 1.4400E 00 | 6.9444E-01 | 1.5278E 00 | 6.5455E-01 | 3.2727E 00 | 6.9444E-01 | 7.8704E-01 | 7.6389E-01 |
| 3 | 1.7280E 00 | 5.7870E-01 | 2.1065E 00 | 4.7473E-01 | 2.3736E 00 | 1.8519E 00 | 7.4460E-01 | 7.0216E-01 |
| 4 | 2.0736E 00 | 4.8225E-01 | 2.5887E 00 | 3.8629E-01 | 1.9313E 00 | 3.2986E 00 | 7.0563E-01 | 6.4718E-01 |
| 5 | 2.4883E 00 | 4.0188E-01 | 2.9906E 00 | 3.3438E-01 | 1.6719E 00 | 4.9061E 00 | 6.6980E-01 | 5.9812E-01 |
| 6 | 2.9860E 00 | 3.3490E-01 | 3.3255E 00 | 3.0071E-01 | 1.5035E 00 | 6.5806E 00 | 6.3678E-01 | 5.5425E-01 |
| 7 | 3.5832E 00 | 2.7908E-01 | 3.6046E 00 | 2.7742E-01 | 1.3871E 00 | 8.2551E 00 | 6.0632E-01 | 5.1494E-01 |
| 8 | 4.2998E 00 | 2.3257E-01 | 3.8372E 00 | 2.6061E-01 | 1.3030E 00 | 9.8831E 00 | 5.7717E-01 | 4.7964E-01 |
| 9 | 5.1598E 00 | 1.9381E-01 | 4.0310E 00 | 2.4808E-01 | 1.2404E 00 | 1.1434E 01 | 5.5211E-01 | 4.4789E-01 |
| 10 | 6.1917E 00 | 1.6151E-01 | 4.1925E 00 | 2.3852E-01 | 1.1926E 00 | 1.2887E 01 | 5.2796E-01 | 4.1925E-01 |
| 11 | 7.4301E 00 | 1.3459E-01 | 4.3271E 00 | 2.3110E-01 | 1.1555E 00 | 1.4233E 01 | 5.0534E-01 | 3.9337E-01 |
| 12 | 8.9161E 00 | 1.1216E-01 | 4.4392E 00 | 2.2526E-01 | 1.1263E 00 | 1.5467E 01 | 4.8467E-01 | 3.6993E-01 |
| 13 | 1.0699E 01 | 9.3464E-02 | 4.5327E 00 | 2.2062E-01 | 1.1031E 00 | 1.6588E 01 | 4.6524E-01 | 3.4867E-01 |
| 14 | 1.2839E 01 | 7.7887E-02 | 4.6106E 00 | 2.1689E-01 | 1.0845E 00 | 1.7601E 01 | 4.4712E-01 | 3.2933E-01 |
| 15 | 1.5407E 01 | 6.4905E-02 | 4.6755E 00 | 2.1388E-01 | 1.0694E 00 | 1.8509E 01 | 4.3019E-01 | 3.1170E-01 |
| 16 | 1.8488E 01 | 5.4088E-02 | 4.7296E 00 | 2.1144E-01 | 1.0572E 00 | 1.9321E 01 | 4.1435E-01 | 2.9560E-01 |
| 18 | 2.6623E 01 | 3.7561E-02 | 4.8122E 00 | 2.0781E-01 | 1.0390E 00 | 2.0680E 01 | 3.8561E-01 | 2.6734E-01 |
| 20 | 3.8338E 01 | 2.6084E-02 | 4.8696E 00 | 2.0536E-01 | 1.0268E 00 | 2.1739E 01 | 3.6025E-01 | 2.4348E-01 |
| 25 | 9.5396E 01 | 1.0483E-02 | 4.9476E 00 | 2.0212E-01 | 1.0106E 00 | 2.3428E 01 | 3.0080E-01 | 1.9790E-01 |
| 30 | 2.3737E 02 | 4.2127E-03 | 4.9789E 00 | 2.0085E-01 | 1.0042E 00 | 2.4263E 01 | 2.6904E-01 | 1.6596E-01 |
| 35 | 5.9067E 02 | 1.6930E-03 | 4.9915E 00 | 2.0034E-01 | 1.0017E 00 | 2.4661E 01 | 2.3816E-01 | 1.4262E-01 |
| 40 | 1.4698E 03 | 6.8034E-04 | 4.9966E 00 | 2.0014E-01 | 1.0007E 00 | 2.4847E 01 | 2.1344E-01 | 1.2491E-01 |
| 45 | 3.6573E 03 | 2.7343E-04 | 4.9986E 00 | 2.0005E-01 | 1.0003E 00 | 2.4932E 01 | 1.9324E-01 | 1.1108E-01 |
| 50 | 9.1004E 03 | 1.0988E-04 | 4.9995E 00 | 2.0002E-01 | 1.0001E 00 | 2.4970E 01 | 1.7847E-01 | 9.9989E-02 |

Discrete compound interest = 25%

| | Single payment | | Uniform annual series | | | Uniform gradient series | Depreciation series | |
| | Compound-interest factor | Present-worth factor | Unacost present-worth factor | Capital-recovery factor | Capitalized-cost factor | Present-worth factor | Sum-of-digits present-worth factor | Straight-line present-worth factor |
| | $(1+i)^n$ | $\dfrac{1}{(1+i)^n}$ | $\dfrac{(1+i)^n - 1}{i(1+i)^n}$ | $\dfrac{(1+i)^n}{(1+i)^n - 1}$ | $\dfrac{(1+i)^n}{(1+i)^n - 1}$ | $\dfrac{F_{RP} - nF_{SP}}{i}$ | $\dfrac{n - F_{RP}}{0.5n(n+1)i}$ | $\dfrac{1}{niF_{PK}}$ |
| | P to S | S to P | R to P | P to R | P to K | G to P | SD to P | SL to P |
| n | $F_{PS}$ | $F_{SP}$ | $F_{RP}$ | $F_{PR}$ | $F_{PK}$ | $F_{GP}$ | $F_{SDP}$ | $F_{SLP}$ |
|---|---|---|---|---|---|---|---|---|
| 1 | 1.2500E 00 | 8.0000E-01 | 8.0000E-01 | 1.2500E 00 | 5.0000E 00 | ........E-01 | 8.0000E-01 | 8.0000E-01 |
| 2 | 1.5625E 00 | 6.4000E-01 | 1.4400E 00 | 6.9444E-01 | 2.7778E 00 | 6.4000E-01 | 7.4667E-01 | 7.2000E-01 |
| 3 | 1.9531E 00 | 5.1200E-01 | 1.9520E 00 | 5.1230E-01 | 2.0492E 00 | 1.6640E 00 | 6.9867E-01 | 6.5067E-01 |
| 4 | 2.4414E 00 | 4.0960E-01 | 2.3616E 00 | 4.2344E-01 | 1.6938E 00 | 2.8928E 00 | 6.5536E-01 | 5.9040E-01 |
| 5 | 3.0518E 00 | 3.2768E-01 | 2.6893E 00 | 3.7185E-01 | 1.4874E 00 | 4.2035E 00 | 6.1619E-01 | 5.3786E-01 |
| 6 | 3.8147E 00 | 2.6214E-01 | 2.9514E 00 | 3.3882E-01 | 1.3553E 00 | 5.5142E 00 | 5.8068E-01 | 4.9190E-01 |
| 7 | 4.7684E 00 | 2.0972E-01 | 3.1611E 00 | 3.1634E-01 | 1.2654E 00 | 6.7725E 00 | 5.4841E-01 | 4.5159E-01 |
| 8 | 5.9605E 00 | 1.6777E-01 | 3.3289E 00 | 3.0040E-01 | 1.2016E 00 | 7.9469E 00 | 5.1901E-01 | 4.1611E-01 |
| 9 | 7.4506E 00 | 1.3422E-01 | 3.4631E 00 | 2.8876E-01 | 1.1550E 00 | 9.0207E 00 | 4.9217E-01 | 3.8479E-01 |
| 10 | 9.3132E 00 | 1.0737E-01 | 3.5705E 00 | 2.8007E-01 | 1.1203E 00 | 9.9870E 00 | 4.6760E-01 | 3.5705E-01 |
| 11 | 1.1642E 01 | 8.5899E-02 | 3.6564E 00 | 2.7349E-01 | 1.0940E 00 | 1.0846E 01 | 4.4507E-01 | 3.3240E-01 |
| 12 | 1.4552E 01 | 6.8719E-02 | 3.7251E 00 | 2.6845E-01 | 1.0738E 00 | 1.1602E 01 | 4.2435E-01 | 3.1043E-01 |
| 13 | 1.8190E 01 | 5.4976E-02 | 3.7801E 00 | 2.6454E-01 | 1.0582E 00 | 1.2262E 01 | 4.0527E-01 | 2.9078E-01 |
| 14 | 2.2737E 01 | 4.3980E-02 | 3.8241E 00 | 2.6150E-01 | 1.0460E 00 | 1.2833E 01 | 3.8765E-01 | 2.7315E-01 |
| 15 | 2.8422E 01 | 3.5184E-02 | 3.8593E 00 | 2.5912E-01 | 1.0365E 00 | 1.3326E 01 | 3.7136E-01 | 2.5728E-01 |
| 16 | 3.5527E 01 | 2.8147E-02 | 3.8874E 00 | 2.5724E-01 | 1.0290E 00 | 1.3748E 01 | 3.5625E-01 | 2.4296E-01 |
| 18 | 5.5511E 01 | 1.8014E-02 | 3.9279E 00 | 2.5459E-01 | 1.0183E 00 | 1.4415E 01 | 3.2917E-01 | 2.1822E-01 |
| 20 | 8.6736E 01 | 1.1529E-02 | 3.9539E 00 | 2.5292E-01 | 1.0117E 00 | 1.4893E 01 | 3.0564E-01 | 1.9769E-01 |
| 25 | 2.6470E 02 | 3.7779E-03 | 3.9849E 00 | 2.5095E-01 | 1.0038E 00 | 1.5562E 01 | 2.5865E-01 | 1.5940E-01 |
| 30 | 8.0779E 02 | 1.2379E-03 | 3.9950E 00 | 2.5031E-01 | 1.0012E 00 | 1.5832E 01 | 2.2370E-01 | 1.3317E-01 |
| 35 | 2.4652E 03 | 4.0565E-04 | 3.9984E 00 | 2.5010E-01 | 1.0004E 00 | 1.5937E 01 | 1.9684E-01 | 1.1424E-01 |
| 40 | 7.5232E 03 | 1.3292E-04 | 3.9995E 00 | 2.5003E-01 | 1.0001E 00 | 1.5977E 01 | 1.7561E-01 | 9.9987E-02 |
| 45 | 2.2959E 04 | 4.3566E-05 | 3.9998E 00 | 2.5001E-01 | 1.0000E 00 | 1.5991E 01 | 1.5845E-01 | 8.8885E-02 |
| 50 | 7.0065E 04 | 1.4272E-05 | 3.9999E 00 | 2.5000E-01 | 1.0000E 00 | 1.5997E 01 | 1.4431E-01 | 7.9999E-02 |

# Table A1.1 (*Continued*)

Discrete compound interest $= 30\%$

| | Single payment | | Uniform annual series | | Uniform gradient series | | Depreciation series | |
|---|---|---|---|---|---|---|---|---|
| | Compound-interest factor | Present-worth factor | Unacost present-worth factor | Capital-recovery factor | Capitalized-cost factor | Present-worth factor | Sum-of-digits present-worth factor | Straight-line present worth factor |
| | $(1+i)^n$ | $\dfrac{1}{(1+i)^n}$ | $\dfrac{(1+i)^n - 1}{i(1+i)^n}$ | $\dfrac{(1+i)^n}{(1+i)^n - 1}$ | $\dfrac{(1+i)^n}{(1+i)^n - 1}$ | $\dfrac{F_{RP} - nF_{SP}}{i}$ | $\dfrac{n - F_{RP}}{0.5n(n+1)i}$ | $\dfrac{1}{niF_{PK}}$ |
| | P to S | S to P | R to P | P to R | P to K | G to P | SD to P | SL to P |
| $n$ | $F_{PS}$ | $F_{SP}$ | $F_{RP}$ | $F_{PR}$ | $F_{PK}$ | $F_{GP}$ | $F_{SDP}$ | $F_{SLP}$ |
| 1 | 1.3000E 00 | 7.6923E-01 | 7.6923E-01 | 1.3000E 00 | 4.3333E 00 | | 7.6923E-01 | 7.6923E-01 |
| 2 | 1.6900E 00 | 5.9172E-01 | 1.3609E 00 | 7.3478E-01 | 2.4493E 00 | 5.9172E-01 | 7.1006E-01 | 6.8047E-01 |
| 3 | 2.1970E 00 | 4.5517E-01 | 1.8161E 00 | 5.5063E-01 | 1.8354E 00 | 1.5020E 00 | 6.5772E-01 | 6.0537E-01 |
| 4 | 2.8561E 00 | 3.5013E-01 | 2.1662E 00 | 4.6163E-01 | 1.5388E 00 | 2.5524E 00 | 6.1125E-01 | 5.4156E-01 |
| 5 | 3.7129E 00 | 2.6933E-01 | 2.4356E 00 | 4.1058E-01 | 1.3686E 00 | 3.6297E 00 | 5.6987E-01 | 4.8711E-01 |
| 6 | 4.8268E 00 | 2.0718E-01 | 2.6427E 00 | 3.7839E-01 | 1.2613E 00 | 4.6656E 00 | 5.3290E-01 | 4.4046E-01 |
| 7 | 6.2749E 00 | 1.5937E-01 | 2.8021E 00 | 3.5687E-01 | 1.1896E 00 | 5.6218E 00 | 4.9975E-01 | 4.0030E-01 |
| 8 | 8.1573E 00 | 1.2259E-01 | 2.9247E 00 | 3.4192E-01 | 1.1397E 00 | 6.4800E 00 | 4.6994E-01 | 3.6559E-01 |
| 9 | 1.0604E 01 | 9.4300E-02 | 3.0190E 00 | 3.3124E-01 | 1.1041E 00 | 7.2343E 00 | 4.4304E-01 | 3.3544E-01 |
| 10 | 1.3786E 01 | 7.2538E-02 | 3.0915E 00 | 3.2346E-01 | 1.0782E 00 | 7.8872E 00 | 4.1869E-01 | 3.0915E-01 |
| 11 | 1.7922E 01 | 5.5799E-02 | 3.1473E 00 | 3.1773E-01 | 1.0591E 00 | 8.4452E 00 | 3.9660E-01 | 2.8612E-01 |
| 12 | 2.3298E 01 | 4.2922E-02 | 3.1903E 00 | 3.1345E-01 | 1.0448E 00 | 8.9173E 00 | 3.7648E-01 | 2.6586E-01 |
| 13 | 3.0288E 01 | 3.3017E-02 | 3.2233E 00 | 3.1024E-01 | 1.0341E 00 | 9.3135E 00 | 3.5812E-01 | 2.4794E-01 |
| 14 | 3.9374E 01 | 2.5398E-02 | 3.2487E 00 | 3.0782E-01 | 1.0261E 00 | 9.6437E 00 | 3.4131E-01 | 2.3205E-01 |
| 15 | 5.1186E 01 | 1.9537E-02 | 3.2682E 00 | 3.0598E-01 | 1.0199E 00 | 9.9172E 00 | 3.2588E-01 | 2.1788E-01 |
| 16 | 6.6542E 01 | 1.5028E-02 | 3.2832E 00 | 3.0456E-01 | 1.0153E 00 | 1.0143E 01 | 3.1169E-01 | 2.0520E-01 |
| 18 | 1.1246E 02 | 8.8924E-03 | 3.3037E 00 | 3.0269E-01 | 1.0090E 00 | 1.0479E 01 | 2.8648E-01 | 1.8354E-01 |
| 20 | 1.9005E 02 | 5.2614E-03 | 3.3158E 00 | 3.0159E-01 | 1.0053E 00 | 1.0702E 01 | 2.6483E-01 | 1.6579E-01 |
| 25 | 7.0564E 02 | 1.4172E-03 | 3.3286E 00 | 3.0043E-01 | 1.0014E 00 | 1.0977E 01 | 2.2227E-01 | 1.3314E-01 |
| 30 | 2.6200E 03 | 3.8164E-04 | 3.3321E 00 | 3.0011E-01 | 1.0004E 00 | 1.1069E 01 | 1.9117E-01 | 1.1107E-01 |
| 35 | 9.7279E 03 | 1.0280E-04 | 3.3330E 00 | 3.0003E-01 | 1.0001E 00 | 1.1098E 01 | 1.6755E-01 | 9.5228E-02 |
| 40 | 3.6119E 04 | 2.7686E-05 | 3.3332E 00 | 3.0001E-01 | 1.0000E 00 | 1.1107E 01 | 1.4905E-01 | 8.3331E-02 |
| 45 | 1.3411E 05 | 7.4567E-06 | 3.3333E 00 | 3.0000E-01 | 1.0000E 00 | 1.1110E 01 | 1.3419E-01 | 7.4074E-02 |
| 50 | 4.9793E 05 | 2.0083E-06 | 3.3333E 00 | 3.0000E-01 | 1.0000E 00 | 1.1111E 01 | 1.2200E-01 | 6.6667E-02 |

## Discrete compound interest = 40%

| | Single payment | | Uniform annual series | | | Uniform gradient series | Depreciation series | |
|---|---|---|---|---|---|---|---|---|
| | Compound-interest factor | Present-worth factor | Unacost present-worth factor | Capital-recovery factor | Capitalized-cost factor | Present-worth factor | Sum-of-digits present-worth factor | Straight-line present-worth factor |
| | $(1+i)^n$ | $\dfrac{1}{(1+i)^n}$ | $\dfrac{(1+i)^n-1}{i(1+i)^n}$ | $\dfrac{(1+i)^n}{(1+i)^n-1}$ | $\dfrac{(1+i)^n}{(1+i)^n-1}$ | $\dfrac{F_{RP}-nF_{SP}}{i}$ | $\dfrac{n-F_{RP}}{0.5n(n+1)i}$ | $\dfrac{1}{niF_{PK}}$ |
| | P to S | S to P | R to P | P to R | P to K | G to P | SD to P | SL to P |
| $n$ | $F_{PS}$ | $F_{SP}$ | $F_{RP}$ | $F_{PR}$ | $F_{PK}$ | $F_{GP}$ | $F_{SDP}$ | $F_{SLP}$ |
| 1 | 1.4000E 00 | 7.1429E-01 | 7.1429E-01 | 1.4000E 00 | 3.5000E 00 | | 7.1429E-01 | 7.1429E-01 |
| 2 | 1.9600E 00 | 5.1020E-01 | 1.2245E 00 | 8.1667E-01 | 2.0417E 00 | 5.1020E-01 | 6.4626E-01 | 6.1224E-01 |
| 3 | 2.7440E 00 | 3.6443E-01 | 1.5889E 00 | 6.2936E-01 | 1.5734E 00 | 1.2391E 00 | 5.8795E-01 | 5.2964E-01 |
| 4 | 3.8416E 00 | 2.6031E-01 | 1.8492E 00 | 5.4077E-01 | 1.3519E 00 | 2.0200E 00 | 5.3769E-01 | 4.6231E-01 |
| 5 | 5.3782E 00 | 1.8593E-01 | 2.0352E 00 | 4.9136E-01 | 1.2284E 00 | 2.7637E 00 | 4.9414E-01 | 4.0703E-01 |
| 6 | 7.5295E 00 | 1.3281E-01 | 2.1680E 00 | 4.6126E-01 | 1.1532E 00 | 3.4278E 00 | 4.5619E-01 | 3.6133E-01 |
| 7 | 1.0541E 01 | 9.4865E-02 | 2.2628E 00 | 4.4192E-01 | 1.1048E 00 | 3.9970E 00 | 4.2296E-01 | 3.2326E-01 |
| 8 | 1.4758E 01 | 6.7760E-02 | 2.3306E 00 | 4.2907E-01 | 1.0727E 00 | 4.4713E 00 | 3.9371E-01 | 2.9132E-01 |
| 9 | 2.0661E 01 | 4.8400E-02 | 2.3790E 00 | 4.2034E-01 | 1.0509E 00 | 4.8585E 00 | 3.6783E-01 | 2.6433E-01 |
| 10 | 2.8925E 01 | 3.4572E-02 | 2.4136E 00 | 4.1432E-01 | 1.0358E 00 | 5.1696E 00 | 3.4484E-01 | 2.4136E-01 |
| 11 | 4.0496E 01 | 2.4694E-02 | 2.4383E 00 | 4.1013E-01 | 1.0253E 00 | 5.4166E 00 | 3.2431E-01 | 2.2166E-01 |
| 12 | 5.6694E 01 | 1.7639E-02 | 2.4559E 00 | 4.0718E-01 | 1.0180E 00 | 5.6106E 00 | 3.0590E-01 | 2.0466E-01 |
| 13 | 7.9371E 01 | 1.2599E-02 | 2.4685E 00 | 4.0510E-01 | 1.0128E 00 | 5.7618E 00 | 2.8933E-01 | 1.8988E-01 |
| 14 | 1.1112E 02 | 8.9993E-03 | 2.4775E 00 | 4.0363E-01 | 1.0091E 00 | 5.8788E 00 | 2.7435E-01 | 1.7696E-01 |
| 15 | 1.5557E 02 | 6.4281E-03 | 2.4839E 00 | 4.0259E-01 | 1.0065E 00 | 5.9688E 00 | 2.6075E-01 | 1.6560E-01 |
| 16 | 2.1780E 02 | 4.5915E-03 | 2.4885E 00 | 4.0185E-01 | 1.0046E 00 | 6.0376E 00 | 2.4837E-01 | 1.5553E-01 |
| 18 | 4.2688E 02 | 2.3426E-03 | 2.4941E 00 | 4.0094E-01 | 1.0023E 00 | 6.1299E 00 | 2.2669E-01 | 1.3856E-01 |
| 20 | 8.3668E 02 | 1.1952E-03 | 2.4970E 00 | 4.0048E-01 | 1.0012E 00 | 6.1828E 00 | 2.0837E-01 | 1.2485E-01 |
| 25 | 4.4999E 03 | 2.2223E-04 | 2.4994E 00 | 4.0009E-01 | 1.0002E 00 | 6.2347E 00 | 1.7308E-01 | 9.9978E-02 |
| 30 | 2.4201E 04 | 4.1320E-05 | 2.4999E 00 | 4.0002E-01 | 1.0000E 00 | 6.2466E 00 | 1.4765E-01 | 8.3330E-02 |
| 35 | 1.3016E 05 | 7.6824E-06 | 2.5000E 00 | 4.0000E-01 | 1.0000E 00 | 6.2493E 00 | 1.2897E-01 | 7.1428E-02 |
| 40 | 7.0004E 05 | 1.4285E-06 | 2.5000E 00 | 4.0000E-01 | 1.0000E 00 | 6.2498E 00 | 1.1433E-01 | 6.2500E-02 |
| 45 | 3.7650E 06 | 2.6561E-07 | 2.5000E 00 | 4.0000E-01 | 1.0000E 00 | 6.2500E 00 | 1.0266E-01 | 5.5556E-02 |
| 50 | 2.0249E 07 | 4.9385E-08 | 2.5000E 00 | 4.0000E-01 | 1.0000E 00 | 6.2500E 00 | 9.3137E-02 | 5.0000E-02 |

**Table A1.1** (*Continued*)

Discrete compound interest = 50%

| | Single payment | | Uniform annual series | | | Uniform gradient series | Depreciation series | |
|---|---|---|---|---|---|---|---|---|
| | Compound-interest factor $(1+i)^n$ | Present-worth factor $\dfrac{1}{(1+i)^n}$ | Unacost present-worth factor $\dfrac{(1+i)^n-1}{i(1+i)^n}$ | Capital-recovery factor $\dfrac{(1+i)^n}{(1+i)^n-1}$ | Capitalized-cost factor $\dfrac{(1+i)^n}{(1+i)^n-1}$ | Present-worth factor $\dfrac{F_{RP}-nF_{SP}}{i}$ | Sum-of-digits present-worth factor $\dfrac{n-F_{RP}}{0.5n(n+1)i}$ | Straight-line present worth factor $\dfrac{1}{niF_{PK}}$ |
| | P to S | S to P | R to P | P to R | P to K | G to P | SD to P | SL to P |
| $n$ | $F_{PS}$ | $F_{SP}$ | $F_{RP}$ | $F_{PR}$ | $F_{PK}$ | $F_{GP}$ | $F_{SDP}$ | $F_{SLP}$ |
| 1 | 1.5000E 00 | 6.6667E-01 | 6.6667E-01 | 1.5000E 00 | 3.0000E 00 | | 6.6667E-01 | 6.6667E-01 |
| 2 | 2.2500E 00 | 4.4444E-01 | 1.1111E 00 | 9.0000E-01 | 1.8000E 00 | 4.4444E-01 | 5.9259E-01 | 5.5556E-01 |
| 3 | 3.3750E 00 | 2.9630E-01 | 1.4074E 00 | 7.1053E-01 | 1.4211E 00 | 1.0370E 00 | 5.3086E-01 | 4.6914E-01 |
| 4 | 5.0625E 00 | 1.9753E-01 | 1.6049E 00 | 6.2308E-01 | 1.2462E 00 | 1.6296E 00 | 4.7901E-01 | 4.0123E-01 |
| 5 | 7.5938E 00 | 1.3169E-01 | 1.7366E 00 | 5.7583E-01 | 1.1517E 00 | 2.1156E 00 | 4.3512E-01 | 3.4733E-01 |
| 6 | 1.1391E 01 | 8.7791E-02 | 1.8244E 00 | 5.4812E-01 | 1.0962E 00 | 2.5953E 00 | 3.9767E-01 | 3.0407E-01 |
| 7 | 1.7086E 01 | 5.8528E-02 | 1.8829E 00 | 5.3108E-01 | 1.0622E 00 | 2.9465E 00 | 3.6550E-01 | 2.6899E-01 |
| 8 | 2.5629E 01 | 3.9019E-02 | 1.9220E 00 | 5.2030E-01 | 1.0406E 00 | 3.2196E 00 | 3.3767E-01 | 2.4025E-01 |
| 9 | 3.8443E 01 | 2.6012E-02 | 1.9480E 00 | 5.1335E-01 | 1.0267E 00 | 3.4277E 00 | 3.1342E-01 | 2.1644E-01 |
| 10 | 5.7665E 01 | 1.7342E-02 | 1.9653E 00 | 5.0882E-01 | 1.0176E 00 | 3.5838E 00 | 2.9217E-01 | 1.9653E-01 |
| 11 | 8.6498E 01 | 1.1561E-02 | 1.9769E 00 | 5.0585E-01 | 1.0117E 00 | 3.6994E 00 | 2.7343E-01 | 1.7972E-01 |
| 12 | 1.2975E 02 | 7.7073E-03 | 1.9846E 00 | 5.0388E-01 | 1.0078E 00 | 3.7842E 00 | 2.5681E-01 | 1.6538E-01 |
| 13 | 1.9462E 02 | 5.1382E-03 | 1.9897E 00 | 5.0258E-01 | 1.0052E 00 | 3.8459E 00 | 2.4198E-01 | 1.5306E-01 |
| 14 | 2.9193E 02 | 3.4255E-03 | 1.9931E 00 | 5.0172E-01 | 1.0034E 00 | 3.8904E 00 | 2.2870E-01 | 1.4237E-01 |
| 15 | 4.3789E 02 | 2.2837E-03 | 1.9954E 00 | 5.0114E-01 | 1.0023E 00 | 3.9224E 00 | 2.1674E-01 | 1.3303E-01 |
| 16 | 6.5684E 02 | 1.5224E-03 | 1.9970E 00 | 5.0076E-01 | 1.0052E 00 | 3.9452E 00 | 2.0593E-01 | 1.2481E-01 |
| 18 | 1.4779E 03 | 6.7664E-04 | 1.9986E 00 | 5.0034E-01 | 1.0007E 00 | 3.9729E 00 | 1.8715E-01 | 1.1104E-01 |
| 20 | 3.3253E 03 | 3.0073E-04 | 1.9994E 00 | 5.0015E-01 | 1.0003E 00 | 3.9868E 00 | 1.7143E-01 | 9.9970E-02 |
| 25 | 2.5251E 04 | 3.9602E-05 | 1.9999E 00 | 5.0002E-01 | 1.0000E 00 | 3.9979E 00 | 1.4159E-01 | 7.9997E-02 |
| 30 | 1.9175E 05 | 5.2151E-06 | 2.0000E 00 | 5.0000E-01 | 1.0000E 00 | 3.9997E 00 | 1.2043E-01 | 6.6666E-02 |
| 35 | 1.4561E 06 | 6.8676E-07 | 2.0000E 00 | 5.0000E-01 | 1.0000E 00 | 3.9999E 00 | 1.0476E-01 | 5.7143E-02 |
| 40 | 1.1057E 07 | 9.0438E-08 | 2.0000E 00 | 5.0000E-01 | 1.0000E 00 | 4.0000E 00 | 9.2683E-02 | 5.0000E-02 |
| 45 | 8.3967E 07 | 1.1909E-08 | 2.0000E 00 | 5.0000E-01 | 1.0000E 00 | 4.0000E 00 | 8.3092E-02 | 4.4444E-02 |
| 50 | 6.3762E 08 | 1.5683E-09 | 2.0000E 00 | 5.0000E-01 | 1.0000E 00 | 4.0000E 00 | 7.5294E-02 | 4.0000E-02 |

# APPENDIX
# 2

## CONTINUOUS
## COMPOUND
## INTEREST

## Table A2.1 Continuous compound interest: Value of

$$F_{PS,\bar{i},n} = e^{in}$$

| $in \rightarrow$ $n \downarrow$ | 0.00 | 0.01 | 0.02 | 0.03 | 0.04 | 0.05 | 0.06 | 0.07 | 0.08 | 0.09 |
|---|---|---|---|---|---|---|---|---|---|---|
| 0 | 1.0000E 00 | 1.0101E 00 | 1.0202E 00 | 1.0305E 00 | 1.0408E 00 | 1.0513E 00 | 1.0618E 00 | 1.0725E 00 | 1.0833E 00 | 1.0942E 00 |
| 0.1 | 1.1052E 00 | 1.1163E 00 | 1.1275E 00 | 1.1388E 00 | 1.1503E 00 | 1.1618E 00 | 1.1735E 00 | 1.1853E 00 | 1.1972E 00 | 1.2092E 00 |
| 0.2 | 1.2214E 00 | 1.2337E 00 | 1.2461E 00 | 1.2586E 00 | 1.2712E 00 | 1.2840E 00 | 1.2969E 00 | 1.3100E 00 | 1.3231E 00 | 1.3364E 00 |
| 0.3 | 1.3499E 00 | 1.3634E 00 | 1.3771E 00 | 1.3910E 00 | 1.4049E 00 | 1.4191E 00 | 1.4333E 00 | 1.4477E 00 | 1.4623E 00 | 1.4770E 00 |
| 0.4 | 1.4918E 00 | 1.5068E 00 | 1.5220E 00 | 1.5373E 00 | 1.5527E 00 | 1.5683E 00 | 1.5841E 00 | 1.6000E 00 | 1.6161E 00 | 1.6323E 00 |
| 0.5 | 1.6487E 00 | 1.6653E 00 | 1.6820E 00 | 1.6989E 00 | 1.7160E 00 | 1.7333E 00 | 1.7507E 00 | 1.7683E 00 | 1.7860E 00 | 1.8040E 00 |
| 0.6 | 1.8221E 00 | 1.8404E 00 | 1.8589E 00 | 1.8776E 00 | 1.8965E 00 | 1.9155E 00 | 1.9348E 00 | 1.9542E 00 | 1.9739E 00 | 1.9937E 00 |
| 0.7 | 2.0138E 00 | 2.0340E 00 | 2.0544E 00 | 2.0751E 00 | 2.0959E 00 | 2.1170E 00 | 2.1383E 00 | 2.1598E 00 | 2.1815E 00 | 2.2034E 00 |
| 0.8 | 2.2255E 00 | 2.2479E 00 | 2.2705E 00 | 2.2933E 00 | 2.3164E 00 | 2.3396E 00 | 2.3632E 00 | 2.3869E 00 | 2.4109E 00 | 2.4351E 00 |
| 0.9 | 2.4596E 00 | 2.4843E 00 | 2.5093E 00 | 2.5345E 00 | 2.5600E 00 | 2.5857E 00 | 2.6117E 00 | 2.6379E 00 | 2.6645E 00 | 2.6912E 00 |
| 1.0 | 2.7183E 00 | 2.7456E 00 | 2.7732E 00 | 2.8011E 00 | 2.8292E 00 | 2.8577E 00 | 2.8864E 00 | 2.9154E 00 | 2.9447E 00 | 2.9743E 00 |
| 1.1 | 3.0042E 00 | 3.0344E 00 | 3.0649E 00 | 3.0957E 00 | 3.1268E 00 | 3.1582E 00 | 3.1899E 00 | 3.2220E 00 | 3.2544E 00 | 3.2871E 00 |
| 1.2 | 3.3201E 00 | 3.3535E 00 | 3.3872E 00 | 3.4212E 00 | 3.4556E 00 | 3.4903E 00 | 3.5254E 00 | 3.5609E 00 | 3.5966E 00 | 3.6328E 00 |
| 1.3 | 3.6693E 00 | 3.7062E 00 | 3.7434E 00 | 3.7810E 00 | 3.8190E 00 | 3.8574E 00 | 3.8962E 00 | 3.9354E 00 | 3.9749E 00 | 4.0149E 00 |
| 1.4 | 4.0552E 00 | 4.0960E 00 | 4.1371E 00 | 4.1787E 00 | 4.2207E 00 | 4.2631E 00 | 4.3060E 00 | 4.3492E 00 | 4.3929E 00 | 4.4371E 00 |
| 1.5 | 4.4817E 00 | 4.5267E 00 | 4.5722E 00 | 4.6182E 00 | 4.6646E 00 | 4.7115E 00 | 4.7588E 00 | 4.8066E 00 | 4.8550E 00 | 4.9037E 00 |
| 1.6 | 4.9530E 00 | 5.0028E 00 | 5.0531E 00 | 5.1039E 00 | 5.1552E 00 | 5.2070E 00 | 5.2593E 00 | 5.3122E 00 | 5.3656E 00 | 5.4195E 00 |
| 1.7 | 5.4739E 00 | 5.5290E 00 | 5.5845E 00 | 5.6407E 00 | 5.6973E 00 | 5.7546E 00 | 5.8124E 00 | 5.8709E 00 | 5.9299E 00 | 5.9895E 00 |
| 1.8 | 6.0496E 00 | 6.1104E 00 | 6.1719E 00 | 6.2339E 00 | 6.2965E 00 | 6.3598E 00 | 6.4237E 00 | 6.4883E 00 | 6.5535E 00 | 6.6194E 00 |
| 1.9 | 6.6859E 00 | 6.7531E 00 | 6.8210E 00 | 6.8895E 00 | 6.9588E 00 | 7.0287E 00 | 7.0993E 00 | 7.1707E 00 | 7.2427E 00 | 7.3155E 00 |
| 2.0 | 7.3891E 00 | 7.4633E 00 | 7.5383E 00 | 7.6141E 00 | 7.6906E 00 | 7.7679E 00 | 7.8460E 00 | 7.9248E 00 | 8.0045E 00 | 8.0849E 00 |

$F_{PS,\bar{i},n} = e^{in}$

| in → <br> ↓ | 0.00 | 0.01 | 0.02 | 0.03 | 0.04 | 0.05 | 0.06 | 0.07 | 0.08 | 0.09 |
|---|---|---|---|---|---|---|---|---|---|---|
| 2.1 | 8.1662E 00 | 8.2482E 00 | 8.3311E 00 | 8.4149E 00 | 8.4994E 00 | 8.5849E 00 | 8.6711E 00 | 8.7583E 00 | 8.8463E 00 | 8.9352E 00 |
| 2.2 | 9.0250E 00 | 9.1157E 00 | 9.2073E 00 | 9.2999E 00 | 9.3933E 00 | 9.4877E 00 | 9.5831E 00 | 9.6794E 00 | 9.7767E 00 | 9.8749E 00 |
| 2.3 | 9.9742E 00 | 1.0074E 01 | 1.0176E 01 | 1.0278E 01 | 1.0381E 01 | 1.0486E 01 | 1.0591E 01 | 1.0697E 01 | 1.0805E 01 | 1.0913E 01 |
| 2.4 | 1.1023E 01 | 1.1134E 01 | 1.1246E 01 | 1.1359E 01 | 1.1473E 01 | 1.1588E 01 | 1.1705E 01 | 1.1822E 01 | 1.1941E 01 | 1.2061E 01 |
| 2.5 | 1.2182E 01 | 1.2305E 01 | 1.2429E 01 | 1.2554E 01 | 1.2680E 01 | 1.2807E 01 | 1.2936E 01 | 1.3066E 01 | 1.3197E 01 | 1.3330E 01 |
| 2.6 | 1.3464E 01 | 1.3599E 01 | 1.3736E 01 | 1.3874E 01 | 1.4013E 01 | 1.4154E 01 | 1.4296E 01 | 1.4440E 01 | 1.4585E 01 | 1.4732E 01 |
| 2.7 | 1.4880E 01 | 1.5029E 01 | 1.5180E 01 | 1.5333E 01 | 1.5487E 01 | 1.5643E 01 | 1.5800E 01 | 1.5959E 01 | 1.6119E 01 | 1.6281E 01 |
| 2.8 | 1.6445E 01 | 1.6610E 01 | 1.6777E 01 | 1.6945E 01 | 1.7116E 01 | 1.7288E 01 | 1.7462E 01 | 1.7637E 01 | 1.7814E 01 | 1.7993E 01 |
| 2.9 | 1.8174E 01 | 1.8357E 01 | 1.8541E 01 | 1.8728E 01 | 1.8916E 01 | 1.9106E 01 | 1.9299E 01 | 1.9492E 01 | 1.9688E 01 | 1.9886E 01 |
| 3.0 | 2.0086E 01 | 2.0287E 01 | 2.0491E 01 | 2.0697E 01 | 2.0905E 01 | 2.1115E 01 | 2.1328E 01 | 2.1542E 01 | 2.1758E 01 | 2.1977E 01 |
| 3.1 | 2.2198E 01 | 2.2421E 01 | 2.2646E 01 | 2.2874E 01 | 2.3104E 01 | 2.3336E 01 | 2.3571E 01 | 2.3807E 01 | 2.4047E 01 | 2.4288E 01 |
| 3.2 | 2.4533E 01 | 2.4779E 01 | 2.5028E 01 | 2.5280E 01 | 2.5534E 01 | 2.5790E 01 | 2.6050E 01 | 2.6311E 01 | 2.6576E 01 | 2.6843E 01 |
| 3.3 | 2.7113E 01 | 2.7385E 01 | 2.7660E 01 | 2.7938E 01 | 2.8219E 01 | 2.8503E 01 | 2.8789E 01 | 2.9079E 01 | 2.9371E 01 | 2.9666E 01 |
| 3.4 | 2.9964E 01 | 3.0265E 01 | 3.0569E 01 | 3.0877E 01 | 3.1187E 01 | 3.1500E 01 | 3.1817E 01 | 3.2137E 01 | 3.2460E 01 | 3.2786E 01 |
| 3.5 | 3.3115E 01 | 3.3448E 01 | 3.3784E 01 | 3.4124E 01 | 3.4467E 01 | 3.4813E 01 | 3.5163E 01 | 3.5517E 01 | 3.5874E 01 | 3.6234E 01 |
| 3.6 | 3.6598E 01 | 3.6966E 01 | 3.7338E 01 | 3.7713E 01 | 3.8092E 01 | 3.8475E 01 | 3.8861E 01 | 3.9252E 01 | 3.9646E 01 | 4.0045E 01 |
| 3.7 | 4.0447E 01 | 4.0854E 01 | 4.1264E 01 | 4.1679E 01 | 4.2098E 01 | 4.2521E 01 | 4.2948E 01 | 4.3380E 01 | 4.3816E 01 | 4.4256E 01 |
| 3.8 | 4.4701E 01 | 4.5150E 01 | 4.5604E 01 | 4.6063E 01 | 4.6525E 01 | 4.6993E 01 | 4.7465E 01 | 4.7942E 01 | 4.8424E 01 | 4.8911E 01 |
| 3.9 | 4.9402E 01 | 4.9899E 01 | 5.0400E 01 | 5.0907E 01 | 5.1419E 01 | 5.1935E 01 | 5.2457E 01 | 5.2985E 01 | 5.3517E 01 | 5.4055E 01 |
| 4.0 | 5.4598E 01 | 5.5147E 01 | 5.5701E 01 | 5.6261E 01 | 5.6826E 01 | 5.7397E 01 | 5.7974E 01 | 5.8557E 01 | 5.9145E 01 | 5.9740E 01 |
| 4.1 | 6.0340E 01 | 6.0947E 01 | 6.1559E 01 | 6.2178E 01 | 6.2803E 01 | 6.3434E 01 | 6.4072E 01 | 6.4715E 01 | 6.5366E 01 | 6.6023E 01 |

## Table A2.1 (Continued)

$F_{PS,i,n} = e^{in}$

| $in \rightarrow$ <br> $i,n \downarrow$ | 0.00 | 0.01 | 0.02 | 0.03 | 0.04 | 0.05 | 0.06 | 0.07 | 0.08 | 0.09 |
|---|---|---|---|---|---|---|---|---|---|---|
| 4.2 | 6.6686E 01 | 6.7357E 01 | 6.8033E 01 | 6.8717E 01 | 6.9408E 01 | 7.0105E 01 | 7.0810E 01 | 7.1522E 01 | 7.2240E 01 | 7.2966E 01 |
| 4.3 | 7.3700E 01 | 7.4440E 01 | 7.5189E 01 | 7.5944E 01 | 7.6708E 01 | 7.7478E 01 | 7.8257E 01 | 7.9044E 01 | 7.9838E 01 | 8.0640E 01 |
| 4.4 | 8.1451E 01 | 8.2269E 01 | 8.3096E 01 | 8.3931E 01 | 8.4775E 01 | 8.5627E 01 | 8.6488E 01 | 8.7357E 01 | 8.8235E 01 | 8.9121E 01 |
| 4.5 | 9.0017E 01 | 9.0922E 01 | 9.1836E 01 | 9.2759E 01 | 9.3691E 01 | 9.4632E 01 | 9.5583E 01 | 9.6544E 01 | 9.7514E 01 | 9.8494E 01 |
| 4.6 | 9.9484E 01 | 1.0048E 02 | 1.0149E 02 | 1.0251E 02 | 1.0354E 02 | 1.0458E 02 | 1.0564E 02 | 1.0670E 02 | 1.0777E 02 | 1.0885E 02 |
| 4.7 | 1.0995E 02 | 1.1105E 02 | 1.1217E 02 | 1.1330E 02 | 1.1443E 02 | 1.1558E 02 | 1.1675E 02 | 1.1792E 02 | 1.1910E 02 | 1.2030E 02 |
| 4.8 | 1.2151E 02 | 1.2273E 02 | 1.2397E 02 | 1.2521E 02 | 1.2647E 02 | 1.2774E 02 | 1.2902E 02 | 1.3032E 02 | 1.3163E 02 | 1.3295E 02 |
| 4.9 | 1.3429E 02 | 1.3564E 02 | 1.3700E 02 | 1.3838E 02 | 1.3977E 02 | 1.4117E 02 | 1.4259E 02 | 1.4403E 02 | 1.4547E 02 | 1.4694E 02 |
|  | .0 | .1 | .2 | .3 | .4 | .5 | .6 | .7 | .8 | .9 |
| 5.0 | 1.4841E 02 | 1.6440E 02 | 1.8127E 02 | 2.0034E 02 | 2.2141E 02 | 2.4469E 02 | 2.7043E 02 | 2.9887E 02 | 3.3030E 02 | 3.6504E 02 |
| 6.0 | 4.0343E 02 | 4.4586E 02 | 4.9275E 02 | 5.4457E 02 | 6.0185E 02 | 6.6514E 02 | 7.3510E 02 | 8.1241E 02 | 8.9785E 02 | 9.9227E 02 |
| 7.0 | 1.0966E 03 | 1.2120E 03 | 1.3394E 03 | 1.4803E 03 | 1.6360E 03 | 1.8080E 03 | 1.9982E 03 | 2.2083E 03 | 2.4406E 03 | 2.6973E 03 |
| 8.0 | 2.9810E 03 | 3.2945E 03 | 3.6410E 03 | 4.0239E 03 | 4.4471E 03 | 4.9148E 03 | 5.4317E 03 | 6.0029E 03 | 6.6342E 03 | 7.3320E 03 |
| 9.0 | 8.1031E 03 | 8.9553E 03 | 9.8971E 03 | 1.0938E 04 | 1.2088E 04 | 1.3360E 04 | 1.4765E 04 | 1.6318E 04 | 1.8034E 04 | 1.9930E 04 |
| 10.0 | 2.2026E 04 | 2.4343E 04 | 2.6903E 04 | 2.9733E 04 | 3.2860E 04 | 3.6316E 04 | 4.0135E 04 | 4.4356E 04 | 4.9021E 04 | 5.4176E 04 |

**Table A2.2 Continuous compound interest: Value of**

$$F_{SP,\bar{i},n} = e^{-in}$$

| in → ↓ | 0.00 | 0.01 | 0.02 | 0.03 | 0.04 | 0.05 | 0.06 | 0.07 | 0.08 | 0.09 |
|---|---|---|---|---|---|---|---|---|---|---|
| 0 | 1.0000E 00 | 9.9005E-01 | 9.8020E-01 | 9.7045E-01 | 9.6079E-01 | 9.5123E-01 | 9.4176E-01 | 9.3239E-01 | 9.2312E-01 | 9.1393E-01 |
| 0.1 | 9.0484E-01 | 8.9583E-01 | 8.8692E-01 | 8.7810E-01 | 8.6936E-01 | 8.6071E-01 | 8.5214E-01 | 8.4366E-01 | 8.3527E-01 | 8.2696E-01 |
| 0.2 | 8.1873E-01 | 8.1058E-01 | 8.0252E-01 | 7.9453E-01 | 7.8663E-01 | 7.7880E-01 | 7.7105E-01 | 7.6338E-01 | 7.5578E-01 | 7.4826E-01 |
| 0.3 | 7.4082E-01 | 7.3345E-01 | 7.2615E-01 | 7.1892E-01 | 7.1177E-01 | 7.0469E-01 | 6.9768E-01 | 6.9073E-01 | 6.8386E-01 | 6.7706E-01 |
| 0.4 | 6.7032E-01 | 6.6365E-01 | 6.5705E-01 | 6.5051E-01 | 6.4404E-01 | 6.3763E-01 | 6.3128E-01 | 6.2500E-01 | 6.1878E-01 | 6.1263E-01 |
| 0.5 | 6.0653E-01 | 6.0050E-01 | 5.9452E-01 | 5.8860E-01 | 5.8275E-01 | 5.7695E-01 | 5.7121E-01 | 5.6553E-01 | 5.5990E-01 | 5.5433E-01 |
| 0.6 | 5.4881E-01 | 5.4335E-01 | 5.3794E-01 | 5.3259E-01 | 5.2729E-01 | 5.2205E-01 | 5.1685E-01 | 5.1171E-01 | 5.0662E-01 | 5.0158E-01 |
| 0.7 | 4.9659E-01 | 4.9164E-01 | 4.8675E-01 | 4.8191E-01 | 4.7711E-01 | 4.7237E-01 | 4.6767E-01 | 4.6301E-01 | 4.5841E-01 | 4.5384E-01 |
| 0.8 | 4.4933E-01 | 4.4486E-01 | 4.4043E-01 | 4.3605E-01 | 4.3171E-01 | 4.2741E-01 | 4.2316E-01 | 4.1895E-01 | 4.1478E-01 | 4.1066E-01 |
| 0.9 | 4.0657E-01 | 4.0252E-01 | 3.9852E-01 | 3.9455E-01 | 3.9063E-01 | 3.8674E-01 | 3.8289E-01 | 3.7908E-01 | 3.7531E-01 | 3.7158E-01 |
| 1.0 | 3.6788E-01 | 3.6422E-01 | 3.6059E-01 | 3.5701E-01 | 3.5345E-01 | 3.4994E-01 | 3.4646E-01 | 3.4301E-01 | 3.3960E-01 | 3.3622E-01 |
| 1.1 | 3.3287E-01 | 3.2956E-01 | 3.2628E-01 | 3.2303E-01 | 3.1982E-01 | 3.1664E-01 | 3.1349E-01 | 3.1037E-01 | 3.0728E-01 | 3.0422E-01 |
| 1.2 | 3.0119E-01 | 2.9820E-01 | 2.9523E-01 | 2.9229E-01 | 2.8938E-01 | 2.8650E-01 | 2.8365E-01 | 2.8083E-01 | 2.7804E-01 | 2.7527E-01 |
| 1.3 | 2.7253E-01 | 2.6982E-01 | 2.6714E-01 | 2.6448E-01 | 2.6185E-01 | 2.5924E-01 | 2.5666E-01 | 2.5411E-01 | 2.5158E-01 | 2.4908E-01 |
| 1.4 | 2.4660E-01 | 2.4414E-01 | 2.4171E-01 | 2.3931E-01 | 2.3693E-01 | 2.3457E-01 | 2.3224E-01 | 2.2993E-01 | 2.2764E-01 | 2.2537E-01 |
| 1.5 | 2.2313E-01 | 2.2091E-01 | 2.1871E-01 | 2.1654E-01 | 2.1438E-01 | 2.1225E-01 | 2.1014E-01 | 2.0805E-01 | 2.0598E-01 | 2.0393E-01 |
| 1.6 | 2.0190E-01 | 1.9989E-01 | 1.9790E-01 | 1.9593E-01 | 1.9398E-01 | 1.9205E-01 | 1.9014E-01 | 1.8825E-01 | 1.8637E-01 | 1.8452E-01 |
| 1.7 | 1.8268E-01 | 1.8087E-01 | 1.7907E-01 | 1.7728E-01 | 1.7552E-01 | 1.7377E-01 | 1.7204E-01 | 1.7033E-01 | 1.6864E-01 | 1.6696E-01 |
| 1.8 | 1.6530E-01 | 1.6365E-01 | 1.6203E-01 | 1.6041E-01 | 1.5882E-01 | 1.5724E-01 | 1.5567E-01 | 1.5412E-01 | 1.5259E-01 | 1.5107E-01 |
| 1.9 | 1.4957E-01 | 1.4808E-01 | 1.4661E-01 | 1.4515E-01 | 1.4370E-01 | 1.4227E-01 | 1.4086E-01 | 1.3947E-01 | 1.3807E-01 | 1.3670E-01 |
| 2.0 | 1.3534E-01 | 1.3399E-01 | 1.3266E-01 | 1.3134E-01 | 1.3003E-01 | 1.2873E-01 | 1.2745E-01 | 1.2619E-01 | 1.2493E-01 | 1.2369E-01 |

# Table A2.2 (Continued)

$$F_{SP,\bar{i},n} = e^{-\bar{i}n}$$

| $\bar{i} \to$ / $n \downarrow$ | 0.00 | 0.01 | 0.02 | 0.03 | 0.04 | 0.05 | 0.06 | 0.07 | 0.08 | 0.09 |
|---|---|---|---|---|---|---|---|---|---|---|
| 2.1 | 1.2246E-01 | 1.2124E-01 | 1.2003E-01 | 1.1884E-01 | 1.1765E-01 | 1.1648E-01 | 1.1533E-01 | 1.1418E-01 | 1.1304E-01 | 1.1192E-01 |
| 2.2 | 1.1080E-01 | 1.0970E-01 | 1.0861E-01 | 1.0753E-01 | 1.0646E-01 | 1.0540E-01 | 1.0435E-01 | 1.0331E-01 | 1.0228E-01 | 1.0127E-01 |
| 2.3 | 1.0026E-01 | 9.9261E-02 | 9.8274E-02 | 9.7296E-02 | 9.6328E-02 | 9.5369E-02 | 9.4420E-02 | 9.3481E-02 | 9.2551E-02 | 9.1630E-02 |
| 2.4 | 9.0718E-02 | 8.9815E-02 | 8.8922E-02 | 8.8037E-02 | 8.7161E-02 | 8.6294E-02 | 8.5435E-02 | 8.4585E-02 | 8.3743E-02 | 8.2910E-02 |
| 2.5 | 8.2085E-02 | 8.1268E-02 | 8.0460E-02 | 7.9659E-02 | 7.8866E-02 | 7.8082E-02 | 7.7305E-02 | 7.6536E-02 | 7.5774E-02 | 7.5020E-02 |
| 2.6 | 7.4274E-02 | 7.3535E-02 | 7.2803E-02 | 7.2078E-02 | 7.1361E-02 | 7.0651E-02 | 6.9948E-02 | 6.9252E-02 | 6.8563E-02 | 6.7881E-02 |
| 2.7 | 6.7206E-02 | 6.6537E-02 | 6.5875E-02 | 6.5219E-02 | 6.4570E-02 | 6.3928E-02 | 6.3292E-02 | 6.2662E-02 | 6.2039E-02 | 6.1421E-02 |
| 2.8 | 6.0810E-02 | 6.0205E-02 | 5.9606E-02 | 5.9013E-02 | 5.8426E-02 | 5.7844E-02 | 5.7269E-02 | 5.6699E-02 | 5.6135E-02 | 5.5576E-02 |
| 2.9 | 5.5023E-02 | 5.4476E-02 | 5.3934E-02 | 5.3397E-02 | 5.2866E-02 | 5.2340E-02 | 5.1819E-02 | 5.1303E-02 | 5.0793E-02 | 5.0287E-02 |
| 3.0 | 4.9787E-02 | 4.9292E-02 | 4.8801E-02 | 4.8316E-02 | 4.7835E-02 | 4.7359E-02 | 4.6888E-02 | 4.6421E-02 | 4.5959E-02 | 4.5502E-02 |
| 3.1 | 4.5049E-02 | 4.4601E-02 | 4.4157E-02 | 4.3718E-02 | 4.3283E-02 | 4.2852E-02 | 4.2426E-02 | 4.2004E-02 | 4.1586E-02 | 4.1172E-02 |
| 3.2 | 4.0762E-02 | 4.0357E-02 | 3.9955E-02 | 3.9557E-02 | 3.9164E-02 | 3.8774E-02 | 3.8388E-02 | 3.8006E-02 | 3.7628E-02 | 3.7254E-02 |
| 3.3 | 3.6883E-02 | 3.6516E-02 | 3.6153E-02 | 3.5793E-02 | 3.5437E-02 | 3.5084E-02 | 3.4735E-02 | 3.4390E-02 | 3.4047E-02 | 3.3709E-02 |
| 3.4 | 3.3373E-02 | 3.3041E-02 | 3.2712E-02 | 3.2387E-02 | 3.2065E-02 | 3.1746E-02 | 3.1430E-02 | 3.1117E-02 | 3.0807E-02 | 3.0501E-02 |
| 3.5 | 3.0197E-02 | 2.9897E-02 | 2.9599E-02 | 2.9305E-02 | 2.9013E-02 | 2.8725E-02 | 2.8439E-02 | 2.8156E-02 | 2.7876E-02 | 2.7598E-02 |
| 3.6 | 2.7324E-02 | 2.7052E-02 | 2.6783E-02 | 2.6516E-02 | 2.6252E-02 | 2.5991E-02 | 2.5733E-02 | 2.5476E-02 | 2.5223E-02 | 2.4972E-02 |
| 3.7 | 2.4724E-02 | 2.4478E-02 | 2.4234E-02 | 2.3993E-02 | 2.3754E-02 | 2.3518E-02 | 2.3284E-02 | 2.3052E-02 | 2.2823E-02 | 2.2596E-02 |
| 3.8 | 2.2371E-02 | 2.2148E-02 | 2.1928E-02 | 2.1710E-02 | 2.1494E-02 | 2.1280E-02 | 2.1068E-02 | 2.0858E-02 | 2.0651E-02 | 2.0445E-02 |
| 3.9 | 2.0242E-02 | 2.0041E-02 | 1.9841E-02 | 1.9644E-02 | 1.9448E-02 | 1.9255E-02 | 1.9063E-02 | 1.8873E-02 | 1.8686E-02 | 1.8500E-02 |
| 4.0 | 1.8316E-02 | 1.8133E-02 | 1.7953E-02 | 1.7774E-02 | 1.7597E-02 | 1.7422E-02 | 1.7249E-02 | 1.7077E-02 | 1.6907E-02 | 1.6739E-02 |
| 4.1 | 1.6573E-02 | 1.6408E-02 | 1.6245E-02 | 1.6083E-02 | 1.5923E-02 | 1.5764E-02 | 1.5608E-02 | 1.5452E-02 | 1.5299E-02 | 1.5146E-02 |

$$F_{SP,\bar{i},n} = e^{-in}$$

| in → ↓ | 0.00 | 0.01 | 0.02 | 0.03 | 0.04 | 0.05 | 0.06 | 0.07 | 0.08 | 0.09 |
|---|---|---|---|---|---|---|---|---|---|---|
| 4.2 | 1.4996E-02 | 1.4846E-02 | 1.4699E-02 | 1.4552E-02 | 1.4408E-02 | 1.4264E-02 | 1.4122E-02 | 1.3982E-02 | 1.3843E-02 | 1.3705E-02 |
| 4.3 | 1.3569E-02 | 1.3434E-02 | 1.3300E-02 | 1.3168E-02 | 1.3037E-02 | 1.2907E-02 | 1.2778E-02 | 1.2651E-02 | 1.2525E-02 | 1.2401E-02 |
| 4.4 | 1.2277E-02 | 1.2155E-02 | 1.2034E-02 | 1.1914E-02 | 1.1796E-02 | 1.1679E-02 | 1.1562E-02 | 1.1447E-02 | 1.1333E-02 | 1.1221E-02 |
| 4.5 | 1.1109E-02 | 1.0998E-02 | 1.0889E-02 | 1.0781E-02 | 1.0673E-02 | 1.0567E-02 | 1.0462E-02 | 1.0358E-02 | 1.0255E-02 | 1.0153E-02 |
| 4.6 | 1.0052E-02 | 9.9518E-03 | 9.8528E-03 | 9.7548E-03 | 9.6577E-03 | 9.5616E-03 | 9.4665E-03 | 9.3723E-03 | 9.2790E-03 | 9.1867E-03 |
| 4.7 | 9.0953E-03 | 9.0048E-03 | 8.9152E-03 | 8.8265E-03 | 8.7386E-03 | 8.6517E-03 | 8.5656E-03 | 8.4804E-03 | 8.3960E-03 | 8.3125E-03 |
| 4.8 | 8.2297E-03 | 8.1479E-03 | 8.0668E-03 | 7.9865E-03 | 7.9071E-03 | 7.8284E-03 | 7.7505E-03 | 7.6734E-03 | 7.5970E-03 | 7.5214E-03 |
| 4.9 | 7.4466E-03 | 7.3725E-03 | 7.2991E-03 | 7.2265E-03 | 7.1546E-03 | 7.0834E-03 | 7.0129E-03 | 6.9431E-03 | 6.8741E-03 | 6.8057E-03 |
|  | .0 | .1 | .2 | .3 | .4 | .5 | .6 | .7 | .8 | .9 |
| 5.0 | 6.7379E-03 | 6.0967E-03 | 5.5166E-03 | 4.9916E-03 | 4.5166E-03 | 4.0868E-03 | 3.6979E-03 | 3.3460E-03 | 3.0276E-03 | 2.7394E-03 |
| 6.0 | 2.4788E-03 | 2.2429E-03 | 2.0294E-03 | 1.8363E-03 | 1.6616E-03 | 1.5034E-03 | 1.3604E-03 | 1.2309E-03 | 1.1138E-03 | 1.0078E-03 |
| 7.0 | 9.1188E-04 | 8.2510E-04 | 7.4659E-04 | 6.7554E-04 | 6.1125E-04 | 5.5308E-04 | 5.0045E-04 | 4.5283E-04 | 4.0973E-04 | 3.7074E-04 |
| 8.0 | 3.3546E-04 | 3.0354E-04 | 2.7465E-04 | 2.4852E-04 | 2.2487E-04 | 2.0347E-04 | 1.8411E-04 | 1.6659E-04 | 1.5073E-04 | 1.3639E-04 |
| 9.0 | 1.2341E-04 | 1.1167E-04 | 1.0104E-04 | 9.1424E-05 | 8.2724E-05 | 7.4852E-05 | 6.7729E-05 | 6.1283E-05 | 5.5452E-05 | 5.0175E-05 |
| 10.0 | 4.5400E-05 | 4.1080E-05 | 3.7170E-05 | 3.3633E-05 | 3.0432E-05 | 2.7536E-05 | 2.4916E-05 | 2.2545E-05 | 2.0400E-05 | 1.8458E-05 |

## Table A2.3 Continuous compound interest: Value of

$$F_{\overline{RP},i,n} = \frac{1 - e^{-in}}{in}$$

| in↓ \ in→ | 0.00 | 0.01 | 0.02 | 0.03 | 0.04 | 0.05 | 0.06 | 0.07 | 0.08 | 0.09 |
|---|---|---|---|---|---|---|---|---|---|---|
| 0   | 1.0000E 00 | 9.9502E-01 | 9.9007E-01 | 9.8515E-01 | 9.8026E-01 | 9.7541E-01 | 9.7059E-01 | 9.6580E-01 | 9.6105E-01 | 9.5632E-01 |
| 0.1 | 9.5163E-01 | 9.4696E-01 | 9.4233E-01 | 9.3773E-01 | 9.3316E-01 | 9.2861E-01 | 9.2410E-01 | 9.1962E-01 | 9.1517E-01 | 9.1074E-01 |
| 0.2 | 9.0635E-01 | 9.0198E-01 | 8.9764E-01 | 8.9333E-01 | 8.8905E-01 | 8.8480E-01 | 8.8057E-01 | 8.7637E-01 | 8.7220E-01 | 8.6806E-01 |
| 0.3 | 8.6394E-01 | 8.5985E-01 | 8.5578E-01 | 8.5175E-01 | 8.4773E-01 | 8.4375E-01 | 8.3979E-01 | 8.3585E-01 | 8.3194E-01 | 8.2806E-01 |
| 0.4 | 8.2420E-01 | 8.2037E-01 | 8.1656E-01 | 8.1277E-01 | 8.0901E-01 | 8.0527E-01 | 8.0156E-01 | 7.9787E-01 | 7.9420E-01 | 7.9056E-01 |
| 0.5 | 7.8694E-01 | 7.8334E-01 | 7.7977E-01 | 7.7622E-01 | 7.7269E-01 | 7.6918E-01 | 7.6570E-01 | 7.6224E-01 | 7.5880E-01 | 7.5538E-01 |
| 0.6 | 7.5198E-01 | 7.4861E-01 | 7.4525E-01 | 7.4192E-01 | 7.3861E-01 | 7.3531E-01 | 7.3204E-01 | 7.2879E-01 | 7.2556E-01 | 7.2235E-01 |
| 0.7 | 7.1916E-01 | 7.1599E-01 | 7.1284E-01 | 7.0971E-01 | 7.0660E-01 | 7.0351E-01 | 7.0044E-01 | 6.9739E-01 | 6.9435E-01 | 6.9134E-01 |
| 0.8 | 6.8834E-01 | 6.8536E-01 | 6.8240E-01 | 6.7946E-01 | 6.7654E-01 | 6.7363E-01 | 6.7074E-01 | 6.6787E-01 | 6.6502E-01 | 6.6218E-01 |
| 0.9 | 6.5937E-01 | 6.5657E-01 | 6.5378E-01 | 6.5102E-01 | 6.4827E-01 | 6.4554E-01 | 6.4282E-01 | 6.4012E-01 | 6.3744E-01 | 6.3477E-01 |
| 1.0 | 6.3212E-01 | 6.2949E-01 | 6.2687E-01 | 6.2427E-01 | 6.2168E-01 | 6.1911E-01 | 6.1655E-01 | 6.1401E-01 | 6.1149E-01 | 6.0898E-01 |
| 1.1 | 6.0648E-01 | 6.0400E-01 | 6.0154E-01 | 5.9909E-01 | 5.9665E-01 | 5.9423E-01 | 5.9182E-01 | 5.8943E-01 | 5.8705E-01 | 5.8469E-01 |
| 1.2 | 5.8234E-01 | 5.8000E-01 | 5.7768E-01 | 5.7537E-01 | 5.7308E-01 | 5.7080E-01 | 5.6853E-01 | 5.6627E-01 | 5.6403E-01 | 5.6181E-01 |
| 1.3 | 5.5959E-01 | 5.5739E-01 | 5.5520E-01 | 5.5302E-01 | 5.5086E-01 | 5.4871E-01 | 5.4657E-01 | 5.4445E-01 | 5.4233E-01 | 5.4023E-01 |
| 1.4 | 5.3815E-01 | 5.3607E-01 | 5.3400E-01 | 5.3195E-01 | 5.2991E-01 | 5.2788E-01 | 5.2587E-01 | 5.2386E-01 | 5.2187E-01 | 5.1988E-01 |
| 1.5 | 5.1791E-01 | 5.1595E-01 | 5.1401E-01 | 5.1207E-01 | 5.1014E-01 | 5.0823E-01 | 5.0632E-01 | 5.0443E-01 | 5.0255E-01 | 5.0068E-01 |
| 1.6 | 4.9881E-01 | 4.9696E-01 | 4.9512E-01 | 4.9329E-01 | 4.9148E-01 | 4.8967E-01 | 4.8787E-01 | 4.8608E-01 | 4.8430E-01 | 4.8253E-01 |
| 1.7 | 4.8077E-01 | 4.7903E-01 | 4.7729E-01 | 4.7556E-01 | 4.7384E-01 | 4.7213E-01 | 4.7043E-01 | 4.6874E-01 | 4.6706E-01 | 4.6539E-01 |
| 1.8 | 4.6372E-01 | 4.6207E-01 | 4.6043E-01 | 4.5879E-01 | 4.5716E-01 | 4.5555E-01 | 4.5394E-01 | 4.5234E-01 | 4.5075E-01 | 4.4917E-01 |
| 1.9 | 4.4760E-01 | 4.4603E-01 | 4.4448E-01 | 4.4293E-01 | 4.4139E-01 | 4.3986E-01 | 4.3834E-01 | 4.3682E-01 | 4.3532E-01 | 4.3382E-01 |
| 2.0 | 4.3233E-01 | 4.3085E-01 | 4.2938E-01 | 4.2791E-01 | 4.2646E-01 | 4.2501E-01 | 4.2357E-01 | 4.2213E-01 | 4.2071E-01 | 4.1929E-01 |

$$F_{\overline{RP},\,\overline{i,n}} = \frac{1 - e^{-in}}{in}$$

| in → ↓ | 0.00 | 0.01 | 0.02 | 0.03 | 0.04 | 0.05 | 0.06 | 0.07 | 0.08 | 0.09 |
|---|---|---|---|---|---|---|---|---|---|---|
| 2.1 | 4.1788E-01 | 4.1647E-01 | 4.1508E-01 | 4.1369E-01 | 4.1231E-01 | 4.1094E-01 | 4.0957E-01 | 4.0821E-01 | 4.0686E-01 | 4.0552E-01 |
| 2.2 | 4.0418E-01 | 4.0285E-01 | 4.0153E-01 | 4.0021E-01 | 3.9890E-01 | 3.9760E-01 | 3.9631E-01 | 3.9502E-01 | 3.9373E-01 | 3.9246E-01 |
| 2.3 | 3.9119E-01 | 3.8993E-01 | 3.8868E-01 | 3.8743E-01 | 3.8618E-01 | 3.8495E-01 | 3.8372E-01 | 3.8250E-01 | 3.8128E-01 | 3.8007E-01 |
| 2.4 | 3.7887E-01 | 3.7767E-01 | 3.7648E-01 | 3.7529E-01 | 3.7411E-01 | 3.7294E-01 | 3.7177E-01 | 3.7061E-01 | 3.6946E-01 | 3.6831E-01 |
| 2.5 | 3.6717E-01 | 3.6603E-01 | 3.6490E-01 | 3.6377E-01 | 3.6265E-01 | 3.6154E-01 | 3.6043E-01 | 3.5932E-01 | 3.5823E-01 | 3.5714E-01 |
| 2.6 | 3.5605E-01 | 3.5497E-01 | 3.5389E-01 | 3.5282E-01 | 3.5176E-01 | 3.5070E-01 | 3.4964E-01 | 3.4859E-01 | 3.4755E-01 | 3.4651E-01 |
| 2.7 | 3.4548E-01 | 3.4445E-01 | 3.4343E-01 | 3.4241E-01 | 3.4140E-01 | 3.4039E-01 | 3.3939E-01 | 3.3839E-01 | 3.3740E-01 | 3.3641E-01 |
| 2.8 | 3.3542E-01 | 3.3445E-01 | 3.3347E-01 | 3.3250E-01 | 3.3154E-01 | 3.3058E-01 | 3.2963E-01 | 3.2868E-01 | 3.2773E-01 | 3.2679E-01 |
| 2.9 | 3.2585E-01 | 3.2492E-01 | 3.2400E-01 | 3.2307E-01 | 3.2215E-01 | 3.2124E-01 | 3.2033E-01 | 3.1943E-01 | 3.1853E-01 | 3.1763E-01 |
| 3.0 | 3.1674E-01 | 3.1585E-01 | 3.1497E-01 | 3.1409E-01 | 3.1321E-01 | 3.1234E-01 | 3.1147E-01 | 3.1061E-01 | 3.0975E-01 | 3.0890E-01 |
| 3.1 | 3.0805E-01 | 3.0720E-01 | 3.0636E-01 | 3.0552E-01 | 3.0469E-01 | 3.0386E-01 | 3.0303E-01 | 3.0221E-01 | 3.0139E-01 | 3.0057E-01 |
| 3.2 | 2.9976E-01 | 2.9895E-01 | 2.9815E-01 | 2.9735E-01 | 2.9655E-01 | 2.9576E-01 | 2.9497E-01 | 2.9419E-01 | 2.9341E-01 | 2.9263E-01 |
| 3.3 | 2.9185E-01 | 2.9108E-01 | 2.9032E-01 | 2.8955E-01 | 2.8879E-01 | 2.8803E-01 | 2.8728E-01 | 2.8653E-01 | 2.8578E-01 | 2.8504E-01 |
| 3.4 | 2.8430E-01 | 2.8357E-01 | 2.8283E-01 | 2.8210E-01 | 2.8138E-01 | 2.8065E-01 | 2.7993E-01 | 2.7922E-01 | 2.7850E-01 | 2.7779E-01 |
| 3.5 | 2.7709E-01 | 2.7638E-01 | 2.7568E-01 | 2.7498E-01 | 2.7429E-01 | 2.7360E-01 | 2.7291E-01 | 2.7223E-01 | 2.7154E-01 | 2.7086E-01 |
| 3.6 | 2.7019E-01 | 2.6951E-01 | 2.6884E-01 | 2.6818E-01 | 2.6751E-01 | 2.6685E-01 | 2.6619E-01 | 2.6554E-01 | 2.6489E-01 | 2.6424E-01 |
| 3.7 | 2.6359E-01 | 2.6294E-01 | 2.6230E-01 | 2.6166E-01 | 2.6103E-01 | 2.6040E-01 | 2.5976E-01 | 2.5914E-01 | 2.5851E-01 | 2.5789E-01 |
| 3.8 | 2.5727E-01 | 2.5665E-01 | 2.5604E-01 | 2.5543E-01 | 2.5482E-01 | 2.5421E-01 | 2.5361E-01 | 2.5301E-01 | 2.5241E-01 | 2.5181E-01 |
| 3.9 | 2.5122E-01 | 2.5063E-01 | 2.5004E-01 | 2.4945E-01 | 2.4887E-01 | 2.4829E-01 | 2.4771E-01 | 2.4714E-01 | 2.4656E-01 | 2.4599E-01 |
| 4.0 | 2.4542E-01 | 2.4485E-01 | 2.4429E-01 | 2.4373E-01 | 2.4317E-01 | 2.4261E-01 | 2.4206E-01 | 2.4150E-01 | 2.4095E-01 | 2.4041E-01 |
| 4.1 | 2.3986E-01 | 2.3932E-01 | 2.3878E-01 | 2.3824E-01 | 2.3770E-01 | 2.3717E-01 | 2.3663E-01 | 2.3610E-01 | 2.3557E-01 | 2.3505E-01 |

**Table A2.3** (*Continued*)

$$F_{\overline{RP},i,n} = \frac{1 - e^{-in}}{in}$$

| $in \rightarrow \downarrow$ | 0.00 | 0.01 | 0.02 | 0.03 | 0.04 | 0.05 | 0.06 | 0.07 | 0.08 | 0.09 |
|---|---|---|---|---|---|---|---|---|---|---|
| 4.2 | 2.3452E-01 | 2.3400E-01 | 2.3348E-01 | 2.3297E-01 | 2.3245E-01 | 2.3194E-01 | 2.3143E-01 | 2.3092E-01 | 2.3041E-01 | 2.2991E-01 |
| 4.3 | 2.2940E-01 | 2.2890E-01 | 2.2840E-01 | 2.2791E-01 | 2.2741E-01 | 2.2692E-01 | 2.2643E-01 | 2.2594E-01 | 2.2545E-01 | 2.2497E-01 |
| 4.4 | 2.2448E-01 | 2.2400E-01 | 2.2352E-01 | 2.2304E-01 | 2.2257E-01 | 2.2209E-01 | 2.2162E-01 | 2.2115E-01 | 2.2068E-01 | 2.2022E-01 |
| 4.5 | 2.1975E-01 | 2.1929E-01 | 2.1883E-01 | 2.1837E-01 | 2.1791E-01 | 2.1746E-01 | 2.1700E-01 | 2.1655E-01 | 2.1610E-01 | 2.1565E-01 |
| 4.6 | 2.1521E-01 | 2.1476E-01 | 2.1432E-01 | 2.1388E-01 | 2.1344E-01 | 2.1300E-01 | 2.1256E-01 | 2.1213E-01 | 2.1169E-01 | 2.1126E-01 |
| 4.7 | 2.1083E-01 | 2.1040E-01 | 2.0998E-01 | 2.0955E-01 | 2.0913E-01 | 2.0870E-01 | 2.0828E-01 | 2.0787E-01 | 2.0745E-01 | 2.0703E-01 |
| 4.8 | 2.0662E-01 | 2.0621E-01 | 2.0580E-01 | 2.0539E-01 | 2.0498E-01 | 2.0457E-01 | 2.0417E-01 | 2.0376E-01 | 2.0336E-01 | 2.0296E-01 |
| 4.9 | 2.0256E-01 | 2.0216E-01 | 2.0177E-01 | 2.0137E-01 | 2.0098E-01 | 2.0059E-01 | 2.0020E-01 | 1.9981E-01 | 1.9942E-01 | 1.9904E-01 |
|  | .0 | .1 | .2 | .3 | .4 | .5 | .6 | .7 | .8 | .9 |
| 5.0 | 1.9865E-01 | 1.9488E-01 | 1.9125E-01 | 1.8774E-01 | 1.8435E-01 | 1.8108E-01 | 1.7791E-01 | 1.7485E-01 | 1.7189E-01 | 1.6903E-01 |
| 6.0 | 1.6625E-01 | 1.6357E-01 | 1.6096E-01 | 1.5844E-01 | 1.5599E-01 | 1.5361E-01 | 1.5131E-01 | 1.4907E-01 | 1.4690E-01 | 1.4478E-01 |
| 7.0 | 1.4273E-01 | 1.4073E-01 | 1.3879E-01 | 1.3689E-01 | 1.3505E-01 | 1.3326E-01 | 1.3151E-01 | 1.2981E-01 | 1.2815E-01 | 1.2654E-01 |
| 8.0 | 1.2496E-01 | 1.2342E-01 | 1.2192E-01 | 1.2045E-01 | 1.1902E-01 | 1.1762E-01 | 1.1626E-01 | 1.1492E-01 | 1.1362E-01 | 1.1234E-01 |
| 9.0 | 1.1110E-01 | 1.0988E-01 | 1.0868E-01 | 1.0752E-01 | 1.0637E-01 | 1.0526E-01 | 1.0416E-01 | 1.0309E-01 | 1.0204E-01 | 1.0101E-01 |
| 10.0 | 9.9995E-02 | 9.9006E-02 | 9.8036E-02 | 9.7084E-02 | 9.6151E-02 | 9.5235E-02 | 9.4337E-02 | 9.3456E-02 | 9.2591E-02 | 9.1741E-02 |
| 11.0 | 9.0908E-02 | 9.0089E-02 | 8.9284E-02 | 8.8494E-02 | 8.7718E-02 | 8.6956E-02 | 8.6206E-02 | 8.5469E-02 | 8.4745E-02 | 8.4033E-02 |
| 12.0 | 8.3333E-02 | 8.2644E-02 | 8.1967E-02 | 8.1300E-02 | 8.0645E-02 | 8.0000E-02 | 7.9365E-02 | 7.8740E-02 | 7.8125E-02 | 7.7519E-02 |
| 13.0 | 7.6923E-02 | 7.6336E-02 | 7.5757E-02 | 7.5188E-02 | 7.4627E-02 | 7.4074E-02 | 7.3529E-02 | 7.2993E-02 | 7.2464E-02 | 7.1942E-02 |
| 14.0 | 7.1429E-02 | 7.0922E-02 | 7.0422E-02 | 6.9930E-02 | 6.9444E-02 | 6.8965E-02 | 6.8493E-02 | 6.8027E-02 | 6.7568E-02 | 6.7114E-02 |
| 15.0 | 6.6667E-02 | 6.6225E-02 | 6.5789E-02 | 6.5359E-02 | 6.4935E-02 | 6.4516E-02 | 6.4103E-02 | 6.3694E-02 | 6.3291E-02 | 6.2893E-02 |
| 16.0 | 6.2500E-02 | 6.2112E-02 | 6.1728E-02 | 6.1350E-02 | 6.0976E-02 | 6.0606E-02 | 6.0241E-02 | 5.9880E-02 | 5.9524E-02 | 5.9172E-02 |
| 17.0 | 5.8824E-02 | 5.8480E-02 | 5.8140E-02 | 5.7803E-02 | 5.7471E-02 | 5.7143E-02 | 5.6818E-02 | 5.6497E-02 | 5.6180E-02 | 5.5866E-02 |
| 18.0 | 5.5556E-02 | 5.5249E-02 | 5.4945E-02 | 5.4645E-02 | 5.4348E-02 | 5.4054E-02 | 5.3763E-02 | 5.3476E-02 | 5.3191E-02 | 5.2910E-02 |
| 19.0 | 5.2632E-02 | 5.2356E-02 | 5.2083E-02 | 5.1813E-02 | 5.1546E-02 | 5.1282E-02 | 5.1020E-02 | 5.0761E-02 | 5.0505E-02 | 5.0251E-02 |
| 20.0 | 5.0000E-02 | 4.9751E-02 | 4.9505E-02 | 4.9261E-02 | 4.9020E-02 | 4.8780E-02 | 4.8544E-02 | 4.8309E-02 | 4.8077E-02 | 4.7847E-02 |

## Table A2.4  Continuous compound interest: Value of

$$F_{SDP,\bar{i},n} = \frac{2}{in}\left(1 - \frac{1 - e^{-in}}{in}\right)$$

| in $\rightarrow$ $\downarrow$ | 0.00 | 0.01 | 0.02 | 0.03 | 0.04 | 0.05 | 0.06 | 0.07 | 0.08 | 0.09 |
|---|---|---|---|---|---|---|---|---|---|---|
| 0 | 1.0000E 00 | 9.9688E-01 | 9.9337E-01 | 9.9007E-01 | 9.8680E-01 | 9.8354E-01 | 9.8030E-01 | 9.7707E-01 | 9.7386E-01 | 9.7066E-01 |
| 0.1 | 9.6748E-01 | 9.6643E-01 | 9.6117E-01 | 9.5804E-01 | 9.5492E-01 | 9.5182E-01 | 9.4873E-01 | 9.4566E-01 | 9.4261E-01 | 9.3956E-01 |
| 0.2 | 9.3654E-01 | 9.3353E-01 | 9.3053E-01 | 9.2755E-01 | 9.2458E-01 | 9.2163E-01 | 9.1869E-01 | 9.1576E-01 | 9.1285E-01 | 9.0995E-01 |
| 0.3 | 9.0707E-01 | 9.0420E-01 | 9.0135E-01 | 8.9851E-01 | 8.9568E-01 | 8.9287E-01 | 8.9007E-01 | 8.8728E-01 | 8.8451E-01 | 8.8175E-01 |
| 0.4 | 8.7900E-01 | 8.7627E-01 | 8.7355E-01 | 8.7084E-01 | 8.6814E-01 | 8.6546E-01 | 8.6279E-01 | 8.6014E-01 | 8.5749E-01 | 8.5486E-01 |
| 0.5 | 8.5225E-01 | 8.4964E-01 | 8.4705E-01 | 8.4446E-01 | 8.4189E-01 | 8.3934E-01 | 8.3679E-01 | 8.3426E-01 | 8.3174E-01 | 8.2923E-01 |
| 0.6 | 8.2673E-01 | 8.2425E-01 | 8.2177E-01 | 8.1931E-01 | 8.1686E-01 | 8.1442E-01 | 8.1199E-01 | 8.0957E-01 | 8.0717E-01 | 8.0477E-01 |
| 0.7 | 8.0239E-01 | 8.0002E-01 | 7.9766E-01 | 7.9530E-01 | 7.9297E-01 | 7.9064E-01 | 7.8832E-01 | 7.8601E-01 | 7.8371E-01 | 7.8143E-01 |
| 0.8 | 7.7915E-01 | 7.7689E-01 | 7.7463E-01 | 7.7239E-01 | 7.7015E-01 | 7.6793E-01 | 7.6572E-01 | 7.6351E-01 | 7.6132E-01 | 7.5914E-01 |
| 0.9 | 7.5696E-01 | 7.5480E-01 | 7.5264E-01 | 7.5050E-01 | 7.4837E-01 | 7.4624E-01 | 7.4413E-01 | 7.4202E-01 | 7.3992E-01 | 7.3784E-01 |
| 1.0 | 7.3576E-01 | 7.3369E-01 | 7.3163E-01 | 7.2958E-01 | 7.2754E-01 | 7.2551E-01 | 7.2349E-01 | 7.2148E-01 | 7.1947E-01 | 7.1748E-01 |
| 1.1 | 7.1549E-01 | 7.1351E-01 | 7.1154E-01 | 7.0958E-01 | 7.0763E-01 | 7.0569E-01 | 7.0375E-01 | 7.0183E-01 | 6.9991E-01 | 6.9800E-01 |
| 1.2 | 6.9610E-01 | 6.9421E-01 | 6.9233E-01 | 6.9045E-01 | 6.8859E-01 | 6.8673E-01 | 6.8488E-01 | 6.8303E-01 | 6.8120E-01 | 6.7937E-01 |
| 1.3 | 6.7755E-01 | 6.7574E-01 | 6.7394E-01 | 6.7214E-01 | 6.7036E-01 | 6.6858E-01 | 6.6680E-01 | 6.6504E-01 | 6.6328E-01 | 6.6153E-01 |
| 1.4 | 6.5979E-01 | 6.5806E-01 | 6.5633E-01 | 6.5461E-01 | 6.5290E-01 | 6.5120E-01 | 6.4950E-01 | 6.4781E-01 | 6.4613E-01 | 6.4445E-01 |
| 1.5 | 6.4278E-01 | 6.4112E-01 | 6.3947E-01 | 6.3782E-01 | 6.3618E-01 | 6.3455E-01 | 6.3292E-01 | 6.3130E-01 | 6.2969E-01 | 6.2808E-01 |
| 1.6 | 6.2648E-01 | 6.2489E-01 | 6.2330E-01 | 6.2172E-01 | 6.2015E-01 | 6.1859E-01 | 6.1703E-01 | 6.1547E-01 | 6.1393E-01 | 6.1239E-01 |
| 1.7 | 6.1085E-01 | 6.0933E-01 | 6.0781E-01 | 6.0629E-01 | 6.0478E-01 | 6.0328E-01 | 6.0179E-01 | 6.0030E-01 | 5.9881E-01 | 5.9733E-01 |
| 1.8 | 5.9586E-01 | 5.9440E-01 | 5.9294E-01 | 5.9149E-01 | 5.9004E-01 | 5.8860E-01 | 5.8716E-01 | 5.8573E-01 | 5.8431E-01 | 5.8289E-01 |
| 1.9 | 5.8148E-01 | 5.8007E-01 | 5.7867E-01 | 5.7728E-01 | 5.7589E-01 | 5.7450E-01 | 5.7312E-01 | 5.7175E-01 | 5.7039E-01 | 5.6902E-01 |
| 2.0 | 5.6767E-01 | 5.6632E-01 | 5.6497E-01 | 5.6363E-01 | 5.6230E-01 | 5.6097E-01 | 5.5964E-01 | 5.5833E-01 | 5.5701E-01 | 5.5570E-01 |

# Table A2.4 (Continued)

$$F_{\text{SDP}, \bar{i}, n} = \frac{2}{in}\left(1 - \frac{1 - e^{-in}}{in}\right)$$

| in → <br> in ↓ | 0.00 | 0.01 | 0.02 | 0.03 | 0.04 | 0.05 | 0.06 | 0.07 | 0.08 | 0.09 |
|---|---|---|---|---|---|---|---|---|---|---|
| 2.1 | 5.5440E-01 | 5.5310E-01 | 5.5181E-01 | 5.5052E-01 | 5.4924E-01 | 5.4797E-01 | 5.4669E-01 | 5.4543E-01 | 5.4416E-01 | 5.4291E-01 |
| 2.2 | 5.4165E-01 | 5.4041E-01 | 5.3916E-01 | 5.3793E-01 | 5.3669E-01 | 5.3547E-01 | 5.3424E-01 | 5.3302E-01 | 5.3181E-01 | 5.3060E-01 |
| 2.3 | 5.2940E-01 | 5.2820E-01 | 5.2700E-01 | 5.2581E-01 | 5.2463E-01 | 5.2345E-01 | 5.2227E-01 | 5.2110E-01 | 5.1993E-01 | 5.1877E-01 |
| 2.4 | 5.1761E-01 | 5.1646E-01 | 5.1531E-01 | 5.1416E-01 | 5.1302E-01 | 5.1188E-01 | 5.1075E-01 | 5.0962E-01 | 5.0850E-01 | 5.0738E-01 |
| 2.5 | 5.0627E-01 | 5.0516E-01 | 5.0405E-01 | 5.0295E-01 | 5.0185E-01 | 5.0076E-01 | 4.9967E-01 | 4.9858E-01 | 4.9750E-01 | 4.9642E-01 |
| 2.6 | 4.9535E-01 | 4.9428E-01 | 4.9321E-01 | 4.9215E-01 | 4.9109E-01 | 4.9004E-01 | 4.8899E-01 | 4.8794E-01 | 4.8690E-01 | 4.8586E-01 |
| 2.7 | 4.8483E-01 | 4.8383E-01 | 4.8277E-01 | 4.8175E-01 | 4.8073E-01 | 4.7972E-01 | 4.7871E-01 | 4.7770E-01 | 4.7669E-01 | 4.7569E-01 |
| 2.8 | 4.7470E-01 | 4.7370E-01 | 4.7271E-01 | 4.7173E-01 | 4.7075E-01 | 4.6977E-01 | 4.6879E-01 | 4.6782E-01 | 4.6685E-01 | 4.6589E-01 |
| 2.9 | 4.6493E-01 | 4.6397E-01 | 4.6302E-01 | 4.6207E-01 | 4.6112E-01 | 4.6018E-01 | 4.5924E-01 | 4.5830E-01 | 4.5737E-01 | 4.5644E-01 |
| 3.0 | 4.5551E-01 | 4.5458E-01 | 4.5366E-01 | 4.5275E-01 | 4.5183E-01 | 4.5092E-01 | 4.5002E-01 | 4.4911E-01 | 4.4821E-01 | 4.4731E-01 |
| 3.1 | 4.4642E-01 | 4.4553E-01 | 4.4464E-01 | 4.4376E-01 | 4.4287E-01 | 4.4200E-01 | 4.4112E-01 | 4.4025E-01 | 4.3938E-01 | 4.3851E-01 |
| 3.2 | 4.3765E-01 | 4.3679E-01 | 4.3593E-01 | 4.3508E-01 | 4.3423E-01 | 4.3338E-01 | 4.3253E-01 | 4.3169E-01 | 4.3085E-01 | 4.3001E-01 |
| 3.3 | 4.2918E-01 | 4.2835E-01 | 4.2752E-01 | 4.2670E-01 | 4.2587E-01 | 4.2505E-01 | 4.2424E-01 | 4.2342E-01 | 4.2261E-01 | 4.2180E-01 |
| 3.4 | 4.2100E-01 | 4.2020E-01 | 4.1940E-01 | 4.1860E-01 | 4.1780E-01 | 4.1701E-01 | 4.1622E-01 | 4.1544E-01 | 4.1465E-01 | 4.1387E-01 |
| 3.5 | 4.1309E-01 | 4.1232E-01 | 4.1154E-01 | 4.1077E-01 | 4.1001E-01 | 4.0924E-01 | 4.0848E-01 | 4.0772E-01 | 4.0696E-01 | 4.0620E-01 |
| 3.6 | 4.0545E-01 | 4.0470E-01 | 4.0395E-01 | 4.0321E-01 | 4.0247E-01 | 4.0173E-01 | 4.0099E-01 | 4.0025E-01 | 3.9952E-01 | 3.9879E-01 |
| 3.7 | 3.9806E-01 | 3.9733E-01 | 3.9661E-01 | 3.9589E-01 | 3.9517E-01 | 3.9446E-01 | 3.9374E-01 | 3.9303E-01 | 3.9232E-01 | 3.9161E-01 |
| 3.8 | 3.9091E-01 | 3.9021E-01 | 3.8951E-01 | 3.8881E-01 | 3.8811E-01 | 3.8742E-01 | 3.8673E-01 | 3.8604E-01 | 3.8536E-01 | 3.8467E-01 |
| 3.9 | 3.8399E-01 | 3.8331E-01 | 3.8263E-01 | 3.8196E-01 | 3.8128E-01 | 3.8061E-01 | 3.7994E-01 | 3.7928E-01 | 3.7861E-01 | 3.7795E-01 |
| 4.0 | 3.7729E-01 | 3.7663E-01 | 3.7597E-01 | 3.7532E-01 | 3.7467E-01 | 3.7402E-01 | 3.7337E-01 | 3.7273E-01 | 3.7208E-01 | 3.7144E-01 |
| 4.1 | 3.7080E-01 | 3.7016E-01 | 3.6953E-01 | 3.6889E-01 | 3.6826E-01 | 3.6763E-01 | 3.6700E-01 | 3.6638E-01 | 3.6575E-01 | 3.6513E-01 |

$$F_{\mathrm{SDP},\bar{i},n} = \frac{2}{in}\left(1 - \frac{1 - e^{-in}}{in}\right)$$

| in ↓ | 0.00 | 0.01 | 0.02 | 0.03 | 0.04 | 0.05 | 0.06 | 0.07 | 0.08 | 0.09 |
|------|------|------|------|------|------|------|------|------|------|------|
| 4.2 | 3.6451E-01 | 3.6389E-01 | 3.6328E-01 | 3.6266E-01 | 3.6205E-01 | 3.6144E-01 | 3.6083E-01 | 3.6023E-01 | 3.5962E-01 | 3.5902E-01 |
| 4.3 | 3.5842E-01 | 3.5782E-01 | 3.5722E-01 | 3.5663E-01 | 3.5603E-01 | 3.5544E-01 | 3.5485E-01 | 3.5426E-01 | 3.5368E-01 | 3.5309E-01 |
| 4.4 | 3.5251E-01 | 3.5193E-01 | 3.5135E-01 | 3.5077E-01 | 3.5019E-01 | 3.4962E-01 | 3.4905E-01 | 3.4848E-01 | 3.4791E-01 | 3.4734E-01 |
| 4.5 | 3.4678E-01 | 3.4621E-01 | 3.4565E-01 | 3.4509E-01 | 3.4453E-01 | 3.4397E-01 | 3.4342E-01 | 3.4287E-01 | 3.4231E-01 | 3.4176E-01 |
| 4.6 | 3.4121E-01 | 3.4067E-01 | 3.4012E-01 | 3.3958E-01 | 3.3904E-01 | 3.3850E-01 | 3.3796E-01 | 3.3742E-01 | 3.3688E-01 | 3.3635E-01 |
| 4.7 | 3.3582E-01 | 3.3529E-01 | 3.3476E-01 | 3.3423E-01 | 3.3370E-01 | 3.3318E-01 | 3.3265E-01 | 3.3213E-01 | 3.3161E-01 | 3.3109E-01 |
| 4.8 | 3.3058E-01 | 3.3006E-01 | 3.2955E-01 | 3.2903E-01 | 3.2852E-01 | 3.2801E-01 | 3.2750E-01 | 3.2700E-01 | 3.2649E-01 | 3.2599E-01 |
| 4.9 | 3.2548E-01 | 3.2498E-01 | 3.2448E-01 | 3.2399E-01 | 3.2349E-01 | 3.2299E-01 | 3.2250E-01 | 3.2201E-01 | 3.2152E-01 | 3.2103E-01 |
| (in ↓) | .0 | .1 | .2 | .3 | .4 | .5 | .6 | .7 | .8 | .9 |
| 5.0 | 3.2054E-01 | 3.1573E-01 | 3.1106E-01 | 3.0651E-01 | 3.0209E-01 | 2.9779E-01 | 2.9360E-01 | 2.8953E-01 | 2.8555E-01 | 2.8169E-01 |
| 6.0 | 2.7792E-01 | 2.7424E-01 | 2.7066E-01 | 2.6716E-01 | 2.6375E-01 | 2.6043E-01 | 2.5718E-01 | 2.5401E-01 | 2.5091E-01 | 2.4789E-01 |
| 7.0 | 2.4494E-01 | 2.4205E-01 | 2.3923E-01 | 2.3647E-01 | 2.3377E-01 | 2.3113E-01 | 2.2855E-01 | 2.2602E-01 | 2.2355E-01 | 2.2113E-01 |
| 8.0 | 2.1876E-01 | 2.1644E-01 | 2.1417E-01 | 2.1194E-01 | 2.0976E-01 | 2.0762E-01 | 2.0552E-01 | 2.0347E-01 | 2.0145E-01 | 1.9947E-01 |
| 9.0 | 1.9753E-01 | 1.9563E-01 | 1.9376E-01 | 1.9193E-01 | 1.9013E-01 | 1.8837E-01 | 1.8663E-01 | 1.8493E-01 | 1.8326E-01 | 1.8162E-01 |
| 10.0 | 1.8000E-01 | 1.7841E-01 | 1.7686E-01 | 1.7532E-01 | 1.7382E-01 | 1.7234E-01 | 1.7088E-01 | 1.6945E-01 | 1.6804E-01 | 1.6665E-01 |
| 11.0 | 1.6529E-01 | 1.6395E-01 | 1.6263E-01 | 1.6133E-01 | 1.6005E-01 | 1.5879E-01 | 1.5755E-01 | 1.5633E-01 | 1.5513E-01 | 1.5394E-01 |
| 12.0 | 1.5278E-01 | 1.5163E-01 | 1.5050E-01 | 1.4938E-01 | 1.4828E-01 | 1.4720E-01 | 1.4613E-01 | 1.4508E-01 | 1.4404E-01 | 1.4302E-01 |
| 13.0 | 1.4201E-01 | 1.4102E-01 | 1.4004E-01 | 1.3907E-01 | 1.3812E-01 | 1.3717E-01 | 1.3625E-01 | 1.3533E-01 | 1.3443E-01 | 1.3353E-01 |
| 14.0 | 1.3265E-01 | 1.3178E-01 | 1.3093E-01 | 1.3008E-01 | 1.2924E-01 | 1.2842E-01 | 1.2760E-01 | 1.2680E-01 | 1.2600E-01 | 1.2522E-01 |
| 15.0 | 1.2444E-01 | 1.2368E-01 | 1.2292E-01 | 1.2218E-01 | 1.2144E-01 | 1.2071E-01 | 1.1999E-01 | 1.1927E-01 | 1.1857E-01 | 1.1788E-01 |
| 16.0 | 1.1719E-01 | 1.1651E-01 | 1.1584E-01 | 1.1517E-01 | 1.1452E-01 | 1.1387E-01 | 1.1322E-01 | 1.1259E-01 | 1.1196E-01 | 1.1134E-01 |
| 17.0 | 1.1073E-01 | 1.1012E-01 | 1.0952E-01 | 1.0892E-01 | 1.0834E-01 | 1.0776E-01 | 1.0718E-01 | 1.0661E-01 | 1.0605E-01 | 1.0549E-01 |
| 18.0 | 1.0494E-01 | 1.0439E-01 | 1.0385E-01 | 1.0332E-01 | 1.0279E-01 | 1.0226E-01 | 1.0175E-01 | 1.0123E-01 | 1.0072E-01 | 1.0022E-01 |
| 19.0 | 9.9723E-02 | 9.9230E-02 | 9.8741E-02 | 9.8258E-02 | 9.7779E-02 | 9.7304E-02 | 9.6835E-02 | 9.6369E-02 | 9.5909E-02 | 9.5452E-02 |
| 20.0 | 9.5000E-02 | 9.4552E-02 | 9.4108E-02 | 9.3669E-02 | 9.3233E-02 | 9.2802E-02 | 9.2374E-02 | 9.1951E-02 | 9.1531E-02 | 9.1115E-02 |

# APPENDIX
# 3

# PROBABILITY

**Table A3.1  Cumulative probability for Z or less for a standard normal distribution**

| Z | 0 | 1 | 2 | 3 | 4 | 5 | 6 | 7 | 8 | 9 |
|---|---|---|---|---|---|---|---|---|---|---|
| −3.9 | .0000 | .0000 | .0000 | .0000 | .0000 | .0000 | .0000 | .0000 | .0000 | .0000 |
| −3.8 | .0001 | .0001 | .0001 | .0001 | .0001 | .0001 | .0001 | .0001 | .0001 | .0001 |
| −3.7 | .0001 | .0001 | .0001 | .0001 | .0001 | .0001 | .0001 | .0001 | .0001 | .0001 |
| −3.6 | .0002 | .0002 | .0001 | .0001 | .0001 | .0001 | .0001 | .0001 | .0001 | .0001 |
| −3.5 | .0002 | .0002 | .0002 | .0002 | .0002 | .0002 | .0002 | .0002 | .0002 | .0002 |
| −3.4 | .0003 | .0003 | .0003 | .0003 | .0003 | .0003 | .0003 | .0003 | .0003 | .0002 |
| −3.3 | .0005 | .0005 | .0005 | .0004 | .0004 | .0004 | .0004 | .0004 | .0004 | .0003 |
| −3.2 | .0007 | .0007 | .0006 | .0006 | .0006 | .0006 | .0006 | .0005 | .0005 | .0005 |
| −3.1 | .0010 | .0009 | .0009 | .0009 | .0008 | .0008 | .0008 | .0008 | .0007 | .0007 |
| −3.0 | .0013 | .0013 | .0013 | .0012 | .0012 | .0011 | .0011 | .0011 | .0010 | .0010 |
| −2.9 | .0019 | .0018 | .0017 | .0017 | .0016 | .0016 | .0015 | .0015 | .0014 | .0014 |
| −2.8 | .0026 | .0025 | .0024 | .0023 | .0023 | .0022 | .0021 | .0021 | .0020 | .0019 |
| −2.7 | .0035 | .0034 | .0033 | .0032 | .0031 | .0030 | .0029 | .0028 | .0027 | .0026 |
| −2.6 | .0047 | .0045 | .0044 | .0043 | .0041 | .0040 | .0039 | .0038 | .0037 | .0036 |
| −2.5 | .0062 | .0060 | .0059 | .0057 | .0055 | .0054 | .0052 | .0051 | .0049 | .0048 |
| −2.4 | .0082 | .0080 | .0078 | .0075 | .0073 | .0071 | .0069 | .0068 | .0066 | .0064 |
| −2.3 | .0107 | .0104 | .0102 | .0099 | .0096 | .0094 | .0091 | .0089 | .0087 | .0084 |
| −2.2 | .0139 | .0136 | .0132 | .0129 | .0126 | .0122 | .0119 | .0116 | .0113 | .0110 |
| −2.1 | .0179 | .0174 | .0170 | .0166 | .0162 | .0158 | .0154 | .0150 | .0146 | .0143 |
| −2.0 | .0228 | .0222 | .0217 | .0212 | .0207 | .0202 | .0197 | .0192 | .0188 | .0183 |

**Table A3.1** (*Continued*)

| Z | 0 | 1 | 2 | 3 | 4 | 5 | 6 | 7 | 8 | 9 |
|---|---|---|---|---|---|---|---|---|---|---|
| -1.9 | .0287 | .0281 | .0274 | .0268 | .0262 | .0256 | .0250 | .0244 | .0238 | .0233 |
| -1.8 | .0359 | .0352 | .0344 | .0336 | .0329 | .0322 | .0314 | .0307 | .0300 | .0294 |
| -1.7 | .0446 | .0436 | .0427 | .0418 | .0409 | .0401 | .0392 | .0384 | .0375 | .0367 |
| -1.6 | .0548 | .0537 | .0526 | .0516 | .0505 | .0495 | .0485 | .0475 | .0465 | .0455 |
| -1.5 | .0668 | .0655 | .0643 | .0630 | .0618 | .0606 | .0594 | .0582 | .0570 | .0559 |
| -1.4 | .0808 | .0793 | .0778 | .0764 | .0749 | .0735 | .0722 | .0708 | .0694 | .0681 |
| -1.3 | .0968 | .0951 | .0934 | .0918 | .0901 | .0885 | .0869 | .0853 | .0838 | .0823 |
| -1.2 | .1151 | .1131 | .1112 | .1093 | .1075 | .1056 | .1038 | .1020 | .1003 | .0985 |
| -1.1 | .1357 | .1335 | .1314 | .1292 | .1271 | .1251 | .1230 | .1210 | .1190 | .1170 |
| -1.0 | .1587 | .1562 | .1539 | .1515 | .1492 | .1469 | .1446 | .1423 | .1401 | .1379 |
| -0.9 | .1841 | .1814 | .1788 | .1762 | .1736 | .1711 | .1685 | .1660 | .1635 | .1611 |
| -0.8 | .2119 | .2090 | .2061 | .2033 | .2005 | .1977 | .1949 | .1922 | .1894 | .1867 |
| -0.7 | .2420 | .2389 | .2358 | .2327 | .2297 | .2266 | .2236 | .2206 | .2177 | .2148 |
| -0.6 | .2743 | .2709 | .2676 | .2643 | .2611 | .2578 | .2546 | .2514 | .2483 | .2451 |
| -0.5 | .3085 | .3050 | .3015 | .2981 | .2946 | .2912 | .2877 | .2843 | .2810 | .2776 |
| -0.4 | .3446 | .3409 | .3372 | .3336 | .3300 | .3264 | .3228 | .3192 | .3156 | .3121 |
| -0.3 | .3821 | .3783 | .3745 | .3707 | .3669 | .3632 | .3594 | .3557 | .3520 | .3483 |
| -0.2 | .4207 | .4168 | .4129 | .4090 | .4052 | .4013 | .3974 | .3936 | .3897 | .3859 |
| -0.1 | .4602 | .4562 | .4522 | .4483 | .4443 | .4404 | .4364 | .4325 | .4286 | .4247 |
| -0.0 | .5000 | .4960 | .4920 | .4880 | .4840 | .4801 | .4761 | .4721 | .4681 | .4641 |

| Z | 0 | 1 | 2 | 3 | 4 | 5 | 6 | 7 | 8 | 9 |
|---|---|---|---|---|---|---|---|---|---|---|
| 0.0 | .5000 | .5040 | .5080 | .5120 | .5160 | .5199 | .5239 | .5279 | .5319 | .5359 |
| 0.1 | .5398 | .5438 | .5478 | .5517 | .5557 | .5596 | .5636 | .5675 | .5714 | .5753 |
| 0.2 | .5793 | .5832 | .5871 | .5910 | .5948 | .5987 | .6026 | .6064 | .6103 | .6141 |
| 0.3 | .6179 | .6217 | .6255 | .6293 | .6331 | .6368 | .6406 | .6443 | .6480 | .6517 |
| 0.4 | .6554 | .6591 | .6628 | .6664 | .6700 | .6736 | .6772 | .6808 | .6844 | .6879 |
| 0.5 | .6915 | .6950 | .6985 | .7019 | .7054 | .7088 | .7123 | .7157 | .7190 | .7224 |
| 0.6 | .7257 | .7291 | .7324 | .7357 | .7389 | .7422 | .7454 | .7486 | .7517 | .7549 |
| 0.7 | .7580 | .7611 | .7642 | .7673 | .7703 | .7734 | .7764 | .7794 | .7823 | .7852 |
| 0.8 | .7881 | .7910 | .7939 | .7967 | .7995 | .8023 | .8051 | .8078 | .8106 | .8133 |
| 0.9 | .8159 | .8186 | .8212 | .8238 | .8264 | .8289 | .8315 | .8340 | .8365 | .8389 |
| 1.0 | .8413 | .8438 | .8461 | .8485 | .8508 | .8531 | .8554 | .8577 | .8599 | .8621 |
| 1.1 | .8643 | .8665 | .8686 | .8708 | .8729 | .8749 | .8770 | .8790 | .8810 | .8830 |
| 1.2 | .8849 | .8869 | .8888 | .8907 | .8925 | .8944 | .8962 | .8980 | .8997 | .9015 |
| 1.3 | .9032 | .9049 | .9066 | .9082 | .9099 | .9115 | .9131 | .9147 | .9162 | .9177 |
| 1.4 | .9192 | .9207 | .9222 | .9236 | .9251 | .9265 | .9278 | .9292 | .9306 | .9319 |
| 1.5 | .9332 | .9345 | .9357 | .9370 | .9382 | .9394 | .9406 | .9418 | .9430 | .9441 |
| 1.6 | .9452 | .9463 | .9474 | .9484 | .9495 | .9505 | .9515 | .9525 | .9535 | .9545 |
| 1.7 | .9554 | .9564 | .9573 | .9582 | .9591 | .9599 | .9608 | .9616 | .9625 | .9633 |
| 1.8 | .9641 | .9648 | .9656 | .9664 | .9671 | .9678 | .9686 | .9693 | .9700 | .9706 |
| 1.9 | .9713 | .9719 | .9726 | .9732 | .9738 | .9744 | .9750 | .9756 | .9762 | .9767 |

**TABLE A3.1**  *(Continued)*

| Z | 0 | 1 | 2 | 3 | 4 | 5 | 6 | 7 | 8 | 9 |
|---|---|---|---|---|---|---|---|---|---|---|
| 2.0 | .9772 | .9778 | .9783 | .9788 | .9793 | .9798 | .9803 | .9808 | .9812 | .9817 |
| 2.1 | .9821 | .9826 | .9830 | .9834 | .9838 | .9842 | .9846 | .9850 | .9854 | .9857 |
| 2.2 | .9861 | .9864 | .9868 | .9871 | .9874 | .9878 | .9881 | .9884 | .9887 | .9890 |
| 2.3 | .9893 | .9896 | .9898 | .9901 | .9904 | .9906 | .9909 | .9911 | .9913 | .9916 |
| 2.4 | .9918 | .9920 | .9922 | .9925 | .9927 | .9929 | .9931 | .9932 | .9934 | .9936 |
| 2.5 | .9938 | .9940 | .9941 | .9943 | .9945 | .9946 | .9948 | .9949 | .9951 | .9952 |
| 2.6 | .9953 | .9955 | .9956 | .9957 | .9959 | .9960 | .9961 | .9962 | .9963 | .9964 |
| 2.7 | .9965 | .9966 | .9967 | .9968 | .9969 | .9970 | .9971 | .9972 | .9973 | .9974 |
| 2.8 | .9974 | .9975 | .9976 | .9977 | .9977 | .9978 | .9979 | .9979 | .9980 | .9981 |
| 2.9 | .9981 | .9982 | .9982 | .9983 | .9984 | .9984 | .9985 | .9985 | .9986 | .9986 |
| 3.0 | .9987 | .9987 | .9987 | .9988 | .9988 | .9989 | .9989 | .9989 | .9990 | .9990 |
| 3.1 | .9990 | .9991 | .9991 | .9991 | .9992 | .9992 | .9992 | .9992 | .9993 | .9993 |
| 3.2 | .9993 | .9993 | .9994 | .9994 | .9994 | .9994 | .9994 | .9995 | .9995 | .9995 |
| 3.3 | .9995 | .9995 | .9995 | .9996 | .9996 | .9996 | .9996 | .9996 | .9996 | .9997 |
| 3.4 | .9997 | .9997 | .9997 | .9997 | .9997 | .9997 | .9997 | .9997 | .9997 | .9998 |
| 3.5 | .9998 | .9998 | .9998 | .9998 | .9998 | .9998 | .9998 | .9998 | .9998 | .9998 |
| 3.6 | .9998 | .9998 | .9999 | .9999 | .9999 | .9999 | .9999 | .9999 | .9999 | .9999 |
| 3.7 | .9999 | .9999 | .9999 | .9999 | .9999 | .9999 | .9999 | .9999 | .9999 | .9999 |
| 3.8 | .9999 | .9999 | .9999 | .9999 | .9999 | .9999 | .9999 | .9999 | .9999 | .9999 |
| 3.9 | 1.000 | 1.000 | 1.000 | 1.000 | 1.000 | 1.000 | 1.000 | 1.000 | 1.000 | 1.000 |

| | | | | | | | | | |
|---|---|---|---|---|---|---|---|---|---|
| 14159 | 26535 | 89793 | 23846 | 26433 | 83279 | 50288 | 41971 | 69399 | 37510 |
| 58209 | 74944 | 59230 | 78164 | 06286 | 20899 | 86280 | 34825 | 34211 | 70679 |
| 82148 | 08651 | 32823 | 06647 | 09384 | 46095 | 50582 | 23172 | 53594 | 08128 |
| | | | | | | | | | |
| 48111 | 74502 | 84102 | 70193 | 85211 | 05559 | 64462 | 29489 | 54930 | 38196 |
| 44288 | 10975 | 66593 | 34461 | 28475 | 64823 | 37867 | 83165 | 27120 | 19091 |
| 45648 | 56692 | 34603 | 48610 | 45432 | 66482 | 13393 | 60726 | 02491 | 41273 |
| | | | | | | | | | |
| 72458 | 70066 | 06315 | 58817 | 48815 | 20920 | 96282 | 92540 | 91715 | 36436 |
| 78925 | 90360 | 01133 | 05305 | 48820 | 46652 | 13841 | 46951 | 94151 | 16094 |
| 33057 | 27036 | 57595 | 91953 | 09218 | 61173 | 81932 | 61179 | 31051 | 18548 |
| | | | | | | | | | |
| 07446 | 23799 | 62749 | 56735 | 18857 | 52724 | 89122 | 79381 | 83011 | 94912 |
| 98336 | 73362 | 44065 | 66430 | 86021 | 39494 | 63952 | 24737 | 19070 | 21798 |
| 60943 | 70277 | 05392 | 17176 | 29317 | 67523 | 84674 | 81846 | 76694 | 05132 |
| | | | | | | | | | |
| 00056 | 81271 | 45263 | 56082 | 77857 | 71342 | 75778 | 96091 | 73637 | 17872 |
| 14684 | 40901 | 22495 | 34301 | 46549 | 58537 | 10507 | 92279 | 68925 | 89235 |
| 42019 | 95611 | 21290 | 21960 | 86403 | 44181 | 59813 | 62977 | 47713 | 09960 |
| | | | | | | | | | |
| 51870 | 72113 | 49999 | 99837 | 29780 | 49951 | 05973 | 17328 | 16096 | 31859 |
| 50244 | 59455 | 34690 | 83026 | 42522 | 30825 | 33446 | 85035 | 26193 | 11881 |
| 71010 | 00313 | 78387 | 52886 | 58753 | 32083 | 81420 | 61717 | 76691 | 47303 |
| | | | | | | | | | |
| 59825 | 34904 | 28755 | 46873 | 11595 | 62863 | 88235 | 37875 | 93751 | 95778 |
| 18577 | 80532 | 17122 | 68066 | 13001 | 92787 | 66111 | 95909 | 21642 | 01989 |

# APPENDIX
# 4

## LEARNING CURVE

# Table A4.1 Values for the exponential learning-curve function

| n | 90% learning ratio | | | 80% learning ratio | | | 70% learning ratio | | |
|---|---|---|---|---|---|---|---|---|---|
| | $E(N)$ | $\Sigma E(N)$ | $\Sigma E(N)/N$ | $E(N)$ | $\Sigma E(N)$ | $\Sigma E(N)/N$ | $E(N)$ | $\Sigma E(N)$ | $\Sigma E(N)/N$ |
| 1 | 1.0000E 00 | 1.0000E 00 | 1.0000E 00 | 1.0000E 00 | 1.0000E 00 | 1.0000E 00 | 1.0000E 00 | 1.0000E 00 | 1.0000E 00 |
| 2 | 9.0000E-01 | 1.9000E 00 | 9.5000E-01 | 8.0000E-01 | 1.8000E 00 | 9.0000E-01 | 7.0000E-01 | 1.7000E 00 | 8.5000E-01 |
| 3 | 8.4621E-01 | 2.7462E 00 | 9.1540E-01 | 7.0210E-01 | 2.5021E 00 | 8.3403E-01 | 5.6818E-01 | 2.2682E 00 | 7.5606E-01 |
| 4 | 8.1000E-01 | 3.5562E 00 | 8.8905E-01 | 6.4000E-01 | 3.1421E 00 | 7.8553E-01 | 4.9000E-01 | 2.7582E 00 | 6.8955E-01 |
| 5 | 7.8299E-01 | 4.3392E 00 | 8.6784E-01 | 5.9564E-01 | 3.7377E 00 | 7.4755E-01 | 4.3685E-01 | 3.1950E 00 | 6.3901E-01 |
| 6 | 7.6159E-01 | 5.1008E 00 | 8.5013E-01 | 5.6188E-01 | 4.2994E 00 | 7.1657E-01 | 3.9773E-01 | 3.5928E 00 | 5.9879E-01 |
| 7 | 7.4395E-01 | 5.8447E 00 | 8.3496E-01 | 5.3449E-01 | 4.8339E 00 | 6.9056E-01 | 3.6740E-01 | 3.9601E 00 | 5.6574E-01 |
| 8 | 7.2900E-01 | 6.5737E 00 | 8.2172E-01 | 5.1200E-01 | 5.3459E 00 | 6.6824E-01 | 3.4300E-01 | 4.3031E 00 | 5.3789E-01 |
| 9 | 7.1606E-01 | 7.2898E 00 | 8.0998E-01 | 4.9295E-01 | 5.8389E 00 | 6.4876E-01 | 3.2283E-01 | 4.6260E 00 | 5.1400E-01 |
| 10 | 7.0469E-01 | 7.9945E 00 | 7.9945E-01 | 4.7651E-01 | 6.3154E 00 | 6.3154E-01 | 3.0579E-01 | 4.9318E 00 | 4.9318E-01 |
| 11 | 6.9455E-01 | 8.6890E 00 | 7.8991E-01 | 4.6211E-01 | 6.7775E 00 | 6.1613E-01 | 2.9116E-01 | 5.2229E 00 | 4.7481E-01 |
| 12 | 6.8543E-01 | 9.3745E 00 | 7.8120E-01 | 4.4935E-01 | 7.2268E 00 | 6.0224E-01 | 2.7841E-01 | 5.5013E 00 | 4.5844E-01 |
| 13 | 6.7714E-01 | 1.0052E 01 | 7.7320E-01 | 4.3792E-01 | 7.6647E 00 | 5.8960E-01 | 2.6717E-01 | 5.7685E 00 | 4.4373E-01 |
| 14 | 6.6955E-01 | 1.0721E 01 | 7.6580E-01 | 4.2759E-01 | 8.0923E 00 | 5.7802E-01 | 2.5718E-01 | 6.0257E 00 | 4.3041E-01 |
| 15 | 6.6257E-01 | 1.1384E 01 | 7.5891E-01 | 4.1820E-01 | 8.5105E 00 | 5.6737E-01 | 2.4821E-01 | 6.2739E 00 | 4.1826E-01 |
| 16 | 6.5610E-01 | 1.2040E 01 | 7.5249E-01 | 4.0960E-01 | 8.9201E 00 | 5.5751E-01 | 2.4010E-01 | 6.5140E 00 | 4.0712E-01 |
| 17 | 6.5008E-01 | 1.2690E 01 | 7.4646E-01 | 4.0166E-01 | 9.3218E 00 | 5.4834E-01 | 2.3273E-01 | 6.7467E 00 | 3.9687E-01 |
| 18 | 6.4446E-01 | 1.3334E 01 | 7.4080E-01 | 3.9436E-01 | 9.7162E 00 | 5.3979E-01 | 2.2598E-01 | 6.9727E 00 | 3.8737E-01 |
| 19 | 6.3918E-01 | 1.3974E 01 | 7.3545E-01 | 3.8755E-01 | 1.0104E 01 | 5.3178E-01 | 2.1978E-01 | 7.1925E 00 | 3.7855E-01 |
| 20 | 6.3422E-01 | 1.4608E 01 | 7.3039E-01 | 3.8121E-01 | 1.0485E 01 | 5.2425E-01 | 2.1405E-01 | 7.4065E 00 | 3.7033E-01 |
| 25 | 6.1307E-01 | 1.7713E 01 | 7.0853E-01 | 3.5478E-01 | 1.2309E 01 | 4.9234E-01 | 1.9083E-01 | 8.4040E 00 | 3.3616E-01 |
| 30 | 5.9631E-01 | 2.0727E 01 | 6.9090E-01 | 3.3456E-01 | 1.4020E 01 | 4.6733E-01 | 1.7375E-01 | 9.3050E 00 | 3.1017E-01 |
| 35 | 5.8250E-01 | 2.3660E 01 | 6.7617E-01 | 3.1836E-01 | 1.5663E 01 | 4.4694E-01 | 1.6050E-01 | 1.0135E 01 | 2.8951E-01 |
| 40 | 5.7080E-01 | 2.6543E 01 | 6.6357E-01 | 3.0497E-01 | 1.7193E 01 | 4.2984E-01 | 1.4984E-01 | 1.0902E 01 | 2.7256E-01 |

**TABLE A4.1**  *(Continued)*

| n | 90% learning ratio | | | 80% learning ratio | | | 70% learning ratio | | |
|---|---|---|---|---|---|---|---|---|---|
| | $E(N)$ | $\Sigma E(N)$ | $\Sigma E(N)/N$ | $E(N)$ | $\Sigma E(N)$ | $\Sigma E(N)/N$ | $E(N)$ | $\Sigma E(N)$ | $\Sigma E(N)/N$ |
| 45 | 5.6067E-01 | 2.9366E 01 | 6.5257E-01 | 2.9362E-01 | 1.8684E 01 | 4.1519E-01 | 1.4103E-01 | 1.1625E 01 | 2.5832E-01 |
| 50 | 5.5176E-01 | 3.2142E 01 | 6.4284E-01 | 2.8383E-01 | 2.0122E 01 | 4.0243E-01 | 1.3358E-01 | 1.2307E 01 | 2.4614E-01 |
| 60 | 5.3668E-01 | 3.7574E 01 | 6.2623E-01 | 2.6765E-01 | 2.2868E 01 | 3.8113E-01 | 1.2162E-01 | 1.3574E 01 | 2.2624E-01 |
| 70 | 5.2425E-01 | 4.2871E 01 | 6.1244E-01 | 2.5469E-01 | 2.5471E 01 | 3.6387E-01 | 1.1235E-01 | 1.4738E 01 | 2.1054E-01 |
| 80 | 5.1372E-01 | 4.8054E 01 | 6.0067E-01 | 2.4397E-01 | 2.7957E 01 | 3.4947E-01 | 1.0489E-01 | 1.5819E 01 | 1.9774E-01 |
| 90 | 5.0460E-01 | 5.3140E 01 | 5.9044E-01 | 2.3490E-01 | 3.0346E 01 | 3.3718E-01 | 9.8719E-02 | 1.6833E 01 | 1.8703E-01 |
| 100 | 4.9659E-01 | 5.8141E 01 | 5.8141E-01 | 2.2706E-01 | 3.2651E 01 | 3.2651E-01 | 9.3509E-02 | 1.7791E 01 | 1.7791E-01 |
| 150 | 4.6690E-01 | 8.2156E 01 | 5.4771E-01 | 1.9928E-01 | 4.3234E 01 | 2.8822E-01 | 7.5900E-02 | 2.1972E 01 | 1.4648E-01 |
| 200 | 4.4693E-01 | 1.0496E 02 | 5.2482E-01 | 1.8165E-01 | 5.2720E 01 | 2.6360E-01 | 6.5456E-02 | 2.5482E 01 | 1.2741E-01 |
| 300 | 4.2021E-01 | 1.4820E 02 | 4.9401E-01 | 1.5942E-01 | 6.9663E 01 | 2.3221E-01 | 5.3130E-02 | 3.1342E 01 | 1.0447E-01 |
| 400 | 4.0223E-01 | 1.8927E 02 | 4.7317E-01 | 1.4532E-01 | 8.4849E 01 | 2.1212E-01 | 4.5819E-02 | 3.6260E 01 | 9.0649E-02 |
| 500 | 3.8882E-01 | 2.2879E 02 | 4.5757E-01 | 1.3525E-01 | 9.8847E 01 | 1.9769E-01 | 4.0849E-02 | 4.0577E 01 | 8.1153E-02 |
| 600 | 3.7819E-01 | 2.6711E 02 | 4.4519E-01 | 1.2754E-01 | 1.1197E 02 | 1.8661E-01 | 3.7191E-02 | 4.4468E 01 | 7.4114E-02 |
| 800 | 3.6201E-01 | 3.4103E 02 | 4.2629E-01 | 1.1626E-01 | 1.3627E 02 | 1.7034E-01 | 3.2074E-02 | 5.1355E 01 | 6.4194E-02 |
| 1000 | 3.4994E-01 | 4.1217E 02 | 4.1217E-01 | 1.0820E-01 | 1.5867E 02 | 1.5867E-01 | 2.8594E-02 | 5.7401E 01 | 5.7401E-02 |

# ANSWERS
# TO SELECTED
# PROBLEMS

**Chapter 2**

**2.1** $500 interest payment first month; $217.35 principal payment first month; $497.83 interest payment second month; $219.52 principal payment second month

**2.3** $23.81

**2.5** $1006.29

**2.8** $22,104

**2.11** 29,180 M$

**2.13** $16,531 net receipt

**2.20** $13,276

**2.23** Choose $4000 cash now plus $7000 in 4 years. The present-value advantage is $1219

**2.27** 79.7 h

**2.28** (a) $128,780; (b) 11.9 years from a selling price of $50,000

**2.29** $59,381

**2.35** $469.32

**2.36** $1845.60

**2.40** 7.64 years. Could be seven payments of $1000 plus one last payment of $363

**2.41** 0.875 percent monthly rate; 10.5 percent nominal rate; 11.02 percent effective rate

**2.44** 14.1 percent per year

**2.49** 5 years

**2.50** (a) $838.35; (b) $7391.10; (c) $1419.60

**2.52** 73.2 percent per year

**2.53** (a) 12.7 percent per year; (b) 23 percent per year

**2.54** A royalty is more economical. The unacost is $80,000 against $85,396.

**2.55** 12.5 years for machine B; machine C never pays off

**2.56** (a) $P_{A,10} = \$57,695$    $P_{B,10} = \$53,029$
      (b) $P_{A,4} = \$28,479$    $P_{B,4} = \$26,176$
      (c) $R_A = \$8598$    $R_B = \$7903$
      (d) $K_A = \$107,480$    $K_B = \$98,790$
      (e) 4.6 years; (f) approximately 15.5 percent per year
**2.61** $125,330 as a first cost now for a 3-year life
**2.62** Plan B is better. The present value for 5 years is 4.659 M$ against 2.839 M$.
**2.65** $315,400

## Chapter 3

**3.2**

| Year | Twice straight-line declining-balance depreciation | Units of production depreciation |
|------|------|------|
| 1 | $22,500 | $2,500 |
| 2 | 11,250 | 10,000 |
| 3 | 5,625 | 20,000 |
| 4 | 625 | 7,500 |

**3.3** Present values are $8223 for straight-line depreciation, $9413 for twice straight-line declining balance with switch, and $8993 for sum-of-the-years-digits depreciation.
**3.4** $F_{DB} = 0.1059$
**3.5** $F_{DB} = 0.27522$, but the maximum permissible value is 0.25.
**3.6** Depreciation by years, in order, is $3333.30, $2667.70, $2000.00, $1333.30, and $666.70, which checks the ordinary calculation for sum-of-the-years-digits depreciation.
**3.11** Decimal depreciation, by years, is

| | |
|---|---|
| 1 | 0.45455 |
| 2 | 0.29091 |
| 3 | 0.16364 |
| 4 | 0.07273 |
| 5 | 0.01818 |

**3.15** The minimum shift is 140 k$, but it can be as high as 160 k$.
**3.27** 12.5 years for machine B; machine C never pays off
**3.32** Machine B is more economical. The after-tax unacost is $4497 against $5614.
**3.33** The lowest after-tax unacost is 37.4 k$/year for 4 years.
**3.38** Minimum for 4 years
**3.39** 1781 k$/year
**3.40** The advantage for an immediate write-off as a present value after taxes is 26,669 − 24,214 = $2455.

## Chapter 4

**4.1**

$$P = \frac{2Q}{(in)^2}[1 - e^{-in}(in + 1)]$$

**4.2**
$$P = \frac{\overline{R}_2 - \overline{R}_1}{i^2 n}[1 - e^{-in}(in + 1)] + n\overline{R}_1\left[\frac{1 - e^{-in}}{in}\right]$$

**4.3**
$$P = Y\frac{1 - e^{-ikn}}{e^{in} - 1}$$

**4.7** $32.95/month

**4.12**

|  | $\psi$ | $1 - 0.34\psi$ |
|---|---|---|
| Simulated depreciation | 0.64278 | 0.78145 |
| End-of-year depreciation | 0.58417 | 0.80138 |

**4.14** (*a*) $1100; (*b*) $1097

**4.15** (*a*) $997,620; (*b*) $1,454,800; (*c*) $1,147,700

## Chapter 5

**5.1** 2.2 years

**5.2** (*a*) 2.6 years; (*b*) 2.8 years

**5.3** NPV = 0.587 M$; DCFRR = 36 percent

**5.4**

|  | *A* |  | *B* |  | *C* |  |
|---|---|---|---|---|---|---|
|  |  | **Rank** |  | **Rank** |  | **Rank** |
| Discounted cash flow | 28% | 1 | 26% | 2 | 23% | 3 |
| Present worth at 10% | 371 | 3 | 394 | 1 | 389 | 2 |

**5.5** $4.15 per part

**5.6** NPV *A* = −2855 k$; NPV *B* = −2742 k$. Choose plan *B*.

**5.7** *P* = $4055

**5.8** Plan *A* NPV = −72.64 k$; plan *B* NPV = 613.60 k$. Build a full-size plant now.

**5.9** 20 percent

**5.10** 11.3 percent

**5.11** 6.9 percent

**5.12** NPV = $111,000 at 0 percent, $24,000 at 4 percent, −$2000 at 6 percent, DCFRR = 5.8 percent

**5.13** (*a*) 13.7 percent; (*b*) 4.4 years

## Chapter 6

**6.2** $231,430

**6.13** 8 percent per year

**6.16** $231,430

**6.17** The productivity rate gain *p* may vary within a system.

## Chapter 7

**7.1** 60,000 m$^3$/year

**7.2** $NV$ = 5.31 M$ at $N$ = 2.5. The incremental variable cost = 1.875 M$ per unit $\Delta N$ at $N$ = 2.5

**7.3** $1.98/t

**7.4** 55 percent

**7.5** 13.5, say, 14

**7.6** (a) $N$ = 32.1 percent; (b) $N$ = 26.5 percent; (c) $N$ = 40.9 percent; (d) $33,325 extra profit

**7.10** (a) At 75 percent capacity, $0.629/kg; at 100 percent capacity, $0.532/kg; (b) $0.24/day/$\Delta$kg; (c) $19.4 × 10$^{-6}$/kg/$\Delta$kg

**7.11** (a) $0.3/kg; (b) $0.2625/kg

**7.12** (a) $40,000; (b) $NS$ = $538,460; (c) $NS$ = $661,540

**7.13** (a) Do not shut down; $Z_g$ = −$180,000 against −$225,000; (b) $77.50

**7.14** 1 − 5000: buy; 5001 − 25,000: make; more than 25,000: buy

**7.15** The maximum profit occurs at a selling price of $15.

**7.16** The maximum profit is $3125 at a sales price of $15, with $3125 for advertising, and selling 1250 units.

**7.19** (a) $N$ = 2037 break-even point; $N$ = 12,681 profit limit point; (b) maximum profit $Z_g$ = $130,940 at $N$ = 7663; (c) the average profit is $13.589/ton; the incremental profit is −13.1 $\Delta$$ per $\Delta$ton.

**7.21** (a) 101 or 9703 parts per period; (b) $N$ = 4902 parts per period with $Z_g$ = $235,100 per period; (c) $N$ = 7071 parts per period with a minimum average cost of $2.828

## Chapter 8

**8.1** Mean 2.187; variance 0.273

**8.5** (a) 0.194; (b) 0.349; (c) 0.93; (d) 0.264

**8.6** (a) 0.083; (b) 0.5; (c) 0.417

**8.7** 0.433

**8.11**

| X | $P_d(X)$ | X | $P_d(X)$ |
|---|---|---|---|
| −5 | 0.0022 | 2 | 0.1760 |
| −3 | 0.0270 | 3 | 0.1210 |
| −1 | 0.1210 | 5 | 0.0270 |
| 0 | 0.1760 | 7 | 0.0022 |
| 1 | 0.1944 | | |

**8.18**

| Total min | Random numbers | Total min | Random numbers |
|---|---|---|---|
| 52.5 | 01–03 | 127.5 | 48–67 |
| 67.5 | 04–08 | 142.5 | 68–86 |
| 82.5 | 09–17 | 157.5 | 87–94 |
| 97.5 | 18–31 | 172.5 | 95–00 |
| 112.5 | 32–47 | | |

**8.19**

| $X$ | $P_c(X)$ | $X$ | $P_c(X)$ |
|-----|----------|------|----------|
| 0   | 0.0228   | 2    | 0.9772   |
| 0.5 | 0.1587   | 2.25 | 0.9938   |
| 1   | 0.5000   | 2.5  | 0.9987   |
| 1.5 | 0.8413   |      |          |

**8.24** The expected cost is 18.48 M$. The probability of a cost between 18.48 M$ and 20 percent over that is 9.1 percent.

**8.25** (*a*) 25.24 k$; (*b*) 1 percent error in *B* leads to a 25.2 percent change in the NPV, 1 percent error in *C* leads to an 11.1 percent change in the NPV, 1 percent error in *D* leads to a 4.3 percent change in the NPV; (*c*) 14 percent

**8.26** 3.62 M$ approximately

## Chapter 9

**9.1** *R* existent = $11,400; *R* displacer = $9623

**9.2** 20 percent per year approximately

**9.3** *R* existent = $8000; *R* displacer = $6848; after taxes

**9.7** $R = \$2204/\text{year}$; $K = \$18,368$

**9.9** The new learning curve will probably have a smaller slope and will have to be reestablished.

**9.10** Minimum at 11 years with $K/C_i = 1.3826$

**9.11** 284 bulbs per week

**9.12** The minimum cost is $217 per period by displacing the group every third period.

**9.13** 132.1 h

**9.14** 93.5 h

**9.15** (*a*) 149.2 h; (*b*) 100.4 h

## Chapter 10

**10.1** 63.0 m

**10.2** 16.8 MW

**10.3** 0.068 m

**10.4** Invest $126,000. The additional investment returns less than 17 percent.

**10.5** 15 stories

**10.6** Build a 10-Mkg plant. This is the only acceptable option.

**10.7** 14,180 parts per batch

**10.8** 11,000 parts per batch, approximately

**10.9** $T = 0.016$ h per stamping; $533,000/year

**10.10** 5

**10.11** 2.25 m; $42,719

**10.12** $0.24/kg of *Z*

**10.13** 3.7 years

**10.14** $N = 2.13$, say, two effects

**10.15** (*a*) 5.5 h, including cleaning time; (*b*) 5139 kg of cake per cycle; (*c*) 6 h, including cleaning time

**10.16** (a) $A_{in} = 900$ kg/day; $A_{out} = 450$ kg/day; $B_{out} = 450$ kg/day; efficiency $= 0.50$

(b) Cumulative amounts for the first 10 days:

$A_{in} = 9500$ kg; $A_{out} = 2333$ kg; $B_{out} = 7167$ kg

Averages for the first 10 days:

$A_{in} = 950$ kg/day; $A_{out} = 233.3$ kg/day; $B_{out} = 716.7$ kg/day; average efficiency 0.754

(c) Amounts for the tenth day only:

$A_{in} = 905$ kg; $A_{out} = 429.5$ kg; $B_{out} = 475.5$ kg; efficiency 0.525

## Chapter 11

**11.1** (a) 100 items; (b) \$16/year; (c) five items

**11.4** 100 items

**11.5** $Q_{opt} = 24,361$ items for a 61-day supply

**11.7** $N = 4500$ approximately

**11.9** 10 days

**11.10** 74.1

**11.11** Five machinists

**11.12** 0.428

**11.13** 400 kg

**11.14** 8.9 Mkg

**11.15** 27 items

**11.17** The minimum cost is \$21.54/week for a stock of three.

**11.20** \$115.26/h

**11.21** 0.142

**11.23** 127 tests per hour capability

**11.24** \$5670/week with three crews

**11.25** 14 or 15

**11.26** 0.55 approximately

**11.27** The minimum cost is for the sequence $ABDC$.

## Chapter 12

**12.3**

$$X_1 + 3X_2 \geq 8$$

$$3X_1 + 4X_2 \geq 19$$

$$3X_1 + X_2 \geq 7$$

$$Z = 50X_1 + 25X_2$$

**12.8** 15 of the most difficult and 75 of the least difficult for maximum grade of 390

**12.9** The minimum cost is 800 k\$ for 40 radio advertisements.

**12.10** 17.25 of $A$ and 30.8 of $B$; $Z = \$155$/day

**12.11** $X_1 = 400$, $X_2 = 500$, $X_3 = 100$

**12.12** 14,000 bbl/day crude $A$ to process $D$; 12,000 bbl/day crude $B$ to process $C$; 6000 bbl/day crude $B$ to process $D$

**12.17** The optimum is either zero item 1, three item 2, and zero item 3; or two item 1, zero item 2, and zero item 3.

**12.18** Overall, the best operation is at 80 percent efficiency; type $A$, unit 1, with unit 2 at a temperature of 430°C for a total 5-year income of \$560,000.

**12.19** The minimum cost is $100,000.

| Into #1 | Out of #1 Into #2 | Out of #2 Into #3 | Out of #3 |
|---------|-------------------|-------------------|-----------|
| 20%     | 50%               | 60%               | 99%       |

**12.20**

|       | Operating cost, % | Fixed cost, % |
|-------|-------------------|---------------|
| (a)   | 23.8              | 76.2          |
| (b)   | 23.8              | 76.2          |
| (c)   | 22.4              | 77.6          |
| (d)   | 22.4              | 77.6          |

**12.21**

$$v = 2.889A^{0.4444}B^{0.2222}C^{0.3333}$$

$$X_1 = 1.284A^{-0.5556}B^{0.2222}C^{0.3333}$$

$$X_2 = 0.9813A^{0.2222}B^{0.1111}C^{-0.3333}$$

**Chapter 13**

**13.1**   1.70 to 1.735

**13.4**

|       | Range for $X$      | $F_r$   |
|-------|--------------------|---------|
| (a)   | 1.8571 to 2.4286   | 0.2857  |
| (b)   | 2      to 2.5      | 0.25    |
| (c)   | 2      to 2.25     | 0.125   |
| (d)   | 2.077  to 2.2308   | 0.0769  |

**13.5**   $X = Y = 0$

**13.7**   Place $U = 5Y$; thus $Z = X^2 + U^2$

**13.11** (a)
$$-2A_m + 1A_{m+1} + 1A_{m+2} = 0$$

(b)
$$\frac{-1}{(1 + 2X)(1 - X)}$$

(c)
$$-\tfrac{1}{3}[(-1)^n 2^{n+1} + 1]$$

(d)
$$\Sigma = \frac{-1 + (-2A_{n-1} + A_n)X^{n+1} - 2A_n X^{n+2}}{1 - X - 2X^2}$$

with $A_n$ and $A_{n-1}$ given by part (c)

**13.12** The scale of relation is

$$-A_m + 3A_{m+1} - 3A_{m+2} + A_{m+3} = 0$$

The next term of the series is $36\ U^7$.

**13.13**
$$\frac{1}{1 - 3U + 3U^2 - U^3}$$

**13.16** (a)
$$\frac{1}{1 - U - U^2}$$

(b)
$$\frac{1}{\sqrt{5}}\left[\left(\frac{1 + \sqrt{5}}{2}\right)^{N+1} - \left(\frac{1 - \sqrt{5}}{2}\right)^{N+1}\right]$$

**13.17** $(1 + \sqrt{5})/2$

**13.18** $\Sigma = F_{N+2} - 1$ where $F_{N+2}$ can be obtained from Prob. 13.16, part (b)

### Chapter 14

**14.9** 7.26 M$

**14.10** 11.3 M$

**14.12** 32.3 M$

**14.14** Low, 19.4 M$; average, 23.1 M$; high, 26.6 M$

### Chapter 15

**15.2** $1,560,000/year or $1.57/lb

### Chapter 16

**16.1** The critical path is 1–3–6–8 at 107 periods.

**16.2** The critical path is 1–2–3–5 at 19 periods.

| Activity | Total float | Free float |
|----------|-------------|------------|
| 1–2 | 0 | 0 |
| 1–3 | 6 | 6 |
| 1–4 | 7 | 0 |
| 2–3 | 0 | 0 |
| 2–4 | 9 | 2 |
| 2–5 | 7 | 7 |
| 3–5 | 0 | 0 |
| 4–5 | 7 | 7 |

**16.3** (a) $1950; (b) $150

**16.6** The critical path is 1–3–6–7 at 12 days.

| Activity | Total float | Free float |
|----------|-------------|------------|
| 1–2 | 3 | 0 |
| 1–3 | 0 | 0 |
| 1–4 | 1 | 0 |
| 2–5 | 3 | 1 |
| 3–5 | 2 | 0 |
| 3–6 | 0 | 0 |
| 4–6 | 1 | 1 |
| 5–7 | 2 | 2 |
| 6–7 | 0 | 0 |

**16.7**  10.5 days

**16.8**  The critical path is 1–3–6–5–7 at 15 days.

| Activity | Total float | Free float |
|----------|-------------|------------|
| 1–2      | 6           | 0          |
| 1–3      | 0           | 0          |
| 1–4      | 1           | 0          |
| 2–5      | 6           | 6          |
| 3–5      | 5           | 5          |
| 3–6      | 0           | 0          |
| 4–6      | 1           | 1          |
| 5–7      | 0           | 0          |
| 6–5      | 0           | 0          |
| 6–7      | 3           | 3          |

**16.9**  Replace 1–3 by $1–3_1–3_2–3_3$, all 1 day apart. Add the dummy activity from $3_3$ to 7. The critical path is 1–4–6–7 at 11 days.

**16.10**  The critical path is 8.5 h.

# NAME INDEX

**621**

# SUBJECT INDEX